Peanut Processing Characteristics and Quality Evaluation

Qiang Wang

Peanut Processing Characteristics and Quality Evaluation

Springer

Qiang Wang
Institute of Food Science and Technology
Chinese Academy of Agricultural Sciences
Beijing, China

ISBN 978-981-10-6174-5 ISBN 978-981-10-6175-2 (eBook)
DOI 10.1007/978-981-10-6175-2

Library of Congress Control Number: 2017950765

Printed on acid-free paper

This Springer imprint is published by Springer Nature
The registered company is Springer Nature Singapore Pte Ltd.
The registered company address is: 152 Beach Road, #21-01/04 Gateway East, Singapore 189721, Singapore

Preface

China is the global leader in peanut production, and peanut is an important component of the Chinese diet. Peanut is also an important crop for the economic sustainability of Chinese farmers.

Peanut Processing Quality is a book that describes not only the ideas formed when the author presided over the research of the National Key R&D Program of China, the Special Fund for Agro-scientific Research in the Public Interest, the National Natural Science Foundation, the National Key Technology Research and Development Program of the Ministry of Science and Technology of China, and a number of major national projects or topics but also the peanut processing quality research achievements obtained by the author and his research team. Over the years, the author has been engaged in the research of peanut processing characteristics and quality formation mechanism and regulation, published more than 50 high-level SCI papers, and obtained some scientific research achievements and awards at the provincial and ministerial levels.

Based on the deep analysis on the research progress of peanut processing quality domestically and abroad, the author systematically studied the sensory quality, physicochemical nutrition quality, and processing characteristics of different peanut varieties; revealed the correlation between raw material quality and peanut protein, peanut oil, export peanut, and other processed product quality; built the processing quality evaluation model; established the processing suitability evaluation methods and standards; selected the specialized peanut processing varieties suitable for high-gel and high-solubility protein and for export and high-stability oil processing; and established the specialized variety characteristic fingerprints. With new contents and full and accurate data, the book is systematical and complete and has its own characteristics and innovation in the relation research between raw material quality and product quality, processing quality evaluation model construction, and other aspects, and it is a rare research work related to peanut processing quality evaluation.

This book provides a systematic review of research on the sensory quality, physicochemical nutrition quality, and processing characteristics of different peanut varieties. A major focus of this book is the relationship between raw material quality and product quality of peanut. The book is an important reference for anyone involved in peanut processing quality in China and also provides an important basis for further strengthening the development and use of peanut resources in China, promoting industrial technology and upgrading and enhancing industrial international competitiveness simultaneously. The publication of this book will provide a reference for the material basis, change mechanism, and regulation technology of peanut processing quality in China and lay the foundation for the development of new subjects of "grain and oil processing quality."

C. Corley Holbrook
President (2017) American Peanut
Research and Education Society

Introduction

Peanut quality is an important index to measure the quality of peanut varieties and peanut products; it can be divided into sensory quality (color, seed kernel size, appearance shape, etc.), physicochemical and nutrition quality (crude protein, crude fat, moisture, ash, composition and content of amino acid, protein, and fatty acid, etc.), and quality characteristics related to processing (pure kernel rate, oil yield, protein extraction rate, oleic acid/linoleic acid, arachin/conarachin, etc.). Peanut quality is relative to its processed products, and different types of processed products have different requirements for raw material quality. Therefore, researching the processing characteristics of different peanut varieties (such as peanut varieties that have good protein functional characteristics and high oil stability and are suitable for export) for different processing purposes; establishing the evaluation technologies, methods, and standards of these high-quality peanut varieties; building the peanut-processing quality evaluation index system; and selecting specialized varieties suitable for peanut processing have guiding significances in promoting the healthy development of the peanut-processing industry of China.

In recent years, my research team and I presided over the special scientific research of the National Key Technologies R&D Program of China "Study on the Change of Component Structure and Quality Control Mechanism in Food Processing Process(2016YFD0400200)", Special Fund for Agro-scientific Research in the Public Interest Program "Major Agricultural Products Processing Characteristics Research and Quality Evaluation Technology (200903043) – Peanut Processing Characteristics Research and Quality Evaluation Technology," the "12th Five-Year Plan" National Key Technologies R&D Program of the Ministry of Science and Technology of China "Edible Agricultural Product Processing Suitability Evaluation and Risk Monitoring Technology Research Demonstration (2012BAD29B00)," the International Scientific and Technological Cooperation Project "Key Technical Cooperation in Preparation of Functional Ingredients in Agricultural and Sideline Products (2012DFA31400)," and many other major national projects or topics; carried out the deep research in the peanut-processing research field for more than 10 years; and won the second prize of the National Technology Invention Award in 2014, the first prize of the China Agricultural Science and Technology Award in 2013, the China Agricultural Science and Technology Award for Outstanding Innovation Team in 2015 (which is equal to scientific research first prize), the Harald Perten Prize (highest academic award) of the International Cereal Science and Technology Association (ICC) in 2012, the first prize of the Science and Technology Achievement Award of the Chinese Academy of Agricultural Sciences in 2011, the second prize of the Science and Technology Achievement Award of the Chinese Academy of Agricultural Sciences in 2009, and the first prize of the Science and Technology Award of the Chinese Cereals and Oils Association in 2012. Six agricultural industry standards were also researched and formulated, including *Grades and Specifications of Peanut for Processing*, *Near-Infrared Rapid Nondestructive Testing Methods of Amino Acid in Peanut*, *Vegetable Protein and Its Product Terms*, *Production Technical Specifications for Oil Preparation Through Low-Temperature Pressing*, etc. More than 100 articles were published in *Food Chemistry*, the *Journal of the Science of Food and Agriculture*, the *International Journal of Food Properties*, *Advanced Materials Research*, *Food Science and Technology*, *Food Research International*, the *Journal of Agricultural Engineering*, *Chinese Oils and Fats*, *Food Science*, the *Journal of the Chinese Cereals and Oils Association*, and the *Journal of Peanut Science*, among others. Twenty relevant national invention patents were obtained or applied, and several books (*Peanuts: Processing Technology and Product Development; Introduction to Peanut Bioactive Substances*) were published. On these bases, the *Peanut-Processing Characteristics and Quality Evaluation* was written after systematic organization. This book includes ten chapters: Chapter 1 is an overview of peanut-processing quality and introduces the distribution and utilization of peanut variety resources at home and abroad, definition and classification of peanut-processing quality, and research progress of peanut-processing quality. Chapter 2 describes the quality characteristics of peanut raw materials and their determination methods; introduces the sensory quality, physicochemical nutrition quality, and processing characteristic index determination method of peanut; and analyzes the quality characteristics of main cultivated peanuts in China. Chapter 3 describes the quality characteristics of peanut products; introduces the edible quality, nutritional quality, and processing quality characteristics of peanut protein and oil; and analyzes their relations. Chapter 4 introduces the relations between peanut raw material quality and product quality; analyzes the correlation among peanut raw materials quality and protein gel property, protein solubility, peanut oil stability, and export peanut quality characteristics; and builds the appropriate evaluation model. Chapter 5 describes the peanut-processing suitability evaluation standards and introduces the high-gel protein, high-solubility protein, high oil stability, suitable export peanut quality evaluation standards, and DNA fingerprint of each special variety. Chapter 6 describes the functional improvement of peanut protein concentrate. Chapter 7 describes the improvement of gelation of peanut protein isolate. Chapter 8 describes the gelation improvement of peanut protein component. Chapter 9 describes the preparation of functional peanut oligopeptide and its biological activity and Chapter 10 describes the oxidation stability improvement of peanut oil. This book was written on the basis of graduation theses of four PhD and eight master's students in my research team with the contents having attached greater prominence to the systematisms, novelty, innovation, cutting edge, and leading functions. This book aims to provide useful reference and guidance for the processing and utilization of peanut-processing characteristics so as to provide technical support for the healthy development of the peanut-processing industry of China.

The members involved in the preparation, collation, and experiment processes include Shi Aimin, Liu Hongzhi, Liu Li, Hu Hui, Wang Li, Zhang Yuhao, Feng Xiaolong, Du Yin, Zhang Jianshu, Shen Minjiang, Gong Ana, Lin Weijing, Liu Lina, Ma Tiezheng, Jiang Nan, He Xuanhui, Li Ning, Zhang Jinchuang, Deng Lei, Yu Hongwei, Wang Jun, Guo Yalong, Jiao Bo, Zhu Song, Bai Wenqiang, Xia Xiaoyong, Zhao Zhihao, Li Jiaxiao, Chen Bingyu, Chen Yan, and other members, who have made their contributions to the research contents and manuscript collation. Also, during the writing, the works and papers of experts and scholars domestic and foreign have been referenced simultaneously. I want to show them my wholehearted thanks.

Due to the limitation of materials, means, research methods, and level of author expertise, some problems and deficiencies in views and conclusions may exist in this book inevitably, and I sincerely hope that readers could give criticism and correction in the reading process.

March 12, 2017 Qiang Wang

Contents

Chapter 1
Overview of Peanut Processing Quality

Peanut is a kind of plant (*Arachis*) which originates in the South American continent and belongs to legume and Faboideae. The cultispecies peanut (*Arachis hypogaea*) is the only species cultivated widely in the peanut, and the others are all wild relatives and are rarely cultivated or used. At present, it is generally believed that Bolivia and Amazonian basin (area) is the original centers of cultispecies peanuts; the cultispecies peanuts were divided into two subspecies and six variants internationally, and the cultispecies peanuts were divided into two groups, five types, eight variety groups, and several variety groups (Yu 2008) in China. The worldwide peanut production is widely distributed in Asia, Europe, Africa, the Americas, Oceania, and other regions in the world. In 2014, the worldwide peanut planting area was 26,541,660 ha, and the production quantity was 43,915,365 tons; the countries whose planting areas were ranked in the top six places were India, China, Nigeria, Sudan, the United Republic of Tanzania, Senegal, and the countries whose production quantity were ranked in the top six places were China, India, Nigeria, the United States, Sudan, and the United Republic of Tanzania. In 2014, the peanut planting area was 4,625,494 ha, and the production quantity was 165,500,213 tons in China, respectively, accounting for 17.4 and 37.69% in the peanut planting area and production quantity of the world and ranking first in the world (FAO statistics).

China has rich peanut variety resources and wide peanut planting area. Except for Qinghai and Ningxia, peanuts are widely cultivated in China, and the cultivation regions are mainly concentrated in the Huang-Huai-Hai basin, the southeast coastal, and the Yangtze river basin, respectively, accounting for about 50%, 20%, and 15% in the planting area and yield of the whole country. At present, about 55% of the peanuts in China are used for oil extraction, 30% for food, 7% for exports, and 8% for seed reservation. In 2010, the total export volume of peanut and its products in the world was 2,680,000 tons and that of China was 530,000 tons, accounting for 19.77% (the shelled peanut volume accounted for 47%) in the total peanut trade volume of the world. Good edible and nutritional quality (large grain, sugar content, O/L) make the peanuts, and their processed products of China to be loved by

Q. Wang, *Peanut Processing Characteristics and Quality Evaluation*,
DOI 10.1007/978-981-10-6175-2_1

foreign consumers and occupy the important position in the international peanut trade.

1 Distribution and Utilization of Peanut Variety Resources

1.1 Distribution and Utilization of Peanut Resources in the World

1.1.1 General of Peanut Germplasm Resources in the World

Peanut is a kind of plant (*Arachis*) which originates in the South American continent and belongs to legume and Faboideae. The arachis is a larger genus and includes more than 70 species. The cultispecies peanut (*Arachis hypogaea*) is the only species cultivated widely in the arachis, and the others are all wild relatives and are rarely cultivated or used in other manners. At present, it is generally believed that Bolivia and Amazonian basin (area) are the original centers of cultispecies peanuts, and others are secondary differentiation centers. In 1980, Gregory proposed the cultispecies peanut classification system (two subspecies (alternative blooming subspecies and continuous blooming subspecies) and four variants (Sabina variants, Rongmao variants, Shuzhi variants, and Zhudou variants)). In 1994, Krapovickas divided the cultispecies peanuts into two subspecies and six variants on the basis of the original classification, and the Peruvian variants and equatorial variants were added newly. Although the above classification method (two subspecies and six variants) has become the latest international cultispecies peanut classification system, the collected Peruvian variants and equatorial variants materials are still fewer, and the classification system (four variants) is still used at present.

According to the incomplete statistics, the world's total collection quantity of peanut germplasm resources is more than 40,000, and the institutions and countries with larger collection quantity of peanut germplasm resources in the word are, in order, the International Crop Research Institute for the Semi-Arid Tropics (ICRISAT) (15,342), the United States (8719), China (7490, Taiwan Province is not included), Argentina (2200), Indonesia (1730), Brazil (1300), Senegal (900), Uganda (900), the Philippines (753), etc. (Yu 2008).

1.1.2 Peanut Planting Area and Production Quantity in the World

The nationwide peanut production and distribution areas are widely spread around Asia, Europe, Africa, the Americas, Oceania, and other regions of the world. According to the statistics in FAO, in 2014, the peanut planting area was 26,541,660 ha, and the production quantity was 43,915,365 tons in the world. The countries whose planting areas were ranked in the top ten positions were

India, China, Nigeria, Sudan, the United Republic of Tanzania, Senegal, Niger, Chad, the United States, and the Democratic Republic of the Congo, and the countries whose production quantity were ranked in the top ten positions were China, India, Nigeria, the United States, Sudan, the United Republic of Tanzania, Argentina, Myanmar, Chad, and Senegal. At present, the peanut planting area of China is ranked second and production quantity ranked first in the world (See Tables 1.1 and 1.2).

1.1.3 International Trade of Peanut and Its Processed Product in the World

FAO statistics show that the trade volume of world peanut and its processed product has been increasing year by year. From 2000 to 2013, the import volume of **shelled peanut** increased from 1,222,654 to 1,680,197 tons with an increase of 37.42%, the import amount increased from 851,129 to 2,466,799 tons with an increase of 189.83%, the export volume increased from 1,199,316 to 167,1318 tons with a decrease of 39.36%, and the export amount increased from 787,404,000 to 2,195,422,000 dollars with an increase of 178.83%; the import volume of **peanut oil** decreased from 261,749 to 208,114 tons with a decrease of 20.49%, the import amount increased from 221,795,000 to 389,713,000 dollars with an increase of 75.71%, the export volume decreased from 260,473 to 186,308 tons with a decrease of 28.47%, and the export amount increased from 197,957,000 to 351,265,000 dollars with an increase of 77.45%; the import volume of peanut butter increased from 41,919 to 93,040 tons with an increase of 121.95%, the import amount increased from 67,191,000 to 312,410,000 dollars with an increase of 364.96%, the export volume increased from 43,825 to 83,641 tons with an increase of 90.85%, and the export amount increased from 75,275,000 to 293,505,000 dollars with an increase of 289.91%; the import volume of peanut cake decreased from 259,702 to 112,356 tons with a decrease of 56.74%, the import amount increased from 38,834,000 to 52,371,000 dollars with an increase of 34.86%, the export volume decreased from 223,901 to 122,103 tons with a decrease of 45.47%, and the export amount increased from 28,278,000 to 49,142,000 dollars with an increase of 73.78% (see Table 1.3 for details).

It can be seen from the trade volume of the continents of the world that the main trade contacts of peanut and its processed product occurred in the Americas, Asia, and Europe. The Americas is the main export region of peanut kernel, peanut oil, and processed peanut. In 2013, the export volumes of peanut kernel, peanut oil; Asia is the main export region of peanut kernel; in 2013, the export volumes of peanut kernel was 553,062 tons, accounting for 33.09% in the total export volume of the world; Europe is the main import region of peanut kernel and peanut oil; in 2013, the import volumes of peanut kernel and peanut oil were 881,888 and 98,700 tons, respectively, accounting for 52.49 and 47.43% in the total import volume of the world (see Tables 1.4, 1.5, 1.6, 1.7, and 1.8).

Table 1.1 The planting area of groundnuts in the world (unit: ha)

Country/ region	Year														
	2000	2001	2002	2003	2004	2005	2006	2007	2008	2009	2010	2011	2012	2013	2014
Algeria	4020	4250	3750	3380	4081	2837	2718	2683	2840	2574	2524	2552	2293	2335	2460
Angola	38,541	81,482	85,766	150,281	175,202	181,778	170,000	231,741	259,081	277,772	285,287	314,232	231,620	334,662	335,580
Argentina	219,260	251,060	222,370	156,426	167,473	210,820	163,677	215,060	227,389	257,430	218,828	264,568	307,166	404,022	409,312
Australia	17,400	14,700	10,254	13,801	14,005	14,000	11,896	12,197	11,000	9954	11,800	7282	7400	7000	5500
Bangladesh	29,150	25,900	26,710	25,900	26,034	28,852	29,394	33,666	31,084	32,540	33,591	31,755	31,182	28,750	29,560
Barbados	30	15	15	30	17	13	11	5	14	24	8	10	10	9	8
Belize	83	84	96	62	57	60	55	71	98	44	59	70	71	70	69
Benin	138,586	145,472	163,744	169,942	160,719	148,494	100,889	121,325	125,025	147,798	175,430	136,699	140,014	162,412	163,009
Bolivia (Plurinational State of)	11,182	10,940	11,194	11,312	11,706	11,915	11,930	11,998	12,042	12,657	12,102	12,290	13,406	11,919	13,066
Botswana	1230	1000	1000	600	1000	1109	1000	293	959	2350	3075	1408	560	641	1239
Brazil	102,914	105,000	97,093	89,174	104,501	136,048	110,777	113,776	121,456	94,325	94,329	106,671	110,366	120,970	142,952
Bulgaria	10,000	10,000	10,000	10,000	9000	10,000	10,000	10,000	10,000	10,000	10,000	10,000	10,000	10,000	10,000
Burkina Faso	236,880	330,904	342,637	404,110	352,528	274,603	310,597	415,171	369,125	458,222	409,922	388,704	397,762	448,767	375,040
Burundi	12,000	12,500	13,000	13,200	13,200	13,000	12,500	13,000	12,900	15,000	16,679	17,751	21,673	23,130	16,708
Cabo Verde												300	318	400	350
Cambodia	10,271	11,271	11,505	14,277	18,009	15,288	13,596	21,460	18,183	16,474	20,041	16,287	18,048	18,000	18,000
Cameroon	204,158	205,000	290,136	295,939	301,858	209,347	236,951	330,802	345,756	366,501	377,496	505,469	422,464	463,209	440,000
Central African Republic	95,568	110,000	116,000	130,000	135,000	140,000	140,686	156,374	159,501	162,691	97,777	99,265	95,715	105,000	100,000
Chad	437,848	477,126	379,498	414,066	369,860	528,174	485,968	453,587	546,375	625,001	1,039,713	675,565	919,918	891,039	774,604
China	4,884,910	5,016,421	4,946,100	5,082,000	4,766,685	4,684,628	3,980,323	3,967,969	4,268,322	4,398,431	4,547,917	4,604,366	4,719,430	4,651,599	4,625,494
Colombia	3141	1948	1314	1546	1573	1807	2262	2278	2310	2885	3604	3182	2222	2696	1566
Comoros	1026	1020	1040	1050	1000	1070	1090	1000	1090	1088	1091	1096	1102	1108	1114
Congo	38,640	39,027	39,416	40,000	42,000	40,162	42,122	45,000	47,000	48,500	55,324	55,786	48,749	49,392	50,011
Costa Rica	170	171	172	174	174	176	179	180	175	182	186	189	200	200	195
Cuba	15,000	10,000	26,000	14,377	12,604	11,838	11,106	10,000	9786	9337	8886	8371	5500	5500	5400
Cyprus	280	410	325	290	260	245	170	74	39	20	14	15	8	61	51
Côte d'Ivoire	79,897	80,920	67,483	69,708	71,974	72,708	72,569	74,622	74,369	76,525	77,782	79,864	80,595	80,000	78,000
Democratic Republic of the Congo	491,003	473,644	456,915	462,262	467,674	473,149	473,959	474,769	475,578	476,388	477,199	473,000	501,515	513,334	524,462

Dominican Republic	1877	2307	3351	2309	2128	2613	2205	2591	1905	3072	2679	2867	2568	4083	4219
Ecuador	11,169	11,170	10,524	12,828	15,368	15,754	15,800	16,000	16,500	17,500	16,895	17,231	17,185	17,334	17,512
Egypt	60,329	63,347	59,288	61,868	60,725	62,260	55,550	65,250	61,401	63,780	66,764	65,050	62,468	62,067	57,321
El Salvador															
Eritrea	1433	1412	1320	1500	1600	2621	2375	5475	2202	2077	2914	2822	2795	2838	2882
Ethiopia	13,550	17,200	16,034	20,000	20,217	27,084	35,468	37,126	40,198	41,579	49,603	64,477	90,156	79,747	64,649
Fiji	284	290	297	303	309	315	300	330	338	346	354	400	425	430	407
Gabon	18,000	17,000	18,000	17,000	16,479	16,463	16,729	17,677	19,602	17,606	17,628	17,887	17,822	17,855	17,888
Gambia	118,100	138,888	105,607	107,937	116,627	105,000	110,000	117,591	133,703	148,331	135,500	111,924	116,507	100,305	81,026
Georgia	405	343	400	190	315	442	100	28	30	78	139	151	150	149	148
Ghana	218,000	254,000	384,000	464,700	431,667	450,000	480,000	341,640	350,660	342,550	353,376	356,780	345,186	328,940	334,000
Greece	600	500	500	500	490	500	494	457	300	590	576	283	668	671	580
Guatemala	1680	1610	1,,680	1700	1610	1700	1800	2795	2795	2865	2878	2829	2926	2986	3045
Guinea	153,427	159,063	167,637	176,672	186,195	196,231	206,806	217,955	203,884	214,356	227,453	239,053	295,758	297,564	215,000
Guinea-Bissau	18,539	18,840	19,213	17,000	16,000	21,373	20,000	23,230	32,283	29,163	35,038	30,421	38,643	35,356	35,082
Guyana	3825	2200	2299	2067	2021	1989	2136	2107	2153	2225	2299	2363	2300	2300	2350
Haiti	26,000	25,000	25,500	26,000	25,000	29,000	25,000	32,000	31,000	25,000	31,424	35,260	41,772	52,844	42,154
Honduras	195	148	134	135	137	139	144	140	145	146	147	147	150	150	155
Hungary				3	6	3	3	2	12	12	12	12	10	6	0
India	6,558,600	6,238,100	5,935,500	5,987,000	6,640,400	6,736,000	5,615,100	6,292,000	6,164,900	5,477,500	5,860,000	5,310,000	4,770,000	5,505,210	4,685,000
Indonesia	683,554	654,800	646,953	683,537	723,434	720,526	706,753	660,480	636,229	622,616	620,563	539,230	559,532	519,056	499,338
Iran (Islamic Republic of)	1311	1321	1328	1337	1400	1500	1400	1411	1426	1415	1410	1407	1400	1450	1443
Iraq	6500	7000	8500	11,000	11,250	11,750	12,500	12,000	1750	6000	4000	5465	1546	1410	1451
Israel	3700	4080	4000	3020	2850	3060	2920	3000	3000	2880	2810	1790	2410	1087	2088
Italy															
Jamaica	2383	3507	2738	2457	2000	2504	2459	2729	1950	1916	1751	2013	2125	2218	2006
Japan	10,800	10,300	9950	9530	9110	8990	8600	8310	8070	7870	7720	7440	7180	6970	6840
Jordan	0	0	0	0	0	0	0	0	0	0	0	0	0	0	0
Kazakhstan	300	150	172	250	150	100	44	41	93	40	100	100	100	100	4
Kenya	17,000	16,114	15,194	15,022	14,932	14,963	17,230	20,329	18,981	20,640	19,291	13,504	16,387	17,311	21,611
Kyrgyzstan	115	116	333	373	273	244	271	298	198	255	98	113	107	59	54

(continued)

Table 1.1 (continued)

Country/ region	Year														
	2000	2001	2002	2003	2004	2005	2006	2007	2008	2009	2010	2011	2012	2013	2014
Lao People's Democratic Republic	12,800	12,100	13,698	14,585	14,605	16,650	18,385	15,965	19,377	19,577	23,570	32,780	21,620	24,595	25,310
Lebanon	2650	2500	2300	2200	2000	1900	2500	2707	3202	2933	2785	2558	2200	2300	2764
Liberia	8000	7496	7285	7000	8000	8150	8500	8700	8000	7000	7331	7500	8000	8000	8455
Libya	12,000	12,006	13,000	13,500	10,500	12,006	10,500	10,500	10,500	9500	10,000	10,000	10,000	10,000	10,000
Luxembourg															
Madagascar	47,205	47,450	47,735	47,950	48,480	54,506	54,800	52,000	54,000	50,000	52,000	53,000	55,000	55,000	53,000
Malawi	169,078	181,337	205,726	229,996	218,028	248,276	244,567	258,111	266,115	266,946	295,236	308,094	353,190	362,824	373,925
Malaysia	379	239	354	300	350	350	238	224	176	190	189	154	119	127	142
Mali	199,735	197,057	273,640	252,524	219,589	259,029	334,671	330,544	395,822	335,031	336,976	340,000	344,000	373,383	351,977
Mauritania	2820	1830	860	868	500	875	1000	1000	1000	1100	1100	1067	1000	1014	1002
Mauritius	123	123	116	255	212	137	178	140	143	241	208	163	266	182	240
Mexico	91,756	79,427	62,010	50,222	65,538	48,014	44,863	52,236	52,026	51,343	52,564	61,344	57,832	56,382	59,024
Morocco	18,200	21,700	20,400	21,500	21,600	19,500	14,900	19,700	20,800	22,700	22,900	17,840	15,490	16,300	16,090
Mozambique	269,442	237,272	326,700	292,537	293,921	424,400	322,900	396,900	458,200	357,000	365,856	288,000	389,266	398,000	416,500
Myanmar	559,895	586,130	567,186	654,438	654,000	684,000	730,000	755,500	815,000	844,267	866,499	887,034	477,200	479,200	484,000
Namibia	536	822	251	93	550	536	650	750	983	772	936	605	585	595	609
Nicaragua	22,326	21,496	16,009	23,009	26,761	30,735	26,028	28,127	38,579	32,290	32,562	35,283	37,980	45,421	40,000
Niger	360,338	192,300	334,600	423,400	349,300	308,800	310,100	375,318	675,477	588,651	795,768	690,853	741,309	760,455	778,994
Nigeria	1,934,000	1,731,000	1,878,000	1,985,000	2,097,000	2,187,000	2,224,000	2,202,638	2,336,400	2,643,330	2,789,180	2,353,680	2,659,800	2,732,700	2,770,100
Occupied Palestinian Territory	4	5	6	5	5	5	5	5	6	8	8	5	5	5	5
Pakistan	81,500	99,400	86,500	102,400	105,816	93,714	93,500	94,900	92,800	87,400	82,905	95,486	81,630	93,799	96,258
Papua New Guinea	1040	1036	1035	1200	960	1200	1207	1250	1300	1300	1294	1300	1324	1342	1359
Paraguay	29,330	30,254	29,059	33,410	34,926	35,000	37,000	37,000	24,113	25,100	24,698	25,819	25,000	24,999	25,000

Peru	5479	4277	2802	2805	3777	3700	3900	4294	4223	5041	4213	3843	3835	4035	3759
Philippines	26,866	26,668	27,057	26,709	26,896	27,472	27,642	28,316	27,726	28,235	27,125	26,902	26,108	25,602	25,048
Portugal	10	10	11	12	10	10	13	15	15	15	14	14	15	16	14
Republic of Korea	4662	4763	5483	4079	3454	3352	2970	3318	3366	4111	5381	4413	4072	4374	4565
Rwanda	13,463	14,767	15,900	16,803	18,883	16,011	16,196	17,000	20,898	23,048	20,558	22,846	20,638	24,160	27,513
Reunion	930	1015	1103	1111	1216	1281	1342	1401	1456	1506	1552	1594	1600	1600	1638
Saint Kitts and Nevis	25	24	28	27	23	28	33	30	24	21	34	33	23	28	16
Saint Vincent and the Grenadines	250	255	255	260	310	310	300	300	310	300	140	81	61	61	69
Saudi Arabia	1015	1018	1000	1011	1006	1000	1000	1000	1000	776	500	546	500	500	499
Senegal	1,095390	920,534	813,725	524,843	747,303	772,305	594,264	607,195	836,843	1,059093	1,195573	865,770	708,950	916,750	878,659
Sierra Leone	19,026	39,000	76,000	92,000	120,000	137,000	150,000	84,482	87,017	100,070	110,077	112,254	100,890	78,585	75,600
Somalia	2100	2700	3580	3580	11,000	10,000	4900	5300	5945	6584	6712	6919	7122	7322	7519
South Africa	82,600	165,000	94,160	49,850	71,500	40,000	48,550	40,770	54,200	54,550	57,450	55,150	45,450	46,900	52,125
South Sudan													190,000	226,000	235,000
Spain	71	56	45	26	42	49	202	33	30	35	19	63	1208	648	952
Sri Lanka	10,540	9680	9110	11,380	9980	10,920	11,660	10,420	10,270	9010	9480	9290	11,610	15,200	14,304
Sudan (former)	1,462,719	1,531,320	1,350,200	1,055,040	1,067,640	961,380	594,583	597,917	953,781	945,417	1,151,640	1,698,480			
Sudan													1,619,520	2,161,740	2,104,000
Suriname	200	215	165	245	215	120	62	51	10	23	21	18	26	30	21
Swaziland	5510	5500	5510	5500	5600	5510	5500	5510	5558	5829	5583	5730	6030	6458	7017
Syrian Arab Republic	10,260	16,446	6661	5200	6540	7800	5740	7240	6241	5108	4688	4610	8771	5045	6032
Tajikistan	800	906	712	448	900	1000	1627	1145	1299	1860	2067	1956	2603	2619	2544
Thailand	82,786	66,332	68,876	46,082	39,993	39,894	38,310	32,519	32,674	32,635	30,500	30,114	29,483	26,337	23,660
Timor-Leste	4000	3570	3327	3083	2917	2751	2766	2543	2523	3255	1154	3753	3583	3043	3062
Togo	53,894	58,841	51,730	63,313	59,286	53,545	52,062	56,084	67,357	67,692	69,621	68,785	67,857	60,813	61,120
Tonga	562	1544	1545	1534	1379	1412	1471	1550	1623	1713	1804	1898	2000	2000	2194
Turkey	28,300	27,000	33,000	28,000	26,000	25,850	22,490	25,942	24,830	25,335	27,440	25,471	37,388	35,943	33,327

(continued)

Table 1.1 (continued)

Country/ region	Year														
	2000	2001	2002	2003	2004	2005	2006	2007	2008	2009	2010	2011	2012	2013	2014
Uganda	199,000	208,000	211,000	216,000	221,000	225,000	230,000	235,000	280,000	369,000	394,000	409,000	421,000	422,000	422,300
United Republic of Tanzania	117,000	247,300	366,940	370,745	374,550	409,320	480,000	560,000	470,670	428,550	482,310	675,226	839,631	943,676	1,619,500
United States	540,670	571,380	524,760	530,950	564,140	659,240	489,270	483,599	609,861	436,660	507,890	437,310	649,120	422,090	535,200
Uruguay	1700	2200	2250	2300	2250	2100	2250	2300	2100	2600	2879	3063	3600	3500	3400
Uzbekistan	9000	9000	3360	4450	3800	4460	4030	5000	5000	5000	5000	5000	5000	5000	5000
Vanuatu	2000	2300	1850	1800	2500	1800	2300	2400	1800	2400	2214	2206	2206	2209	2212
Venezuela (Bolivarian Republic of)	560	520	511	400	335	293	392	350	442	220	405	470	1274	1063	1461
Vietnam	244,900	244,600	246,700	243,800	263,700	269,600	246,700	254,500	255,300	245,000	231,400	223,744	219,265	216,215	208,149
Zambia	132,284	108,478	88,000	88,000	88,000	108,245	96,400	124,578	121,375	204,073	254,566	209,237	176,162	207,249	237,423
Zimbabwe	268,100	275,076	258,610	105,052	133,339	200,592	176,196	260,518	180,000	170,000	319,608	329,803	207,120	158,607	200,000
Total	23,264,694	23,102,346	22,926,420	22,862,973	23,583,013	24,083,379	21,534,224	22,708,928	24,216,990	24,148,563	26,142,267	25,105,921	25,194,119	26,880,761	26,541,660

Data: FAOSTAT

Table 1.2 The production quantity of groundnuts in the world (unit: tons)

Country (region)	Year 2000	2001	2002	2003	2004	2005	2006	2007	2008	2009	2010	2011	2012	2013	2014
Algeria	3894	4621	4616	3842	4269	3304	2995	3319	3222	3057	3385	3037	2778	2753	2909
Angola	12,750	27,055	31,447	64,075	49,997	66,001	57,000	66,660	91,925	110,828	115,164	161,116	66,616	192,028	252,480
Argentina	419,600	394,800	361,900	220,900	292,950	444,800	347,323	600,035	625,349	605,491	611,040	701,535	685,722	1,025857	1,165924
Australia	35,100	36,600	27,568	43,633	32,451	32,000	24,508	17,587	24,300	19,346	27,400	18,392	22,600	21,837	15,900
Bangladesh	32,000	27,000	34,000	34,240	34,075	38,880	37,980	45,910	44,268	47,160	53,467	53,664	51,944	49,791	56,439
Barbados	60	30	30	63	28	23	21	9	24	50	16	19	19	17	15
Belize	115	119	139	82	77	113	102	98	99	63	91	99	100	99	98
Benin	121,159	125,377	146,214	124,979	151,666	140,329	99,382	114,460	115,562	132,069	154,403	131,790	129,109	149,762	144,731
Bolivia (Plurinational State of)	12,708	12,210	12,701	13,096	13,738	14,227	14,210	13,900	13,769	13,315	12,809	12,874	16,751	20,045	18,100
Botswana	1230	147	137	578	483	576	570	256	1160	1539	1826	833	200	112	553
Brazil	184,487	201,748	195,284	187,719	236,488	315,239	249,916	263,440	312,802	255,662	261,455	311,409	334,224	389,783	402,626
Bulgaria	8000	6000	7000	7000	7000	7000	7000	7000	7000	7000	7000	7000	7000	7000	8571
Burkina Faso	169,146	301,092	323,642	358,121	245,307	220,525	215,447	244,922	346,292	330,624	340,166	265,322	310,759	349,688	335,223
Burundi	8764	8800	9200	9400	9800	9700	9200	9500	9200	8000	9227	9470	9963	10,243	9296
Cabo Verde												200	223	280	250
Cambodia	7490	8913	9738	18,483	21,543	22,629	23,811	30,000	25,470	21,812	21,957	22,836	30,299	27,400	25,900
Cameroon	196,702	197,631	210,712	218,027	225,720	346,448	414,046	449,123	484,199	503,175	536,187	564,230	633,799	635,947	614,000
Central African Republic	104,500	121,900	127,800	133,600	139,500	145,400	146,100	156,374	159,501	162,691	140,000	160,000	149,264	162,000	157,000
Chad	358,791	448,089	379,968	414,868	385,103	486,255	420,016	464,320	548,342	517,449	1,103,314	569,929	1,297,712	965,162	791,088
China	14,515,755	14,471,835	14,895,099	13,493,462	14,410,302	14,395,479	12,809,561	13,079,363	14,341,175	14,764,870	15,709,036	16,114,231	16,856,845	17,018,965	16,550,213
Colombia	4504	3204	1994	2050	1834	2111	2814	2586	2634	3158	5171	4385	3082	2150	1757
Comoros	829	840	860	870	840	880	900	840	900	904	915	919	926	933	940
Congo	23,184	23,416	23,650	23,700	23,000	22,900	23,400	23,900	28,200	29,100	29,100	28,744	26,283	26,138	25,967
Costa Rica	210	210	211	213	214	215	217	225	210	216	218	219	220	220	221
Cuba	15,000	10,000	26,000	14,469	12,695	11,933	11,206	10,000	9938	9539	9151	8695	5800	5800	5700
Cyprus	1100	1750	1300	1200	1330	1340	1020	910	269	238	243	248	234	271	251
Czechoslovakia	71,853	73,394	61,207	63,226	65,280	67,239	69,239	69,256	81,015	84,952	90,227	91,844	93,490	95,166	117,692

(continued)

Table 1.2 (continued)

Country (region)	Year														
	2000	2001	2002	2003	2004	2005	2006	2007	2008	2009	2010	2011	2012	2013	2014
Democratic Republic of the Congo	382,000	368,495	355,480	359,640	363,850	368,110	368,740	369,370	370,000	370,631	387,734	393,901	405,277	413,342	421,568
Dominican Republic	2268	2780	3436	2496	2846	2951	2522	3147	2454	3917	3830	4252	3899	5279	5307
Ecuador	6211	9702	6739	11,078	19,667	16,651	16,700	17,000	17,500	19,000	18,388	19,182	19,492	20,051	20,651
Egypt	187,169	205,066	191,037	195,869	191,846	199,560	183,970	217,580	208,835	198,012	202,906	206,574	205,419	204,796	183,438
El Salvador															
Eritrea	1300	1369	1188	1200	754	2485	2311	6327	550	1262	2013	2176	2172	2199	2226
Ethiopia	11,929	15,210	13,285	29,284	20,715	29,053	34,150	51,080	44,685	46,425	71,607	103,478	124,419	112,088	103,706
Fiji	258	264	270	276	281	286	300	295	298	301	303	300	325	350	340
Gabon	19,600	16,300	16,200	17,000	17,000	16,700	17,000	18,000	20,000	18,000	18,058	18,360	18,330	18,401	18,472
Gambia	138,000	151,069	71,526	92,937	135,697	107,000	114,500	72,557	109,641	121,950	137,631	83,858	119,614	93,862	80,650
Georgia	567	348	200	72	654	352	88	23	20	59	111	121	121	121	121
Ghana	209,000	258,000	520,000	439,000	389,649	420,000	520,000	301,770	470,100	526,040	530,887	465,103	475,056	408,814	426,280
Greece	3000	2000	1000	1000	1736	1733	1662	1516	1000	2000	2000	1000	2465	2507	2098
Guatemala	1510	1529	1533	1540	1551	1550	1580	3378	3387	3387	3280	3224	3390	3492	3593
Guinea	199,661	209,946	224,647	240,378	257,210	275,222	294,494	304,805	315,107	300,098	332,081	352,779	357,867	363,028	297,000
Guinea-Bissau	20,584	21,011	21,528	20,000	19,000	24,379	22,000	24,709	46,460	30,696	36,177	35,431	45,214	41,297	41,098
Guyana	3300	1900	1992	1798	1768	1745	1879	1858	1904	1973	2044	2106	2056	2061	2111
Haiti	21,000	21,000	20,000	22,000	22,000	24,300	24,300	27,000	26,000	21,500	27,254	30,603	36,353	46,154	36,923
Honduras	110	84	76	76	77	78	81	80	81	81	81	81	80	80	84
Hungary				5	10	5	5	3	10	15	14	14	10	7	0
India	6,480,300	7,027,500	4,121,100	8,126,500	6,774,400	7,993,300	4,863,500	9,182,500	7,168,100	5,428,500	8,265,000	6,964,000	4,695,000	9,472,000	6,557,000
Indonesia	1,292,000	1,245,000	1,259,000	1,378,000	1,469,000	1,467,000	1,470,000	1,384,400	1,353,600	1,364,700	1,267,054	958,524	864,458	701,680	638,896
Iran (Islamic Republic of)	3444	3480	3486	3587	3600	3000	3600	3626	3723	3634	3620	3610	3700	4000	3748
Iraq	15,000	16,000	18,000	21,000	34,000	34,000	32,000	28,000	6000	19,000	10,000	10,888	5448	5834	5477
Israel	23,840	29,040	23,225	25,100	24,100	25,100	25,600	25,646	16,737	16,123	15,861	11,618	12,983	17,799	15,430
Italy															
Jamaica	2766	4436	3015	3121	2572	3175	3413	3728	2825	2704	2007	2643	2701	2832	2465
Japan	26,700	23,100	24,000	22,000	21,300	21,400	20,000	18,800	19,400	20,300	16,200	20,300	17,300	16,200	16,100

Jordan	0	0	0	0	0	0	0	0	0	0	0	0	0	0	0
Kazakhstan	340	400	162	200	200	200	84	82	179	120	130	130	140	30	8
Kenya	30,000	21,974	21,000	21,000	21,000	21,000	24,896	29,908	17,126	21,454	10,894	12,803	24,639	94,072	56,149
Kyrgyzstan	130	152	513	596	368	322	326	306	283	427	167	194	206	100	111
Lao People's Democratic Republic	13,201	16,779	16,377	16,019	12,404	26,990	27,600	35,070	32,690	35,163	50,945	70,195	46,020	54,805	59,250
Lebanon	7700	7000	6200	6000	5700	5500	7000	7668	8556	8031	8222	8450	9000	10,000	10,091
Liberia	4800	4533	4428	4276	4895	5009	5100	5300	5200	4800	5015	5100	6000	6500	6638
Libya	20,000	20,637	22,000	24,000	23,000	26,000	23,000	23,000	23,000	26,000	22,000	17,000	17,000	17,000	17,000
Luxembourg															
Madagascar	35,030	35,240	35,455	35,610	34,600	38,053	41,858	32,000	34,000	26,000	30,000	31,000	33,000	33,000	32,000
Malawi	122,281	155,183	157,874	190,112	153,414	141,078	203,071	261,810	243,215	275,176	297,487	325,215	368,081	380,800	296,498
Malaysia	1937	1336	1320	1581	1696	1700	803	823	706	657	578	507	615	534	1564
Mali	193,073	139,832	191,548	214,727	161,044	279,503	265,549	324,187	388,383	300,624	314,458	524,000	478,870	515,333	509,363
Mauritania	2256	1466	690	690	400	700	800	800	800	860	870	864	815	825	817
Mauritius	408	323	284	893	610	231	381	290	320	587	556	499	723	431	618
Mexico	142,216	119,520	74,643	91,616	98,940	72,853	68,242	82,809	80,735	85,502	81,485	79,827	114,846	99,849	96,346
Morocco	38,820	44,550	40,340	21,110	51,520	47,000	38,240	49,610	41,010	51,690	49,560	39,710	36,348	38,790	38,376
Mozambique	124,290	109,175	101,074	87,463	90,232	93,000	85,977	102,932	102,520	97,000	157,685	95,700	112,913	117,000	140,124
Myanmar	633,805	731,094	756,622	877,700	946,290	1,039,350	1,024,000	1,087,900	1,202,200	1,304,829	1,362,452	1,399,625	843,500	853,000	865,900
Namibia	250	398	118	45	242	47	145	329	370	400	334	199	205	201	199
Nicaragua	97,004	100,806	86,234	134,091	149,097	173,523	145,654	156,750	198,951	167,371	180,250	184,285	200,000	214,285	221,340
Niger	113,216	82,000	149,600	185,600	159,000	139,100	135,900	147,676	308,510	253,497	406,245	395,669	291,763	342,772	403,422
Nigeria	2,901,000	2,683,000	2,855,000	3,037,000	3,250,000	3,478,000	3,825,000	2,847,373	2,872,740	2,977,620	3,799,240	2,962,627	3,313,500	2,474,530	3,413,100
Occupied Palestinian Territory	19	20	27	24	25	25	25	25	30	40	41	25	25	24	23
Pakistan	91,400	101,100	90,300	114,700	76,448	69,129	73,900	83,400	85,500	53,200	67,803	87,894	81,330	100,788	86,663
Papua New Guinea	1000	1000	1000	1200	960	1200	1200	1250	1300	1310	1319	1341	1368	1395	1423
Paraguay	22,045	27,601	26,630	30,644	33,180	34,000	36,000	36,000	29,988	22,590	26,564	24,032	25,700	29,997	25,000
Peru	9458	7327	4977	5105	6171	6154	6817	6715	6541	8461	6860	6035	6006	6422	6246
Philippines	26,827	26,151	26,246	26,125	27,087	28,437	29,151	31,205	30,247	30,978	29,624	29,734	29,134	29,091	29,196
Portugal	24	25	25	25	21	21	27	30	31	31	30	30	33	34	31
Republic of Korea	8918	9620	11,212	7177	8257	6604	6354	6960	7459	10,201	13,991	10,854	9939	10,875	12,402

(continued)

Table 1.2 (continued)

Country (region)	Year														
	2000	2001	2002	2003	2004	2005	2006	2007	2008	2009	2010	2011	2012	2013	2014
Rwanda	7032	9635	10,414	10,275	10,785	10,099	9020	9921	11,472	15,353	14,369	14,756	11,638	14,414	10,181
Reunion	853	944	1035	1128	1264	1361	1459	1557	1651	1742	1829	1910	2000	2000	2042
Saint Kitts and Nevis	28	30	35	34	29	32	56	45	42	35	57	53	37	45	25
Saint Vincent and the Grenadines	250	255	255	260	310	310	300	300	310	300	255	211	210	240	237
Saudi Arabia	3991	3985	4000	3970	3961	4000	4000	4000	4000	3108	2000	2193	2000	2050	2043
Senegal	1,061,540	887,356	260,723	440,709	602,621	703,373	460,481	331,195	731,210	1,032,651	1,286,856	527,528	692,572	677,456	669,329
Sierra Leone	14,704	30,000	58,400	70,500	91,128	104,730	115,200	57,448	59,720	70,049	81,457	83,068	84,748	86,443	86,000
Somalia	2256	3240	4182	4180	9460	8500	5000	5500	6374	7077	7146	7428	7711	7993	8276
South Africa	136,128	222,319	133,730	66,934	115,000	64,000	74,000	58,000	88,800	99,500	88,000	64,250	59,000	41,500	74,500
South Sudan													105,000	120,000	135,000
Spain	188	162	115	90	122	140	249	75	99	100	58	224	1449	2051	2676
Sri Lanka	7070	6460	5730	6580	7930	9040	9820	9840	10,250	13,070	14,350	16,900	21,410	27,490	25,383
Sudan (former)	947,000	990,000	1,267,000	790,000	790,000	520,000	555,000	564,000	716,000	942,000	762,500	1,185,000			
Sudan													1,032,000	1,767,000	1,767,000
Suriname	254	265	207	308	268	146	88	64	17	27	32	26	38	33	31
Swaziland	4055	4100	4200	4100	4300	4018	4100	3620	3249	3293	2945	2808	2728	2680	2649
Syrian Arab Republic	28,106	29,574	20,485	16,200	18,300	24,500	16,230	22,380	18,770	14,574	13,037	11,892	13,101	15,389	17,067
Tajikistan	489	617	529	390	740	1115	3022	1539	1342	4913	4757	5416	6914	7063	7165
Thailand	131,897	107,252	112,149	76,070	64,868	66,595	64,643	53,559	53,015	45,715	45,344	47,837	47,680	43,764	39,670
Timor-Leste	4000	3876	3919	3971	4236	4302	5128	4665	4497	6258	1567	4071	3998	4083	4091
Togo	25,976	38,248	35,680	38,248	34,870	33,448	39,285	35,950	42,648	44,528	46,496	47,369	43,636	38,903	40,663
Tonga	307	846	850	848	766	787	823	850	914	968	1023	1081	1100	1300	1300
Turkey	78,000	72,000	90,000	85,000	80,000	85,000	77,454	86,409	85,274	90,081	97,310	90,416	122,780	128,265	123,600
Uganda	139,000	146,000	148,000	220,000	221,000	225,000	230,000	235,000	229,636	257,967	275,767	327,000	295,000	295,306	295,601
United Republic of Tanzania	52,000	206,800	346,790	339,225	331,660	293,870	350,000	408,058	340,770	347,970	465,290	651,397	810,000	1,425,000	1,635,335
United States	1,481,210	1,939,880	1,506,150	1,879,750	1,945,090	2,208,930	1,575,980	1,696,728	2,341,630	1,674,500	1,885,510	1,659,510	3,063,510	1,892,920	2,353,540
Uruguay	1100	1100	1200	1250	1180	1200	1180	1200	1300	1500	1586	1685	2000	1800	1700

Uzbekistan	10,000	11,000	6910	7220	7220	9040	9388	7000	7000	8000	7000	7000	7000	7000	5000
Vanuatu	2000	2200	2450	2500	2600	2500	2450	2550	2200	2550	2531	2581	2573	2589	2606
Venezuela (Bolivarian Republic of)	1603	1502	1521	1179	963	801	1315	1094	1326	651	1013	1010	4528	3535	3209
Vietnam	355,300	363,100	400,400	406,200	469,000	489,300	462,500	510,000	530,200	510,900	487,200	468,418	468,402	492,005	453,332
Zambia	51,971	41,420	60,000	60,000	60,000	74,218	84,010	55,215	70,527	120,564	163,733	278,775	113,025	106,792	143,591
Zimbabwe	190,890	168,749	58,378	86,494	64,157	57,754	83,170	126,768	80,000	78,570	136,719	97,504	72,623	68,918	84,000
Total	34,741,095	35,882,568	33,021,718	36,225,796	36,401,601	38,577,519	33,334,836	37,397,928	38,652,075	37,315,750	43,421,648	40,860,028	41,311,240	45,836,231	43,915,365

Data: FAOSTAT

Table 1.3 World's peanut trade

	Shelled peanuts				Peanut oil				Peanut butter				Peanut cake			
Year	Import quantity (t)	Import value (1000$)	Export quantity (t)	Export value (1000$)	Import quantity (t)	Import value (1000$)	Export quantity (t)	Export value (1000$)	Import quantity (t)	Import value (1000$)	Export quantity (t)	Export value (1000$)	Import quantity (t)	Import value (1000$)	Export quantity (t)	Export value (1000$)
2013	1,680,197	2,466,799	1,671,318	2,195,422	208,114	389,713	186,308	351,265	93,040	312,410	83,641	293,505	112,356	52,371	122,103	49,142
2012	1,611,702	2,705,754	1,627,011	2,450,856	193,012	438,792	199,459	438,081	88,766	308,773	92,043	319,757	88,143	34,308	82,570	25,562
2011	1,657,880	2,253,753	1,687,264	2,302,260	226,191	390,876	194,176	320,026	77,337	211,765	56,626	159,604	191,668	70,117	154,080	47,887
2010	1,551,848	1,702,481	1,255,622	1,372,877	260,989	341,297	220,184	283,382	70,246	170,266	55,213	138,984	193,301	58,262	180,165	50,698
2009	1,440,488	1,533,836	1,279,128	1,219,344	211,366	308,620	222,597	280,146	61,086	155,883	66,186	161,288	123,269	35,533	93,193	25,986
2008	1,526,165	1,787,396	1,196,346	1,320,946	204,877	398,809	153,135	302,119	68,706	170,976	72,270	179,482	176,158	60,808	117,227	36,387
2007	1,482,913	1,291,059	1,193,951	1,119,011	210,206	281,120	242,524	307,738	65,573	125,003	72,378	144,566	196,107	42,583	189,021	45,142
2006	1,357,856	999,416	1,070,947	800,810	251,123	260,481	249,820	231,884	70,204	118,348	68,514	124,496	262,666	46,703	220,016	33,274
2005	1,368,405	1,017,510	1,130,123	811,795	221,739	247,218	204,307	221,306	54,800	90,394	60,878	104,178	185,676	34,701	220,009	32,549
2004	1,258,561	973,121	1,009,793	787,933	240,619	294,481	207,726	234,860	54,189	90,386	54,102	97,367	291,276	66,145	217,305	38,522
2003	1,222,522	824,696	1,063,473	750,973	233,810	264,350	250,606	267,263	46,418	75,457	47,714	81,979	133,142	25,556	233,925	39,320
2002	1,296,141	787,284	1,134,517	656,334	254,161	191,476	275,241	193,058	43,912	68,980	49,735	81,884	323,187	52,516	231,890	31,098
2001	1,243,023	800,585	1,083,994	666,037	250,039	192,870	265,126	192,332	43,770	70,686	47,004	78,833	349,634	56,475	310,037	42,010
2000	1,222,654	851,129	1,199,316	787,404	261,749	221,795	260,473	197,957	41,919	67,191	43,825	75,275	259,702	38,834	223,901	28,278

Data: FAOSTAT

Table 1.4 African peanut trade

Year	Shelled peanuts				Peanut oil				Peanut butter				Peanut cake			
	Import quantity (t)	Import value (1000$)	Export quantity (t)	Export value (1000$)	Import quantity (t)	Import value (1000$)	Export quantity (t)	Export value (1000$)	Import quantity (t)	Import value (1000$)	Export quantity (t)	Export value (1000$)	Import quantity (t)	Import value (1000$)	Export quantity (t)	Export value (1000$)
2000	59,953	59,953	104,653	48,697	4010	3053	123,421	84,968	21	39	11	3	3869	752	143,781	16,518
2001	76,314	76,314	105,208	38,080	7361	4163	133,289	86,099	62	61	0	0	72,706	8555	159,862	19,578
2002	81,769	81,769	93,345	29,844	2916	1493	124,538	78,594	112	99	393	325	25,006	2703	136,170	17,316
2003	93,130	93,130	63,791	31,198	6291	4980	54,161	47,468	40	88	1	2	4057	840	36,203	8443
2004	92,015	92,015	61,606	35,114	7197	4803	41,594	40,841	71	163	27	10	5021	1246	19,390	4728
2005	86,540	86,540	64,163	30,991	13,020	8447	32,750	34,708	214	428	43	11	2009	359	11,563	1726
2006	87,649	87,649	51,650	25,550	3338	2597	80,718	68,524	262	626	133	310	3199	521	31,150	4564
2007	102,382	102,382	81,981	42,767	8807	5786	85,256	85,870	889	870	179	386	14,281	1496	37,625	7512
2008	105,494	105,494	79,683	47,617	15,786	11,631	24,092	31,575	617	1158	308	291	5520	735	7505	1648
2009	127,703	127,703	90,483	79,083	18,531	11,867	44,564	51,655	428	914	292	279	14,543	3358	25,155	6984
2010	96,243	96,243	79,919	55,299	24,906	12,270	64,096	73,642	436	964	21	17	8340	1576	36,559	7928
2011	109,384	109,384	98,646	98,516	19,732	9642	66,186	88,462	423	1005	29	27	23,488	10,520	43,621	10,957
2012	176,133	176,133	171,047	140,579	11,597	8416	21,366	35,851	1573	3914	1578	1441	6453	995	8388	2584
2013	121,226	121,226	142,657	147,624	12,408	10,956	23,410	45,975	1499	3982	732	373	2045	398	50,040	20,039

Data: FAOSTAT

Table 1.5 American peanut trade

Year	Shelled peanuts				Peanut oil				Peanut butter				Peanut cake			
	Import quantity (t)	Import value (1000$)	Export quantity (t)	Export value (1000$)	Import quantity (t)	Import value (1000$)	Export quantity (t)	Export value (1000$)	Import quantity (t)	Import value (1000$)	Export quantity (t)	Export value (1000$)	Import quantity (t)	Import value (1000$)	Export quantity (t)	Export value (1000$)
2013	239,387	324,067	694,870	890,600	21,682	38,025	125,453	209,578	39,219	136,621	63,991	224,220	40,060	20,900	40,132	13,944
2012	258,351	476,099	519,899	829,254	11,078	21,805	122,737	253,326	33,955	121,164	51,930	186,846	48,464	19,904	53,357	15,721
2011	260,480	373,499	518,275	708,766	16,837	26,209	79,666	125,169	25,583	64,427	41,109	108,956	25,193	10,314	24,018	7081
2010	231,260	259,325	495,052	496,988	27,178	34,879	113,334	131,720	22,813	49,328	40,525	98,972	28,241	10,103	46,668	10,634
2009	221,865	241,200	509,237	460,205	34,756	55,733	129,546	133,970	24,034	53,163	36,463	90,023	32,404	9532	22,596	5748
2008	216,931	299,772	501,862	537,015	39,285	65,856	78,392	143,599	30,713	65,406	39,853	96,832	34,039	8946	23,255	6305
2007	218,348	209,388	474,999	428,623	16,676	20,630	66,274	75,034	31,025	52,164	37,347	74,035	41,577	8271	26,636	4794
2006	200,698	155,486	457,263	322,504	60,133	56,974	114,252	93,611	33,485	50,201	40,660	75,312	57,213	9031	46,782	5683
2005	189,188	154,040	369,891	267,280	18,671	24,089	90,965	86,053	31,396	47,319	36,680	65,785	47,921	7212	42,633	5403
2004	187,797	161,794	322,628	265,887	68,652	81,592	59,483	60,100	27,690	42,024	31,600	55,877	29,292	4435	20,140	2849
2003	159,183	130,645	281,916	214,939	9267	9118	87,383	81,483	22,029	32,315	26,831	47,585	37,334	5287	33,909	4331
2002	202,480	149,640	377,583	256,572	33,909	23,317	84,923	51,556	20,886	31,302	28,698	51,007	31,652	4450	40,233	5446
2001	215,570	176,184	345,670	264,208	35,163	27,322	55,266	36,952	22,223	36,500	29,634	54,847	22,968	3147	38,243	5286
2000	266,164	220,684	476,555	365,516	21,859	17,718	66,743	45,027	21,776	35,207	28,748	54,241	16,481	2448	38,047	4706

Data: FAOSTAT

Table 1.6 Asian peanut trade

Year	Shelled peanuts				Peanut oil				Peanut butter				Peanut cake				
	Export value (1000$)	Import quantity (t)	Import value (1000$)	Export quantity (t)	Export value (1000$)	Import quantity (t)	Import value (1000$)	Export quantity (t)	Export value (1000 $)	Import quantity (t)	Import value (1000 $)	Export quantity (t)	Export value (1000 $)	Import quantity (t)	Import value (1000 $)	Export quantity (t)	Export value (1000 $)
2000	292,365	304,475	154,933	553,062	316,427	42,423	48,107	25,532	26,883	13,918	19,988	10,105	12,836	38,814	7253	32,133	5517
2001	319,666	297,215	137,535	571,915	307,460	39,073	37,011	26,494	23,996	14,332	20,113	10,856	12,903	67,951	12,578	98,389	14,666
2002	305,476	345,967	154,518	593,177	305,685	27,390	24,694	18,293	16,825	14,351	19,804	11,907	14,368	112,401	18,363	46,224	6682
2003	360,925	310,772	152,922	642,499	422,652	25,678	29,294	51,056	56,737	15,915	25,148	12,555	16,466	26,787	5835	153,122	24,809
2004	584,598	297,029	161,304	520,350	363,269	18,548	22,994	57,659	62,565	16,130	25,695	15,677	25,013	212,429	48,704	172,816	30,089
2005	819,077	336,023	179,729	596,239	403,350	22,546	26,554	29,472	33,955	12,144	17,737	15,483	21,736	121,536	24,811	160,705	24,698
2006	668,103	313,508	174,105	468,050	350,214	16,841	21,066	17,116	20,169	22,057	35,702	18,888	28,593	137,013	24,900	138,550	22,443
2007	758,706	377,037	234,070	533,217	510,220	30,886	44,956	51,786	81,453	15,776	29,015	21,373	35,537	67,056	15,065	123,553	32,479
2008	1,066,077	406,578	302,246	509,191	554,304	24,213	55,161	14,758	37,570	15,598	42,269	19,303	43,621	118,710	44,285	83,106	27,217
2009	841,361	377,158	307,024	573,132	505,107	36,965	65,585	13,498	29,242	12,507	30,010	19,782	42,045	48,358	13,606	35,091	9719
2010	1,100,422	427,308	408,525	547,863	613,025	90,536	126,983	15,537	30,265	19,233	47,978	1782	6550	66,002	21,569	73,922	24,881
2011	1,357,027	449,303	483,582	898,883	1,168,252	78,453	145,057	19,987	44,222	20,436	57,438	2641	10,807	50,407	18,307	47,990	14,978
2012	1,379,219	383,792	485,823	779,523	1,132,007	76,895	186,895	31,775	84,748	21,337	72,803	24,531	79,004	14,655	6923	10,956	3754
2013	1,433,529	422,589	518,528	665,379	823,582	73,653	140,183	14,668	39,274	19,514	63,678	1893	9954	39,071	15,750	23,294	11,530

Data: FAOSTAT

Table 1.7 European peanut

Year	Shelled peanuts				Peanut oil				Peanut butter				Peanut cake			
	Import quantity (t)	Import value (1000$)	Export quantity (t)	Export value (1000$)	Import quantity (t)	Import value (1000$)	Export quantity (t)	Export value (1000$)	Import quantity (t)	Import value (1000$)	Export quantity (t)	Export value (1000$)	Import quantity (t)	Import value (1000$)	Export quantity (t)	Export value (1000$)
2013	881,888	1,473,110	167,189	330,975	98,700	195,259	22,192	54,921	32,561	107,362	16,996	58,866	31,180	15,323	8383	3307
2012	779,675	1,558,086	155,562	346,756	91,944	216,993	23,299	62,766	31,646	110,123	14,002	52,455	18,571	6484	9812	3411
2011	824,738	1,289,315	170,728	324,938	109,756	206,278	28,301	61,964	30,706	88,324	12,842	39,804	92,580	30,976	38,450	14,866
2010	780,663	942,351	130,890	204,180	117,129	164,399	27,012	46,832	27,566	71,478	12,878	33,421	90,718	25,014	23,016	7255
2009	703,479	904,759	103,775	171,322	119,631	171,667	34,780	64,359	23,922	71,367	9636	28,890	27,964	9037	10,351	3535
2008	782,474	1,110,784	101,608	174,724	124,208	261,882	35,776	88,832	21,486	61,453	12,802	38,718	17,889	6842	3356	1184
2007	759,082	767,938	100,851	132,479	152,446	207,058	39,045	64,513	17,648	42,434	13,470	34,577	73,193	17,751	1207	3,56
2006	743,739	614,501	91,849	100,009	169,665	177,486	37,576	48,766	14,184	31,377	8832	20,275	65,241	12,251	3534	5,83
2005	748,038	635,204	96,572	106,423	164,765	184,203	50,868	65,426	10,841	24,479	8670	16,636	14,210	2319	5108	722
2004	668,200	594,636	90,049	107,008	143,970	181,538	48,728	70,750	9610	22,219	6797	16,463	44,534	11,760	4959	856
2003	634,921	480,914	71,370	78,341	189,907	216,955	57,476	80,999	8283	17,590	8327	17,926	64,964	13,594	10,691	1735
2002	639,292	432,258	66,531	61,094	187,465	139,427	47,374	45,790	8551	17,728	8737	16,184	154,128	27,000	9260	1630
2001	645,121	445,709	57,044	52,684	166,624	122,247	49,925	45,001	7140	13,957	6514	11,083	184,991	31,777	13,534	2402
2000	580,839	442,247	62,752	54,731	191,070	150,277	44,601	40,775	6191	11,902	4962	8195	200,408	28,261	9920	1514

Data: FAOSTAT

Table 1.8 Oceanian peanut

Year	Shelled peanuts				Peanut oil				Peanut butter				Peanut cake			
	Import quantity (t)	Import value (1000$)	Export quantity (t)	Export value (1000$)	Import quantity (t)	Import value (1000$)	Export quantity (t)	Export value (1000$)	Import quantity (t)	Import value (1000$)	Export quantity (t)	Export value (1000$)	Import quantity (t)	Import value (1000$)	Export quantity (t)	Export value (1000$)
2013	15,107	24,232	1223	2641	1671	5290	585	1517	247	767	29	92	0	0	254	322
2012	13,751	30,400	980	2260	1498	4683	282	1390	255	769	2	11	0	2	57	92
2011	13,975	23,915	732	1788	1413	3690	36	209	189	571	5	10	0	0	1	5
2010	16,374	20,023	1898	3385	1240	2766	205	923	198	518	7	24	0	0	0	0
2009	10,283	12,642	2501	3627	1483	3768	209	920	195	429	13	51	0	0	0	0
2008	14,688	19,818	4002	7286	1385	4279	117	543	292	690	4	20	0	0	5	33
2007	26,064	30,084	2903	4922	1391	2690	163	868	235	520	9	31	0	0	0	1
2006	12,262	10,167	2135	2533	1146	2358	158	814	216	442	1	6	0	0	0	1
2005	8616	7969	3258	3751	2737	3925	252	1164	205	431	2	10	0	0	0	0
2004	13,520	12,459	15,160	16,655	2252	3554	262	604	688	285	1	4	0	0	0	0
2003	24,516	19,412	3897	3843	2667	4003	5,30	576	151	316	0	0	0	0	0	2
2002	26,633	18,296	3881	3139	2481	2545	113	293	12	47	0	0	0	0	3	24
2001	8803	7035	4157	3605	1818	2127	152	284	13	55	0	0	1018	418	9	78
2000	11,224	9434	2294	2034	2387	2640	176	304	13	55	0	0	130	121	20	22

Data: FAOSTAT

It can be seen from the trade in various countries. In 2013, the countries whose exports of *shelled peanut* were ranked in the top five positions were India, the United States, Argentina, the Netherlands, and China; their export volumes were 541,337, 318,046, 188,954, 137,533, and 99,417 tons, and their export amount was 623,337,000, 416,161,000, 240,932,000, 277,297,000, and 170,505,000 dollars, respectively. The countries whose imports of shelled peanut were ranked in the top five positions were the Netherlands, Indonesia, Germany, Mexico, and the Russian Federation, and their import volumes were 354,057, 177,031, 104,516, 101,784, and 89,441 tons; their import amount was 591,724,000, 206,047,000, 183,825,000, 138,987,000, and 135,928 dollars, respectively (see Table 1.9 for details).

In 2013, the countries whose exports of peanut butter were ranked in the top five positions were the United States, Canada, the Netherlands, the United Kingdom, and China; their export volumes were 45,891, 18,043, 10,505, 3337, and 799 tons; and their export amount was 154,106, 69,780, 33,911, 13,376, and 34,700 dollars, respectively. The countries whose imports of peanut butter were ranked in the top five positions were Canada, the United States, the United Kingdom, China, and Saudi Arabia; their import volumes were 19,314, 15,567, 11,112, 5762, and 5674 tons, and their import amount was 62,127, 62,642, 38,011, 14,613, and 17,096 dollars, respectively (see Table 1.10 for details).

In 2013, the countries whose exports of *peanut oil* were ranked in the top five positions were Brazil, Argentina, Nicaragua, Senegal, and Belgium; their export volumes were 63,250, 40,804, 16,433, 16,418, and 11,981 tons, and their export amount was 100,900,000, 66,061,000, 30,138,000, 36,870,000, and 28,766,000 dollars, respectively. The countries whose imports of peanut oil were ranked in the top five positions were China, Italy, France, the United States, and Belgium; their import volumes were 68,325, 33,992, 19,422, 19,123, and 18,939 tons, and their import amount was 126,749,000, 57,744,000, 40,711,000, 31,522,000, and 34,851,000. See Table 1.11 for details.

In 2013, the countries whose exports of *peanut cake* were ranked in the top five positions were Spain, the Netherlands, Rwanda, Armenia, and Hungary; their export volumes were 31,783, 18,018, 17,949, 13,666, and 8551 tons, and their export amount was 0, 2,219,000, 0, 4,935,000, and 0 dollars, respectively. The countries whose imports of peanut cake were ranked in the top five positions were China, France, Chile, Honduras, and Guatemala; their import volumes were 36,807, 17,023, 14,107, 7837, and 6077 tons, and their import amount was 14,702,000, 8,486,000, 6,882,000, 3,636,000, and 2,785,000 dollars, respectively. See Table 1.12 for details.

Table 1.9 World's import and export trade conditions of shelled peanut in 2013

Country/region	Import quantity(t)	Import value (1000$)	Export quantity(t)	Export value (1000$)
Albania	45	73	0	0
Angola	1241	61,295	0	0
Argentina	83	94	188,954	240,932
Australia	13,422	20,802	1197	2520
Austria	3287	5009	343	897
Azerbaijan	1019	947	0	0
Bahamas	11	36	3	4
Bahrain	161	301	0	0
Bangladesh	10	16	101	38
Barbados	236	362	0	0
Belarus	4702	6404	1568	598
Belgium	12,618	21,562	7381	12,804
Benin	55	1	217	44
Bolivia (Plurinational State of)	1832	27	6112	6135
Bosnia and Herzegovina	507	651	0	0
Botswana	0	898	0	0
Brazil	142	176	80,691	112,539
Brunei Darussalam	150	266	0	0
Bulgaria	8097	10,998	428	832
Burkina Faso	12	1	350	506
Burundi	1981	923	0	0
Cambodia	0	7		
Cameroon	0	0	30	7
Canada	84,197	105,548	19	25
Central African Republic	0	0	0	0
Chad	0	0	36	63
Chile	10,674	18,182	0	0
China	16,494	22,352	99,417	170,505
Colombia	8656	11,853	1	6
Congo	30	8	0	0
Costa Rica	2394	2930	0	0
Croatia	1453	2565	38	74
Cyprus	946	2021	19	39
Czechia	2616	4851	70	129
Denmark	6521	12,197	817	1932
Ecuador	433	650	0	1
Egypt	4854	6575	13,482	20,417
El Salvador	4577	3961	0	0
Estonia	484	952	1	4
Ethiopia	91	190	281	301
Fiji	18	65	0	0
Finland	3013	5071	9	42

(continued)

Table 1.9 (continued)

Country/region	Import quantity(t)	Import value (1000$)	Export quantity(t)	Export value (1000$)
France	27,861	48,997	1040	3737
Gabon	0	0	1	3
Gambia	0	0	4148	2224
Georgia	831	1277	5	8
Germany	104,516	183,825	10,920	19,676
Ghana	56	9	304	901
Greece	10,782	18,779	619	895
Guatemala	3465	1769	0	0
Hungary	6266	10,907	26	46
India	114	62	541,337	623,337
Indonesia	177,031	206,047	168	145
Iran (Islamic Republic of)	21,400	18,000	18	12
Ireland	1101	1448	3	6
Israel	3289	6187	0	0
Italy	18,625	35,422	144	575
Japan	26,667	55,718	0	16
Jordan	7492	11,527	409	427
Kazakhstan	3678	4297	391	162
Kenya	7603	5100	223	53
Kuwait	1239	2311	39	86
Kyrgyzstan	0	0	120	102
Latvia	945	1638	186	326
Lebanon	8465	18,156	51	57
Libya	2231	3409	63	55
Lithuania	3488	5466	1331	2444
Luxembourg	5105	9321	0	0
Madagascar	0	0	9955	4868
Malawi	0	0	46,933	58,806
Malaysia	15,661	18,439	248	329
Mali	28	2	1436	211
Malta	321	588	0	0
Mauritius	0	0	0	0
Mexico	101,784	138,987	2530	6621
Morocco	248	333	0	0
Mozambique	1483	1176	16,161	13,982
Myanmar	20	26	0	0
Namibia	19	45	54	284
Netherlands	354,057	591,724	137,533	277,297
New Zealand	1667	3364	26	121
Nicaragua	75	61	92,646	102,899
Niger	80	15	1	0
Nigeria	7800	6400	1193	870

(continued)

Table 1.9 (continued)

Country/region	Import quantity(t)	Import value (1000$)	Export quantity(t)	Export value (1000$)
Norway	7602	15,341	51	99
Occupied Palestinian Territory				
Pakistan	11,481	5730	3	3
Paraguay	57	0	5839	5265
Peru	6053	6828	2	1
Philippines	12,351	4185	0	0
Poland	35,396	60,526	374	693
Portugal	4122	6589	2454	3665
Qatar	193	274	0	0
Republic of Korea	232	371	0	0
Republic of Moldova	609	1050	41	65
Romania	5191	7355	31	66
Russian Federation	89,441	135,928	144	282
Rwanda	0	0	0	0
Saint Kitts and Nevis	0	0	0	0
Saint Lucia	0	0	0	0
Samoa	0	0	0	0
Saudi Arabia	3134	3980	233	279
Senegal	0	0	26,635	21,405
Serbia	7210	12,880	41	75
Singapore	4	9	4717	7598
Slovakia	1988	3379	66	140
Slovenia	174	477	0	0
South Africa	18,763	20,058	7922	13,046
Spain	31,047	55,794	639	1374
Sri Lanka	1023	1111	0	0
Sudan	362	450	5251	4901
Suriname	260	456	0	0
Swaziland	415	571	745	246
Sweden	4204	7702	298	816
Switzerland	645	1383	1	0
Syrian Arab Republic	5000	4000	110	143
Thailand	62,855	69,593	44	147
The former Yugoslav Republic of Macedonia	2887	4217	55	105
Togo	0	0	0	0
Trinidad and Tobago	985	1618	0	1
Tunisia	4736	2947	0	0
Turkey	3600	8904	308	676
Uganda	5368	1011	220	264
Ukraine	31,479	50,856	20	47
United Arab Emirates	21,300	30,000	11,300	13,000

(continued)

Table 1.9 (continued)

Country/region	Import quantity(t)	Import value (1000$)	Export quantity(t)	Export value (1000$)
United Kingdom	83,444	130,761	517	1234
United Republic of Tanzania	8919	2477	4081	2284
United States	6896	18,274	318,046	416,161
Uruguay	490	675	0	0
Uzbekistan	0	0	933	1145
Venezuela (Bolivarian Republic of)	4109	7971	0	0
Yemen	3393	3671	67	61
Zambia	6862	7881	828	635
Zimbabwe	3753	3081	0	0

Data: FAOSTAT

1.2 Distribution and Utilization of Peanut Resources in China

1.2.1 General of Peanut Germplasm Resources in China

China's peanut germplasm can be divided into two groups (continuous blooming and alternative blooming which are equivalent to Shuzhi subspecies and Mizhi subspecies), four types (common type, Longsheng type, multi-grain type, and pearl type), eight species, and a quantity of variety groups. As the parent (subspecies) hybridization between two major groups is widely adopted in the peanut breeding research, the new type with characteristics of two groups occurs, and it is classified as "intermediate type," so the current peanut varieties of China can be divided into five types.

In more than 7490 copies of peanut germplasm resource materials collected by China, there are 4638 copies for domestic local varieties which are, respectively, from 22 provinces, and there are 2852 copies of resources introduced from foreign countries which are, respectively, from ICRISAT, the United States, Thailand, and other 31 countries (units) (Yu 2008). In the domestic resources, the varieties of common type and pearl type are the most, followed by that of the Longsheng type and intermediate type and that of multi-grain type is fewer. The major variety of peanuts are mainly distributed in Hebei, Shandong, Jiangsu, Henan, Anhui, and other provinces; the pearl-type peanuts are mostly concentrated in Guangdong, Guangxi, Jiangxi, Hubei, Hunan, Sichuan, Yunnan, Guizhou, and other provinces; the Longsheng-type peanuts are more in Guangxi, Sichuan, Jiangxi, and other provinces; the intermediate-type peanut varieties are mainly distributed in the middle and lower reaches of the Yangtze River and the large peanut area in the north; the multi-grain varieties are mainly distributed in the northeastward very early-maturing area and northwest inland peanut area.

Table 1.10 World's import and export trade conditions of peanut butter in 2013

Country/reigon	Import quantity(t)	Import value (1000$)	Export quantity(t)	Export value (1000$)
Argentina	0	0	0	0
Armenia	2	10	–	–
Aruba	79	375	0	0
Austria	157	720	15	129
Bahamas	447	920	–	–
Bahrain	314	2027	3	5
Barbados	0	0	0	0
Belarus	19	72	–	–
Belgium	1861	2883	271	1238
Bermuda	141	528	0	1
Bosnia and Herzegovina	40	170	–	–
Bulgaria	246	708	2	8
Cabo Verde	13	43	–	–
Cameroon	261	752	0	1
Canada	19,314	62,127	18,043	69,780
Chile	1943	3945	41	249
China	5762	14,613	799	4700
Colombia	89	343	2	15
Cook Islands	0	0		
Costa Rica	212	737	0	1
Croatia	155	626	0	1
Cyprus	27	132	0	0
Czechia	213	879	61	283
Denmark	406	1771	69	354
Dominica	11	50	–	–
Estonia	17	77	0	0
Fiji	242	747	29	92
Finland	160	693	1	7
France	3299	12,239	518	1687
Germany	5399	15,873	566	2901
Ghana	41	77	459	270
Greece	571	1548	89	325
Grenada	1	9	0	0
Guyana	28	98	0	0
Haiti	80	328		
Hungary	117	466	23	61
Ireland	797	3047	17	71
Italy	660	2678	16	92
Jamaica	0	0	0	0
Japan	5058	16,215	–	–
Kiribati	5	20	–	–
Latvia	17	107	3	18

(continued)

Table 1.10 (continued)

Country/reigon	Import quantity(t)	Import value (1000$)	Export quantity(t)	Export value (1000$)
Libya	110	348		
Lithuania	33	162	1	2
Luxembourg	16	76	0	0
Malaysia	0	0	0	0
Maldives	0	0	–	–
Malta	52	217	–	–
Namibia	713	1883	10	17
Netherlands	1473	5243	10,505	33,911
Norway	837	3222	11	69
Oman	293	1462	0	0
Panama	715	2518	–	–
Philippines	38	80	217	531
Poland	939	3204	385	1178
Portugal	296	1180	11	37
Qatar	1067	6479	–	–
Republic of Korea	0	0	0	0
Republic of Moldova	4	20	–	–
Romania	169	637	8	35
Russian Federation	481	1924	2	20
Saint Lucia	10	56	0	1
Saint Pierre and Miquelon	3	10	363	1203
Saudi Arabia	5674	17,096	–	–
Slovakia	480	1003	0	0
Slovenia	43	206	0	2
Spain	1351	3337	717	1675
Sudan	13	65	0	0
Suriname	534	1724	3	8
Sweden	1085	4175	68	274
Switzerland	0	0	0	0
The former Yugoslav Republic of Macedonia	56	188	300	1112
Togo	2	1	–	–
Trinidad and Tobago	35	165	11	59
Turkey	0	0	0	0
Ukraine	0	0	0	0
United Arab Emirates	1279	5564	511	3515
United Kingdom	11,112	38,011	3337	13,376
United States	15,567	62,642	45,891	154,106
Uzbekistan	0	0	–	–
Venezuela (Bolivarian Republic of)	10	46	0	0
Yemen	0	0	–	–

(continued)

Table 1.10 (continued)

Country/reigon	Import quantity(t)	Import value (1000$)	Export quantity(t)	Export value (1000$)
Zambia	330	769	261	83
Zimbabwe	0	0	0	0

Data: FAOSTAT

1.2.2 Peanut Planting Area and Yield in China

China has a broad range of peanut planting, and peanuts are planted around the country except for Tibet, Qinghai, and other provinces and regions. East China and middle-of-south regions are the main peanut production areas of China. According to the statistics of China Statistical Yearbook, in 2015, the peanut planting area in East China and middle-of-south regions would reach 40,528,500 mu, accounting for 58.53% in the total peanut area of China (69,240,000 mu), and the yield would reach 11,131,000.1 tons, accounting for 67.71% in the peanut yield of China (16,440,000 tons). See Table 1.13 for details.

The statistics of China Statistical Yearbook shows that, in 2000, the peanut production area was 72,829,600 mu and the yield was 14,436,600 tons in China; in 2015, the production area was 69,240,000 mu, the yield was 16,440,000 tons (FAO statistics show that the peanut planting area was 4,625,494 ha and the yield was 16,550,213 tons in China in 2015), the planting area decreased by 4.93%, and the yield increased by 13.88%. The planting area increased significantly in Liaoning, Jilin, Inner Mongolia, and other regions, and the yield increased significantly in Liaoning, Jilin, Heilongjiang, Inner Mongolia, and Gansu. In the development of peanut industry, peanut planting has been gradually shifted from north to south. See Table 1.14 for details.

1.2.3 International Trade of Peanut and Its Processed Product in China

In the past decade, although there have been fluctuations in China's peanut exports, the overall situation has been still optimistic with stronger competitive advantage in the international market and broad market development prospect. China is the largest exporter of peanuts in the world; the main export categories of peanut products are shelled peanuts, peanut oil, and peanut butter which are mainly exported to Europe, Southeast Asia, Japan, the Middle East, and more than 120 countries and regions; the export volume accounts for about 30% in the peanut market share of the world, and China peanut is one of the competitive products internationally in the grain and oil commodities of China currently. In 2013, the annual export volume of peanut and its product was about 109,009 tons, and the foreign exchange income of 202,874,000 dollars was created. According to FAO statistics, from 2000 to 2013, the export volume of shelled peanuts of China decreased from 331,577 to 99,417 tons year by year, the export volume of peanut

Table 1.11 World's import and export trade conditions of peanut oil in 2013

Country/region	Import quantity(t)	Import value (1000$)	Export quantity(t)	Export value (1000$)
Albania	5	11	0	0
Algeria	0	3	0	1
Angola	24	84	0	0
Argentina	6	36	40,804	66,061
Armenia	1	4	0	0
Australia	1023	2867	579	1500
Austria	541	1440	140	730
Azerbaijan	1	1	–	–
Bahamas	2	13	–	–
Bahrain	22	40	–	–
Bangladesh	9	55	–	–
Barbados	1	9	0	0
Belarus	3	12	–	–
Belgium	18,939	34,851	11,981	28,766
Belize	1	2	–	–
Benin	8034	6291	300	429
Bermuda	3	15	–	–
Bhutan	461	647	–	–
Bolivia (Plurinational State of)	1	2	–	–
Bosnia and Herzegovina	5	13	–	–
Botswana	18	73	9	8
Brazil	63	157	63,250	100,900
Brunei Darussalam	0	0	0	0
Bulgaria	5	25	0	1
Burkina Faso	926	173	0	0
Burundi	1	1	–	–
Cabo Verde	1	1	–	–
Cambodia	0	0	0	0
Cameroon	2	9	0	0
Canada	2172	5565	10	31
Central African Republic	25	5	0	0
Chad	1	3	0	0
Chile	2	14	–	–
China	68,325	126,749	8793	27,669
Colombia	15	97	0	0
Comoros	1	2		
Congo	10	33	0	0
Cook Islands	4	10	–	–
Costa Rica	1	10	–	–
Croatia	6	17	0	0
Cuba	0	0	0	0
Cyprus	14	42	–	–

(continued)

Table 1.11 (continued)

Country/region	Import quantity(t)	Import value (1000$)	Export quantity(t)	Export value (1000$)
Czechia	40	140	10	70
Democratic Republic of the Congo	16	62	0	0
Denmark	23	154	1	8
Djibouti	4	15	–	–
Dominican Republic	1	7	10	17
Ecuador	3	12	–	–
Egypt	26	123	105	157
El Salvador	184	206	–	–
Estonia	2	6	1	3
Ethiopia	100	330	0	0
Faroe Islands	0	0	–	–
Fiji	31	67	2	6
Finland	41	294	0	0
France	19,422	40,711	5605	13,455
French Polynesia	329	971	0	0
Gabon	12	46	–	–
Gambia	52	27	0	0
Georgia	2	15	22	37
Germany	4403	11,336	462	2199
Ghana	13	8	1	1
Greece	0	1	0	0
Grenada	0	0	–	–
Guatemala	0	0	206	222
Guinea	0	0	–	–
Guinea-Bissau	0	0	0	0
Guyana	2	3	–	–
Haiti	10	20	–	–
Honduras	33	49	0	0
Hungary	45	110	0	0
Iceland	11	35	–	–
India	58	272	3479	5564
Indonesia	3	15	12	18
Iran (Islamic Republic of)	0	0	0	0
Iraq	0	0	0	2
Ireland	27	107	–	–
Israel	33	145	26	129
Italy	33,992	57,744	625	1851
Jamaica	2	14	1	2
Japan	573	2070	0	0
Jordan	4	22	0	0
Kazakhstan	1	5	–	–

(continued)

Table 1.11 (continued)

Country/region	Import quantity(t)	Import value (1000$)	Export quantity(t)	Export value (1000$)
Kenya	2	7	29	134
Kiribati	0	0	–	–
Kuwait	24	75	1	2
Kyrgyzstan	0	0	–	–
Lao People's Democratic Republic	0	0	–	–
Latvia	2	11	0	0
Lebanon	56	173	5	21
Liberia	7	27	–	–
Libya	1	17	–	–
Lithuania	5	22	2	6
Luxembourg	795	1929	15	62
Madagascar	1	2	0	0
Malawi	0	0	0	0
Malaysia	1490	4651	24	25
Maldives	0	1	–	–
Mali	0	4	0	0
Malta	7	17	0	0
Mauritania	387	335	–	–
Mauritius	3	12	0	0
Mexico	15	110	85	274
Mongolia			0	0
Montenegro	2	10	–	–
Morocco	0	0	0	0
Mozambique	2	3	–	–
Myanmar	0	3	–	0
Namibia	5	36	0	0
Nepal	0	0	0	0
Netherlands	6914	13,029	2617	5488
New Caledonia	63	186	0	0
New Zealand	215	1173	4	11
Nicaragua	0	0	16,433	30,138
Niger	2	4	0	0
Nigeria	200	210	3700	4050
Norway	1454	3894	3	17
Oman	83	180	0	0
Pakistan	32	109	8	26
Panama	5	11	45	78
Papua New Guinea	1	4	–	–
Paraguay	0	0	–	–
Peru	12	75	1	14
Philippines	3	21	151	178

(continued)

Table 1.11 (continued)

Country/region	Import quantity(t)	Import value (1000$)	Export quantity(t)	Export value (1000$)
Poland	42	162	0	0
Portugal	299	731	147	219
Qatar	454	516	0	0
Republic of Korea	18	139	0	0
Republic of Moldova	0	1	–	–
Romania	32	70	1	7
Russian Federation	55	109	0	0
Rwanda	2	3	–	–
Saint Kitts and Nevis	0	0	–	–
Saint Lucia	0	0	–	–
Saint Pierre and Miquelon	2	7	–	–
Saint Vincent and the Grenadines	0	0	–	–
Samoa	0	0	–	–
Sao Tome and Principe	43	44	–	–
Saudi Arabia	0	0	100	142
Senegal	1651	2516	16,418	36,870
Serbia	2	6	0	0
Seychelles	0	0	–	–
Singapore	1909	3927	1934	5284
Slovakia	5	27	0	0
Slovenia	95	254	9	52
Solomon Islands	0	2	–	–
Somalia	1	3	–	–
South Africa	32	46	391	692
Spain	1603	2819	413	1109
Sri Lanka	7	12	0	1
Sudan	11	6	2428	3600
Suriname	0	4	–	–
Swaziland	131	149	0	0
Sweden	57	228	3	16
Switzerland	7559	18,655	19	102
Thailand	25	93	1	3
The former Yugoslav Republic of Macedonia	0	1	–	–
Timor-Leste	0	0	–	–
Togo	620	168	0	0
Tonga	1	2	–	–
Trinidad and Tobago	1	1	0	0
Tunisia	0	0	0	0
Turkey	11	71	0	0
Uganda	0	0	3	1

(continued)

Table 1.11 (continued)

Country/region	Import quantity(t)	Import value (1000$)	Export quantity(t)	Export value (1000$)
Ukraine	5	30	4	35
United Arab Emirates	31	124	112	174
United Kingdom	2257	6247	134	723
United Republic of Tanzania	1	7	22	13
United States	19,123	31,522	4608	11,841
Uruguay	0	0	0	0
Vanuatu	4	8	–	–
Venezuela (Bolivarian Republic of)	0	0	0	0
Vietnam	0	0	0	1
Yemen	3	6	0	0
Zambia	23	11	0	0
Zimbabwe	4	14	0	0

Data: FAOSTAT

oil decreased from 23,755 to 8793 tons, the export amount increased from 24,655,000 to 27,669,000 dollars, and the export volume of peanut butter decreased from 9537 to 799 tons, and the export amount increased from 11,752,000 to 4,700,000 dollars. See Table 1.15. The peanut export varieties of China increased year by year overall, and the export structure was developed from raw materials based to balance between export raw materials and finished products; the export volume of food made and processed with peanuts as raw materials increased, and the export range was expanded year by year. The peanut export products were developed from the original single kernel grade to the grading kernels, processed peanuts, peanut oil, and other varieties.

1.2.4 Peanut Utilization in China

The continuous improvement of peanut yield has promoted the increase of the total peanut processing and utilization volume of China, and the methods and range of utilization have been gradually widened. In the 1990s, the average annual processing volume of peanuts increased by nearly 40% compared with that in the 1980s in China. The peanut oil preparation is the main method to use peanuts, and the oil yield and peanut oil quality have been improved continuously due to the continuous improvement of processing process and peanut quality. With the increase of peanut processing methods, a large number of peanut food occur; the proportion of peanuts used for oil extraction in the total peanut yield has been decreased year by year, while the proportion of peanuts for food processing and direct eating has increased year by year. In the 1990s, the annual average proportion of peanuts for peanut oil preparation in China accounted for 58% in the total domestic peanut utilization volume and decreased by 6% compared with that in

Table 1.12 World's import and export trade conditions of peanut cake in 2013

Country	Import quantity(t)	Import value (1000$)	Export quantity(t)	Export value (1000$)
Albania	0	0	0	0
Algeria	0	0	0	0
Argentina	0	0	0	0
Armenia	0	0	13,666	4935
Austria	0	0	0	0
Austria	–	1	0	0
Bahamas	0	0	–	–
Bahrain	0	1	–	–
Bangladesh	0	0	0	0
Belarus	0	0	–	–
Belgium	3886	2000	1367	711
Benin	0	0	–	–
Bhutan	5	2	–	–
Bolivia (Plurinational State of)	0	0	–	–
Bosnia and Herzegovina	0	0	–	–
Botswana	0	0	0	0
Brazil	0	0	0	0
Brunei Darussalam	0	0	296	95
Bulgaria	0	0	–	–
Burkina Faso	40	3	–	0
Burundi	0	0	–	–
Cabo Verde	0	0	–	–
Cambodia	0	0	–	–
Cameroon	0	0	10	2
Canada	2156	577	1	8
Central African Republic	0	0	0	0
Chad	–	–	21	2
Chile	14,107	6882	0	0
China	36,807	14,702	854	558
Colombia	0	0	–	–
Comoros	0	0	–	–
Congo	0	0	0	0
Costa Rica	145	67	0	0
Cyprus	0	0	0	0
Czechia	0	0	14	0
Democratic People's Republic of Korea	0	0	0	1
Denmark	0	0	–	0
Djibouti	0	0	–	–
Dominican Republic	700	260	19	–
Ecuador	0	0	–	–
Egypt	0	0	0	1
El Salvador	4173	1381	0	0

(continued)

Table 1.12 (continued)

Country	Import quantity(t)	Import value (1000$)	Export quantity(t)	Export value (1000$)
Estonia	0	0	70	0
Ethiopia	0	0	–	–
Faroe Islands	0	0	–	–
Fiji	0	0	–	–
Finland	0	0	–	–
France	17,023	8486	0	16
Gabon	1	0	–	–
Georgia	0	0	0	0
Germany	3853	1215	0	0
Ghana	0	0	–	–
Greece	0	0	0	0
Guatemala	6077	2785	0	0
Guinea	10	1	0	0
Guyana	0	0	–	–
Honduras	7837	3636	0	0
Hungary	0	0	8551	0
India	196	103	658	3816
Indonesia	0	0	0	1185
Iran (Islamic Republic of)	0	0	0	0
Ireland	13	6	0	0
Israel	0	0	0	0
Italy	0	0	–	–
Jamaica	0	0	0	0
Japan	51	39	0	0
Jordan	482	299	0	0
Kazakhstan	0	0	–	–
Kenya	0	0	0	0
Latvia	0	0	–	–
Lebanon	0	0	0	0
Liberia	0	0	0	0
Luxembourg	0	0	0	0
Madagascar	–	–	7453	0
Malawi	0	0	135	0
Malaysia	560	182	0	4180
Mali	108	111	0	19
Malta	0	0	–	–
Mauritania	1462	195	–	–
Mauritius	18	7	0	0
Mexico	3333	1570	3	0
Morocco	0	0	3000	0
Mozambique	0	0	0	1
Myanmar	–	–	16	640
Namibia	0	0	6446	0
Nepal	0	0	0	7

(continued)

Table 1.12 (continued)

Country	Import quantity(t)	Import value (1000$)	Export quantity(t)	Export value (1000$)
Netherlands	1241	682	18,018	2219
New Zealand	0	0	0	0
Nicaragua	0	0	35	6944
Niger	35	2	200	0
Nigeria	25	21	1640	2
Norway	5150	2921	0	253
Oman	0	0	–	–
Pakistan	0	0	0	583
Papua New Guinea	0	0	0	0
Peru	130	43	0	0
Philippines	2	1	0	0
Poland	0	0	0	0
Portugal	0	0	0	0
Qatar	362	193	0	0
Republic of Korea	35	19	0	0
Romania	0	2	362	0
Russian Federation	0	0	–	–
Rwanda	277	19	17,949	0
Saint Kitts and Nevis	0	0	–	–
Saudi Arabia	40	101	1	393
Senegal	0	0	0	8403
Singapore	0	0	66	0
Slovakia	0	0	–	–
Slovenia	0	0	–	–
Solomon Islands	0	0	0	322
South Africa	3	1	16	35
Spain	6	2	31,783	0
Sri Lanka	125	55	0	4
Sudan	–	–	0	11,571
Swaziland	66	38	0	0
Sweden	7	4	0	0
Switzerland	0	0	16	1
Thailand	0	0	–	11
Togo	0	0	0	0
Turkey	0	0	–	0
Uganda	0	0	–	0
Ukraine	–	–	3	–
United Arab Emirates	406	53	0	–
United Kingdom	1	4	8447	11
United Republic of Tanzania	0	0	0	0
United States	1402	3699	0	2057
Uruguay	0	0	728	0
Venezuela (Bolivarian Republic of)	0	0	0	0

(continued)

Table 1.12 (continued)

Country	Import quantity(t)	Import value (1000$)	Export quantity(t)	Export value (1000$)
Vietnam	–	–	5	153
Zambia	0	0	854	0
Zimbabwe	0	0	–	2

Data: FAOSTAT

the 1980s; the annual average proportion of peanuts for peanut food processing and direct eating accounted for 42% in the total domestic peanut utilization volume and increased by 6% compared with that in the 1980s (Zhou 2005). In recent years, the domestic peanut consumption has been mainly concentrated in peanut oil, peanut protein powder, roasted peanuts and peanut kernels, peanut beverages, and other peanut products, and the peanut oil accounts for 55% in the domestic peanut consumption and has become the main method to use peanuts. See Fig. 1.1.

In the edible peanut consumption, the peanut butter accounted for 37%, roasted peanuts for 32%, peanut protein for 6%, peanut milk for 7%, and raw peanut eating for 18% (Yang 2009). Relative to the status of the first large peanut exporting countries, the development rate of peanut food processing industry of China severely lags behind the development of foreign countries. At present, the peanut varieties in China are mixed for application, the special varieties suitable for different processing purposes are insufficient, the comprehensive research on the processing characteristics of different variety resources is carried out less, and the peanut processing quality evaluation technologies, methods, and standards are deficient, which have seriously restricted the development of peanut processing industry.

2 Peanut Processing Quality Definition and Classification

The quality of agricultural products is not only the pursuing goal of agricultural production but also the urgent need of consumers' nutrition and health. The quality of peanut and its processed products directly affects the vital interests and physical and mental health of producers, processors, and consumers, so the peanut quality has become the hotspot in recent years.

For different processing and consumption modes, the requirements for the quality of peanut varieties or raw materials are different. For example, the peanuts are mainly used for oil extraction and export in China; the oil content, proportion of fatty acids and contents of V_E, sterols, and other active components in raw materials are the important quality indicators which determine the peanut oil product quality; and the grain form, skin color, sugar content, and O/L ratio are the important quality indicators which affect the export peanut quality. As another example, the peanuts in the United States have four types, including Lanna (Runners), Valencia

Table 1.13 Peanut planting area and yield in different regions of China in 2015

Regions	Planting area (mu)	Yield (tons)	Accounting for the proportion of the total area of the country (%)	Accounting for the proportion of the total yield of the country (%)
Northeast	7,021,500	1,059,000	10.14	6.44
North China	5,535,000	1,336,000	8.00	8.13
East China	19,650,000	5,294,000	28.38	32.20
South China	9,288,000	1,808,000	13.41	11.00
Northwest	552,000	119,000	0.80	0.72
Southwest	6,310,500	984,000	9.11	5.99
Middle-of-south	20,878,500	5,837,000	30.15	35.50
Total	69,240,000	16,440,000	100	100

Data sources: China Statistical Yearbook

(Valencias), Spanish (Spanish), and Virginia (Virginias), and different types of peanut varieties have different processing characteristics and quality. Lanna peanuts are popular due to their unique seed kernel size and high yield, and about more than half of such peanuts are used to make peanut butter; Valencia peanuts are mainly used for peanut baking, selling with shell, and stewing; Spanish peanuts have high oil content which makes such peanut an ideal oilseed peanut variety, while Spanish peanuts are also used for processing desserts and peanut butter; with large grains, Virginia peanuts are mainly used for peanut baking with shells, salted peanuts, or appetizer peanut processing (Licher 2005). It can be seen that the quality of the varieties or raw materials can be inseparable from the quality of processed products, different processed products have different requirements for raw materials, and it is difficult to process high-quality processed products without suitable varieties or raw materials. Therefore, in recent years, the research on the processing quality of peanut varieties, raw materials, and their products has become a hotspot in this field.

However, the definition, connotation, and classification on the peanut processing quality have not been combed systematically, and there has been no uniform statement currently. Misra (2004) divided the quality characteristics of peanuts into physical quality and chemical quality. Wan Shubo (Wan 2007) divided the peanut quality into nutrition quality, commercial quality, and processing quality, with the nutrition quality referred to protein, fat, sugar compound, vitamins, minerals, and other nutritional indicators; the commercial quality referred to the standards and specifications specified from the commercial view; the processing quality referred to the quality indicators directly related to product processing, but these three kinds of qualities were described for the variety or raw material, and the product quality was not involved. From the perspective of agricultural product quality research of many years, the author thinks that the agricultural product

Table 1.14 Changes of peanut area and yield in China in 2000 and 2015

	2000		2015		Increase (%)	
Province	Area(mu)	Yield(tons)	Area(mu)	Yield(t)	Area (mu)	Yield(t)
Liaoning	2,142,000	256,200	4,167,000	448,000	94.54	74.86
Jilin	812,300	130,900	2,601,000	559,000	220.20	327.04
Heilongjiang	278,700	21,100	253,500	52,000	−9.04	146.45
Beijing	192,000	33,600	27,000	5000	−85.94	−85.12
Tianjin	58,500	8200	15,000	3000	−74.36	−63.41
Hebei	6,949,500	1,325,900	5,143,500	1,274,000	−25.99	−3.91
Shanxi	315,700	41,400	99,000	12,000	−68.64	−71.01
Inner Mongolia	78,000	8800	250,500	42,000	221.15	377.27
Shanghai	21,000	6600	10,500	2000	−50.00	−69.70
Jiangsu	3,418,500	797,400	1,359,000	351,000	−60.25	−55.98
Zhejiang	214,800	36,100	273,000	53,000	27.09	46.81
Anhui	5,010,000	1,111,500	2,866,500	944,000	−42.78	−15.07
Fujian	1,590,000	238,200	1,572,000	286,000	−1.13	20.07
Jiangxi	2,698,500	403,800	2,463,000	464,000	−8.73	14.91
Shandong	13,852,500	3,501,400	11,106,000	3,194,000	−19.83	−8.78
Guangdong	4,965,000	776,800	5,488,500	1,090,000	10.54	40.32
Guangxi	3,609,000	495,500	3,214,500	607,000	−10.93	22.50
Hainan	733,500	95,200	585,000	111,000	−20.25	16.60
Shaanxi	501,750	73,300	489,000	97,000	−2.54	32.33
Gansu	10,500	1300	19,500	5000	85.71	284.62
Xinjiang	70,500	20,000	43,500	17,000	−38.30	−15.00
Chongqing	632,400	75,500	853,500	119,000	34.96	57.62
Sichuan	3,594,000	542,900	3,945,000	678,000	9.77	24.88
Guizhou	664,500	74,500	769,500	105,,000	15.80	40.94
Yunnan	618,000	53,500	741,000	82,000	19.90	53.27

Henan	14,772,000	3,358,800	16,119,000	4,853,000	9.12	44.49
Hubei	2,901,000	657,100	2,986,500	679,000	2.95	3.33
Hunan	2,125,500	290,800	1,773,000	305,000	−16.58	4.88
Total	72,829,600	14,436,600	69,240,000	16,440,000	−4.93	13.88

Data sources: China Statistical Yearbook

Table 1.15 China's peanut and its products import and export trade during 2000–2013

Year	Groundnuts, shelled				Oil, groundnut				Peanut butter				Cake, groundnuts			
	Import quantity (t)	Import value (1000$)	Export quantity (t)	Export value (1000$)	Import quantity (t)	Import value (1000$)	Export quantity (t)	Export value (1000$)	Import quantity (t)	Import value (1000$)	Export quantity (t)	Export value (1000$)	Import quantity (t)	Import value (1000$)	Export quantity (t)	Export value (1000$)
2013	16,494	22,352	99,417	170,505	68,325	126,749	8793	27,669	5762	14,613	799	4700	36,807	14,702	854	558
2012	25,420	33,395	103,699	206,010	71,674	173,299	10,742	34,809	6616	21,262	23,546	75,814	10,329	3353	2280	675
2011	61,566	71,897	116,003	194,373	73,959	133,733	11,966	29,452	6607	16,520	1256	6671	36,814	12,444	3850	976
2010	20,823	21,207	126,801	172,819	85,349	117,145	13,378	25,710	6157	12,259	695	3488	37,843	13,662	2839	691
2009	10,232	7367	174,822	165,358	33,151	58,494	12,034	26,693	5977	12,683	18,872	39,619	36,374	10,409	6036	1263
2008	15,543	16,695	167,257	232,456	18,750	41,864	12,872	33,499	4426	10,981	18,651	41,798	99,236	37,110	3034	703
2007	9509	8446	204,535	210,826	24,881	35,338	13,683	22,642	5283	7574	20,042	32,355	49,141	10,734	4230	658
2006	12,194	8563	234,558	185,605	11,139	13,875	13,869	17,097	6271	8514	17,618	25,673	111,543	19,897	8162	890
2005	7088	4489	363,716	263,765	17,869	19,922	21,197	23,958	4358	5651	14,920	20,480	83,408	16,329	8413	758
2004	7199	4705	325,757	248,944	13,177	15,953	15,273	18,097	5572	7609	13,098	18,438	90,879	21,456	3564	295
2003	5196	3217	400,403	268,968	20,758	22,769	27,453	31,664	5438	7341	12,031	15,323	6357	1132	6457	980
2002	6207	2838	429,275	222,691	22,492	19,360	15,622	14,200	5564	7218	11,648	13,667	9308	1443	11,129	1591
2001	1994	1122	408,010	223,568	34,098	30,981	22,704	20,442	5172	6891	10,433	12,020	8314	1699	25,698	4277
2000	4280	2595	331,577	197,567	35,854	37,147	23,755	24,655	5513	6907	9537	11,752	21,101	4472	14,727	1925

Data sources

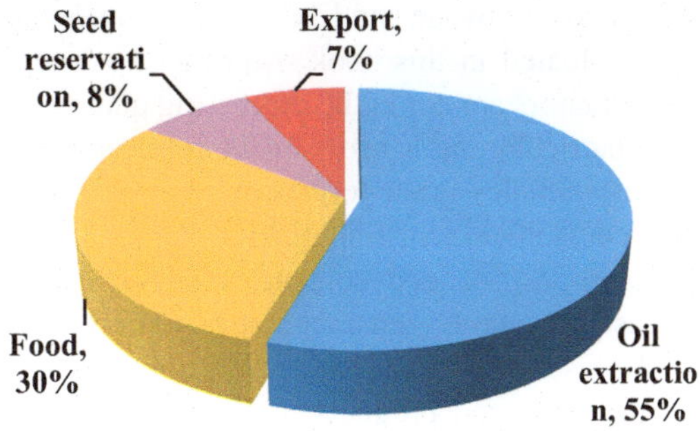

Fig. 1.1 Peanut consumption structure of China

(varieties or raw materials) quality should include sensory quality, physicochemical nutrition quality, and processing quality (characteristic) for the varieties or raw materials. The sensory quality mainly includes color, smell, taste, texture, and other quality properties which can be felt directly by human organs; the physicochemical nutrition quality refers to the protein, fat, carbohydrate, vitamins, minerals, other nutritive characteristics, protease inhibitors, phytohemagglutinin, and other anti-nutritional factors; the processing quality referred to the characteristic indicators and quality properties directly related to the quality of processed products in varieties or raw materials. As for processed products, the processing quality of products includes edible quality, nutritional quality, and safety quality. The edible quality is different from the sensory quality; it is defined for the processed products which can be eaten directly, and its meaning also refers to color, smell, taste, shape, texture, and other quality properties which can be felt directly by human organs. While the sensory quality is defined mainly for raw materials because some raw materials of agricultural products could be eaten directly after processing. The nutritional quality is also different from the physicochemical nutrition quality of raw materials because the anti-nutritional factors and other physicochemical indicators that are not conducive to human health in raw materials must be removed through the processing process. The safety quality refers to the quality properties harmful to the human body in raw materials (such as pesticide residues, heavy metals, and mycotoxins) and those produced in the processing process (such as trans-fatty acids, chloropropanol, and benzopyrene).

According to the analysis above, the quality of peanut varieties or raw materials was divided into three aspects in this book: sensory quality, physicochemical nutrition quality, and processing quality (characteristic); and the processing quality of peanut products was also divided into three aspects: edible quality, nutritional quality, and safety quality. The aflatoxin was not the original composition of peanut, and it was caused by fungi infection before and after peanut ripening (Sargeant et al. 1961), so the aflatoxin would not be considered as an evaluation indicator of peanut processing quality in this book except for the export peanuts. Meanwhile, the peanut protein could cause allergic reactions and bring unsafe factors for processed products, but it had been clearly indicated in peanut products that the peanut allergy consumers could choose peanut products appropriately

(Misra 2004; Savage and Keenan 1994); the peanut allergens would not be analyzed and evaluated in this book. And the heavy metals, pesticide residues, and other safety factors are not in the processing quality evaluation range of this book, either.

In particular, the sensory quality of peanut varieties or raw materials in this book mainly includes grain form, color, seed kernel size, taste, smell, and so on; the physicochemical nutrition quality mainly includes crude protein, crude fat, water, ash, composition and content of amino acid, protein, and fatty acids, and the processing quality (characteristic) mainly includes pure kernel rate, oil yield, protein extraction rate, oleic acid/linoleic acid, and arachin/conarachin. As for the peanut processed products in China currently, the processing quality of peanut products is mainly for peanut oil, peanut protein, and export peanuts.

2.1 Sensory Quality

The sensory quality of peanut mainly includes imperfect fruit, immature fruit, broken fruit, insect-damaged fruit, damaged fruit, pod size, pod shape, fruit waist, imperfect grain (insect-damaged grain, diseased grain, germinative grain, broken grain, immature grain, and damaged grain), metamorphic grain (moldy grain, thermal loss grain, discolored grain, oily grain, spotty grain, and other damaged grains), different variety rates, thousand seed weight, impurity, smell, whole/half peanut kernel, kernel size, color, taste, rancid kernel, peanut pod, kernel appearance characteristics, and other various quality properties which directly affect the peanut quality.

Luan et al. (1986) researched 1072 peanuts and found that the variation range of hundred kernel weight was 30–143.2 g; the variation coefficient of hundred fruit weight of pearl variety was the largest, the variation coefficient of hundred kernel weight of multi-grain variety was the largest, and the order of average hundred kernel weight of various varieties were intermediate type> ordinary type> Longsheng type> pearl type> multi-grain type. Wu et al. (2005) identified the appearance quality and taste quality of 50 peanut varieties in sensory quality, and as a result, a high-tenderness variety and a high-sweetness with strong-flavor variety were selected. Chen et al. (2004) carried out the quantitative grading for the kernel shape, color, and taste in sensory quality of 265 peanut varieties preserved at home and abroad. The results showed that there were significant differences in the performance of kernel shape, color, and taste among the five types of cultivated peanut; the taste of multi-grain-type peanut was excellent, and the kernel shape of common-type peanut was better, while the kernel color, shape, and taste of Longsheng-type peanut were the worst, and the variation range of kernel color, shape, and taste of intermediate-type peanut was larger.

2.2 *Physicochemical and Nutritional Quality*

The physicochemical and nutritional quality refers to the various chemical compositions and varieties of essential nutrients for human body in peanuts, as well as the number and quality of various chemical compositions and nutrients. It mainly includes crude protein, crude fat, crude fiber, water, ash, total sugar, amino acids, fatty acids, vitamins, minerals, phytosterol, resveratrol, squalene, and other nutritional components, as well as protease inhibitors, phytolectin, and other antinutritional factors.

2.2.1 Protein

The protein content in peanut is 24–36% which is second only to that in soybean and higher than that in sesame and rape, compared with several major oilseed crops (Table 1.16). About 10% peanut protein was water-soluble, it was called albumin, and the remaining 90% was salt-soluble proteins (Gao et al. 1995). The salt-soluble protein mainly included arachin and conarachin; the arachin accounted for 73% and the conarachin accounted for 27% (Yamada 1979). The research on the nutritional value of peanut protein showed that the biological value (BV) of peanut protein was 59, the net utilization rate (NPD) of protein was 51, and the pure digestibility was up to 90% (Wan et al. 2004). The nutritional value of peanut protein powder is similar to that of animal protein; the protein content is 2.8 times of that in lean beef, 3.3 times of that in lean pork, 3.8 times of that in egg, and 16 times of that in milk (the nutritional components of peanut protein powder and meat, egg, and milk food are shown in Table 1.17). The nutritional function of peanut protein is close to that of soy protein, but peanut protein can be absorbed more easily than soybean protein; the contents of indigestible sugar, cottonseed sugar, and stachyose in peanut protein are only equivalent to one-seventh of soy protein, and the abdominal distension and belch gas phenomena which occur after eating soy protein will not occur after eating the peanut protein. Therefore, the peanut protein was considered as a great potential kind of alternative of protein base and milk and other animal milk for the lactose intolerance consumers (Wan et al. 2004; Zhang et al. 2008). In addition, the nitrogen solubility index (NSI) of peanut protein was high, and if it was added to animal food or plant food, it could play a role in improving the food quality and strengthening the food nutrition, and also its application prospect was very extensive due to its inherent flavor of peanut (Wan et al. 2004). The peanut protein contains eight kinds of essential amino acids of human body, except that its methionine content is lower; the other amino acid contents reach the standards of United Nations Food and Agriculture Organization (FAO); meantime, the peanut was rich in arginine, leucine, phenylalanine, glutamic acid, and aspartic acid (the amino acid of peanut and other oilseed crops is shown in Table 1.18) (Wang 2010); the lysine content was higher than that in rice, wheat flour, and corn; the effective

Table 1.16 Chemical components of peanut and other oil crop seeds (unit: %)

Crop	Fat	Protein	Carbohydrate	Crude fiber	Ash	Water
Peanut	44.27–53.86	23.94–36.35	9.89–23.62	2.67–6.40	1.75–2.58	5.33–9.16
Soybean	14.95–22.14	41.18–53.61	17.81–30.47	4.22–6.40	3.89–5.72	5.71–12.50
Rape	28.15–48.08	19.13–27.17	16.61–38.86	4.58–11.22	3.34–7.84	6.53–10.53
Sesame	45.17–57.16	19.87–24.25	9.59–19.91	4.00–7.52	4.49–6.87	4.35–8.50
Cottonseed	17.46–23.07	24.27–37.66	19.14–33.33	1.12–3.56	5.12–6.12	9.42–12.09

Table 1.17 Nutritional components of peanut protein powder and meat, egg, and milk food

Name	Protein (g)	Fat (g)	Carbohydrate (g)	Calcium (mg)	Magnesium (mg)	Iron (mg)
Peanut protein powder (100 g)	55.0	4.5	23.0	93	510	2.6
Crucian (100 g)	13.0	1.1	0.1	54	203	2.5
Dry shrimp (100 g)	58.1	2.1	4.6	577	614	13.1
Lean pork (100 g)	16.7	28.8	1.0	11	177	2.4
Beef (100 g)	20.1	10.2	0.1	7	170	0.9
Milk (100 g)	3.3	4.0	5.0	600	465	1.0
Chicken (100 g)	21.5	2.5	0.7	11	190	1.5
Egg (100 g)	14.6	11.6	1.6	55	210	2.7

utilization rate was up to 98.8%; and the effective utilization of lysine in soybean protein was only 78% (Chen et al. 2007).

The research found that there were significant differences in the protein contents of different varieties of peanuts. Jiang et al. (2006) collected 6390 cultivated peanut variety resources in China, and the analysis results showed that the protein contents of peanut resources in Fujian and Jiangxi were high, and the genetic diversity of peanut resources in Hubei, Henan, and Guangxi was higher than that in other areas. The variation amplitude of protein content in peanut varieties collected by Teng et al. (2003) was 13.60–34.82%, and the difference between peanut varieties was significant. The protein content variation in common-type peanut was the largest, followed by that in the pearl-type peanut, that in the intermediate-type peanut was the smallest, and that in other types of peanuts was in the middle. The protein content in common-type peanut varieties was up to 34.75%, the average protein content in pearl-type peanuts was 29.28%, and the protein content in Longsheng-type peanuts was the highest (21.06%).

Andersen et al. (1998) analyzed the amino acid contents of six peanut varieties with high oleic acid content and ten peanut varieties with ordinary oleic acid content. The results showed that the differences between histidine and methionine of various peanut varieties were 1.65 and 2.1 times, respectively. Analyzed the amino acid contents of 379 peanut varieties in Shandong, and the results showed that the variation range of arginine of various peanut varieties was 0.96–6.35% and the variation coefficient was 15.07%; the variation range of methionine was 0.26–1.44%, the variation coefficient was 23.27%, and it could be seen that the amino acid contents of different peanut varieties were significantly different.

Arachin, conarachin I, and conarachin II were the major components of peanut protein, and the physicochemical properties of these three components largely determined the functional characteristics of peanut protein (Govindaraju and Srinivas 2006). Li et al. (1998) obtained four major types of protein composition modes through the SDS-PAGE electrophoresis and two-dimensional electrophoresis analysis for 46 peanut varieties. The main difference in the composition of

Table 1.18 Amino acid composition in peanut and other grain and oil crops (unit: g/16gN)

Amino acid	Peanut protein	Soy protein	Rapeseed protein	Cottonseed protein	Rice protein	Wheat protein	Corn protein	Modes recommended by FAO/WHO
Lys	3.05	6.01	5.81	4.48	3.52	2.44	3.67	5.50
His	2.31	2.25	2.72	2.90	2.32	2.23	3.03	–
Arg	11.30	7.55	6.64	12.45	9.15	4.29	4.70	–
Asp	14.12	10.38	7.12	9.11	8.46	7.22	7.05	–
Thr	2.54	3.66	4.50	3.00	3.85	3.04	4.40	4.00
Ser	4.91	4.61	4.71	4.12	5.02	5.97	3.62	–
Glu	19.92	18.42	17.92	20.47	19.68	29.94	13.25	–
Pro	4.40	6.20	6.14	3.60	3.95	7.56	5.63	–
Gly	5.61	4.62	5.33	4.15	3.27	4.46	6.65	–
Ala	4.12	4.50	4.90	3.73	5.43	4.02	4.62	–
Cys	1.35	1.63	2.64	0.82	2.21	2.54	2.40	>3.50
Met	0.91	1.56	2.38	1.31	1.73	1.41	1.83	
Val	4.50	5.30	5.29	4.60	5.43	4.22	4.95	5.00
Iso	4.14	5.02	4.20	3.47	3.54	3.58	3.28	4.00
Leu	6.72	7.72	7.31	5.81	8.40	7.11	15.20	7.00
Tyr	4.11	3.91	3.10	3.15	3.80	3.20	5.20	>6.00
Phe	5.22	5.00	4.14	5.51	4.75	4.53	4.96	
Try	1.02	1.20	1.45	1.15	1.68	1.14	0.78	1.00

peanut seed protein was rooted in the difference in the composition of arachin subunit; the arachin of type I mainly contained 41 kDa, 38.5 kDa, and 2 18 kDa subunits; the arachin of type II mainly contained 41 kDa, 38.5 kDa, 37.5 kDa, and 3 18 kDa subunits; the arachin of type III mainly contained 41 kDa, 38.5 kDa, 36.5 kDa, and 3 18 kDa subunits; the arachin of type IV mainly contained 41 kDa, 38.5 kDa, 37.5 kDa, 36.5 kDa, and 3 18 kDa subunits; Shokarii et al. (1991), Krishna et al. (1986), and Du et al. (2013) have also reported the similar conclusion.

The protein was a kind of high-molecular polymer composed of various amino acids with a spatial structure, and its physicochemical properties (molecular size and shape, amino acid composition and sequence, charge distribution, intramolecular and molecular actions, and effective hydrophobic effect) and functional properties were closely related (Yuan et al. 2005). Monteiro and Pralash (1994) analyzed the amino acids of different peanut protein components (Table 1.19), and the results showed that each component contained 18 amino acids. In the total protein, the contents of aspartic acid, glutamic acid, and arginine were higher, and the contents of cysteine, methionine, tyrosine, and lysine were lower. In general, the content levels of serine, glycine, and lysine in conarachin I were higher, and the content level of glycine was six times higher than that in the total protein; however, the contents of aspartic acid, proline, alanine, valine, isoleucine, and arginine were lower than that of other protein components. The amino acid contents of different components of peanut protein were significantly different, and the subunits of components of peanut protein of different varieties were different; therefore, there would be a large gap in the functional properties of peanut protein of different varieties, and the research on the relationship between these three was less and needed to be further analyzed. Lu et al. (2000) analyzed the amino acid composition of three major components of peanut protein from two different sources (Table 1.20). The results showed that they contained 17 kinds of amino acids, the contents of aspartic acid, glutamic acid, and arginine were the highest, the contents of methionine and cysteine were very low, and the results were consistent with that of Monteiro et al. (1994).

With certain functional properties, nutritional characteristics, and good flavor, the peanut protein had been widely used in many foods (Mouecoucou et al. 2004). At present, the peanut protein product is mainly peanut protein powder in China. As the raw material for food processing, it can be used in the meat product processing, grain baking product processing, vegetable protein drink and cold drink and cold food production, and non-staple food, spice, and leisure food production; the functional properties of protein has been increased through the modification method, which can improve the product texture and structure and enhance the nutritional value of product. Due to the different consumption habits, Europe and America and other developed countries had much research on peanut butter, peanut candies, and peanut-related slimming foods and selected the suitable varieties for peanut candy processing, but their research on the peanut protein products was relatively less (Wu 2009). Therefore, studying the protein component and subunit composition, protein structure, modification, and functional properties of different

Table 1.19 Amino acid composition of various peanut protein components (unit: g/16gN)

Amino acid	Total protein	Arachin	Conarachin II	Conarachin I
Asp	12.91	12.29	12.68	7.95
Thr	2.50	2.68	2.97	3.06
Ser	5.22	4.60	5.50	7.97
Glu	23.01	21.43	22.24	8.28
Pro	5.01	5.84	3.11	2.63
Gly	5.15	3.99	4.16	29.73
Ala	3.74	3.87	3.99	2.87
Cys	0.33	0.33	0.19	0.23
Val	3.65	4.77	5.73	3.47
Met	0.52	0.71	0.78	0.34
Ile	2.83	3.61	4.12	2.40
Leu	6.25	6.42	7.22	4.26
Tyr	4.12	4.13	2.38	5.69
Phe	5.41	6.87	6.27	6.07
Lys	3.12	3.06	5.46	7.13
His	2.22	3.50	2.81	2.54
Arg	12.43	10.69	10.19	4.82
Trp	1.59	1.21	0.91	0.59

Table 1.20 Analysis on amino acid composition of various stored protein components of two peanut varieties (unit: $mg \cdot g^{-1}$ amino acid)

Amino acid	Shanyou 523			Haihua No. 1		
	Arachin	Conarachin I	Conarachin II	Arachin	Conarachin I	Conarachin II
Asp	133.35	126.82	138.27	131.48	129.13	137.34
Thr[a]	26.27	32.57	21.71	25.53	25.61	24.96
Ser	33.79	33.52	41.20	37.01	31.73	38.39
Glu	208.52	183.70	283.80	210.58	217.55	213.85
Gly	58.32	51.76	33.67	39.31	46.92	59.64
Ala	48.53	53.52	21.07	46.47	41.70	43.97
Val[a]	55.22	67.43	–	52.81	55.02	49.81
Cys	–	–	–	2.60	–	8.69
Met[a]	9.72	13.88	25.64	–	5.71	9.15
Ile[a]	42.27	49.28	29.43	42.20	46.92	37.69
Leu[a]	73.38	80.96	68.51	74.97	67.27	66.03
Tyr	47.85	42.08	7.71	45.92	13.66	36.85
Phe[a]	64.96	69.06	25.47	69.29	56.82	25.38
Lys[a]	25.78	42.45	29.25	25.33	60.48	38.17
His	18.52	19.26	12.64	24.02	29.16	25.57
Arg	108.55	90.63	158.52	126.44	133.09	138.72
Pro	49.88	44.59	33.14	45.97	44.82	43.80

[a]Essential amino acid

peanut varieties and establishing the evaluation method of functional properties of peanut protein have become the research hotspot in this field.

2.2.2 Fat

The fat content in peanut kernel is 44–54%, and compared with several oilseed crops, the fat content in peanut is second only to that in sesame and higher than that in rape, soybean, and cottonseed (Table 1.16). The research of Teng (2003) showed that the fat content of peanut in different areas was different, the oil contents of peanuts in Henan and Zhejiang were high, and the oil contents of peanuts in Sichuan and Guangxi were low; the research of Moore and Knauft (1989) and Ókeefe et al. (1993) found that the fat contents of different variety types were different, the fat content of multi-grain variety was the highest (the average content was 51.09%), and the fat content of intermediate variety was lower (the average content was 46.28%).

The fatty acid in peanut includes saturated fatty acid (palmitic acid 6–18%, stearic acid 1.3–6.5%, anthralycolic acid 1.0–3.0%, behenic acid, wood tar acid, and myristic acid) and unsaturated fatty acid (oleic acid 35–72%, linoleic acid 20–45%, and arachidonic acid), and the total amount of unsaturated fatty acids was over 85% (Liu et al. 2008; Bockisch 1998; Jiang et al. 1998). Its fatty acid composition was similar to that of olive oil (the contents of palmitic acid, stearic acid, oleic acid, linoleic acid, and arachidic acid were 6.9%, 2.3%, 84.4%, 4.6%, and 0.1%, respectively), and the olive oil could reduce the incidence of cardiovascular disease, so the peanut oil was called "Chinese economic olive oil" (Yao 2005). It was reported that oleic acid was a monounsaturated fatty acid that had a good resistance to rancidity, such as high oleic acid content and low polyunsaturated fatty acid content of oleic acid sunflower seed oil (only 2–4% unsaturated fatty acid content), so it had excellent oxidation stability (Shi 1999). The sum of oleic acid and linoleic acid in peanuts was about 80%, and it was essentially constant (Bovi 1983; Basha 1992). The oleic acid content of common variety is the highest in the various types of peanut varieties with an average of 49.32%, and the oleic acid contents of the multi-grain and pearl varieties are the lowest with an average of about 38.0%. The linoleic acid content in various types of varieties was opposite to the oleic acid content (Luan et al. 1990; Han and Luan 1998). Table 1.21 shows the fatty acid composition and contents of several common vegetable oils; it could be seen that from the table, compared with other vegetable oils, the peanut oil contained long-chain fatty acids (C20: 0–C24: 0) which did not exist or had little content in other vegetable oils.

2.2.3 Carbohydrate

The carbohydrate components in peanuts are complex; they can be divided into monosaccharide, oligosaccharide, and polysaccharide from the structural

Table 1.21 Fatty acid contents of several common vegetable oils (unit: %)

Source	C14:0	C16:0	C17:0	C18:0	C20:0	C22:0	C24:0	C16:1	C18:1	C20:1	C22:1	C18:2	C18:3
Almond oil	–	2.65	0.02	0.18	0.02	–	–	–	86.79	–	–	9.82	0.03
Tea oil	1.79	–	–	3.55	–	–	–	11.99	74.41	–	–	8.26	0.97
Cottonseed oil	0.50	20.40	–	1.40	–	–	–	0.30	14.60	–	–	62.80	
Olive oil	1.30	14.41	–	3.26	0.52	–	–	–	71.25	—	–	9.77	1.15
Peanut oil	–	10.90	–	2.70	1.10	1.70	0.60	–	46.50	0.70	–	35.40	0.10
Soybean oil	–	19.05	—	3.87	–	–	–	–	16.99	–	–	51.39	8.70
Sesame oil	–	8.77	–	5.30	—	0.10	–	0.15	39.70	0.19	–	44.69	0.96
Sunflower seed oil	–	4.44	–	3.82	0.28	1.06	0.31	–	38.11	0.16	–	51.82	–
Corn oil	–	11.60	–	1.30	–	–	–	–	30.60	–	–	55.80	–
Wheat germ oil	–	27.89	–	1.48	2.92	–	–	–	15.06	–	–	52.64	–
Rice bran oil	0.61	22.14	0.03	–	–	0.93	–	–	32.70	1.54	–	35.30	–
Rapeseed oil	–	4.50	–	1.40	0.50	0.30	0.20	–	32.00	4.40	24.80	23.60	8.10
Perilla oil	0.21	3.94	–	1.62	0.31	–	–	0.61	12.14	–	–	15.32	62.91

composition, and the polysaccharide can be divided into homopolysaccharide and heteropolysaccharide according to the different compositions. Homopolysaccharide is composed of the same glycosyl and includes starch and cellulose; the heteropolysaccharide is composed of two or more different monosaccharide units. The peanut kernel contains 10–23% carbohydrates, the starch accounts for about 4%, and the rest is free sugar which can be divided into soluble and insoluble sugar. Soluble sugar mainly included sucrose (sucrose based), fructose, glucose, as well as a small amount of stachyose, raffinose, verbascose, and other oligosaccharides; insoluble sugar included galactose, xylose, arabinose, and glucosamine (Yao 2005). The total carbohydrate in the defatted peanut flour was 38.0%, the starch content was 12.5%, the hemicellulose content was 40%, the monosaccharide content was 1.2%, and the oligosaccharide content was 18.0% (Kan 2008). In the carbohydrates, the sucrose was the most important sugar, and the sweet taste of raw peanut was mainly derived from sucrose; the unique flavor (Newell et al. 1967) and nut flavor (Mason et al. 1969) of baked peanuts were generated due to the Maillard reaction (Savage 1994) of monosaccharide (glucose and fructose) and free amino acids in the peanut baking process. The research on the relationships between sweetness and taste of baked peanuts and between seed hardness and sugar content showed that the taste quality was better with the increase of seed hardness and sucrose content, and the relationship between taste quality and seed sweetness of baked peanuts was closer than the relationship between seed hardness and sugar content (Wan 2007).

The chemical composition of dietary fiber in peanut mainly included cellulose, hemicellulose, pectin, and lignin; the dietary fiber mainly existed in peanut shell, peanut kernel (cake), and peanut stems and leaves; the amount of dietary fiber in peanut shells was the largest, about 65–80%, that in peanut cake was 4–6% and that in peanut stems and leaves was about 21.8% (Liao 2004). The dietary fiber components in peanut are complex, so there are a variety of classification methods. According to the different solubility in water, the dietary fiber can be divided into water-soluble dietary fiber and water-insoluble dietary fiber; according to different quality, it can be divided into ordinary dietary fiber and high-quality dietary fiber; according to the different fermentation degrees by coliform bacteria, it can be divided into partial fermented dietary fiber and fully fermented dietary fiber. The dietary fiber in peanut has good water-holding capacity, cationic binding and exchange and adsorption, and other physicochemical properties, as well as many physiological functions, such as blood sugar regulation, cholesterol lowering, blood pressure lowering, and obesity prevention. The dietary fiber in peanut can be added to dairy products, meat products, drinks, baked food, and other foods to improve the sensory quality and nutritional function of food, and it can also be made into snack food or slimming food. The dietary fiber in the peanut has a wide range of sources, many physiological functions, and a wide range of application, so it has a broad market development prospect.

2.2.4 Vitamin

Peanut is rich in vitamins, including niacin; vitamin E; vitamins B1, B2, and B6; pantothenic acid; folic acid; etc. Among them, the contents of nicotinic acid, vitamin E, and pantothenic acid are higher and account for 85% in the total vitamin content, followed by the contents of vitamins B1, B2, and B6; the contents of these three vitamins are between 0.1–1 mg/100 g and account for about 10% in the total vitamin content; the contents of vitamin K, folic acid, and biotin in peanut are lower, less than 0.1 mg/100 g, and account for less than 5% in the total vitamin content.

The V_E in peanut mainly contains three isomers: α-V_E, γ-V_E, and δ-V_E, and their approximate contents are 20 mg/100 g, 7 mg/100 g, and 0.5 mg/100 g, respectively. It is reported that the natural V_E (especially α-V_E) has the effects to enhance immunity, delay senescence, and reduce the incidence of cardiovascular disease and cancer. V_E has a significant impact on the stability of the oil. Tong et al. (2009) measured the oxidation stability difference of small packed vegetable oil free from antioxidant by using the oil oxidation stability tester and pointed out that the oxidation stability of corn oil was better than that of peanut oil, because the deodorant corn oil contained 80–120 mg/100 g of vitamin E, and the content of vitamin E in peanut oil was 26.8–51 mg/100 g which was significantly lower than that of corn oil (Table 1.22).

There have been some reports on differences in the total V_E content between different peanut types in China. On this basis, Zhang Jianshu and Wang Qiang (2012) carried out measurement and statistics analysis for the three isomers (α-V_E, γ-V_E, and δ-V_E) of 45 peanut varieties collected from three main peanut-producing areas (Shandong, Guangdong, and Henan). The results showed that, in the 45 peanut varieties, the content of α-V_E in the peanut varieties from Guangdong was the highest and the contents of γ-V_E and δ-V_E in the peanut varieties from Fujian was the highest, which provided the theoretical basis for selecting V_E-rich peanut varieties.

2.2.5 Phytosterol and Squalene

The phytosterol content in peanuts was 150–250 mg/100 g (Zhang et al, 2008); it mainly contained campesterol, stigmasterol, and β-sitosterol, and their contents were about 22 mg/100 g, 63 mg/100 g, and 77 mg/100 g, respectively (Liu and Yao 2006; Clough 1992); Table 1.23 shows the phytosterol contents in several common foods.

Squalene is widely distributed in the roots, stems, and leaves of plants, especially in vegetable oil, and it is one of the main components of unsaponifiable matter; squalene could be used as singlet oxygen quencher and free radical scavenger (Finotti et al. 2000) to play a role of antioxidation in oil, but it would become pro-oxidant after oxidation and promote the oxidation of oil (Xu et al. 2007); the

Table 1.22 Content and composition of vitamin E in different vegetable oils (unit: mg/100 g)

Food name	Total V_E content	V_E composition				Data source
		α-V_E	β-V_E	γ-V_E	δ-V_E	
Peanut	42.61	33.98	–	16.93	1.3	Zhang (2012)
Rapeseed oil	69.37	23.16	0.19	38.93	7.09	Yang (2009)
Soybean oil	105.91	11.90	4.44	65.48	24.09	Yang (2009)
Corn oil	88.65	31.71	2.44	51.64	2.86	Yang (2009)
Sunflower seed oil	74.45	64.12	5.06	4.52	0.75	Yang (2009)
Raw sesame oil	117.10	–	–	113.60	3.50	Huang et al. (2001)
Olive oil	22.03	19.75	0.64	1.55	0.09	Yang (2009)
Rice bran oil	32.27	29.95	1.47	0.85	0	Yang (2009)
Walnut oil	20.94	1.39	0	16.92	2.63	Yang (2009)
Lycium seed oil	162.44	151.40	11.04		–	Wang et al. (2000)
Raw linseed oil	101.44	1.83	94.03		5.58	Wang et al. (2000)
Almond oil	55.32	3.55	–	49.38	2.38	Ma et al. (2009)

squalene content in olive oil was the highest and reached 383 mg/100 g, which enhanced the oxidation stability of olive oil; the squalene content in peanut oil was less, about 16.00 mg/100 g (Xiao 2006), so it needed to be further studied about whether squalene had antioxidant effect in peanut oil. The foreign research reported that there were differences in the composition and content of phytosterols in peanut varieties in South America, but there were few reports on the differences in content of phytosterol (such as campesterol, stigmasterol, β-sitosterol, squalene, and other functional active components) in peanuts in different areas and of different varieties in China with the world's first peanut yield and planting area (Peng et al. 2006; Nelson and Carlos 1995, 1997). At present, only Zhang et al. (2012) analyzed the contents of campesterol, stigmasterol, β-sitosterol, and squalene in 67 peanut varieties in five main peanut production areas (Shandong, Guangdong, Henan, Fujian, and Jiangsu), and the results showed that, in the 67 analyzed peanut varieties, the order of total sterol content in peanuts in different provinces was Henan> Shandong> Guangdong> Jiangsu > Fujian, and the total sterol content in peanuts in Henan was higher than that in Fujian ($P < 0.05$); the order of squalene content in peanuts in various provinces was Henan> Guangdong> Shandong> Fujian> Jiangsu, and the squalene content in peanuts in Henan was significantly higher than that in Jiangsu ($P < 0.01$). These studies could not only promote the research on functional active components in peanut in China but also provide a theoretical basis for selecting the peanut varieties rich in phytosterols and squalene.

2.2.6 Other Nutritional Components

In addition to rich protein, fat, and carbohydrates, the peanut also contains mineral, saponin, resveratrol, proanthocyanidin, flavonoid, and other bioactive components. The mineral content in peanut seed kernel is only 2–3%, and from the nutrition

Table 1.23 Content and composition of phytosterol in common food (unit: mg/100 g)

Food name	Sterol content	Sterol composition						Data source
		Sitosterol	Campesterol	Stigmasterol	Δ-5 oat sterol	Δ-7 oat sterol	Sitostanol	
Peanut oil	200.1	100	58.3	19.5	–	–		Zhang (2012)
Sunflower seed oil	372.26	268.00	53.51	31.81	–	–	18.95	Feng (2006)
Rapeseed oil	570.16	341.50	155.01	8.04	–	–	11.36	Feng (2006)
Soybean oil	307.34	175.6	58.05	56.1	–	–	16.08	Feng (2006)
Sesame oil	559.27	350.73	102.80	45.03	–	–	60.71	Feng (2006)
Olive oil	265.42	12.87	4.98	43.69	–	–	326.96	Feng (2006)
Corn oil	1032.07	661.70	195.72	50.45	112.44	–	–	Feng (2006)
Palm oil	38.98	23.56	12.25	3.17	–	–	–	Bao et al. (2002)
Wheat	60.3–69.0	30.9	8.8	1.5	2.3	1.4	–	Sheng et al. (2002)
Oat	32.9–52.0	16.7–28.2	2.4–4.0	1.2–1.5	3.2–9.1	2.4	–	Sheng et al. (2002)
Barley	58.6–83.0	28.9	11.1	15.0	14.5	1.7	–	Sheng et al. (2002)
Corn	178.0	120.0	32.0	21.0	–	–	–	Sheng et al. (2002)
Almond	154.7	132.2	5.4	0.7	–	–	16.4	Han et al. (2009)
Radish	34.4	27.34	–	7.06	–	–	–	Han (2001)
Tomato	6.77	3.02	3.25	0.50	–	–	–	Han (2001)
Grape	3.31	2.24	0.70	0.37	–	–	–	Han (2001)
Apple	12.7	12.35	–	0.35	–	–	–	Han (2001)
Strawberry	12.1	11.42	0.51	0.17	–	–	–	Han (2001)
Longan	9.3	2.7	2.7	–	–	–	0.7	Han et al. (2009)
Pawpaw	16.9	8.6	5.4	2.9	–	–	–	Han et al. (2009)
Hawthorn	23.4	20.7	1.1	0.4	–	–	1.2	Han et al. (2009)
Sweet potato	12.1	8.62	0.80	2.67	–	–	–	Han (2001)

Table 1.24 Resveratrol content in each part of peanut (unit: μg/g)

Part	Resveratrol content	Reference	Part	Resveratrol content	Reference
Peanut root (Fall crops)	130–1330	Chen et al. (2002)	Peanut shell	2.5	Zhang et al. (2003)
Peanut root (Spring crops)	15–630	Chen et al. (2002)	Peanut shell	8	Chen (2004)
Peanut root	94.3	Zhang et al. (2003)	Peanut stem	8	Chen (2004)
Peanut root	46.7–58.8	Xiang et al. (2005)	Peanut stem	88.8	Zhang et al. (2003)
Peanut root	179.5	Chen (Chen 2004)	Peanut leaf	0	Zhang et al. (2003)
Peanut skin	650	Sanders (2000)	Peanut leaf	0	Chen (2004)
Peanut skin	45.3	Zhang et al. (2003)	Peanut kernel	2.63	Liu and Yao (2006)
Peanut skin	360	Liu et al. (2006)	Peanut kernel	30–149	Sanders (2000)
Peanut flower	29.3	Zhang et al. (2003)	Peanut kernel	0	Zhang et al. (2003)

point of view, the peanut kernel is rich in zinc, potassium, phosphorus, and magnesium contents and lacks of calcium, iodine, and iron contents. Zn^{2+}, Mg^{2+}, and other metal ions are compositions or activators of many metalloenzymes, and the Zn^{2+} content in 100 g of peanut oil reaches 8.48 mg which is 37 times than that in salad oil, 16 times than that in rapeseed oil, and 7 times than that in soybean oil. The resveratrol content in peanut roots was the most, followed by that in stems, red skins, leaves, and shells, and that in peanut kernels was the least (see Table 1.24) (Zhang et al. 2009). The luteolin was mainly distributed in the peanut shells, but the contents were different due to the differences in the production area, variety, and maturity, and the variation range of contents was between 0.25 and 1.12% (Tang et al. 2005). The proanthocyanidin was rich in peanut skin and accounted for 17% in the dry weight of peanut skin, and about 50% were higher bioactive oligomers (Karchesy and Hemingway 1986).

2.2.7 Anti-nutritional Factors in Peanut

The anti-nutritional factors in peanut mainly included trypsin inhibitors, lectins, phytates, condensed tannins, and α-amylase inhibitors (Ahmed et al. 1998; Ejigui et al. 2005).

Trypsin inhibitor is a substance that binds to trypsin to inhibit this enzyme action. Yang et al. (1998) found that the 2s protein component of peanut has trypsin inhibitory activity through experiments, and the nutritional value of peanut protein products could be enhanced by extracting and purifying the trypsin inhibitor of peanut seed and using a variety of mercapto reductant and protease to passivate the activity of trypsin inhibitor at low temperature. Tian et al. (2009) verified that there were at least three trypsin inhibitor components in peanut with a relative molecular mass of 30–70 kD and an isoelectric point between pH 5.0 and 5.8 through research. The content of trypsin inhibitor in baked and unshelled peanuts was reduced from 16.31 TIU/mg to 12.6TIU/100 g.

Agglutinin mainly exists in legume grains, peanuts, and their cakes. Dev Sagarika et al. (2006) thought that the peanut agglutinin (PNA) was a unique fourth-level open-structure agglutinin of the pod, a kind of homotetramer protein and a single-dominant group consisting of three β-sheet belts. Sun et al. (2011) found that the agglutinin crude extract of peanut seed could not agglutinate A, B, and O blood cells of humans, but could condense A, B, and O blood cells of humans after sialidase treatment through experiments. The agglutination of agglutinin for erythrocytes could be inhibited by lactose, melibiose, raffinose, and D-galactose; it was stable in the pH range of 5.0–11.0, and the activity completely disappeared after thermal insulation of 10 min at 55 °C. Among the six different varieties of peanuts tested, the activity of agglutinin in Hebei Gaoyou seeds was the strongest. At the early growth stage of peanuts, the parts where the agglutinin exists were cotyledons and roots.

For the phytic acid (phytate) (i.e., inositol-6-phosphate), its phosphates can be chelated with a variety of metal ions (such as Zn^{2+}, Ca^{2+}, Cu^{2+}, Fe^{2+}, Mg^{2+}, Mn^{2+}, Mo^{2+}, and Co^{2+}) into the corresponding insoluble complex, then the stable phytates are formed, and they are not easily absorbed by intestinal tracts, which reduces their use by animal bodies. The phytate content in peanut is 1.05–1.76% (by dry weight) and that in soybean is 1–2.22% (by dry weight). Compared with the phytate content in untreated peanuts, the phytate content in the peanuts through germination, baking, and unshelling treatment significantly decreased, and the phytate content decreased from 14.66 to 8.45 mg/g after germination, baking, and unshelling.

Condensed tannins (polyphenols) include tannins, phenolic acids, gossypol, and sinapine, and they mainly exist in the grains, bean grains, cotton rapeseeds and their cakes, and certain root tuber feed. Tannin is also known as tannic acid, it is a water-soluble polyphenols substance with a bitter taste, and it can be divided into condensed tannins with anti-nutritional effect and hydrolytic tannins with toxic effect. The condensed tannins are formed through the condensation of some flavonoid compounds in plants. The condensed tannins cannot be generally hydrolyzed and can be dissolved in water with an extreme polarity. The tannins react with trypsin and amylase or its substrates (proteins and carbohydrates) to reduce the utilization rate of proteins and carbohydrates; the tannins also bind to the gastrointestinal mucosal proteins to form insoluble complexes on the surface of intestinal mucosa which can damage the intestinal wall, interfere with the absorption of certain minerals (such as iron ions), and affect the growth and development of animals. The tannins can not only be combined with calcium, iron, and zinc and other metal ions to form precipitation but also be combined with vitamin B12 to form complexes to reduce their utilization rate. The tannin content in peanuts can be decreased from 0.09 mg/100 g to 0 mg/100 g after unshelling treatment.

α-Amylase inhibitors are mainly found in wheat, kidney beans, taro, mango and immature banana, and other foods. The amylase inhibitors can come into play after much raw or insufficiently cooked and heated foods are ingested, so that the starch contained in the foods cannot be digested and absorbed and utilized by the bodies and most of them are directly excreted. If this phenomenon exists in the long term, the absorption of nutrients will be reduced and the growth and development will be affected. After the bean and wheat foods are sufficiently heated and treated, the activity of digestive enzyme protein inhibitor can be completely removed. The content of α-amylase inhibitor in the peanut can be decreased from 62.5 to 18.0 AIU/g after germination, baking, and unshelling treatment.

2.3 *Processing Quality*

The processing quality (characteristic) mainly includes pure kernel rate, sugar content, oil yield, protein extraction rate, oleic acid/linoleic acid (O/L), unsaturated fatty acid/saturated fatty acid, arachin/conarachin, and other quality properties closely related to processing, and the peanut processing characteristics are closely related to the main processing products of peanut.

2.3.1 Pure Kernel Rate

The pure kernel rate refers to the fraction of the quality of seed kernel gotten and net sample after the shelled peanut net sample is unshelled, and it is also the important indicator to measure the kernel production rate during the peanut kernel production. Luan (1986) analyzed the pure kernel rate of 1072 peanut varieties in China and found that the variation range of pure kernel rate was between 52.0 and 86.1%. The research of Luan (1988) showed that there were significant differences in the pure kernel rates of different varieties and the order of average pure kernel rate of various types of varieties was pearl type >intermediate type >common type >Longsheng type >multi-grain type.

2.3.2 Sugar Content

The sugar content is an important evaluation indicator for export peanut kernels (it is required that the sugar content in peanut is more than 6%), it has a great influence on the peanut taste, and the sugar is also one of the main factors affecting the sweet flavor of baked peanuts at the same time. Bryant et al. (2004) analyzed the content and composition of oligosaccharides in 33 peanut varieties in the United States by using the high-performance liquid chromatography (HPLC Method). The research results showed that only three oligosaccharides (sucrose, raffinose, and stachyose) were detected in the oligosaccharide extracts of 33 peanut varieties. The contents of sucrose and stachyose were 2.84–6.74% and 0.27–0.61%, and the mean values were 4.62% and 0.42%, respectively; the content of raffinose was the lowest, the raffinose was not detected in some samples, the maximum content was 0.12%, and the mean value was 0.06%. The degrees of variation (CV value) of these three oligosaccharides in different peanut varieties were very high and that of raffinose was the highest (97.97%) (Table 1.25). However, there were fewer studies on the differences in sugar content in different peanut varieties in China.

2.3.3 Oil Yield

This refers to the percentage of oil content gotten through extraction accounted for the fat quality contained in peanut. In the oilseed crops, the oil yield of sunflower is 51–54%, peanut is about 45%, rapeseed is 32–38%, domestic soybean is about 18%, and import soybean is about 22%; the oil yield of peanut is second only to that of sunflower and more than double that of import soybean. The fat content in peanut is 44–54%, while there are fewer reports on the oil yield of different varieties of peanuts; Wang (2012) analyzed the oil yield of 45 peanut varieties in main peanut production areas of China and found that there were significant differences and the variation range was 24.04–54.60%. This research provided data support for selecting the peanut varieties with higher oil yield and provided a reference for actual production of peanut oil processing enterprises.

2.3.4 Protein Extraction Rate

The protein extraction rate refers to the percentage of extracted protein quality accounted for protein quality contained in peanut, and it is one of the important measurement indicators of peanut protein preparation. The research conducted by Li (2009) showed that in the actual production, the extraction rate was an important indicator which was necessary to consider when the protein isolate was prepared by soybeans. The current domestic peanut protein production has just started, so there have been fewer articles and reports on the peanut protein extraction rate. Wang (2012) extracted the protein in 66 peanut varieties by using the alkali extraction and acid precipitation method, and the results showed that the variation range of protein extraction rate was 59.51–88.97%, which indicated that there were greater differences in the extraction rate of protein in different varieties.

2.3.5 Oleic Acid/Linoleic Acid

The oleic acid/linoleic acid (O/L) ratio is an important indicator to measure the storability of peanut raw materials and their products. The higher the ratio, the better the stability of peanuts and their products (Savage 1994; Lopez et al. 2001). Jiang and Duan (1993) and Liu and Liang (1993) carried out the quality test for the

Table 1.25 Composition of oligosaccharides in 33 peanut varieties in the United States (unit: %)

Variety	Sucrose	Raffinose	Stachyose	Total sugar
Tx798736–1	3.51	0.05	0.39	3.95
Tx872551	3.63	0.05	0.40	4.08
Tx883772	3.54	0.05	0.60	4.19
Tx872561	3.94	0.05	0.49	4.48
ICGS(E)-56	3.23	ND*	0.61	3.84
ICGS(E)-49	2.84	0.05	0.31	3.20
ICGS(E)-108	3.23	0.05	0.40	3.68
SN55–437	3.67	0.11	0.36	4.14
TP175–3	4.45	0.03	0.39	4.87
TP178–3	5.00	0.05	0.55	5.60
TP172-#2–2	5.48	0.05	0.43	5.96
TP171–2	5.45	0.03	0.55	6.03
TP178–1	5.14	0.05	0.52	5.71
TP175–6	5.45	0.05	0.51	6.01
STARR89 WALLER	3.65	0.05	0.37	4.07
PI 1174	3.53	0.05	0.33	3.91
FLORUNNER INTERSPECIFIC LINE	5.22	0.03	0.43	5.68
BRYAN UPPT 1989 TAMRUN 88	4.26	0.06	0.42	4.74
BRYAN UPPT 1989 Tx 835,741	5.01	0.09	0.27	5.37
BRYAN UPPT 1989 FLORUNNER	5.00	0.05	0.48	5.53
BRYAN UPPT 1989 Tx 835,841	4.75	ND	0.43	5.18
GAINES Co. LP 1989 STARR	5.75	0.05	0.44	6.24
GAINES Co. Tx 798,736–1 1989	6.74	0.12	0.49	7.35
FRIO Co. STARR	3.91	0.05	0.39	4.25
FRIO Co. Tx 798,736–1 1989	4.88	ND	0.37	5.25
FRIO Co. LP LATE MATURING NC 343	4.74	0.07	0.50	5.31
FRIO Co. 1989 FLORUNNER	4.84	0.12	0.48	5.44
FRIO Co. EARLY MATURING LP 0.31PRONTO	4.74	0.06	0.41	5.21
Tx A0.47 g-3	5.08	0.03	0.47	5.58
RMP-12	4.38	0.05	0.31	4.74
FLORUNNER	5.28	ND	0.47	5.75
SPANISH	6.05	ND	0.37	6.42
VIRGINIA	6.15	ND	0.40	6.55
Mean	4.62	0.06	0.42	5.09
COEFF. OF VARIATION	20.52	97.97	25.58	19.64
FISHER PLSD ($p < 0.05$)	0.20	0.06	0.24	

ND indicate that this kind of sugar is not detected

peanut germplasm resources in China, and the results showed that the average O/L value of Longsheng-type peanut was 2.05, common-type peanut was 1.37, pearl-type peanut was 1.36, multi-grain-type peanut was 0.99, and intermediate-type peanut was 1.08. Norden et al. (1987) first discovered that F435 was a natural mutant with high oleic acid content; its oleic acid content was more than 80%, linoleic acid content was 2%, and O/L value was 40:1. Gorbet et al. of University of Florida (1997) selected and cultivated two Runners varieties (SunOleic 95R and Sun Oleic 97R); their O/L values were 15.8 and 27.0, respectively, and the shelf lives of their processed products were 3–15 times longer than that of common varieties. At present, the United States had cultivated more than ten high oleic peanut varieties, and the oleic acid contents of some varieties had reached about 85% (Lu 2005); the breeding experts in China have been cultivating the oil-used peanut varieties with high oleic acid and low linoleic acid contents, which has great significance for improving the economic benefits of peanut processing industry. All the import countries had relevant regulations on the O/L value of peanut currently, for example, the required O/L value of large peanut should be 1.6 or more and that of small peanut should be more than 1.2 (Wan 2007). Some studies also showed that the O/L value had a significant correlation with the flavor of baked peanut, for example, the research conducted by Pattee (2002) showed that the oleic acid content had a significant positive correlation with the flavor intensity of baked peanut.

2.3.6 Unsaturated Fatty Acid/Saturated Fatty Acid

The oxidation of vegetable oil is closely related to fatty acids, and the different vegetable oils have different oxidation stability due to the differences in the types and contents of fatty acids contained. Li et al. (2010) analyzed the fatty acid composition of eight natural edible vegetable oils (such as peanut oil, rapeseed oil, and soybean oil) by using the gas chromatography method and compared their oxidative stability. The results showed that the oxidation stability in an order from strong to weak was peanut oil, camellia oil, sesame oil, rapeseed oil, corn oil, blend oil, soybean oil, and sunflower seed oil. The polyunsaturated fatty acid had larger impacts on the oxidative stability, and the corresponding stability of oil with higher content of polyunsaturated fatty acids was smaller; for example, the contents of polyunsaturated fatty acids in sunflower seed oil, soybean oil, and blend oil were high (they were 60.37%, 58.14%, and 54.57%, respectively) and their stability was unstable. The oxidative stability of vegetable oil was not only affected by the content of polyunsaturated fatty acids but also affected by the total amount of unsaturated fatty acids. The content of polyunsaturated fatty acids in sesame oil (47.25%) was higher than that in rapeseed oil (28.34%), but the rapeseed oil was unstable than sesame oil, because the total amount of unsaturated fatty acids in rapeseed oil (88.53%) was significantly higher than that in sesame oil (81.92%) (Li et al. 2010). Therefore, the unsaturated fatty acid/saturated fatty acid ratio directly affected the oxidation stability of oil, and the corresponding stability of oils with high unsaturated fatty acid/saturated fatty acid ratio was also small; for example, the contents of unsaturated fatty acid/saturated fatty acid in sunflower seed oil and camellia oil were high (they were 7.69 and 5.88, respectively), so their oxidation was the most unstable; the unsaturated fatty acid/saturated fatty acid ratio of peanut oil was low (3.85), so its oxidation stability was the best (Table 1.26).

The FAO/WHO Expert Committee considered that different saturated fatty acids had different impacts on lipid levels. For example, the lauric acid (C12:0), myristic acid (C14:0), and palmitic acid (C16:0) could increase the low-density lipoprotein (LDL) level, while the stearic acid (C18:0) had no significant effect on LDL. According to the epidemiological and randomized controlled trial research of coronary heart disease, the minimum recommended intake of polyunsaturated fatty acids should be 6% in order to reduce LDL and total cholesterol level, increase the high-density lipoprotein (HDL) level, and decrease the risk of coronary heart disease. Some research found that the excessive intake (more than 11%) of polyunsaturated fatty acids would increase the risk of lipid peroxidation in the case of low tocopherol intake level and, thus, the acceptable range of polyunsaturated fatty acids ($n-3$ and $n-6$) was 6–11%. Therefore, the appropriate proportion pattern of unsaturated fatty acids/saturated fatty acids could not only guarantee the oxidation stability of vegetable oil but also reduce the incidence of coronary heart disease. At present, there have been fewer reports on saturated fatty acid/unsaturated fatty acids in different peanut varieties. Zhang and Wang (2012) analyzed the unsaturated fatty acids/saturated fatty acids in 45 peanut varieties in China and found that the variation range of saturated fatty acid/unsaturated fatty acid was 2.22–4.55, the average value was 4.00, and the peanut varieties with the lowest and highest contents of unsaturated fatty acid/saturated fatty acid were black peanut and Longhua 243, respectively. This conclusion provided a useful reference for selecting the peanut varieties with reasonable unsaturated fatty acid/saturated fatty acid.

2.3.7 Arachin/Conarachin

The arachin and conarachin are the main components of peanut protein, and the arachin/conarachin is an important indicator to measure the functional properties of peanut protein, but there are fewer researches on the differences in arachin/conarachin in different peanut varieties and their relationships with functional properties at present. Saio (1969) reported that the hardness, springiness, and viscosity of tofu gel made from soybean 11S globulin were better than that of tofu made from 7S globulin, and 11S/7S had a positive correlation with the quality of tofu. Mujoo et al. (2003) and Cheng et al. (2006) also reported that the 11S/7S ratio of soybean protein was closely related to the thermal gel of protein, and 11S/7S had a significantly positive correlation with the hardness of tofu gel ($r = 0.86$). In recent years, with the breakthrough of low-temperature cold-pressing technology and low modified peanut protein production technology, the development and utilization of peanut protein have attracted great attentions of production enterprises; the research related to the peanut protein components and functional properties has increasingly become the research hotspot in academic world. Wang et al. (2013) analyzed the main components of proteins and contents of subunits in

Table 1.26 Composition and content of fatty acids in different vegetable oils

Vegetable oil	Peanut oil	Rapeseed oil	Camellia oil	Soybean oil	Sunflower seed oil	Blend oil	Corn oil	Sesame oil
SFA	20.93	11.46	14.20	19.41	13.47	19.76	18.72	18.07
MUFA	37.90	60.19	71.29	22.45	26.16	25.67	32.77	34.67
PUFA	41.17	28.34	14.51	58.14	60.37	54.57	48.51	47.25
UFA/SFA	3.85	7.69	5.88	4.17	6.25	4.00	4.35	4.55
n-3PUFA	2.60	9.66	1.07	7.12	0.99	6.37	0.71	1.59
n-6PUFA	38.58	18.68	13.43	51.02	59.37	48.20	47.80	45.66

170 peanut varieties in China and studied the relationship between arachin/conarachin and functional properties of peanut protein. The results showed that the range of arachin/conarachin ratio was between 0.8 and 1.68, the variation coefficient was 15.23, and the ratios of arachin/conarachin in different peanut varieties were different greatly. At the same time, the preliminary research showed that the arachin/conarachin was closely related to the functional properties of peanut protein; the higher the arachin/conarachin, the better the protein solubility; the lower the arachin/conarachin, the better the gel property.

References

Ahmed EM, Ali T, Sitren HS. Extraction and purification of lectin from florunner peanut seeds [J]. Peanut Sci. 1998;15:44–7.

Andersen PC, Hill K, Gorbet DW, et al. Fatty acid and amino acid profiles of selected peanut cultivars and breeding lines [J]. J Food Compos Anal. 1998;11:100–11.

Bao ZD, Xu RN, Zhang SH. Determination of phytosterol and cholesterol in oil by capillary gas chromatography [J]. Chin J Anal Chem. 2002;30(12):1490–3.

Basha SM. Effect of location and season on peanut seed protein and polypeptide composition [J]. J Agric Food Chem. 1992;40:1784–8.

Bockisch M. Vegetable fats and oils[J]. Fats and oils handbook. Champaign: AOCS Press; 1998. p. 174–326.

Bovi MLA. Genotypic and environmental effects on fatty acid composition, iodine value, and oil content of peanut (*Arachis hypogaea* L.)[D]. NT: University of Florida; 1983. p. 44–406.

Bryant RJ, Rao DR, Ogutu S. α and β-galactosidase activities and oligosaccharide content in peanuts[J]. Plant Foods Hum Nutr. 2004;58:213–23.

Chen XX. Studies on the contents of resveratrol, polydatin and peanut resveratrol in Polygonum Cuspidatum [D]. Fuzhou: Fujian Normal University; 2004.

Chen TW, Kang BB, Su LQ, et al. The analysis on nutritional qualities of peanut cultivars which host planted in Fujian province [J]. Chin Agric Sci Bull. 2007;23(11):141–5.

Chen RS, Wu PL, Robin YYC. Peanut roots as a source of resveratrol [J]. J Agric Food Chem. 2002;50:1665–7.

Chen J, Wu LR, Zhang CS. Sensory quality of peanut (Arachis hypogaea L.) [J]. J Peanut Sci. 2004;33(1):24–7.

Cheng CL, Wang ZY, Shi YG, et al. Chin Oils Fats. 2006;31(4):16–9.

Clough A. The determination of tocopherols in vegetable oils by square-wave voltammetry[J]. J Am Oil Chem Soc. 1992;69:456–60.

Du Y, Wang Q, Liu HZ, Wang L, Liu L. Major protein fractions and subunit contents in peanut from different cultivars. Food Sci. 2013;34(9):42–6.

Ejigui J, Savoie L, Marin J, Desrosiers T. Influence of traditional processing methods on the nutritional composition and antinutritional factors of red peanuts (*Arachis hypogea*) and small red kidney beans (*Phaseolus vulgaris*)[J]. J Biol Sci. 2005;5(5):597–605.

FAOSTAT. http://faostat.fao.org/site/291/default.aspx (n.d.)

Feng SY. Study on the method and application of GC determination of phytosterols in common foods [D]. Nanchang: Nanchang University; 2006.

Finotti E, Ambrosio MD, Paoletti F, et al. Synergistic effects of a-tocopherol, b-sitosterol and squalene [J]. Nahrung. 2000;44(5):373–4.

Gao HC, Li WY, Lv XL. Study on protein stability and processing conditions of peanut [J]. J Chin Foreign Tech Inf. 1995;12:39–40.

Gorbet DW, Knauft DA. Registration of 'SunOleic 95R' peanut [J]. Crop Sci. 1997;37(4):1392.

Govindaraju K, Srinivas H. Studies on the effects of enzymatic hydrolysis on functional and physic-chemical properties of arachin [J]. Swiss Soc Food Sci Technol. 2006;39:54–62.

Han JH. The characteristic, function and application of phytosterols [J]. J Environ Hyg. 2001;28 (5):285–91.

Han JH, He M, Zhou SS, et al. Analysis of phytosterol contents in food plant materials and Chinese traditional medicines [J]. J Hyg Res. 2009;38(2):188–91.

Han SP, Luan WQ. Preliminary analysis on protein fat and fatty acid content of introduced peanut varieties in foreign countries [J]. Shandong Agric Sci. 1998;6:30–4.

Huang HN, Liu W, Zhang WZ. Application study on stripe picture of natural vitamin E in vegetable oil by HPLC [J]. Strait J Prev Med. 2001;7(3):10–3.

Jiang HF, Duan NX. Correlation analysis of fat quality and oil content, oleic acid and linoleic acid content in peanut oil [J]. J Peanut Sci. 1993;2:4–5.

Jiang HF, Duan NX, Xiaoping R. Comprehensive evaluation of peanut germplasm resources [J]. Chin J Oil Crop Sci. 1998;20(3):32–3.

Jiang HF, Ren XP, Liao BS, et al. Genetic diversity of peanut genotypes with resistance to bacterial wilt based on seed characters [J]. Chin J Oil Crop Sci. 2006;28(2):144–50.

Karchesy JJ, Hemingway RW. Condensed tannins: (4β-8;2β-O-7)-Linked Procyanidins in Arachis hypogeal L[J]. J Agric Chem. 1986;34:966–70.

Krishna TG, Pawar SE, Mitra R. Variation and inheritance of the arachin polypeptides of groundnut (*Arachis hypogaea* L.) [J]. Theor Appl Genet. 1986;73:82–7.

Li JJ. Screening and suitability of soybean processing special varieties in China [D]. Shijiazhuang: Hebei Agricultural Universit; 2009.

Li HY, Deng ZY, Li J. Study on oxidative stability of plant oil with different fatty acid composition[J]. Science and Technology of Food Industry, 2010, 31(1): 173–182.

Li Y, Huang SZ, Fu JR. Electrophoretic analysis of protein from different peanut seeds [J]. J Integr Plant Biol. 1998;40(6):534–41.

Liao ZD. Chemical utilization of peanut shell [J]. Technol Deve Chem Ind. 2004; 33(02):22–35.

Licher MK. Strategic options for the Virginia Peanut Industry after the 2002 farm bill: a linear programming model [D]. Master of Science, 2005.

Liu GH, Liang ZP. Identification of main quality traits of peanut germplasm resources in China [J]. Chin J Oil Crop Sci. 1993;1:18–21.

Liu DC, Sun W, Yu BQ, et al. Simultaneous preparation of peanut oil and defatted protein powder by low-temperature prepressing, extraction and low-temperature desolventizing [J]. Chin Oils Fats. 2008;33(12):13–5.

Liu HS, Wang JH, Xu XN, et al. Determination of resveratrol in peanut, peanut oil and raisins by RP – HPLC [J]. Chin Tradit Herbal Drug. 2006;37(8):1188–9.

Liu Q, Yao HY. Antioxidants in oilseeds [J]. Chin Food Addit. 2006;2:95–9.

Lopez Y, Smith OD, Senseman SA, et al. Genetic factors influencing high oleic acid content in Spanish market-type peanut cultivars [J]. Crop Sci. 2001;41:51–6.

Lu Y. Peanut and peanut products nutrition should cause universal attention [J]. Chin Food Drug Adm. 2005;9:59.

Lu CB, Huang SZ, Tang XJ, et al. Composition of amino acid in storage proteins and changes in 17.5 kDa polypeptide synthesis during peanut seed development [J]. J Trop Subtrop Bot. 2000;8(4):339–45.

Luan WQ, Feng HS, Wang JS. Study on the main characters of peanut varieties – differences in performance and types [J]. Chin Seed Ind. 1986;2:3–7.

Luan WQ, Han SP, et al. Chin Seed Ind. 1990;2:22–5.

Ma YH, Zhao Z, Li KY, et al. Oil content and composition of almond from different producing Area [J]. J Chin Cereals Oils. 2009;24(11):70–3.

Mason ME, Newell JA, Johnson BR, et al. Non volatile flavour components of groundnut [J]. J Agric Food Chem. 1969;17:732–82.

Misra JB. A mathematical approach to comprehensive evaluation of quality in groundnut [J]. J Food Compos Anal. 2004;17:69–79.

Monteiro PV, Pralash V. Effect of proteases on Arachin, conarachin I and conarachin II from peanut (*Arachis hypogaea* L.) [J]. J Agric Food Chem. 1994;42:268–73.

Moore KM, Knauft DA. The inheritance of high oleic acid in peanut [J]. J Hered. 1989;80:252–3.

Mouecoucou J. β-Lactoglobulin/polysaccharide interactions during in vitro gastric and pancreatic hydrolysis assessed in dialysis bags of different molecular weight cut-offs. Biochim Biophys Acta Gen Subj. 2004;1670(2):105–12.

Mujoo R, Trinh DT, Ng PKW. Characterization of storage proteins in different soybean varieties and their relationship to tofu yield and texture [J]. Food Chem. 2003;82:265–73.

Nelson RG. Proximate, Fatty acid and sterol composition of Aboriginal peanut (*Arachis hypogaea* L) seeds from Bolivia [J]. J Sci Food Agric, 1997, 73: 349–356.

Nelson RG, Carlos AG. Lipid, protein, and ash contents, and fatty acid and sterol composition of peanut (*Arachis hypogaea* L.) seeds from Ecuador[J]. Peanut Sci. 1995;22:84–9.

Newell JA, Mason ME, Matlock RS. Precursors of typical and atypical roasted peanut flavor [J]. J Agric Food Chem. 1967;15:767–72.

Norden AJ, Gorbet DW, Knauft DA, et al. Variability for oil quality among peanut genotypes in the Florida breeding programme [J]. Peanut Sci. 1987;14:7–11.

Ókeefe SF, Wiley VA, Knauft DA. Comparison of oxidative stability of high and normal oleic peanut oils [J]. J Am Oil Chem Soc. 1993;70:489–92.

Peng H, Chen WQ, Deng BW. Analysis of the plant sterol component and content of edible rape oil with gas chromatography [J]. J Anhui Agric Sci. 2006;34(19):4830–1.

Sagarika D, Khan R, Avadhesha S. 2,2,2- Trifluoroethanol-Induced structural change of peanut agglutinin at different pH: A comparative account [J]. IUBMB Life. 2006;58(8):473–9.

Saio K. Food processing characteristics of soybean 11S and 7S protein Part I. Effect of difference of protein component among soybean varieties on formation of tofu-curd [J]. Agric Biol Chem. 1969;33:1301–8.

Sanders TH, Mcmichael RW, Hendrix KW. Occurrence of resveratrol inedible peanuts [J]. J Agric Food Chem. 2000;48:1243–6.

Sargeant K, Sheridan A, O'Kelly J. Toxicity associated with certain samples of groundnuts [J]. Nature. 1961;192:1096–7.

Savage GP, Keenan JI. The composition and nutritive value of groundnut kernels. In: Smartt J, editor. The groundnut crop: a scientific basis for Improvement. Chapman & Hall, London, 1994, 173–213.

Sheng Y, Hua W, Gu WY. The distribution of phytosterol in food sources [J]. Cereals Oils. 2002;5:40–1.

Shi HS. Highly resistant to oxidation of vegetable oil [J]. J Daily Chem Sci. 1999;104(1):23–6.

Shokarii EH, Esen A, Mozingo RW. Immunological Characterization of a 36kD Polypeptide in Peanuts (Arachis hypogaea L.) [J]. Peanut Sci. 1991;18(1):11–5.

Sun J, Lv JJ, Lu FS, Yang QL, et al. Study on properties of lectin in seed of peanut(Arachis hypogaea L.) [J]. Food Sci Technol. 2011;08:134–7.

Tang LP, Gong YQ, Wu XY, et al. HPLC determination of luteolin of peanut hull from different regions [J]. J Peanut Sci. 2005;34(2):1–4.
Teng Z, Liu Q, Guo DL. The content statistics and classify analyse for major nutrition natters of peanut [J]. Food Res Dev. 2003;24(4):84–5.
Tian J, Su CM, Zhou XW. Preliminary study on the trypsin inhibitors in the sweet potato and peanut [J]. Guihaia. 2009;29(1):70–3.
Tong YW, Chen FX, Yang BT. Investigation on oxidation stability of different edible vegetable oil [J]. Chin Oils Fats. 2009; 34(2):31–34.
Wan S, Wang C, Zhao P, et al. Optimized disposition of Intercropping date and planting density for wheat-peanut cropping system [J]. Chin J Oil Crop Sci. 2004;26:55–8.
Wan SB. Peanut quality [M]. Beijing: China Agricultural Science and Technology Press; 2007.
Wang LZ. Soybean research for 50 years [M]. Beijing: China Agricultural Science and Technology Press; 2010.
Wang L. Study on processing characteristics and quality evaluation of peanut peanut [D]. Beijing: Chinese Academy of Agricultural Sciences; 2012.
Wang J, Zhang N, SJ W, et al. Analyzing vitamin E contents of ll plant oils from Ningxia market by HPLC [J]. J Ningxia Med Univ. 2000;22(1):11–2.
Wang S, Wang W, Jin Z, et al. Screening and diversity of plant growth promoting endophytic bacteria from peanut. In: Afr J Microbiol Res; 2013.
Wu HW. Preparation of peanut concentrate protein, mechanism of gel formation and its application [D]. Beijing: Chinese Academy of Agricultural Sciences; 2009.
Wu LR, Chen J, Wang XZ, et al. Main value index of peanut sensory quality [J]. Chin J Oil Crop Sci. 2005;27(1):1–3.
Xiang HY, Zhou CS, Zhong SA, et al. Synthesis of resveratrol imprinted polymer and its application in separation of active ingredient in polygonum cuspidatum Sieb. et Zucc. Extracts [J]. Chin J Appl Chem. 2005;7:739–43.
Xiao YG. Investigation and prospect of bio-active components in vegetable oil [J]. Cereal Food Ind. 2006;13(4):1–5.
Xu L, Wang R, Cao XB, et al. Study on the change of antioxidant components in olive Oil.[J]. J Henan Univ Technol (Nat Sci Ed). 2007;28(2):38–41.
Yamada T. Dissociation-association behavior of Arachin between dimeric and monomeric forms [J]. Agric Biol Chem. 1979;43(12):2549–56.
Yang QL. Current situation of peanut processing in China, 2009.
Yang XQ, Zhang SH, Li Y. Extraction and isolation of peanut 2S protein and its partial properties [J]. J South China Univ Technol (Nat Sci Ed). 1998;26(4):1–5.
Yao YY. Comparison of peanut oil and olive oil in nutritional value [J]. Chin Oils Fats. 2005; 30(4):66–8.
Yu SL. Chinese peanut varieties and their pedigree [M]. Shanghai: Shanghai Science and Technology Press; 2008.
Yuan DQ, Liang LQ, Wang ZF. Application of modificated vegetable proteins in food industry [J]. Food Res Dev. 2005;26(6):13–5.
Zhang JS. Effects of fatty acid composition and endogenous antioxidants on the stability of different peanut oils [D]. Beijing: Chinese Academy of Agricultural Sciences; 2012.
Zhang JQ, Chen J, Li ZY, et al. Advances in functional constituents of peanut [J]. J Grain Oil Process. 2008:73–6.
Zhang J, Gao HY, Ge LM, et al. Factors of resveratrol content in peanut roots and stems [J]. J Agric Eng (Addenda). 2009;25:251–3.
Zhang H, Hu JP, Han M, et al. Determination of resveratrol content different in parts of peanut by HPLC. [J]. J Xinjiang Med Univ. 2003;26(5):440–1.

Zhang JS, Wang Q, Liu HZ, et al. Comparative analysis of tocopherol and phytosterol composition of peanut cultivars from different regions [J]. Food Sci. 2012;33(22):191–5.

Zhou RB. Current situation and development suggestions of peanut production and processing industry in China [J]. Chin Oils Fats. 2005;30(2):5–9.

Chapter 2
Quality Characteristics and Determination Methods of Peanut Raw Materials

As one of the major oilseed crops in the world, peanut is planted in more than 100 countries, but the sensory quality, nutritional quality, and processing characteristics of peanuts planted in different regions and of different varieties are significantly different. Some scholars have reported the basic composition and vitamin E and phytosterol contents of peanuts in China, the United States, India, Canada, and other regions (Dean et al. 2009; Özcan 2010; Wang et al. 2009, 2010). The research conducted by Shin et al. (2010a) found that among the 221 peanut varieties of the United States, the phytosterol content in the Spain peanuts (144.1 mg/100 g $\pm$ 5.3 mg/100 g) was significantly higher than that in Runner peanuts (127.5 mg/100 g $\pm$ 6.3 mg/100 g) and Virginia peanuts (129.3 mg/100 g $\pm$ 6.9 mg/100 g) ($P < 0.05$); the research conducted by Shin et al. (2010b) showed that there were significant differences in the composition of fatty acid in peanut varieties with normal oleic acid, medium oleic acid, and high oleic acid contents ($P < 0.05$); in addition, some varieties were selected due to their high protein content, high oleic acid content, or other processing characteristics.

In recent years, with the rapid development of peanut processing industry, the research on the peanut processing quality and functional characteristics has attracted widespread attention from the academic field, and the high protein and characteristic amino acid contents, scientific and reasonable fatty acid ratio modes, different protein components and subunit contents, high and low glycan contents, and other processing characteristics and functional indicators of peanut varieties have become important research fields. Wang et al. (2011) reported the protein, fat, and sucrose contents in various peanut varieties in different regions of China; Shokarii et al. (1991) reported the protein components and subunit composition of different peanut varieties. There are more than 7490 peanut germplasm resources and varieties in China, but limited systematic research has been conducted on the sensory quality, nutritional quality, and processing characteristics of peanuts of different varieties and regions. Therefore, it is of great significance to analyze the processing quality and functional characteristics of peanut varieties and make clear the relationship between peanut raw material quality and product quality for

Q. Wang, *Peanut Processing Characteristics and Quality Evaluation*,
DOI 10.1007/978-981-10-6175-2_2

scientific and reasonable utilization of peanut resources and improvement of product quality.

1 Determination of Peanut Quality Indicators

To analyze the quality characteristics of representative peanut varieties in China comprehensively and systematically, this research team collected and analyzed 111 main cultivars in 12 major peanut planting provinces in China (Table 2.1). Among them, 59 cultivars were from Shandong, 13 from Henan, 10 from Guangdong, 9 from Fujian, 6 from Jiangsu, 6 from Hubei, 3 from Guangxi, 1 from Liaoning, 1 from Hebei, 1 from Hunan, 1 from Anhui, and 1 from Jiangxi.

A total of 80 indicators (7 sensory quality indicators, 60 physicochemical and nutritional quality indicators, and 13 processing quality indicators) (Table 2.2) of peanut were determined with reference to domestic and foreign literature (Hariprasanna et al. 2008; Misra 2004) and related standards. The means, variation coefficients, upper quartiles, medians, and lower quartiles of sensory quality, physicochemical and nutritional quality, and processing quality indicators were analyzed. The variation range showed the breadth of data coverage; the variation coefficient was the statistics to measure the degree of variation in the data; the upper quartile, median, and lower quartile could reflect the distribution of data. The above research ensured the accuracy and comprehensiveness of analytical data and reflected the quality characteristics of peanuts truly. The analysis results of specific data are described in the following sections, respectively.

2 Analysis of Sensory Quality

The sensory quality of peanut refers to the appearance characteristics of pod and seed kernel, including imperfect grain, metamorphic kernel, thousand grain weight, impurity, color, smell, rancid kernel, and other indicators which directly affect the consumption and utilization of peanuts. Based on the summary of domestic and foreign test methods for various quality indicators, this book determined and analyzed the sensory quality of 66 peanut varieties in China by using the standard determination method and defined the sensory quality situations of different peanut varieties in China basically.

Table 2.1 Name and source of 111 peanut varieties

No.	Name	Source	No.	Name	Source	No.	Name	Source
1	Zhonghua 8	Hubei	38	034-256-1	Shandong	75	Minhua 10	Fujian
2	Shanhua 7	Shandong	39	Ji 9814	Hebei	76	Huayu 36	Shandong
3	Silihong	Liaoning	40	Yuhua 15	Henan	77	Yuhua 14	Henan
4	Luhua 11	Shandong	41	Yuhua 9326	Henan	78	Xiangxiang	Shandong
5	Bianhua 3	Henan	42	Yuhua 9327	Henan	79	Luhua 2	Shandong
6	Haihua 1	Shandong	43	Kainong 30	Henan	80	Huayu HH017	Shandong
7	Shanhua 9	Shandong	44	Kainong 37	Henan	81	Yueyou 551	Guangdong
8	Shuangji 2	Shandong	45	Yuanza 9102	Henan	82	Quanhua 10	Fujian
9	Fenghua 5	Shandong	46	Zhongnong 108	Hubei	83	Quanhua 464	Fujian
10	Yueyou 14	Guangdong	47	Quanhua 551	Fujian	84	Quanhua 327	Fujian
11	Yueyou 40	Guangdong	48	Hua Guanwang	Shandong	85	Xiaopuyang	Shandong
12	Yueyou 45	Guangdong	49	Honghua 1	Shandong	86	Zhongkaihua 4	Guangdong
13	Yueyou 86	Guangdong	50	Qinglan 8	Shandong	87	Pingdu Dalidun	Shandong
14	Minhua 9	Fujian	51	Huayu 8	Shandong	88	Yuhua 11	Henan
15	Guihua 771	Guangxi	52	Luhua 14	Shandong	89	Heyou 4#	Shandong
16	Zhanhua 82	Fujian	53	Xuhua 15	Jiangsu	90	Shuangji 22	Shandong
17	Shanyou 250	Guangdong	54	Huayu 16	Shandong	91	Qinglan 2	Shandong
18	Longhua 243	Fujian	55	Zhonghua 4	Hubei	92	Qixia Laobaoji	Shandong
19	Heyou 11	Guangxi	56	Zhonghua 15	Hubei	93	Wenshangmansheng	Shandong
20	Pearl Red	Guangdong	57	Haiyu 6	Shandong	94	Guihua 95	Guangxi
21	White Peanut	Fujian	58	Lufeng 2	Shandong	95	Xuhua 3	Jiangsu
22	Fenghua 1	Shandong	59	Hongguan	Shandong	96	Qingdao Manshengdali	Shandong
23	Fenghua 3	Shandong	60	Yuanza 9307	Henan	97	Zhenghua 5	Henan
24	Fenghua 4	Shandong	61	Zhengnong 7	Henan	98	Luhua 10	Shandong
25	Xuhua 5	Jiangsu	62	Fenghua 6	Shandong	99	Qingdao Liman	Shandong

(continued)

Table 2.1 (continued)

No.	Name	Source	No.	Name	Source	No.	Name	Source
26	Yuanhua 8	Jiangsu	63	Luhua 9	Shandong	100	Qufu Doupeng	Shandong
27	Xuhua 13	Jiangsu	64	Luhua 15	Shandong	101	Zhufeng 1	Shandong
28	Xuhua 14	Jiangsu	65	Xianghua 509-77	Hunan	102	Dingtao Banmanyang Peanut	Shandong
29	Huayu 19	Shandong	66	Hua 17	Shandong	103	Zhaoyuan Banman	Shandong
30	Huayu 20	Shandong	67	Hua 27	Shandong	104	Pingdu Paman	Shandong
31	Huayu 22	Shandong	68	Luhua 12	Shandong	105	Yueyou 25	Guangdong
32	Huayu 23	Shandong	69	Baisha 101	Henan	106	Qixia Bankangpi	Shandong
33	Huayu 28	Shandong	70	Huayu 33	Shandong	107	Hua 55	Shandong
34	Huayu 31	Shandong	71	Huayu 25	Shandong	108	Jimo Paman	Shandong
35	Baisha 1016	Shandong	72	Luhua 10	Anhui	109	Pearl Black	Guangdong
36	Colorful Peanut	Hubei	73	Lanhua 2	Shandong	110	Shuanghong 2	Shandong
37	Black Peanut	Hubei	74	Huayu 32	Shandong	111	Ganhua 7	Jiangxi

Table 2.2 Peanut quality indicators

Sensory quality	Physicochemical and nutritional quality	Processing quality
Fruit shape, red skin, grain shape, shape, smell, hundred fruit weight, and hundred kernel weight	Water, crude fat, crude protein, total sugar, ash, crude fiber, total amino acids, 18 kinds of amino acids (aspartic acid, threonine, serine, glutamic acid, proline, glycine, alanine, cysteine, valine, methionine, isoleucine, leucine, tyrosine, phenylalanine, lysine, histidine, and arginine), protein composition content and subunit content (arachin, conarachin, conarachin I, conarachin II, 40.5 kDa, 37.5 kDa, 35.5 kDa, 23.5 kDa, 18 kDa, 17 kDa, 15.5 kDa), fatty acids (13, including oleic acid, linoleic acid, etc.), V_E, sterol, and squalene	Pure kernel rate, oil yield, protein extraction rate, oleic acid/linoleic acid, unsaturated fatty acids/saturated fatty acids, and arachin/conarachin

2.1 Determination Methods

2.1.1 Imperfect Fruit, Immature Fruit, Broken Fruit, Insect-Damaged Fruit, Damaged Fruit, and Imperfect Grain (Insect-Damaged Grain, Diseased Grain, Germinative Grain, Broken Grain, Immature Grain, and Damaged Grain)

1. Domestic standard: GB/T 5494-2008 Determination of foreign matter and unsound kernels of grain and oilseeds
2. Methods of American Peanut Council: Shelling and grading

2.1.2 Metamorphic Kernel (Moldy Grain, Thermal Loss Grain, Discolored Grain, Oily Grain, Spotty Grain, and Other Damaged Grains)

1. Domestic standard: SN/T 0798-1999 Inspection of cereals, oils, and feedstuffs for import and export – Terminology for inspection
2. Standard of International Code Council: Codex Stan 200-1995 Codex Standard for Peanuts

2.1.3 Thousand Grain Weight

1. Domestic standard: GB/T 5519-2008 Cereals and pulses – Determination of the mass of 1000 grains
2. Foreign standard: None

2.1.4 Impurity

1. Domestic standard: GB/T 5494-2008 Determination of foreign matter and unsound kernels of grain and oilseeds
2. Foreign standard: None

2.1.5 Color

1. Domestic standard: GB/T 5492-2008 Inspection of grain and oilseeds – Methods for identification of color, odor, and taste
2. Methods of American Peanut Council: Shelling and grading

2.1.6 Smell

1. Domestic standard: GB/T 5492-2008 Inspection of grain and oilseeds – Methods for identification of color, odor, and taste
2. Foreign standard: No standard

2.1.7 Whole Half Peanut Kernel

1. Domestic standard: GB/T 1532-2008 Peanut
2. Foreign standard: None

2.1.8 Different Variety Rate

1. Domestic standard: SN/T 0803.4-1999 Oil-bearing materials for import and export – Method for the inspection of type purity and their mixture
2. Foreign standard: None

2.1.9 Peanut Kernel Size

1. Domestic standard: NY/T 1893-2010 Grades and specifications of peanuts for processing
2. Methods of American Peanut Council: Shelling and grading

2.1.10 Rancid Kernel

1. Domestic standard: None
2. Methods of American Peanut Council: Shelling and grading

2.2 Data Analysis

This research team collected 66 peanut varieties in Table 2.1 for analysis. These 66 varieties were from 10 provinces, 30 from Shandong, 9 from Henan, 6 from Hubei, 6 from Guangdong, 5 from Fujian, 5 from Jiangsu, 2 from Guangxi, 1 from Liaoning, 1 from Hunan, and 1 from Hebei (Table 2.3).

The research and analysis (Table 2.4) found that there were no significant differences in the shape and smell of different varieties of peanuts, while the differences of grain shape, fruit shape, hundred fruit weight, and other two indicators were significant. The maximum variation coefficient of peanut grain shape was 74.54%, which indicated that the difference of grain shape of each variety was large. Through comparison between the mean and median, it was found that the data variations of all indicators were very small, except that of grain shape, which indicated that these indicators of various varieties were evenly distributed and had no extreme value basically; while the data variation of grain shape was large, which indicated that the differences of grain shapes of different peanut varieties were large and several varieties had extreme shapes, for example, the oval peanut varieties accounted for 40% of the varieties analyzed and only Fenghua 6 peanuts were cocoon-shaped peanuts. The maximum hundred fruit weight was 285.00 g and the minimum was 114.80 g. After the average value of hundred fruit weight of peanuts in different areas was analyzed, it was found that peanut varieties in Shandong had the highest value of hundred fruit weight; among the top ten, six varieties were from Shandong, and they were Fenghua 5, Shuangji 2, Huayu 19, Shanhua 9, Shanhua 7, and Huayu 31. The peanut varieties in Guangdong, Fujian, and the south had the lowest average value of hundred fruit weight; the peanut varieties whose value of hundred fruit weight was ranked in the last ten were Yueyou 86, Colorful Peanut, Zhanhua 82, Minhua 9, Shanyou 250, Yueyou 14, Yueyou 20, Longhua 243, Guihua 771, and Pearl Red.

3 Analysis of Physicochemical and Nutritional Quality

The physicochemical and nutritional quality of peanut is the intrinsic quality characteristic of peanut which is closely related to the nutritional value and functional characteristics of peanut, and it specifically includes crude protein, crude fat, crude fiber, ash, total sugar, amino acid, fatty acid, vitamin E, phytosterol, squalene, resveratrol, and other indicators. Based on the summary of domestic and foreign

Table 2.3 Name and source of 66 peanut varieties

No.	Variety name	Variety source	No.	Variety name	Variety source	No.	Variety name	Variety source
1	Zhonghua 8	Hubei	23	Fenghua 3	Shandong	45	Yuanza 9102	Henan
2	Shanhua 7	Shandong	24	Fenghua 4	Shandong	46	Zhongnong 108	Hubei
3	Silihong	Liaoning	25	Xuhua 5	Jiangsu	47	Quanhua 551	Fujian
4	Luhua 11	Shandong	26	Yuanhua 8	Jiangsu	48	Huaguanwang	Shandong
5	Bianhua 3	Henan	27	Xuhua 13	Jiangsu	49	Honghua 1	Shandong
6	Haihua 1	Shandong	28	Xuhua 14	Jiangsu	50	Qinglan 8	Shandong
7	Shanhua 9	Shandong	29	Huayu 19	Shandong	51	Huayu 8	Shandong
8	Shuangji 2	Shandong	30	Huayu 20	Shandong	52	Luhua 14	Shandong
9	Fenghua 5	Shandong	31	Huayu 22	Shandong	53	Xuhua 15	Jiangsu
10	Yueyou 14	Guangdong	32	Huayu 23	Shandong	54	Huayu 16	Shandong
11	Yueyou 40	Guangdong	33	Huayu 28	Shandong	55	Zhonghua 4	Hubei
12	Yueyou 45	Guangdong	34	Huayu 31	Shandong	56	Zhonghua 15	Hubei
13	Yueyou 86	Guangdong	35	Baisha 1016	Shandong	57	Haiyu 6	Shandong
14	Minhua 9	Fujian	36	Colorful Peanut	Hubei	58	Lufeng 2	Shandong
15	Guihua 771	Guangxi	37	Black Peanut	Hubei	59	Hongguan	Shandong
16	Zhanhua 82	Fujian	38	034-256-1	Shandong	60	Yuanza 9307	Henan
17	Shanyou 250	Guangdong	39	Ji 9814	Hebei	61	Zhengnong 7	Henan
18	Longhua 243	Fujian	40	Yuhua 15	Henan	62	Fenghua 6	Shandong
19	Heyou 11	Guangxi	41	Yuhua 9326	Henan	63	Luhua 9	Shandong
20	Pearl Red	Guangdong	42	Yuhua 9327	Henan	64	Luhua 15	Shandong
21	White Peanut	Fujian	43	Kainong 30	Henan	65	Xianghua 509-77	Hunan
22	Fenghua 1	Shandong	44	Kainong 37	Henan	66	Huayu 17	Shandong

Table 2.4 Descriptive analysis of sensory quality of peanut

Factor	Variation range	Mean	Variation coefficient/%	Upper quartile	Median	Lower quartile	Data variation/%
Fruit shape	1.00~8.00	5.07 ± 1.86	36.78	4.00	5.00	7.00	1.38
Red skin	1.00~9.00	5.47 ± 1.46	26.62	5.00	6.00	6.00	9.69
Grain shape	1.00~5.00	2.40 ± 1.79	74.54	1.00	1.00	5.00	58.33
Hundred fruit weight	114.80~285.00	183.07 ± 43.42	23.72	149.50	183.50	213.80	0.23
Hundred kernel weight	38.6~120	72.16 ± 18.64	25.83	57.75	71.90	85.20	0.36

test methods for physicochemical and nutritional quality of peanut, this book determined and analyzed the physicochemical and nutritional quality of 66 peanut varieties in China by using the standard determination methods and then preliminarily defined the physicochemical and nutritional quality of different peanut varieties in China.

3.1 Determination Method

3.1.1 Crude Protein

1. Domestic standard: GB 5009.5-2010 Determination of protein in foods
2. AOAC standard: AOAC Official Method 950.48 Protein (Crude) in Nuts and Nut Products

3.1.2 Crude Fat

1. Domestic standard: GB/T 5009.6-2016 Determination of fat in foods
2. AOAC standard: AOAC Official Method 948.22 Fat (Crude) in Nuts and Nut Products

3.1.3 Crude Fiber

1. Domestic standard: GB/T 5515-2008/ISO 6865:2000 Determination of crude fiber content in grain – Method with intermediate filtration
2. (2) AOAC standard: AOAC Official Method 962.09 Fiber (Crude) in Animal Feed and Pet Food

3.1.4 Water

1. Domestic standard: GB 5009.3-2016 Determination of moisture in food
2. AOAC standard: AOAC Official Method 925.40 Loss on Drying (Moisture) in Nuts and Nut Products

3.1.5 Ash

1. Domestic standard: GB 5009.4-2016 Determination of ash in foods
2. AOAC standard: AOAC Official Method 950.49 Ash of Nuts and Nut Products

3.1.6 Total Sugar

1. Domestic standards:

   Reducing sugar: GB/T 5009.7-2016 Determination of reducing sugar in foods
   Sucrose: GB/T 5009.8-2016 Determination of fructose, glucose, sucrose, maltose, and lactose in foods
   Starch: GB/T 5009.9-2016 Determination of starch in foods
   Total sugar: There is no standard and the phenol sulfuric acid method is adopted mostly. Colorimetric method for determination of sugars and related substances

2. AOAC standard:

   Reducing sugar: AOAC Official Method 950.50 Sugars (Reducing) in Nuts and Nut Products
   Sucrose: AOAC Official Method 950.51 Sucrose in Nuts and Nut Products
   Starch: AOAC Official Method 920.40 Starch in Animal Feed
   Total sugar: There is no standard and the phenol sulfuric acid method is adopted mostly. Colorimetric method for determination of sugars and related substances

3.1.7 Amino Acid

1. Domestic standard: GB/T 5009.124-2016 Determination of amino acids in foods
2. AOAC standard: AOAC Official Method 994.12 Amino Acids in Feeds

3.1.8 Fatty Acid

1. Domestic standard: GB/T 5009.168-2016 Determination of fatty acids in foods
2. AOAC standard: AOAC Official Method 996.06 Fat (Total, Saturated, and Unsaturated)

3.1.9 Vitamin E

1. Domestic standard: GB 5009.82-2016 Determination of VE, VA, and VD in foods
2. AOAC standard: AOAC Official Method 971.30 α-Tocopherol and α-Tocopheryl Acetate in foods and Feeds

3.1.10 Phytosterol

1. Domestic standard: GB/T 25223-2010/ISO 12228: 1999 Animal and vegetable fats and oils – Determination of individual and total sterols contents – Gas chromatographic method
2. AOAC Standard: AOAC Official Method 967.18 Beta-Sitosterol in Butter Oil

3.1.11 Squalene

1. Domestic standard: None
2. AOAC standard: AOAC Official Method 943.04 Squalene in Oils and Fats

3.1.12 Resveratrol

1. Domestic standard: GB/T 24903-2010 Inspection of grain and oils – Determination of resveratrol in peanut by high-performance liquid chromatography
2. AOAC standard: None

3.2 Data Analysis

3.2.1 Basic Component Analysis

Crude fat, crude protein, total sugar, ash, crude fiber, and water are the main components of peanut, and they are also known as the basic components of peanut. The contents and variation ranges of the various peanut varieties researched and analyzed by this team are listed in Table 2.5. It could be seen that the water contents of the selected peanut varieties were all within the safe range (9%) (Codex Stan 200-1995), the variation range of total sugar was the widest (2.87–12.59%), the variation range of crude fat was 42.11–58.59%, and the variety with the highest content was Yuhua 9327. The variation range of crude protein was 21.42–31.40%, and the variety with the widest variation range was Longhua 243. The research results of Cobb and Johnson (1973) showed that the average crude fat content of peanuts in the United States was 50% and the variation range was 44–56%. Jonnala et al. (2005) found that the variation range of peanut fat was 42–49%, that of protein was 25–29% and that of crude fiber was 9–12%. Teng et al. (2003) found that the variation range of peanut fat content was 39.96–58.64%. Luo et al. (2004) analyzed the sugar content of 13 peanut varieties and found that the variation range of sucrose was 5.1–5.9%. From the comparison between domestic and foreign research results, it could be found that the coverage of data variation range of peanut varieties in the research conducted by this team was wider, and the research results showed that the data variation of six basic components of peanut was less than 4%, which indicated that the data of selected peanut varieties was distributed evenly.

Table 2.5 Analysis of physicochemical and nutritional quality of peanut

Factor	Variation range	Mean	Variation coefficient/ %	Upper quartile	Median	Lower quartile	Data variation/ %
Water	3.71–7.41	5.47 ± 0.95	17.43	4.71	5.36	6.18	2.01
Crude fat	42.11–58.59	51.22 ± 3.40	6.63	49.29	51.24	53.59	0.04
Crude protein	21.42–31.40	25.79 ± 2.06	7.97	24.37	25.78	27.09	0.04
Total sugar	2.87–12.59	7.30 ± 2.56	35.08	5.02	7.03	9.59	3.70
Ash	2.19–3.46	2.57 ± 0.20	7.86	2.45	2.56	2.65	0.39
Crude fiber	1.50–6.90	2.53 ± 0.82	32.28	2.10	2.50	2.80	1.19

From the analysis of basic components of selected peanuts in different areas (Table 2.6), it was found that there were significant differences in water, ash, crude fat, crude protein, and total sugar, respectively. The peanut varieties with the highest water content were from Fujian and Guangdong, and the means were 6.78% and 6.17%, respectively. The varieties with high crude fat content were from Henan and Jiangsu, and the means were 53.52% and 53.00%, respectively. The peanut varieties ranked in the first five places were Yuhua 9327, Xuhua 13, Yuhua 15, Yuanza 9102, and Xuhua 14. The peanut varieties from Fujian and Guangdong had high protein content, and the average contents were 29.79% and 30.34%, respectively. The peanut varieties ranked in the first five places were Longhua 243, Pearl Red, Shanyou 250, Yueyou 45, and Yueyou 86, which indicated that the differences of basic components of peanut varieties in different areas were large, and the result provided the basis for reasonable utilization of various peanut varieties.

3.2.2 Analysis of Amino Acid Content

The amino acid content in peanuts is usually determined by using the amino acid automatic analyzer or high-performance liquid chromatography, but these two methods are not suitable for the determination of mass samples and selection of breeding materials due to slow analysis speed and high cost; the near-infrared spectroscopy technology has been widely used in nondestructive test of agricultural products, especially in the analysis of crop quality, but the reports on determination of amino acid in peanuts using this technology do not exist at home and abroad. Therefore, based on the preliminary investigation and mass varieties' collection, this research team built the prediction model of amino acid content in peanuts by using the near-infrared analysis technology, carried out sufficient validation, and thus established near-infrared rapid test methods for amino acid in peanuts to provide rapid and nondestructive peanut amino acid test methods to peanut

Table 2.6 Analysis of physicochemical and nutritional quality of peanuts in different provinces

Producing area	Water	Ash	Crude fat	Crude protein	Crude fiber	Total sugar
Shandong	4.96 ± 0.66 b	2.58 ± 0.16 ab	51.07 ± 3.08 ab	27.03 ± 2.13 c	2.65 ± 1.14 a	7.85 ± 2.45 ab
Henan	5.27 ± 0.83 b	2.71 ± 0.32 a	53.00 ± 3.83 a	26.43 ± 1.60 c	2.40 ± 0.37 a	7.23 ± 3.09 ab
Guangdong	6.17 ± 0.65 a	2.55 ± 0.10 ab	49.41 ± 2.23 b	29.79 ± 1.07 ab	2.55 ± 0.64 a	6.28 ± 2.31 ab
Jiangsu	5.20 ± 1.10 b	2.39 ± 0.20 b	53.52 ± 2.43 a	28.07 ± 1.38 bc	2.78 ± 0.13 a	5.54 ± 2.58 b
Fujian	6.78 ± 0.64 a	2.45 ± 0.08 b	48.21 ± 3.15 b	30.34 ± 2.59 a	2.20 ± 0.67 a	8.69 ± 2.27 a

Note: a, b and c refer to the different levels in the same column ($p < 0.05$)

processing enterprises and breeding experts so as to promote better development of peanut industry in China.

3.2.2.1 Amino Acid Content

Amino acid is the basic composition unit of protein, and the difference of amino acid types and contents in different peanut varieties may lead to the difference of functional characteristics of protein. This team analyzed the amino acid content in 111 different peanut varieties by using the amino acid automatic analyzer; the analysis atlas is shown in Fig. 1, Appendix 4, and the analysis results are shown in Table 2.7. It could be seen that the mean of total amount of amino acids in 111 peanut varieties analyzed was 26.44 g/100 g (the number of g of amino acids in 100 g of peanuts) and the variation range was 19.08–45.53 g/100 g; the variation range of glutamic acid was the widest (2.05–6.12 g/100 g), and the mean was 4.23 g/100 g. The variation range of tryptophan was the smallest (0.16–0.42 g/100 g), and the mean was 0.25 g/100 g. The variation range of lysine was 0.77–1.60 g/100 g, that of methionine was 0.09–0.71 g/100 g and that of threonine was 0.40–1.15 g/100 g. The research results of Dean et al. (2009) showed that the variation range of lysine was 0.49–1.08 g/100 g, that of methionine was 0.09–0.57 g/100 g and that of threonine was 0.12–1.01 g/100 g, which were consistent with the results of this research.

The variation range of arginine content in peanuts was 2.38–5.45 g/100 g with a mean of 3.14 g/100 g (Table 2.7). Arginine was a nonessential amino acid, but it was closely related to the vascular health (Gornik and Creager 2004; Moriguti et al. 2005). Andersen et al. (1998) found that peanut was the main source of arginine, and the variation range was 1.50–4.32 g/100 g. The research conducted by Young and Mason (1972) showed that the total arginine content (free and non-free) could play an important role in selecting other more potential amino acid varieties, while free arginine was used as a sign of maturity of peanut seed, which might be largely related to the particular gene.

3.2.2.2 Characteristic Amino Acids

From the average content of amino acids, it was found that the contents of aspartic acid (3.07 g/100 g $\pm$ 0.60 g/100 g), glutamic acid (4.23 g/100 g $\pm$ 0.64 g/100 g), and arginine (3.14 g/100 g $\pm$ 0.53 g/100 g) were significantly higher than the contents of other amino acids (Fig. 2.1). The results were consistent with those reported in the literature (Dawson et al. 1971; Basha and Cherry 1976).

From the analysis of contents of 18 kinds of amino acids in soybean, peanut, rapeseed, rice, wheat, corn, and other crops (Fig. 2.2), it could be found that the glutamic acid content in various samples was high, the contents of aspartic acid and arginine in peanuts were significantly higher than those in other crops, and the

Table 2.7 Amino acid content in peanut. Unit: g/100 g

	Number of samples	Variation range	Mean ± standard deviation	Variation coefficient/%
Total amino acid	111	19.08–45.53	26.44 ± 4.88	18.440
Aspartic acid	111	2.22–5.42	3.07 ± 0.60	19.60
Threonine	111	0.40–1.15	0.70 ± 0.14	19.52
Serine	111	0.81–2.21	1.25 ± 0.27	21.79
Glutamic acid	111	2.05–6.12	4.23 ± 0.64	15.14
Proline	111	0.79–1.73	1.22 ± 0.20	16.12
Glycine	111	1.11–2.58	1.52 ± 0.30	19.68
Alanine	111	0.63–1.38	1.08 ± 0.35	18.58
Cystine	111	0.35–1.14	0.70 ± 0.23	32.22
Valine	111	0.90–1.76	1.19 ± 0.16	13.05
Methionine	111	0.09–0.71	0.33 ± 0.12	36.62
Isoleucine	111	0.71–1.53	0.97 ± 0.16	16.28
Leucine	111	1.28–3.05	1.77 ± 0.33	18.76
Tyrosine	111	0.46–1.89	1.08 ± 0.33	30.64
Phenylalanine	111	0.80–2.45	1.49 ± 0.25	16.53
Lysine	111	0.77–1.60	1.02 ± 0.14	13.37
Histidine	111	0.47–1.00	0.65 ± 0.11	17.26
Tryptophan	111	0.16–0.42	0.25 ± 0.05	19.51
Arginine	111	2.38–5.45	3.14 ± 0.53	16.97

contents of these two amino acids in peanuts were high, so these two amino acids were the characteristic amino acids of peanut.

3.2.2.3 Near-Infrared Fingerprint

Place the samples for spectral acquisition in the same laboratory with the near-infrared spectrometer for more than 24 h, so that the environmental conditions of samples are consistent with that of instrument to reduce the impact of temperature on the samples. Clean each sample to remove impurities and broken particles without further pretreatment. Turn on the near-infrared spectrometer to preheat for 30 min under 25 °C, place the samples evenly directly by hands using the natural loading method, fill with the sample cups, scan each sample twice and repeat loading three times, and take the average spectral acquisition method to overcome the sample heterogeneity and collect as much sample information as possible. Store the calculated mean of near-infrared absorption spectrum in the computer for further use of establishment of amino acid calibration model.

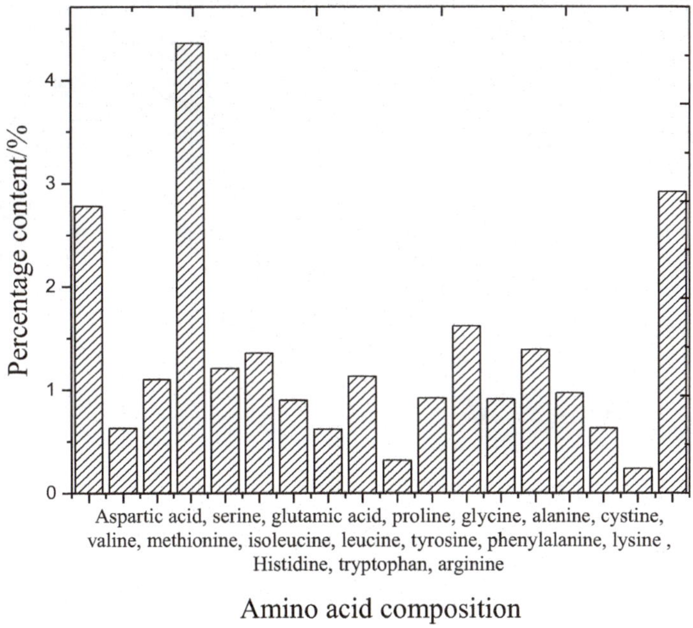

Fig. 2.1 Analysis of amino acid content in peanut

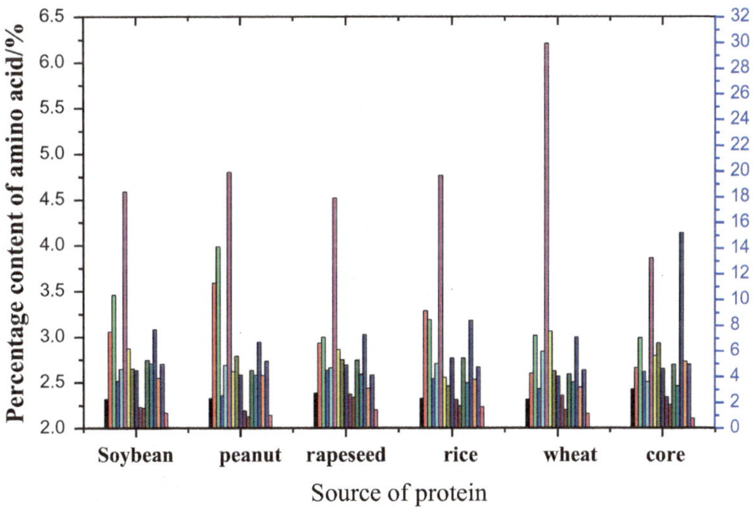

Fig. 2.2 Analysis of amino acid in peanut and other crops (lysine, histidine, aspartic acid, threonine, serine, glutamic acid, proline, glycine, alanine, cystine, methionine, valine, isoleucine, bright ammonia acid, tyrosine, phenylalanine, and tryptophan from *left* to *right*)

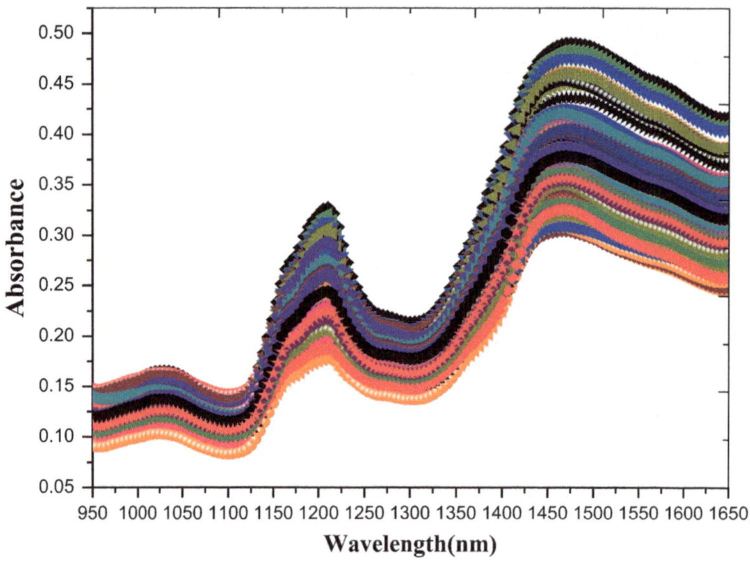

Fig. 2.3 Original absorption spectrums of peanut samples

1. Near-Infrared Characteristic Spectrum

Figure 2.3 shows the original near-infrared absorption spectrums of 111 peanut samples in the 950–1650 nm spectral region. The figure showed that there were three absorption peaks: the first peak was in 975–1025 nm, and it referred to the secondary frequency multiplication of O-H and N-H bond stretching vibration and third-level frequency multiplication of C-H bond stretching; there was a strong absorption band at 1210 nm, and it might be a secondary frequency multiplication absorption band of C-H bond stretching vibration; there was a strong absorption peak at 1460 nm, and it might be a first-level frequency multiplication absorption band of N-H bond stretching vibration, which were caused by a large amount of fat and protein in the peanut kernels (Song et al. 2011; Hourant et al. 2000).

2. Analysis of Principal Components of Infrared Spectrum Data

The near-infrared spectrum data represents a series of multivariates which contain many overlapping information. The multivariate statistical analysis is to extract some useful information in the original data to reduce the dimensionality of data. The principal component obtained after principal component analysis is a linear combination of original data. The cumulative contribution rate of the first two principal components in this research was 98.96% which was greater than 95%, so they could reflect the information of original data completely (Serudo et al. 2007). Therefore, the first two principal components were used to establish the regression model.

In order to understand the near-infrared spectrum classification of peanut samples in different regions preliminarily, a scatter diagram had been made by using the scores of the first and second principal components of near-infrared spectrum data

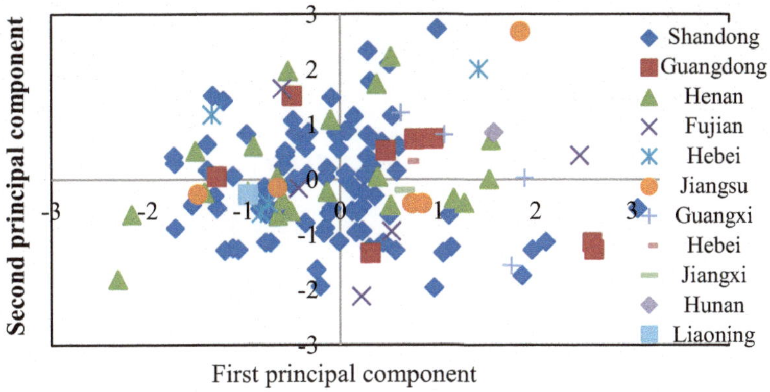

Fig. 2.4 Scatter diagram of first two principal components of near-infrared spectrum data of peanut

of peanut (Fig. 2.4). From the diagram, it could be seen that the spatial distribution limit of peanut samples in various provinces was not very obvious, but the samples in different provinces occupied their own spaces, which initially indicated that the peanut samples in different regions were very different.

3. Spectrum Data Preprocessing

In the near-infrared spectrum analysis, in order to eliminate the errors caused by sample heterogeneity, high-frequency random noise, baseline drift, stray light, and other factors, different spectrum preprocessing methods were used to process the spectrums and then establish the model, which could improve the accuracy and reliability of the model (Chu et al. 2004). Figure 2.5 is the second-derivative diagram of peanut samples. It could be seen from the figure that the fineness of spectrum had been improved significantly and the spectrum contour was clearer. A referred to the second overtone of O–H and N–H and the third overtone of C–H; B, C, and D were generated by C–H bond stretching vibration, O–H bond bonding vibration, and O–H bond denaturation overtone, respectively; E referred to the first overtone of O–H bond and N–H bond in the amino group.

4. Near-Infrared Prediction Model of Amino Acids

Seventy-four varieties were randomly selected as the modeling sample set from the 111 varieties in Table 2.1, 37 varieties were selected as the modeling validation sample set, the model was established by using PCA and PLS, and the internal cross validation and external validation were conducted to verify the model reliability. The internal cross validation is to remove one or more samples crosswise each time, predict the samples to be removed by modeling with other samples in turn, and measure the model quality by comparing the predicted values and chemical values R^2 and RESECV of samples. The external validation is to verify the model by using the samples that are not involved in modeling.

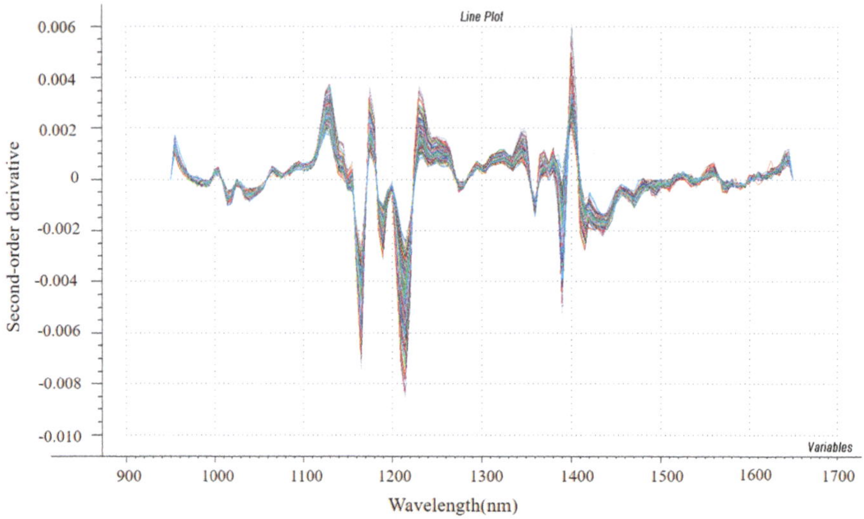

Fig. 2.5 Second-derivative diagram of peanut sample

The internal cross validation and external validation results of NIRS prediction models of eight kinds of amino acids (Table 2.8) showed that the variation range of different amino acids R^2 was 0.83–0.96; the variation range of RPD was 2.35–7.50 (R^2 of other amino acids was less than 0.80, and RPD values were less than 2 or more than 10, which were not listed). The research results of Williams and Norris (2001) showed that the model whose variation range of R^2 of NIRS prediction model was 0.83–0.95 and RPD value was 2–10 was reliable.

The scatter diagram of predicted values and true values of eight kinds of amino acids (Asp, Thr, Ser, Glu, Gly, Leu, Arg, and Cys) through the external validation of 37 varieties is shown in Fig. 2.6. The results showed that the correlation coefficient of predicted value and true value of each amino acid was greater than 0.90.

3.2.3 Analysis of Relative Contents of Protein Component and Subunit (SDS-PAGE Method)

3.2.3.1 Protein Extraction Method

The peanut protein was extracted by using alkali extraction and acid precipitation method, phosphate buffer method, and Tris-HCl method; the extraction rate and purity of protein extracted by using these three methods were calculated, and the polyacrylamide gel electrophoresis (SDS-PAGE) (Table 2.9) analysis was carried out for the protein prepared by using these three methods. From the table, it could be seen that the obtained electrophoretograms of peanut protein extracted by using these three methods were consistent; the extraction rate and purity of protein extracted by alkali extraction and acid precipitation method were higher than

Table 2.8 Prediction and validation results of amino acid in peanut

Parameter	Predicted set					Verified set				
	R_c^2	RMSE	R^2	RMSEV	SD/RMSE	SEP	R_v^2	Bias	Slope	RPD
Asp	0.88	0.18	0.6	0.34	1.56	0.21	0.84	−0.033	1.106	2.52
Thr	0.83	0.03	0.77	0.04	2.25	0.03	0.82	0.006	1.023	3.00
Ser	0.86	0.09	0.54	0.17	1.41	0.1	0.82	−0.011	1.11	2.40
Glu	0.87	0.45	0.54	0.86	1.47	0.49	0.85	0.063	1.085	2.57
Gly	0.88	0.09	0.55	0.18	1.44	0.11	0.82	−0.003	1.047	2.36
Leu	0.88	0.09	0.61	0.17	0.88	0.05	0.81	−0.033	0.983	3.00
Arg	0.89	0.16	0.66	0.29	1.69	0.17	0.86	−0.028	1.059	2.88
Cys	0.96	0.01	0.94	0.02	3.50	0.004	0.995	0.000	0.992	7.50

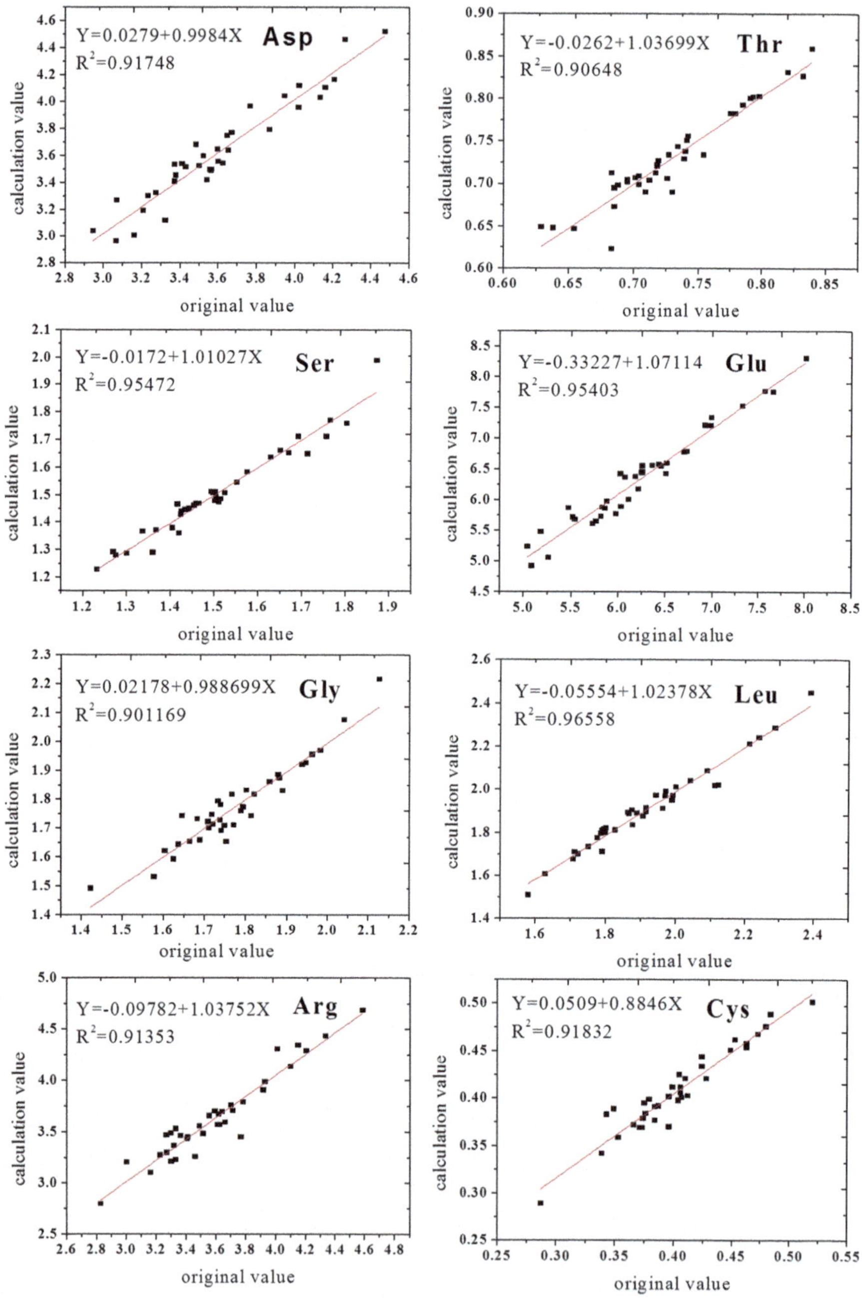

Fig. 2.6 Scatter diagram of true values and predicted values of amino acids in peanut

Table 2.9 Extraction rate, purity, and electrophoretogram of protein extracted by different extraction methods

Extraction method	Extraction rate	Purity	Electrophoretogram
Alkali extraction and acid precipitation	65%	92%	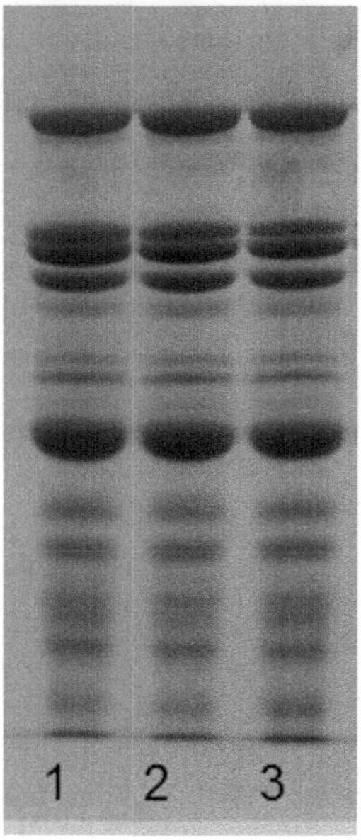
Phosphate buffer	62%	88%	(1) Alkali extraction and acid precipitation method, (2) phosphate buffer method, (3) Tris-HCl method
Tris-HCl	61%	55%	

those of protein extracted by the other two methods, and this method was convenient for industrialized production. From the above, the alkali extraction and acid precipitation method was used to prepare protein for the further research.

3.2.3.2 Main Component and Relative Content of Protein

Main Components and Relative Contents of Protein in Different Varieties of Peanuts

The peanut protein included arachin and conarachin, and the conarachin contains conarachin I and conarachin II (Yamada 1979; Prakash and Rao 1986). From Fig. 2.7, it could be seen that each peanut variety contained arachin, conarachin I, and conarachin II.

The analysis of relative contents of protein components and subunits of 111 peanut varieties is shown in Tables 2.10 and 2.11. It could be seen from the tables that the variation coefficients of conarachin I and conarachin II contents were large, and they were 11.51% and 13.40%, respectively.

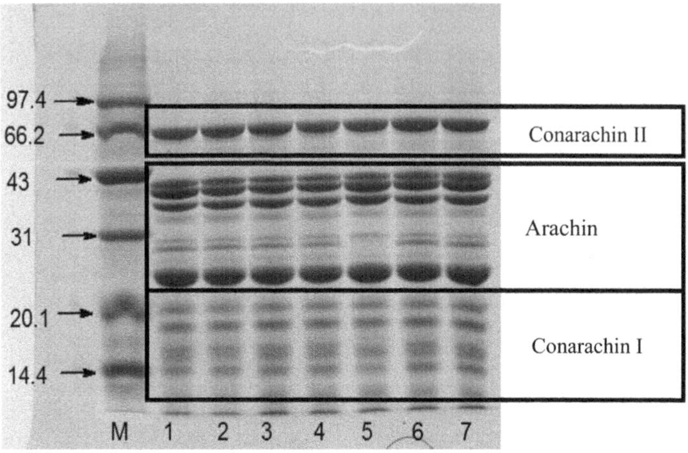

Fig. 2.7 Protein components of different peanut varieties

Table 2.10 Descriptive analysis of contents of peanut protein components

		Variation range/%	Mean/% ± standard deviation	Variation coefficient/%
Arachin	Total content	46.40–62.70	56.00 ± 3.79	6.77
Conarachin	Total content	37.30–53.50	43.98 ± 3.81	8.67
	Conarachin I	20.90–33.40	25.16 ± 2.90	11.51
	Conarachin II	13.40–25.30	18.82 ± 2.52	13.40

Table 2.11 Analysis of contents of peanut protein subunits

		Variation range/%	Mean/% ± standard deviation	Variation coefficient/%
Arachin	40.5 kDa	7.70–14.50	10.61 ± 1.74	16.38
	37.5 kDa	10.50–17.90	13.99 ± 1.68	11.99
	35.5 kDa	0.00–19.20	9.43 ± 5.86	62.14
	23.5 kDa	18.70–26.50	21.97 ± 1.78	8.11
Conarachin	61 kDa	13.40–25.30	18.82 ± 2.52	13.40
	18 kDa	6.60–11.40	8.71 ± 1.34	15.41
	17 kDa	6.90–13.20	9.56 ± 1.40	14.63
	15.5 kDa	3.70–11.90	6.90 ± 1.37	19.89

Peanut protein is usually composed of eight subunits. The arachin has four subunits with the molecular weights of 40.5, 37.5, 35.5, and 23.5 kDa, respectively; the conarachin II has one subunit with the molecular weight of 61 kDa; the conarachin I has three subunits with the molecular weights of 15.5, 17, and 18 kDa, respectively (Fig. 2.7).

The SDS-PAGE atlas of 111 peanut varieties is shown in Fig. 2, Appendix 4. It could be seen from the figure that the number of conarachin subunits of different peanut varieties was consistent without differences, but there were differences in the number of arachin subunits. The arachin of some peanut varieties lacks 35.5 kDa subunits, such as Shuangji 2, Yueyou 14, Minhua 9, Zhanhua 82, Shanyou 250, Longhua 243, Heyou 11, Pearl Red, Yuanhua 8, Baisha 1016, Quanhua 551, Huaguanwang, Qinglan 8, Huayu 8, Huayu 16, Zhonghua 4, and other 26 varieties, accounting for 23.42% of the tested varieties. Although it has been unclear how the deficiency of 35.5 kDa subunits affects the functional properties of peanut proteins, some research showed that the functional properties (such as solubility, foamability, gelation, thermal coherence, and emulsibility) of plant protein were closely related to the molecular weight distribution, size/composition of subunit, dissociation/polymerization property of subunit, amount of disulfide bond and its thermal stability, and hydrophilicity and hydrophobicity of plant protein during processing (Yang et al. 2001). Therefore, it could be inferred that the deficiency of 35.5 kDa subunits was certain to affect the functional properties of peanut protein.

There were significant differences in the thickness of strips and shade of dyeing in Fig. 2, Appendix 4, which indicated that there were significant differences in the content of each subunit. The statistical analysis (Table 2.11) showed that the component in arachin with the maximum variance coefficient (62.14%) was 35.5 kDa subunit with significant differences among varieties; the variation coefficient of relative content of 40.5 kDa subunit was 16.38 % which was only second to that of 35.5 kDa subunit, and the component with the minimum variance coefficient (8.11%) was 23.5 kDa subunit. The component in conarachin with the maximum variance coefficient (19.89%) was 15.5 kDa subunit, and the variation coefficients of the other subunits were more than 10% with significant differences. The content of 23.5 kDa subunits in arachin was much higher than that of the other subunits, and the relative content was $21.97 \pm 1.78\%$. The content of 61 kDa subunits in conarachin was much higher than that of the other subunits, and the relative content was $18.82 \pm 2.52\%$.

Utsumi and Kinsella (1985a, b) found that the acidic subunit As-III was more important than As-IV in 11S protein during the formation of soybean protein gel; three subunits in 7S protein were involved in the formation of gel, the β subunit in 7S globulin and basic subunit in 11S globulin in the soybean protein isolate gel interacted with each other selectively and affected the gelation of SPI, and the acidic subunit in 11S protein played an important role in the formation of gel network structure. Tay (2004) and Cheng Cuilin et al. (Cheng et al. 2006) found that the subunits (α′, α, and β) were positively correlated with emulsibility, while the subunit $A_{1, 2, 4}$ and subunit β were negatively correlated with emulsibility. Liu et al. (2008) analyzed the SPI prepared by four protein subunit variation types of soybean varieties, and the results showed that the deficiency of single subunit in 11S component did not affect the solubility, emulsibility, emulsifying stability, and thermal stability of soybean protein significantly, and the significant reduction or deficiency of 11S component content could improve the emulsibility and emulsifying stability of soy protein. Salleh et al. (2004) researched the impacts of deficiency of α and α′ subunits in 7S globulin on soybean protein gel. It was found that the influence relations of components on the gel hardness were deficient α′ subunit>7S whole subunit>deficient α subunit under the same conditions. It is a good point to cultivate special soybean varieties for food with peculiar 11S and 7S protein component contents and subunit germplasm for the research of protein subunit based on molecular genetics, combined with food science research of effect of soybean protein subunits on its functions. Since the 1990s, in the research of soybean protein subunit composition and its processing characteristics, it has gotten more and more attention to change the subunit composition of soybean protein by some ways to obtain the soybean varieties that lack some subunits so as to make these varieties become the special soybean raw materials suitable for processing need characteristics, while there has been no report on the impacts of deficiency of peanut protein subunits on functional properties of protein until now. Therefore, the research on the relationship between the deficiency of different protein component subunits and protein functions will be the emphasis of the further research of this team. This research will provide important basis for the improvement of quality and functional properties of peanut protein and cultivation of peanut varieties with different subunit types.

3.2.4 Analysis of Relative Contents of Protein Component and Subunit (Near-Infrared Spectroscopy)

3.2.4.1 Near-Infrared Spectrum Data

1. Near-infrared characteristic spectrum

   Analyze the relationship between various absorption peaks and absorption groups and see Sect. 3.2.2.3 (1).

2. Analysis of main components of near-infrared spectrum data

   Analyze the differences in near-infrared spectrum data of various varieties by using the main component analysis method and see Sect. 3.2.2.3 (2).

3. Preprocessing of near-infrared spectrum data

   Preprocess the near-infrared spectrum data by using the second-derivative analysis method and see Sect. 3.2.2.3 (3).

3.2.4.2 Near-Infrared Prediction Model

Seventy-four varieties were randomly selected as the modeling sample set from the 111 varieties in Table 2.1, 37 varieties were selected as the modeling validation sample set, the model was established by using PCA and PLS, and the internal cross validation and external validation were conducted to verify the model reliability. See the specific method in 2.2.2.3 (4). The internal cross validation and external validation results of NIRS prediction model are shown in Table 2.12. The scatter diagram of predicted values and true values of three kinds of protein components (arachin, conarachin, and conarachin I) through external validation of 37 varieties is shown in Fig. 2.8. The data in the table and figure showed that the established model was reliable.

3.2.5 Analysis of Fatty Acid Content (Gas Chromatography)

Zhang (2012) determined the composition and contents of fatty acids in 45 peanut varieties in Table 2.1. The results are shown in Table 2.13. There were 13 major fatty acids in peanut, including myristic acid (C14:0), palmitic acid (C16:0), heptadecanoic acid (C17:0), stearic acid (C18:0), arachidic acid (C20:0), behenic acid (C22:0), lignoceric acid (C24:0), palmitoleic acid (C16:1), heptadecenoic acid (C17:1), oleic acid (C18:1), arachidonic acid (C20:1), linoleic acid (C18:2), and linolenic acid (C18:3). Through the calculation by the means, the contents of 13 fatty acids in peanuts were sorted from high to low: C18:1 (37.70%), C18:2 (34.21%), C16:0 (12.37%), C18:0 (3.63%), C18:3 (2.20%), C22:0 (1.77%), C16:1 (1.43%), C20:0 (1.10%), C20:1 (0.70%), C24:0 (0.65%), C17:0 (0.07%), C14:0 (0.03%), and C17:1 (0.03%). Oleic acid and linoleic acid are two fatty acids with high contents in peanuts, and they determine the quality of peanuts and their products to a great extent. Compared with other vegetable oils, the peanut oil has a unique long-chain fatty acid (C20–C24). Among the 45 peanut varieties tested, the long-chain fatty acids content accounted for 2.20–5.83% of the total fatty acid component content. Some research pointed out that the degree of impact of oxidation on long-chain fatty acids was higher than that of short-chain fatty acids, so the unique long-chain fatty acid components in peanut oil were also the causes for oxidative deterioration (Zhang and Wang 2010; Nelson 1997).

The contents and variation coefficients of saturated fatty acids (SFA), unsaturated fatty acids (UFA), polyunsaturated fatty acids (PUFA), monounsaturated fatty acids (MUFA), and other fatty acids in different peanut varieties were analyzed. The results are shown in Table 2.14. Through the calculation by the mean, the contents of SFA, UFA, PUFA, and MUFA in peanuts were 19.63%, 76.28%, 36.41%, and 39.87%, respectively, which indicated that unsaturated fatty acid was the main component in peanuts and accounted for 76% of the total fatty acid content, so the storability of peanut and its products was poor and they were easy to be oxidized. The SFA in peanuts was mainly oriented by palmitic acid (C16:0) with an average content of 12.37%; the variety with the highest content of palmitic acid

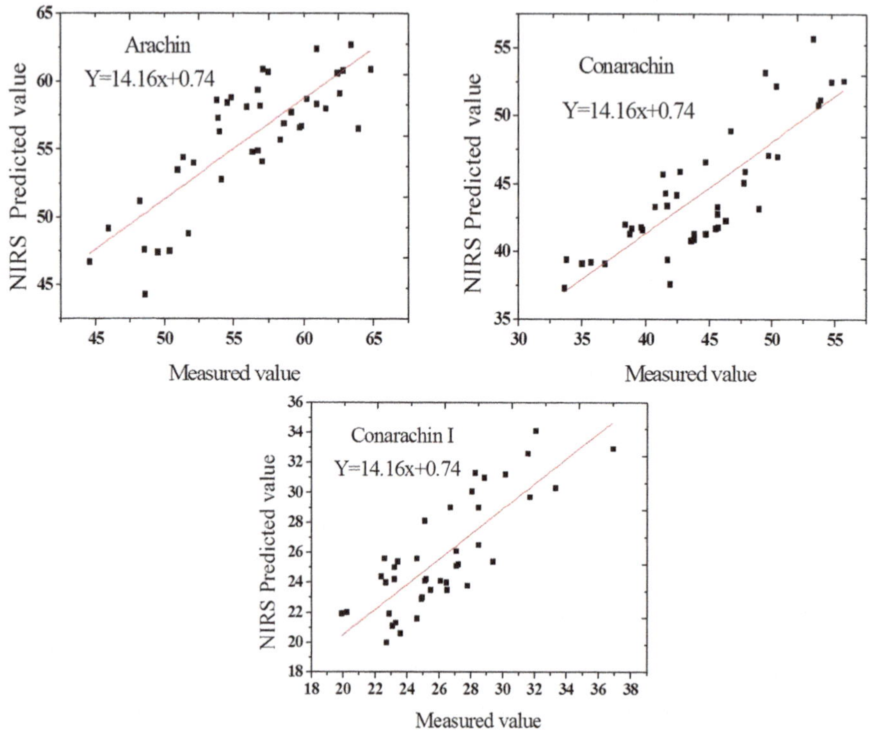

Fig. 2.8 Scatter diagram of true values and predicted values of peanut protein component content

was Fenghua 3 with a content of 15.40%, and the variety with the lowest content was Shanhua 7 with a content of 10.10% and variation coefficient of 8.30%. The variation range of UFA was 64.13–80.05%, and the mean was 76.28%; the varieties with the highest and lowest contents of UFA were Colorful Peanut and Black Peanut; UFA was oriented by oleic acid and linoleic acid, the varieties with the highest and lowest contents of oleic acid content were Kainong 37 (46.90%) and Yueyou 86 (30.50%) with the variation coefficient of 11.35%, and the varieties with the highest and lowest contents of linoleic acid were Zhonghua 8 (40.30%) and Shanhua 7 (24.30%) with the variation coefficient of 11.18%. The peanuts and their products with high contents of oleic acid were storable, and the peanuts with high contents of linoleic acid had medicinal values (Li et al. 2004; Ma et al. 2009); the variation range of PUFA content was 24.48–42.46% with the mean of 36.41%, and the varieties with the highest and lowest PUFA contents were Fenghua 4 and Shanhua 7. The varieties with high PUFA content had high nutritional value, and the varieties with high UFA content had poor oil oxidative stability (Guo et al. 2010). There were high UFA and PUFA contents in peanuts, so peanut oil had high nutritional value and it was easy to be oxidated.

It could be seen from Table 2.14 that the fatty acid components with more than 10% of average contents in the peanut fatty acid composition included C16:0,

Table 2.12 Predication and validation results of amino acids in peanut

No.	Indicator	Calibration set		Internal cross validation set		External validation					
		R^2	RMSE	R^2	RMSE	RSD	RMSD	SEP	R^2	Bias	Slope
1	Arachin	0.64	0.38	0.59	0.44	0.29	0.37	0.31	0.62	0.06	2.023
2	Conarachin	0.61	0.45	0.57	0.54	0.31	0.43	0.33	0.60	−0.33	2.106
3	Conarachin I	0.66	0.33	0.60	0.42	0.28	0.31	0.28	0.62	0.21	2.319

Table 2.13 Fatty acid contents of peanut varieties (unit:%)

Variety	C14:0	C16:0	C17:0	C18:0	C20:0	C22:0	C24:0	C16:1	C17:1	C18:1	C20:1	C18:2	C18:3	SFA	PUFA	UFA	MUFA
Zhonghua 8	0.04	11.60	0.06	2.80	1.12	2.34	1.13	0.50	0.03	37.00	0.90	40.30	1.20	19.09	41.50	79.93	38.43
Shanhua 7	0.04	10.10	0.13	3.10	1.06	1.04	0.51	3.19	0.22	39.50	0.63	24.30	0.18	15.98	24.48	68.02	43.54
Luhua 11	0.00	11.40	0.08	2.85	0.88	1.21	0.55	1.56	0.05	43.20	0.53	30.50	2.30	16.97	32.80	78.14	45.34
Haihua 1	0.03	12.30	0.08	6.29	1.36	2.25	1.03	0.58	0.05	36.20	0.66	36.30	1.29	23.34	37.59	75.08	37.49
Shuangji 2	0.05	11.80	0.08	3.50	0.94	1.29	0.54	3.46	0.16	32.50	0.48	30.70	5.93	18.2	36.63	73.23	36.60
Shanhua 9	0.04	11.10	0.13	3.24	1.14	1.75	0.69	1.22	0.08	37.70	0.64	36.60	2.55	18.09	39.15	78.79	39.64
Fenghua 5	0.03	12.10	0.06	7.42	1.14	1.92	0.76	0.46	0.03	33.40	0.69	39.40	1.14	23.43	40.54	75.12	34.58
Fenghua 1	0.03	12.00	0.07	2.77	0.98	1.65	0.63	1.16	0.05	37.30	0.65	38.00	2.02	18.13	40.02	79.18	39.16
Fenghua 3	0.02	11.40	0.07	2.37	0.77	1.36	0.30	3.46	0.00	32.90	0.53	32.50	5.60	16.29	38.10	74.99	36.89
Fenghua 4	0.03	12.00	0.07	2.91	0.87	1.45	0.23	1.79	0.00	33.10	0.56	39.40	3.06	17.56	42.46	77.91	35.45
Huayu 19	0.05	11.80	0.08	2.86	1.06	1.54	0.70	0.94	0.04	46.30	0.71	27.60	1.19	18.09	28.79	76.78	47.99
Huayu 20	0.03	12.30	0.08	4.66	1.59	2.40	0.71	0.66	0.02	40.70	0.62	32.40	1.10	21.77	33.50	75.50	42.00
Huayu 22	0.03	12.10	0.08	3.01	1.24	2.21	1.02	0.34	0.03	44.20	0.84	32.50	0.52	19.69	33.02	78.43	45.41
Huayu 23	0.04	11.9	0.08	2.39	1.01	1.85	0.87	0.75	0.05	45.40	0.96	30.60	1.05	18.14	31.65	78.81	47.16
Huayu 28	0.03	12.01	0.04	3.79	1.43	2.27	0.82	0.42	0.03	38.80	0.60	36.30	1.42	20.39	37.72	77.57	39.85
Huayu 31	0.03	12.10	0.08	3.06	1.26	2.07	0.94	0.55	0.03	39.90	0.83	35.70	1.13	19.54	36.83	78.14	41.31
Baisha 1016	0.04	11.90	0.08	4.17	1.52	2.20	0.76	0.50	0.03	40.00	0.70	34.70	0.94	20.67	35.64	76.87	41.23
Bianhua 3	0.00	13.20	0.09	3.10	0.96	1.54	0.63	1.70	0.08	34.50	0.56	36.50	2.80	19.52	39.30	76.14	36.84
White Peanut	0.04	13.10	0.06	4.09	1.13	1.76	0.35	0.6	0.00	35.60	0.44	40.30	0.98	20.53	41.28	77.92	36.64
Yuhua 15	0.03	13.10	0.08	3.70	1.33	1.73	0.60	0.76	0.02	38.40	0.61	35.50	1.21	20.57	36.71	76.50	39.79
Yuhua 9326	0.04	13.40	0.08	3.64	1.17	1.82	0.49	1.03	0.03	35.70	0.64	36.70	1.62	20.64	38.32	75.72	37.40
Yuhua 9327	0.03	12.70	0.08	3.50	1.28	1.84	0.15	1.47	0.01	36.80	0.80	33.90	2.13	19.58	36.03	75.11	39.08

Kainong 30	0.03	13.50	0.08	4.04	1.45	1.95	0.48	0.78	0.03	39.30	0.82	33.80	1.27	21.53	35.07	76.00	40.93
Kainong 37	0.04	11.40	0.08	3.23	1.22	1.97	0.57	0.83	0.03	46.90	0.93	27.70	1.22	18.51	28.92	77.61	48.69
Yuanza 9102	0.04	14.50	0.08	3.97	1.31	1.89	0.54	0.56	0.02	35.20	0.66	38.50	0.73	22.33	39.23	75.67	36.44
Black Peanut	0.02	11.40	0.04	15.8	1.39	1.76	0.81	0.58	0.02	31.30	0.65	30.50	1.08	31.22	31.58	64.13	32.55
Xuhua 5	0.04	11.50	0.08	5.51	1.38	2.18	1.07	0.28	0.03	36.90	0.79	38.30	0.66	21.76	38.96	76.96	38.00
Yuanhua 8	0.04	12.00	0.08	4.00	1.30	1.81	0.90	0.56	0.03	38.30	0.70	36.80	1.22	20.13	38.02	77.61	39.59
Xuhua 13	0.03	12.20	0.07	3.73	1.31	1.55	0.60	0.83	0.05	39.20	0.62	35.50	1.65	19.49	37.15	77.85	40.70
Xuhua 14	0.04	13.50	0.06	3.76	1.23	1.72	0.54	1.77	0.02	33.50	0.64	34.80	2.62	20.85	37.42	73.35	35.93
Colorful Peanut	0.02	12.70	0.06	2.55	0.89	1.76	0.58	0.46	0.02	41.30	1.02	36.30	0.95	18.56	37.25	80.05	42.80
Yueyou 14	0.00	11.50	0.06	2.79	1.00	1.92	0.85	2.20	0.00	39.90	0.70	30.40	3.49	18.12	33.89	76.69	42.80
Yueyou 40	0.00	13.70	0.00	1.87	0.67	1.47	0.94	2.70	0.00	35.50	0.84	32.90	3.94	18.65	36.84	75.88	39.04
Yueyou 45	0.00	15.40	0.00	1.90	0.62	1.25	0.00	2.00	0.00	34.40	0.76	37.80	3.15	19.17	40.95	78.11	37.16
Yueyou 86	0.00	14.50	0.11	1.86	0.59	0.98	0.00	4.46	0.00	30.50	0.63	32.20	5.89	18.04	38.09	73.68	35.59
Guihua 771	0.00	12.90	0.08	2.86	1.01	1.84	0.77	1.65	0.00	34.70	0.65	37.50	2.81	19.46	40.31	77.31	37.00
Pearl Red	0.04	11.50	0.08	2.87	0.93	1.73	0.82	2.26	0.00	39.80	0.67	30.90	2.91	17.97	33.81	76.54	42.73
Minhua 9	0.00	13.00	0.09	2.54	1.03	2.07	0.98	1.54	0.00	38.50	0.90	34.60	2.08	19.71	36.68	77.62	40.94
Zhanhua 82	0.00	13.90	0.11	2.71	0.75	1.03	0.36	2.51	0.00	34.30	0.59	35.80	4.26	18.86	40.06	77.46	37.40
Shanyou 250	0.03	12.10	0.07	2.97	1.09	2.61	1.15	0.48	0.00	42.50	0.98	34.10	0.74	20.02	34.84	78.80	43.96
Longhua 243	0.00	11.70	0.08	2.27	0.71	1.03	0.00	5.31	0.00	30.80	0.52	28.40	7.96	15.79	36.36	72.99	36.63
Silihong	0.00	11.50	0.06	2.87	0.94	1.99	0.88	2.20	0.09	31.00	0.89	35.40	4.90	18.24	40.30	74.48	34.18
Heyou 11	0.03	12.80	0.07	3.43	1.10	2.17	0.77	0.51	0.02	37.4	0.73	38.70	1.02	20.37	39.72	78.38	38.66
Jihua 9814	0.04	12.60	0.08	3.50	1.22	1.84	0.59	1.61	0.05	39.90	0.57	30.70	1.78	19.87	32.48	74.61	42.13
034-256-1	0.03	11.40	0.08	3.30	1.32	1.72	0.84	0.99	0.07	46.50	0.81	27.10	1.31	18.69	28.41	76.78	48.37
Mean	0.03	12.37	0.07	3.63	1.10	1.77	0.65	1.43	0.03	37.70	0.70	34.21	2.20	19.63	36.41	76.28	38.09

Table 2.14 Analysis of fatty acid contents of peanut varieties (unit:%)

	Maximum	Minimum	Mean	Variation amplitude	Variation coefficient		Maximum	Minimum	Mean	Variation amplitude	Variation coefficient
C14:0	0.05	0.00	0.03	0.05	62.66	C17:1	0.22	0.00	0.03	0.22	124.54
C16:0	15.40	10.10	12.37	5.30	8.30	C18:1	46.90	30.50	37.70	16.40	11.35
C17:0	0.13	0.00	0.07	0.13	32.05	C20:1	1.02	0.44	0.70	0.58	19.88
C18:0	15.80	1.86	3.63	13.94	58.71	C18:2	40.30	24.30	34.21	16.00	11.18
C20:0	1.59	0.59	1.10	1.00	21.98	C18:3	7.96	0.18	2.20	7.78	76.98
C22:0	2.61	0.98	1.77	1.63	21.87	SFA	31.22	15.79	19.63	15.43	12.53
C24:0	1.15	0.00	0.65	1.15	44.49	PUFA	42.46	24.48	36.41	17.98	10.74
C16:1	5.31	0.28	1.43	5.03	80.27	UFA	80.05	64.13	76.28	15.92	3.74
						MUFA	48.69	32.55	39.87	16.14	9.81

Table 2.15 Analysis of main fatty acid composition of peanuts in different provinces

	Variation range/%	Mean/%	Standard deviation	Variation amplitude/%	Variation coefficient/%	Are there significant differences in fatty acid contents of peanuts in different provinces ($P < 0.05$)
C16:0	11.76–13.25	12.60	0.57	1.49	4.50	Yes
C18:1	35.80–38.71	37.19	1.13	2.91	3.04	No
C18:2	33.09–36.34	34.40	1.23	3.11	3.59	No
SFA	18.57–21.60	19.61	1.28	3.03	6.54	No
MUFA	38.71–40.71	39.52	0.77	2.00	1.94	No
PUFA	35.91–37.76	36.85	0.76	1.85	2.05	No
UFA	74.98–77.16	76.37	0.83	2.18	1.08	No

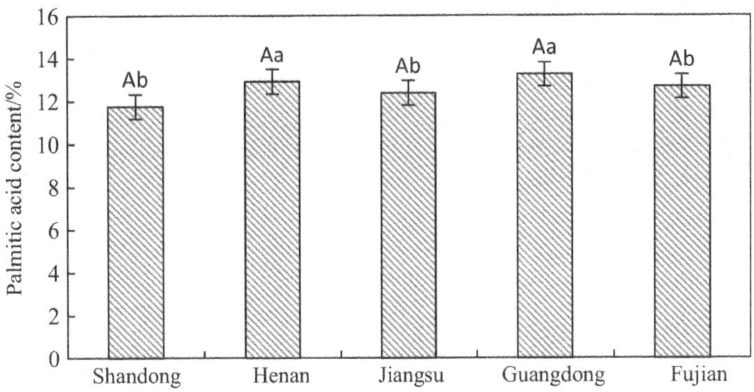

Fig. 2.9 Comparison of palmitic acid contents of peanut varieties in different provinces. Note: a, b, different letters mean significant differences ($P < 0.05$)

C18:1, C18:2 SFA, MUFA, PUFA, and UFA. After the statistical analysis was carried out for above fatty acid contents according to different provinces, it was found that there were differences in the fatty acid contents of peanuts in different provinces. The results are shown in Table 2.15.

In the range of selected peanut varieties, the variation ranges of above fatty acid contents in peanuts in different provinces were C16:0 (11.76–13.25%), C18:1 (35.80–38.71%), C18:2 (33.09–36.34%), SFA (18.57–21.60%), MUFA (38.71–40.71%), PUFA (35.91–37.76%), and UFA (74.98–77.16%). There were significant differences ($P < 0.05$) in the contents of C16:0 in peanuts in different provinces, as shown in Fig. 2.9; the contents of C16:0 in peanuts in Henan and Guangdong were significantly higher than those in Shandong, Jiangsu, and Fujian ($P < 0.05$), and the differences of fatty acid contents in peanuts in other provinces did not reach the significant level.

3.2.6 Analysis of Vitamin E Content (HPLC Method)

3.2.6.1 Standard Curve

The high-performance liquid chromatography (HPLC) was used to analyze the V_E contents in peanuts. Analysis conditions: chromatographic column, Waters C18 (4.6 × 250 mm, 5 μm); mobile phase, methanol-water (98:2, V/V) mixing; UV detection wavelength, 300 nm; injection volume, 20 μL; flow rate, 1.2 mL/min; column temperature, 30 °C.

Accurately weigh a certain amount of α-V_E, γ-V_E, and δ-V_E standard samples, and dissolve and mix them with anhydrous ethanol so that the mass concentrations of α-V_E were 0.1, 0.2, 0.4, 0.6, 0.8, 1.0, and 1.2 mg/mL, that of γ-V_E were 0.1, 0.2, 0.3, 0.4, 0.5, 0.6, and 0.7 mg/mL and that of δ-V_E were 0.01, 0.03, 0.05, 0.07, 0.09, and 0.10 mg/mL, respectively; under the corresponding chromatographic conditions, introduce the samples, respectively, take the peak area X as the abscissa, and take the concentration C (mg/mL) as the ordinate for plotting; the linear regression equations of α-V_E, γ-V_E, and δ-V_E were obtained by the regression fitting of Microsoft Excel software.

Under the determined chromatographic conditions of V_E, the chromatograms of V_E mixed standard samples are shown in Fig. 2.10 and the linear regression equations are shown in Table 2.16.

3.2.6.2 Sample Determination

Zhang (2012) determined the V_E contents in 45 varieties of peanuts in Table 2.1 (Table 2.17). After oil was extracted from various peanut varieties, 5.00 g of peanut oil was weighed and 30 mL of absolute ethyl alcohol, 10 mL of potassium hydroxide solution (1+1), and 5 mL of 10% ascorbic acid solution were added. After full mixing, the mixture was boiled and refluxed for 60 min. The unsaponifiable compounds were extracted, and the extractives were dissolved with 2 mL of ethanol and filtered through a 0.45 μm filter membrane. 20 μL of sample solution was taken and filtered through a 0.45 μm organic phase filter membrane and then HPLC analysis was carried out. The chromatogram of Yuhua 15 is shown in Fig. 2.10b, and the chromatograms of other varieties are shown in Fig. 4, Appendix 4. The retention time and peak area of each chromatographic peak were obtained according to the chromatograms, the quality was determined according to the consistency of retention time, the quantity was determined, and the V_E isomer content in peanut was calculated according to the peak area and standard curve:

$$V_E \text{ isomer content in peanut oil (mg/100g)} = \frac{c \times V}{m} \times 100$$

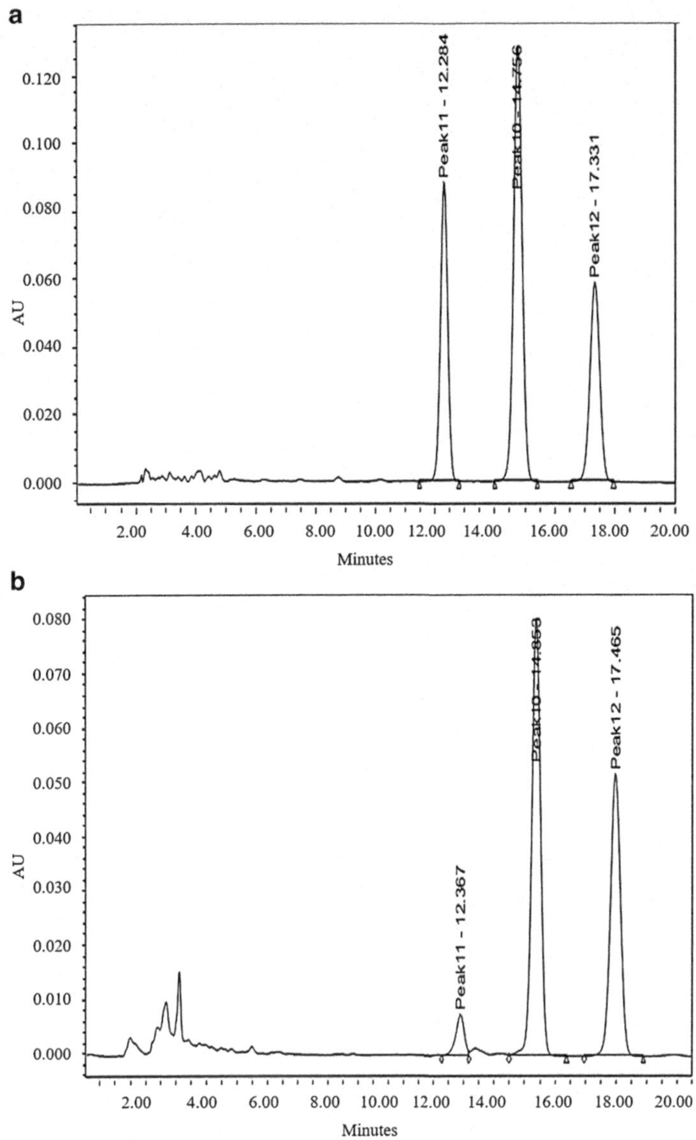

Fig. 2.10 Chromatograms of V_E mixed standard samples and peanut samples. Note: (**a**) standard sample; (**b**) Yuhua 15; 1, δ-V_E; 2, γ-V_E; 3, α-V_E

Table 2.16 V_E linear regression equation

Standard sample	Linear regression equation	Linear range	R^2
α-V_E	$C = 3 \times 10^{-7} \times +0.089$	0.10–1.2 mg/mL	0.994
γ-V_E	$C = 10^{-7} \times +0.001$	0.10–0.70 mg/mL	0.997
δ-V_E	$C = 10^{-7} \times +0.002$	0.01–0.10 mg/mL	0.995

Table 2.17 Contents of V_E and its isomers in different peanut varieties (unit: mg/100 g)

Variety	VE			
	α-V_E	γ-V_E	δ-V_E	Total V_E
Zhonghua 8	27.75 ± 0.28	6.74 ± 0.11	0.36 ± 0.01	34.85 ± 0.18
Shanhua 7	16.80 ± 0.42	6.72 ± 0.20	0.63 ± 0.02	24.15 ± 0.25
Luhua 11	22.96 ± 0.27	9.95 ± 0.30	0.82 ± 0.02	33.72 ± 0.58
Haihua 1	20.08 ± 0.06	9.09 ± 0.03	0.72 ± 0.03	29.88 ± 0.11
Shuangji 2	19.73 ± 0.24	5.89 ± 0.11	0.46 ± 0.02	26.08 ± 0.37
Shanhua 9	16.65 ± 0.28	7.25 ± 0.07	0.50 ± 0.15	24.40 ± 0.21
Fenghua 5	18.13 ± 0.03	5.71 ± 0.04	0.36 ± 0.01	24.20 ± 0.02
Fenghua 1	22.35 ± 0.28	7.41 ± 0.07	0.49 ± 0.07	30.25 ± 0.14
Fenghua 3	19.25 ± 0.35	6.25 ± 0.08	0.41 ± 0.00	25.91 ± 0.27
Fenghua 4	9.36 ± 0.10	7.61 ± 0.03	0.57 ± 0.01	17.54 ± 0.08
Huayu 19	23.34 ± 0.27	8.53 ± 0.06	0.80 ± 0.04	32.67 ± 0.17
Huayu 20	17.67 ± 0.11	6.52 ± 0.04	0.69 ± 0.06	24.87 ± 0.01
Huayu 22	25.25 ± 0.04	9.73 ± 0.01	0.86 ± 0.00	35.92 ± 0.05
Huayu 23	12.16 ± 0.23	4.88 ± 0.14	0.43 ± 0.01	17.47 ± 0.38
Huayu 28	14.75 ± 0.34	4.19 ± 0.06	0.34 ± 0.01	19.28 ± 0.27
Huayu 31	15.09 ± 0.13	6.69 ± 0.17	0.50 ± 0.04	22.28 ± 0.09
Baisha 1016	21.06 ± 0.24	9.49 ± 0.07	0.76 ± 0.02	31.31 ± 0.19
Bianhua 3	19.47 ± 0.01	4.02 ± 0.06	0.29 ± 0.02	23.78 ± 0.06
White Peanut	23.73 ± 0.20	16.93 ± 0.03	0.97 ± 0.02	41.63 ± 0.19
Yuhua 15	21.10 ± 0.21	6.00 ± 0.16	0.62 ± 0.01	27.71 ± 0.36
Yuhua 9326	14.33 ± 0.01	3.47 ± 0.06	0.33 ± 0.01	18.13 ± 0.03
Yuhua 9327	16.00 ± 0.20	4.74 ± 0.04	0.50 ± 0.08	21.24 ± 0.07
Kainong 30	33.98 ± 0.11	6.84 ± 0.07	0.46 ± 0.01	41.28 ± 0.03
Kainong 37	28.28 ± 0.03	9.00 ± 0.08	0.61 ± 0.02	37.89 ± 0.13
Yuanza 9102	27.72 ± 0.10	13.59 ± 0.06	1.30 ± 0.01	42.61 ± 0.04
Black Peanut	14.78 ± 0.28	6.16 ± 0.07	0.41 ± 0.06	21.35 ± 0.42
Xuhua 5	15.42 ± 0.34	5.15 ± 0.06	0.36 ± 0.04	20.93 ± 0.44
Yuanhua 8	20.73 ± 0.24	7.73 ± 0.10	0.57 ± 0.02	29.03 ± 0.16
Xuhua 13	13.61 ± 0.27	4.55 ± 0.06	0.39 ± 0.03	18.55 ± 0.36
Xuhua 14	18.74 ± 0.04	8.03 ± 0.01	0.73 ± 0.10	27.50 ± 0.05
Colorful Peanut	13.52 ± 0.08	9.70 ± 0.20	0.56 ± 0.00	23.78 ± 0.28
Yueyou 14	26.82 ± 0.42	5.73 ± 0.06	0.45 ± 0.06	33.00 ± 0.54
Yueyou 40	24.88 ± 0.41	6.87 ± 0.01	0.58 ± 0.01	32.33 ± 0.43
Yueyou 45	25.50 ± 0.06	9.73 ± 0.06	0.62 ± 0.02	35.85 ± 0.13
Yueyou 86	26.61 ± 0.01	9.41 ± 0.03	0.60 ± 0.10	36.62 ± 0.14
Guihua 771	20.08 ± 0.11	5.08 ± 0.04	0.40 ± 0.02	25.56 ± 0.17
Pearl Red	23.45 ± 0.07	4.29 ± 0.17	0.36 ± 0.02	28.10 ± 0.12
Minhua 9	24.21 ± 0.14	5.71 ± 0.07	0.45 ± 0.02	30.37 ± 0.23
Zhanhua 82	24.00 ± 0.35	6.45 ± 0.08	0.52 ± 0.03	30.97 ± 0.47
Shanyou 250	19.84 ± 0.16	3.72 ± 0.06	0.35 ± 0.01	23.91 ± 0.11
Longhua 243	15.82 ± 0.47	2.88 ± 0.13	0.26 ± 0.01	18.96 ± 0.58

(continued)

Table 2.17 (continued)

	VE			
Variety	α-V_E	γ-V_E	δ-V_E	Total V_E
Silihong	16.76 ± 0.31	6.33 ± 0.23	0.33 ± 0.02	23.42 ± 0.10
Heyou 11	29.07 ± 0.14	7.38 ± 0.07	0.52 ± 0.01	36.97 ± 0.22
Jihua 9814	17.71 ± 0.11	4.78 ± 0.06	0.40 ± 0.01	22.89 ± 0.15
034-256-1	18.57 ± 0.16	5.95 ± 0.03	0.48 ± 0.03	25.00 ± 0.15

c is the content of vitamin E (mg/mL) found on the standard curve, *V* is the concentrated and constant volume of sample (mL), and *m* is the sample mass (g).

There were mainly three V_E isomers (α-V_E, γ-V_E, and δ-V_E) in peanuts. The content of α-V_E was the highest and accounted for 73% of the total V_E content calculated according to the mean, followed by the content of γ-VE which accounted for 25% of the total V_E, and the content of δ-V_E was the lowest and accounted for 2% of the total V_E, which were consistent with the research results of Yang et al. (2009a) and Huang et al. (2001). The physiological activity order of V_E isomers in vivo was δ-V_E<γ-V_E<α-V_E (Kan 2002), so the physiological functions of different peanut varieties might be different if the contents and proportions of V_E and its isomers in different peanut varieties were different.

The maximums, minimums, means, variation amplitudes, and variation coefficients of contents of total vitamin E and its isomers in peanuts were analyzed. The results are shown in Table 2.18. In the 45 peanut varieties, the total V_E content was in the range of 17.47–42.61 mg/100 g and the mean was 27.87 mg/100 g; the variety with the highest V_E content was "Yuanza 9102" and the variety with the lowest V_E content was "Huayu 23"; the α-V_E content was in the range of 9.36–33.98 mg/100 g and the mean was 20.38 mg/100 g; the variety with the highest α-V_E content was "Kainong 30" and the variety with the lowest α-V_E content was "Fenghua 4"; the γ-V_E content was in the range of 2.88–16.93 mg/100 g and the mean was 6.95 mg/100 g; the variety with the highest γ-V_E content was "White Peanut" and the variety with the lowest γ-VE content was "Longhua 243"; the δ-V_E content was in the range of 0.26–1.30 mg/100 g and the mean was 0.54 mg/100 g; the variety with the highest δ-V_E content was "Yuanza 9102" and the variety with the lowest δ-V_E content was "Longhua 243." The variation coefficients of total V_E and α- V_E, γ- V_E, and δ- V_E contents in different peanut varieties were 24.46%, 25.55%, 37.60%, and 37.38%, respectively. The variation coefficients of contents of V_E and its isomers in peanut varieties were large, which could provide the basis for the production of functional peanut products and selection of high-quality varieties.

Statistical analysis was carried out for the contents of V_E and its isomers in different peanut varieties by provincial sources (Fig. 2.11, Table 2.19). In the range of selected peanut varieties, the variation range of α-V_E contents in peanuts in different provinces was 16.40–24.56 mg/100 g, the order of contents from high to low was Guangdong > Henan> Fujian> Shandong> Jiangsu, and the α-V_E content in peanuts in Guangdong was significantly higher than that in peanuts in Jiangsu ($P < 0.01$); the variation range of γ-V_E content was 4.69–7.86 mg/100 g, the order of contents from high to low was Fujian>Guangdong>Jiangsu>Shandong>Henan, and

Table 2.18 Variation analysis of contents of V_E and its isomers in different peanut varieties (unit: %)

Endogenous antioxidant	Maximum	Minimum	Mean	Variation amplitude	Variation coefficient
α-V_E	33.98	9.36	20.38	24.62	25.55
γ-V_E	16.93	2.88	6.95	14.05	37.60
δ-V_E	1.30	0.26	0.54	1.04	37.38
Total V_E	42.61	17.47	27.87	25.14	24.46

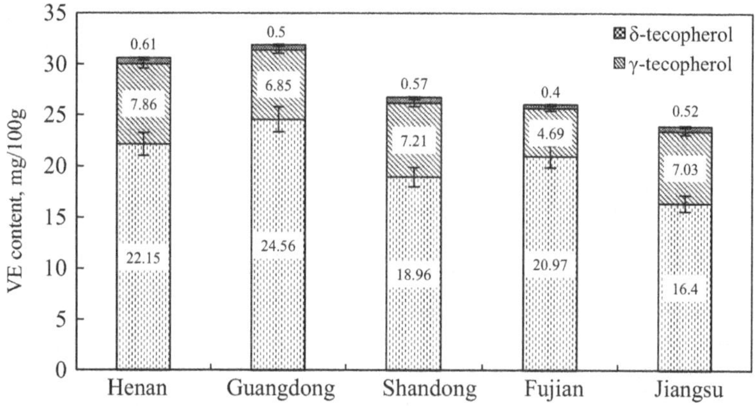

Fig. 2.11 Statistical analysis of contents of V_E and its isomers of peanut varieties in different provinces

Table 2.19 Statistical analysis of contents of V_E and its isomers in peanuts in different provinces

	Shandong	Henan	Jiangsu	Guangdong	Fujian
Total V_E	Aa	Aa	Aa	Aa	Aa
α-V_E	ABab	ABab	Bb	Aa	ABab
γ-V_E	Aa	Aa	Aa	Aa	Aa
δ-V_E	Aa	Aa	Aa	Aa	Aa

Note: A, B: different letters mean highly significant differences ($P < 0.01$); a, b: different letters mean significant differences ($P < 0.05$)

there was no significant difference in the γ-V_E content in peanuts in different provinces ($P < 0.05$); the variation range of δ-V_E content was 0.40–0.61 mg/100 g, the order of contents from high to low was Fujian>Guangdong>Jiangsu>Shandong>Henan, and there was no significant difference in the δ-V_E content in peanuts in different provinces ($P < 0.05$); the variation range of total V_E content was 31.91–23.96 mg/100 g, the order of contents from high to low was Guangdong>Henan>Shandong>Fujian>Jiangsu, and there was no significant difference in the total V_E content in peanuts in different provinces ($P < 0.05$).

In the isomers of V_E, the physiological activity in vivo of α-V_E was the strongest and the oil antioxidant ability of γ-V_E and γ-V_E was the strongest (Kan 2002). In the above five provinces, the α-V_E content in peanuts in Guangdong was the highest and the γ-V_E and δ-V_E contents in peanuts in Fujian were the highest. Therefore, it could be inferred that the varieties with high α-V_E contents could be selected from Guangdong peanut varieties as the varieties with physiological activity and high V_E content in vivo, and the varieties with high γ-V_E and δ- V_E contents could be selected from Fujian peanut varieties as the varieties with oil antioxidant ability and high V_E content.

3.2.7 Analysis of Sterol and Squalene Contents (HPLC Method)

3.2.7.1 Standard Curve

The contents of phytosterol and squalene in peanuts were analyzed by using the high-performance liquid chromatography (HPLC). Analysis conditions: chromatographic column,, Waters SunFire C18 (4.6 × 250 mm, 5 μm); mobile phase: acetonitrile/water (90:10, v/v), mixing and degassing; UV detection wavelength. 210 nm; injection volume, 20 μL; flow rate, 1.5 mL/min; column temperature, 30 °C. A certain amount of stigmasterol, campesterol, and β-sitosterol standard samples were accurately weighed and dissolved and mixed with anhydrous ethanol, so that the mass concentrations of campesterol were 0.2, 0.6, 0.8, 1.0, and 1.2 mg/mL, that of stigmasterol were 0.05, 0.15, 0.25, 0.40, and 0.50 mg/mL and that of β-sitosterol were 0.5, 1.0, 1.5, 2.0, and 3.0 mg/mL, respectively. Under the corresponding chromatographic conditions, introduce the samples, respectively, take the peak area X as the abscissa, and take the concentration C (mg/mL) as the ordinate for plotting; the linear regression equations of stigmasterol, campesterol, and β-sitosterol were obtained by the regression fitting of Microsoft Excel software. Under the determined chromatographic conditions of phytosterol and squalene, the chromatograms of phytosterol and squalene mixing standard samples are shown in Fig. 2.12 and the linear regression equations are shown in Table 2.20.

3.2.7.2 Sample Determination

Zhang et al. (2012) tested the phytosterol and squalene contents in 45 different peanut varieties in Table 2.1 simultaneously by using the high-performance liquid chromatography (Table 2.21). 5.00 g of peanut oil was accurately weighed, and 30 mL of absolute ethyl alcohol, 10 mL of potassium hydroxide solution (1 + 1), and 5 mL of 10% ascorbic acid solution were added. After mixing, the mixture was boiled and refluxed for 60 min. The unsaponifiable compounds were extracted, and the extractives were dissolved with 2 mL of ethanol and filtered through a 0.45 μm of filter membrane. 20 μL of sample solution was taken and filtered through a 0.45 μm organic phase filter membrane and then the HPLC

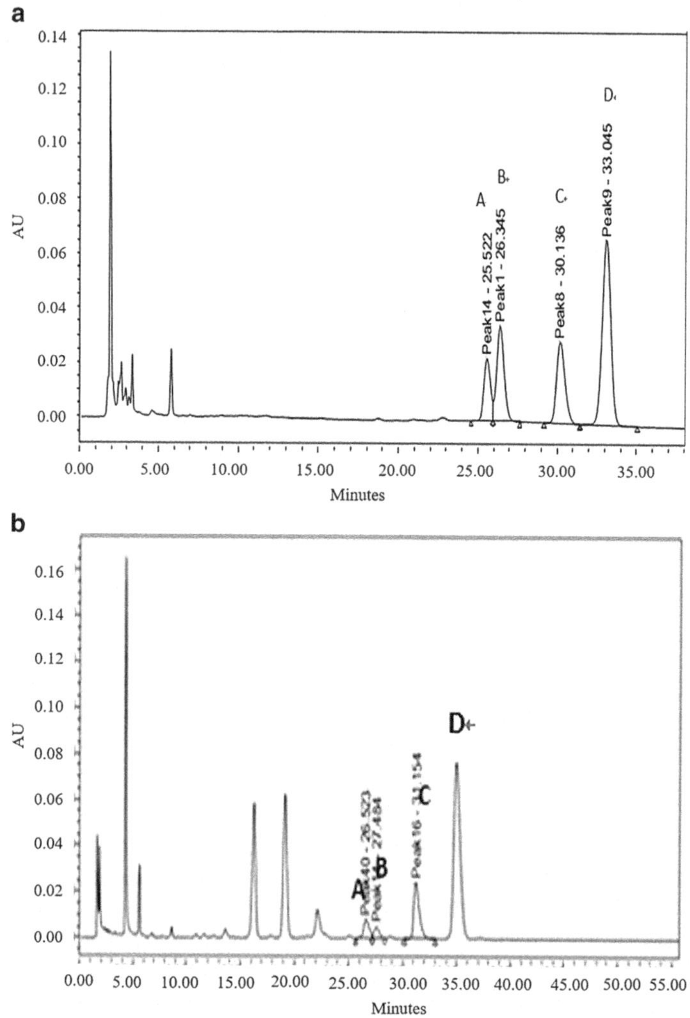

Fig. 2.12 Chromatograms of phytosterol and squalene mixing standard samples and peanut samples. Note: (**a**) standard sample; (**b**) Luhua 11; *A*, campesterol; *B*, stigmasterol; *C*, β-Sitosterol; *D*, squalene

Table 2.20 Linear regression equations of phytosterol and squalene

Standard sample	Linear regression equation	Linear range	R^2
Campesterol	$C = 4 \times 10^{-7} - 0.412$	0.20–1.20 mg/mL	0.996
Stigmasterol	$C = 3 \times 10^{-7} - 0.138$	0.05–0.50 mg/mL	0.997
β-Sitosterol	$C = 3 \times 10^{-7} - 0.119$	0.50–3.00 mg/mL	0.998
Squalene	$C = 10^{-8} + 0.003$	0.10–0.50 mg/mL	1.000

Table 2.21 Phytosterol and squalene contents of different peanut varieties. Unit: mg/100 g

	Phytosterol			Squalene
Variety	Stigmasterol	Campesterol	β-sitosterol	Squalene
Zhonghua 8	15.32 ± 0.33	30.60 ± 1.92	100.05 ± 0.07	7.60 ± 0.01
Shanhua 7	3.14 ± 0.24	22.04 ± 0.40	73.76 ± 0.18	8.59 ± 0.08
Luhua 11	15.42 ± 1.06	32.91 ± 1.16	101.23 ± 0.49	12.81 ± 0.06
Haihua 1	12.76 ± 0.95	28.68 ± 0.07	96.89 ± 0.86	13.56 ± 0.07
Shuangji 2	11.18 ± 0.59	18.54 ± 0.03	67.62 ± 0.66	6.68 ± 0.28
Shanhua 9	8.11 ± 0.88	17.92 ± 1.17	74.01 ± 0.20	9.56 ± 0.28
Fenghua 5	10.58 ± 1.29	19.04 ± 0.53	80.08 ± 1.63	9.25 ± 0.35
Fenghua 1	13.48 ± 0.57	27.61 ± 0.64	100.63 ± 0.20	10.63 ± 0.55
Fenghua 3	6.25 ± 1.03	20.48 ± 0.66	73.65 ± 0.18	9.67 ± 0.55
Fenghua 4	11.79 ± 1.29	27.82 ± 0.59	89.23 ± 0.06	6.40 ± 0.17
Huayu 19	17.41 ± 0.17	34.98 ± 0.14	100.65 ± 0.48	8.98 ± 0.30
Huayu 20	9.03 ± 0.04	20.29 ± 0.11	58.65 ± 0.93	6.09 ± 0.13
Huayu 22	17.38 ± 0.23	37.13 ± 1.06	108.29 ± 0.57	10.68 ± 0.27
Huayu 23	3.18 ± 0.13	4.20 ± 0.16	44.09 ± 0.28	3.77 ± 0.78
Huayu 28	9.11 ± 0.48	21.37 ± 0.18	57.45 ± 0.37	9.27 ± 0.38
Huayu 31	8.17 ± 0.41	16.22 ± 0.57	65.72 ± 0.49	7.61 ± 0.59
Baisha 1016	14.39 ± 0.86	33.21 ± 0.03	77.34 ± 0.08	9.10 ± 0.18
Bianhua 3	9.62 ± 0.88	16.28 ± 0.25	65.44 ± 0.61	7.63 ± 0.10
White Peanut	14.93 ± 1.39	42.53 ± 0.21	87.17 ± 1.19	16.45 ± 0.17
Yuhua 15	15.89 ± 0.01	26.61 ± 0.34	72.93 ± 0.13	9.26 ± 0.21
Yuhua 9326	9.83 ± 0.65	15.31 ± 0.44	60.51 ± 0.48	8.04 ± 0.18
Yuhua 9327	9.23 ± 0.33	19.74 ± 0.45	65.09 ± 0.03	7.70 ± 0.44
Kainong 30	17.16 ± 0.28	44.00 ± 1.47	85.86 ± 1.22	12.71 ± 0.17
Kainong 37	22.62 ± 0.07	48.64 ± 0.04	110.88 ± 0.13	14.35 ± 0.11
Yuanza 9102	17.02 ± 0.08	67.39 ± 0.11	115.70 ± 0.44	13.88 ± 0.13
Black Peanut	7.16 ± 0.06	17.49 ± 0.27	68.40 ± 0.42	10.13 ± 0.23
Xuhua 5	6.16 ± 0.23	11.48 ± 0.75	54.00 ± 0.06	5.80 ± 0.03
Yuanhua 8	2.95 ± 0.81	36.59 ± 1.50	92.64 ± 0.06	8.61 ± 0.04
Xuhua 13	7.66 ± 0.47	10.73 ± 0.91	46.49 ± 0.55	5.71 ± 0.27
Xuhua 14	7.76 ± 0.18	30.74 ± 1.20	75.18 ± 0.38	7.80 ± 0.18
Colorful Peanut	10.25 ± 0.25	17.76 ± 0.44	53.04 ± 0.06	5.28 ± 0.14
Yueyou 14	12.65 ± 1.65	27.48 ± 0.57	87.61 ± 1.12	9.00 ± 0.11
Yueyou 40	8.97 ± 1.09	19.41 ± 0.47	80.03 ± 0.24	12.60 ± 0.04
Yueyou 45	3.42 ± 0.31	26.49 ± 0.71	80.66 ± 0.66	10.96 ± 0.13
Yueyou 86	1.04 ± 0.06	28.29 ± 0.04	78.64 ± 1.53	11.09 ± 0.11
Guihua 771	6.15 ± 0.58	11.78 ± 0.11	48.37 ± 0.01	4.05 ± 0.17
Pearl Red	10.33 ± 0.47	16.53 ± 0.57	55.15 ± 0.14	8.18 ± 0.24
Minhua 9	10.82 ± 0.64	17.18 ± 0.61	68.24 ± 0.33	9.34 ± 0.33
Zhanhua 82	9.43 ± 0.07	18.94 ± 0.98	67.81 ± 0.10	9.59 ± 0.17
Shanyou 250	9.61 ± 0.86	15.60 ± 0.57	69.40 ± 0.28	7.46 ± 0.27
Longhua 243	3.73 ± 0.11	3.89 ± 0.18	38.82 ± 0.33	5.39 ± 0.16

(continued)

Table 2.21 (continued)

	Phytosterol			Squalene
Variety	Stigmasterol	Campesterol	β-sitosterol	Squalene
Silihong	0.31 ± 0.10	14.48 ± 1.17	94.87 ± 0.54	5.39 ± 0.16
Heyou 11	11.12 ± 0.40	28.00 ± 0.64	85.55 ± 0.42	13.09 ± 0.03
Jihua 9814	11.23 ± 0.08	16.72 ± 0.44	52.75 ± 0.81	7.09 ± 0.03
034-256-1	11.03 ± 0.74	21.26 ± 0.41	66.56 ± 0.64	8.00 ± 0.66

analysis was carried out. The retention time and peak area of each chromatographic peak were obtained, the quality was determined according to the consistency of retention time, the quantity was determined, and the phytosterol and squalene contents in peanuts were calculated according to the peak area and standard curve. The chromatogram of Luhua 11 is shown in Fig. 2.16 and the chromatograms of other varieties are shown in Fig. 5, Appendix 4.

Peanut mainly contained β-sitosterol, campesterol, and stigmasterol, and the order of their contents from high to low was β-sitosterol>campesterol>stigmasterol, which were consistent with the research results of Feng Shuyuan (2006), Nelson and Carlos (1995), and Ramakanth et al. (2006). After calculation according to the means, the average contents of these three phytosterols in peanuts were 75.49 mg/100 g, 24.10 mg/100 g, and 10.11 mg/100 g, respectively, and the average content of squalene was 9.00 mg/100 g.

Variation analysis was carried out on the phytosterol and squalene contents in 45 peanut varieties (Table 2.22). It could be seen that the campesterol content was in the range of 3.89–67.39 mg/100 g and the mean was 24.10 mg/100 g; the varieties with the highest and lowest campesterol contents were "Yuanza 9102" and "Longhua 243," respectively; the stigmasterol content was in the range of 0.31–22.62 mg/100 g and the mean was 10.11 mg/100 g; the varieties with the highest and lowest stigmasterol contents were "Kainong 37" and "Silihong," respectively; the β-sitosterol content was in the range of 38.82–115.70 mg/100 g and the mean was 75.49 mg/100 g; the varieties with the highest and lowest β-sitosterol contents were "Yuanza 9102" and "Longhua 243," respectively. The variation coefficients of total phytosterol and campesterol, stigmasterol, and β-sitosterol contents in different peanut varieties were 29.56%, 48.94%, 47.51%, and 24.99%, respectively; the squalene content was in the range of 3.77–16.45 mg/100 g and the mean was 9.00 mg/100 g; the varieties with the highest and lowest squalene contents were "White Peanut" and "Huayu 23," respectively, with the variation coefficient of 31.55%. There were significant differences in the contents of phytosterol components and squalene in peanut varieties, and the variation coefficient was large, which laid a theoretical foundation for development of peanut products with high phytosterols and squalene contents and selection of peanut varieties with high content and quality.

Statistical analysis was carried out on the contents of phytosterol and its isomers of different peanut varieties by provincial sources (Fig. 2.13, Table 2.23). In the range of selected peanut varieties, the variation range of total sterol content in

Table 2.22 Variation analysis of phytosterol and squalene contents of different peanut varieties

Endogenous antioxidant	Maximum/ mg/100 g	Minimum/ mg/100 g	Mean/mg/ 100 g	Variation amplitude	Variation coefficient/%
Stigmasterol	22.62	0.31	10.11	22.31	48.94
Campesterol	67.39	3.89	24.10	63.50	29.56
β-Sitosterol	115.70	38.82	75.49	76.88	47.51
Squalene	16.45	3.77	9.00	12.68	31.55

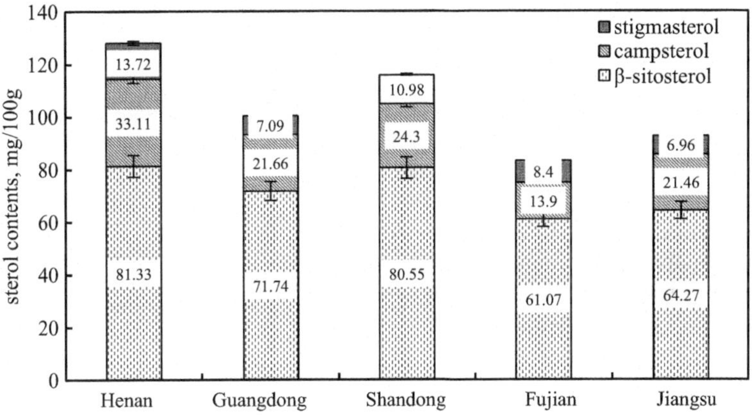

Fig. 2.13 Statistical analysis of phytosterol contents of peanut varieties in different provinces

Table 2.23 Analysis of difference significance of phytosterol contents in peanuts in different provinces

	Shandong	Henan	Jiangsu	Guangdong	Fujian
Stigmasterol	Aab	Aa	Ab	Ab	Ab
Campesterol	ABab	Aa	ABab	ABab	Bb
β-Sitosterol	Aa	Aa	Aa	Aa	Aa
Total sterol	Aab	Aa	Aab	Aab	Ab

Note: A, B, different letters mean highly significant differences ($P < 0.01$); a, b, different letters mean significant differences ($P < 0.05$)

peanuts in different provinces was 83.37–128.16 mg/100 g, the order of contents from high to low was Henan>Shandong>Guangdong>Jiangsu>Fujian, and the total sterol content in peanuts in Henan was significantly higher than that in peanuts in Fujian ($P < 0.05$); the variation range of stigmasterol content was 7.09–13.72 mg/100 g, the order of contents from high to low was Henan>Shandong>Fujian>Guangdong>Jiangsu, and the stigmasterol content in peanuts in Henan was significantly higher than that in peanuts in Jiangsu, Guangdong, and Fujian ($P < 0.05$); the variation range of campesterol content was 13.90–33.11 mg/100 g, the order of contents from high to low was

Henan>Shandong>Guangdong>Jiangsu>Fujian, and the campesterol content in peanuts in Henan was extremely significantly higher than that in peanuts in Fujian ($P < 0.01$); the variation range of β-sitosterol content was 61.07–81.33 mg/100 g, the order of contents from high to low was Henan>Shandong>Guangdong>Jiangsu>Fujian, and there was no significant difference in the β-sitosterol content in peanuts in different provinces ($P < 0.05$).

Phytosterol had several physiological functions such as serum cholesterol reduction, prostate disease prevention, anticancer, anti-inflammatory, immune regulation, and so on, and among the above three kinds of phytosterols (stigmasterol, campesterol, and β-sitosterol), the above physiological functions of β-sitosterol were the strongest (Li et al. 2011). Among the above peanuts in five provinces, the total sterol and β-sitosterol contents in peanuts in Henan were the highest, so it could be inferred that the varieties with high total sterol or β-sitosterol contents could be selected from the peanut varieties in Henan as the raw materials for development of functional peanut products.

The statistical analysis was carried out for the squalene content in different peanut varieties by provincial source (Fig. 2.14). In the range of selected peanut varieties, the variation range of squalene content in peanuts in different provinces was 6.64–11.13 mg/100 g, the order of contents from high to low was Henan>Guangdong>Shandong>Fujian>Jiangsu, and the squalene content in peanuts in Henan was significantly higher than that in peanuts in Jiangsu ($P < 0.01$). Squalene was a kind of natural antioxidant that could prevent the peroxidation of unsaturated fatty acids, protect cells from damage of free radical, and enhance cellular immune systems in cells (Qiao et al. 2011). The squalene content in peanuts in Henan was the highest among the five provinces, so it could be initially inferred that the varieties with high squalene content could be selected from the peanut varieties in Henan.

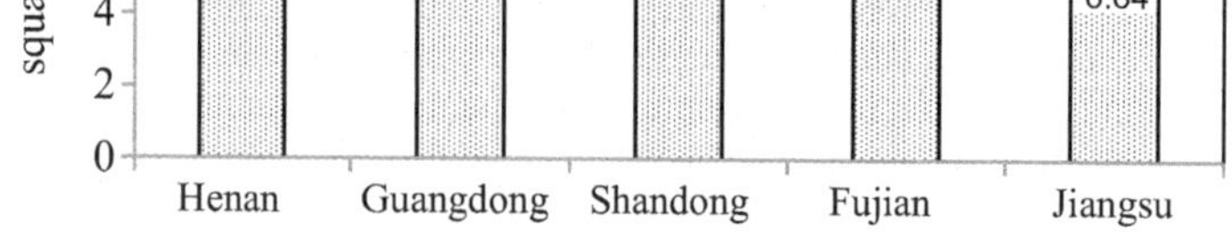

Fig. 2.14 Statistical analysis of squalene content of peanut varieties in different provinces. Note: *A*, *B*: different letters mean highly significant differences ($P < 0.01$); *a*, *b*: different letters mean significant differences ($P < 0.05$)

4 Analysis of Processing Quality

The processing characteristics of peanut varieties or raw materials are closely related to the quality of processed peanut products; if the quality of different varieties or raw materials is different, the quality of their processed products is different and there are no suitable processing varieties or raw materials, and it is difficult to produce high-quality peanut products. Therefore, the processing quality of peanut varieties directly affects the product quality, and it is of great significance to comprehensively analyze the processing quality of peanut varieties or raw materials; research the processing quality testing technologies, methods, and standards; and select the appropriate special varieties for processing for effectively promoting the healthy development of peanut processing industry in China. In this section, based on the summary of domestic and international testing technologies, the normative determination methods were used to conduct analytic determination and statistical determination for the processing quality of peanut varieties in China, and the processing quality characteristics of peanut varieties in China were cleared initially in this book.

4.1 Determination Method

4.1.1 Pure Kernel Rate

1. Domestic standard: GB/T 5499-2008 Determination of pure kernel yield of unhulled oilseeds
2. Foreign standard: There is no standard and the oil seeds are weighed after shelling by shelling machine.

4.1.2 Oil Yield

1. Domestic method: Quality of extracted oil/quality of oil in raw materials
2. Foreign methods: Quality of extracted oil/quality of oil in raw materials

4.1.3 Protein Extraction Rate

1. Domestic method: Quality of extracted protein/quality of protein in raw materials
2. Foreign method: Quality of extracted protein/quality of protein in raw materials

4.1.4 Oleic Acid/Linoleic Acid

1. Domestic standards: GB/T 5009.168-2016 Determination of fatty acids in foods

2. AOAC standard: AOAC Official Method 996.06 Fat (Total, Saturated, and Unsaturated)

4.1.5 Unsaturated Fatty Acids/Saturated Fatty Acids

1. Domestic standards: GB/T 5009.168-2016 Determination of fatty acids in foods
2. AOAC standard: AOAC Official Method 996.06 Fat (Total, Saturated, and Unsaturated)

4.1.6 Relative Contents of Protein Component and Subunit

1. Domestic method: There is no standard and SDS-PAGE is used with optical density analysis currently.
2. Foreign method: There is no standard and SDS-PAGE is used with optical density analysis currently.

4.2 Data Analysis

4.2.1 Pure Kernel Rate, Protein Extraction Rate, and Oil Yield

Through the analysis of pure kernel rate, protein extraction rate, and oil yield of different peanut varieties (Table 2.24), it could be found that there were significant differences in that of different varieties. The pure kernel rate refers to the ratio of kernel and shell of peanut; the greater the value, the higher utilization rate of peanut. In this study, the variation range of pure kernel rate of peanut was large (50.31–79.94%), which indicated that the plumpness of various varieties was different and might affect the properties of peanut and its products. The variation range of protein extraction rate was 59.51–88.97%, which indicated that there were differences in the contents, composition, and properties of protein in different peanut varieties. The variation coefficient of oil yield was 13.80%, which indicated that there were differences in the contents, composition, and properties of fat in different peanut varieties.

Table 2.24 Statistical analysis of pure kernel rate, protein extraction rate, oil yield, and O/L of different peanut varieties

Factor	Variation range	Mean	Variation coefficient/%
Pure kernel rate	50.31–79.94	69.93 ± 5.94	8.50
Oil yield	24.04–54.60	38.25 ± 5.28	13.80
Protein extraction rate	59.51–88.97	74.00 ± 5.78	7.81
O/L	0.84–1.72	1.12 ± 0.07	20.61%

4.2.2 Oleic Acid/Linoleic Acid

After the oleic acid/linoleic acid (O/L) in different tested peanut varieties was analyzed, it was found that there were six varieties with the O/L ratio of more than 1.4; they were Shanhua 7, Luhua 11, Huayu 19, Huayu 23, Kainong 37, and 034-256-1, and their O/L values were 1.63, 1.42, 1.68, 1.48, 1.69, and 1.72, respectively. From the statistical results, it could be seen that the variation range of O/L values of various peanut varieties was 0.84–1.72, the mean was 1.12, the variation coefficient was 20.61% which was large, and the O/L values were widely distributed; O/L was an important indicator to measure the nutritional quality and storage quality of peanuts, and the developed countries had considered the O/L value as one of the main quality indicators of peanut breeding (Ding 2011); from the aspect of storability, the higher the O/L value, the longer the shelf life, the better the storability (Guo et al. 2010); from the aspect of nutritional quality, the lower the O/L, the higher the relative content of linoleic acid, the higher the nutritional value (Wan 2007); if the peanut processing purposes were different, the requirements of O/L value were different; it was required that the O/L value of export peanut was higher than 1.4 (Wan 2007) and the O/L value of fresh peanut was low (Wan 2007; Wu et al. 2008). Therefore, the storable type varieties, nutritional type varieties, and other special varieties for different processing purposes could be selected according to the difference of O/L value.

4.2.3 Unsaturated Fatty Acids/Saturated Fatty Acids

The statistical analysis was carried out for the unsaturated fatty acids/saturated fatty acids (UFA/SFA) and their variation coefficients of different peanut varieties. The results are shown in Tables 2.25 and 2.26. It could be seen that the variation range of UFA/SFA value of peanut was 2.05–4.62 and the mean was 3.94; the varieties with the highest and lowest UFA/SFA values were Longhua 243 and Black Peanut, respectively.

The difference analysis was carried out for UFA/SFA of peanut varieties in different provinces. The results are shown in Table 2.26 and Fig. 2.15. In the range of selected peanut varieties, the UFA/SFA range of peanuts in different provinces was 3.57–4.15, and there were different degrees of differences in the peanuts in different provinces. Through the comparison of UFA/SFA diagrams of peanuts in different provinces, it was found that the UFA/SFA of peanuts in Guangdong and Fujian was significantly higher than that in Henan ($P < 0.05$), and there was no significant difference in UFA/SFA of peanuts in Shandong, Jiangsu, Guangdong, and Fujian.

Table 2.25 UFA/SFA of peanut variety

Variety	UFA/SFA	Variety	UFA/SFA	Variety	UFA/SFA
Zhonghua 8	4.19	Huayu 31	4.00	Colorful Peanut	4.31
Shanhua 7	4.26	Baisha 1016	3.72	Yueyou 14	4.23
Luhua 11	4.60	Bianhua 3	3.90	Yueyou 40	4.07
Haihua 1	3.22	White Peanut	3.80	Yueyou 45	4.07
Shuangji 2	4.02	Yuhua 15	3.72	Yueyou 86	4.08
Shanhua 9	4.36	Yuhua 9326	3.67	Guihua 771	3.97
Fenghua 5	3.21	Yuhua 9327	3.84	Pearl Red	4.26
Fenghua 1	4.37	Kainong 30	3.53	Minhua 9	3.94
Fenghua 3	4.60	Kainong 37	4.19	Zhanhua 82	4.10
Fenghua 4	4.44	Yuanza 9102	3.39	Shanyou 250	3.94
Huayu 19	4.24	Black Peanut	2.05	Longhua 243	4.62
Huayu 20	3.47	Xuhua 5	3.54	Silihong	4.08
Huayu 22	3.98	Yuanhua 8	3.86	Heyou 11	3.85
Huayu 23	4.34	Xuhua 13	3.99	Jihua 9814	3.75
Huayu 28	3.80	Xuhua 14	3.52	034-256-1	4.11
Mean 3.94					

Table 2.26 Statistical analysis of contents of main fatty acid composition of peanuts in different provinces

	Variation range	Mean	Standard deviation	Variation amplitude	Variation coefficient	Are there significant differences in fatty acid contents of peanuts in different provinces ($P < 0.05$)
UFA/SFA	3.57–4.15	3.94	0.24	0.58	6.11	Yes

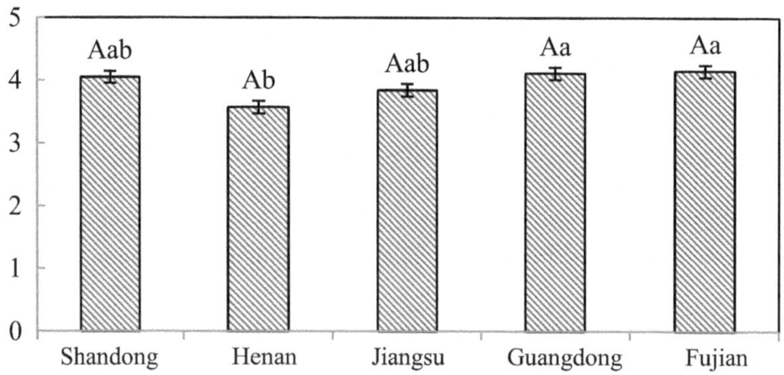

Fig. 2.15 Comparison of UFA/SFA of peanut varieties in different provinces. Note: *a*, *b*: different letters mean significant differences ($P < 0.05$)

4.2.4 Arachin/Conarachin

The types, structures, and contents of various components of peanut protein were closely related to the functional properties of protein. The research showed that arachin/conarachin was an important indicator to measure the functional properties of peanut protein. This team analyzed the contents and proportions of arachin and conarachin in 170 varieties of peanuts by using SDS-PAGE. The results were shown in Table 2.27. It could be seen that the variation coefficients of arachin I and arachin II contents of 170 peanut varieties were large and they were 11.59% and 13.02%, respectively, which indicated that the differences among various varieties were large.

The variation range of ratio of arachin and conarachin was between 0.80 and 1.68 with a mean of 1.24 ± 0.19; there were 85 copies of samples with the ratio higher than the mean, accounting for half of the total number of samples (Table 2.28 and Fig. 2.16). The variation coefficient was 15.23% and exceeded 10%, which indicated that there was a large genetic variation in the protein composition among different peanut varieties.

The ratios of arachin/conarachin of different peanut varieties were different. Among the 170 varieties analyzed, the variety with the highest ratio was Xuhua 5 with the ratio of 1.68 and the variety with the lowest ratio was Luhua 8 with the ratio of 0.80. The research showed that 11S, 7S, and 11S/7S of soybean were closely related to the functional properties, the disulfide bond content of 11S was higher than that of 7S, and the structure of 11S was compact; therefore, the formed gelation hardness of 11S was larger than that of 7S, but the emulsibility and solubility of 7S component were higher than that of 11S component (Song et al. 2010). Saio (1969) reported that the hardness, springiness, and viscosity of tofu gel

Table 2.27 Variation analysis of peanut protein component contents

Subunit	Sample capacity	Variation amplitude (%)	Mean±standard deviation (%)	Variation coefficient (%)
Arachin	170	44.30–62.70	55.00 ± 3.90	7.08
Conarachin I	170	20.00–34.10	25.79 ± 2.99	11.59
Conarachin II	170	13.40–25.70	19.21 ± 2.50	13.02
Conarachin I and II	170	37.30–55.70	45.00 ± 3.90	8.67

Table 2.28 Ratio of arachin and conarachin

Ratio of arachin/conarachin	Number of varieties	Average ratio	Proportion/%
<1.0	19	0.90	11.18
1.0–1.2	47	1.12	27.65
1.2–1.4	72	1.29	42.35
1.4–1.6	26	1.47	15.29
>1.6	6	1.65	3.53

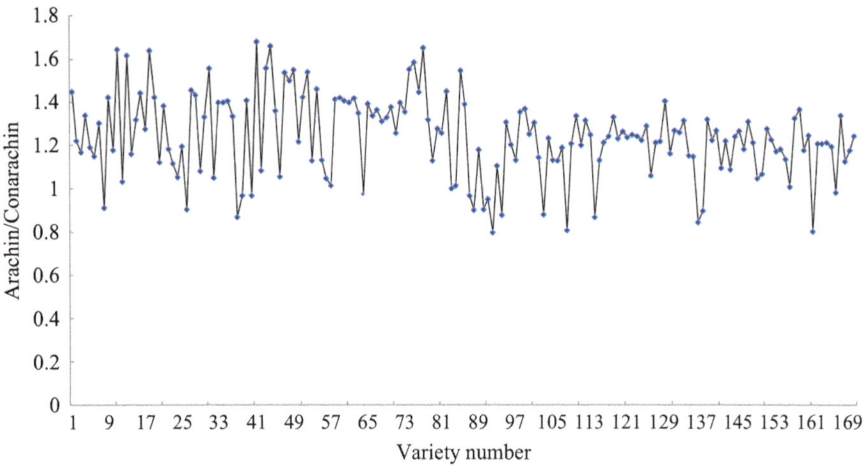

Fig. 2.16 Distribution diagram of ratios of arachin and conarachin

made from 11S globulin were significantly better than that of tofu made from 7S globulin. The research conducted by Arrese et al. (1991) showed that the 11S/7S ratio of soybean protein was closely related to the thermal gel of protein. There was a significant positive correlation between 11S/7S and hardness of tofu gel ($r = 0.86$) (Cheng et al. 2006; Mujoo et al. 2003). The difference in arachin/conarachin ration among different peanut varieties was large, which provided the basis for selecting the peanut varieties with high-quality functional properties.

5 Correlation in Quality Characteristic Indicators of Peanut Raw Materials

The sensory quality, physicochemical and nutritional quality, processing characteristics, and other quality properties of peanut varieties or raw materials are formed constantly in the growth and development process of peanuts. The formation of quality in the planting process is determined by genetic factors and nongenetic factors. The genetic factors refer to the genetic modes and genetic characteristics that determine the characteristics of varieties, and the nongenetic factors refer to all factors other than genetic factors, such as ecological and environmental conditions, cultivation measures, mineral nutrients, and other factors. The quality property of peanuts after harvest has been determined, and the analysis shows that there is a certain correlation between the quality characteristics of different peanut varieties or raw materials. The research on the correlation between the quality characteristics of peanut raw materials is of practical guiding significance to deep understanding of processing characteristics of raw materials, mutual influence in processing and their interaction rules, and further production of high-quality processed peanut products.

5.1 Correlation in Sensory Quality

The analysis of correlation among peanut sensory quality indicators was carried out (Table 1, Appendix 3), and it was found that the peanut fruit shape (the order was hockey shape<hump shape<bead shape<common shape<bee waist shape<gourd shape<cocoon shape<axe shape) was significantly positively correlated with the grain shape (the order was elliptical shape<triangular shape<peach shape <conical shape<cylindrical shape) ($r = 0.438$), which indicated that the peanut fruit shape was closely related to the grain shape and the change of fruit shape would affect the change of grain shape significantly. At the same time, the results showed that the grain shape (the order was oval shape<triangle shape<peach shape<conical shape<cylindrical shape) was negatively correlated with the hundred kernel weight ($r = -0.464$), there was a highly significant positive correlation between the hundred fruit weight and hundred kernel weight ($r = 0.923$), and there was a positive correlation between the hundred kernel weight and pure kernel rate ($r = 0.422$), which indicated that the higher the hundred kernel weight, the higher the pure kernel rate.

5.2 Correlation in Physicochemical and Nutritional Quality

The research and analysis of physicochemical and nutritional quality indicators of different peanut varieties (Table 1, Appendix 3) showed that there were different correlations among the indicators. There was a significant negative correlation between crude fat and crude protein which was consistent with the results of Jiang and Duan (1992) and Liu and Liang (1993), which indicated that it was feasible to cultivate the peanut varieties with high protein or high fat. There were significant negative correlations between crude fat and many amino acids, such as total amino acid ($r = -0.563$), aspartic acid ($r = -0.513$), threonine ($r = -0.488$), serine ($r = -0.538$), glutamate ($r = -0.556$), glycine ($r = -0.550$), cystine ($r = -0.573$), valine ($r = -0.613$), isoleucine ($r = -0.436$), leucine ($r = -0.497$), tyrosine ($r = -0.514$), phenylalanine ($r = -0.545$), lysine ($r = -0.594$), and arginine ($r = -0.403$), which indicated that the reduction of fat content could increase the relative content of amino acids. This conclusion provided a basis for the cultivation of new peanut varieties with high amino acid content. There were significant positive correlations between the crude protein and total amino acid ($r = 0.408$), aspartic acid ($r = 0.409$), proline ($r = 0.676$), valine ($r = 0.431$), isoleucine ($r = 0.536$), leucine ($r = 0.449$), histidine ($r = 0.528$), and arginine ($r = 0.533$), and the result was consistent with that of Yang et al. (2009b), which indicated that the content and quality of peanut protein could be improved by increasing the content of certain amino acid.

The correlation analysis of relative contents of peanut protein components and their subunits (Table 1, Appendix 3) showed that there were correlations among arachin, conarachin, and arachin/conarachin and various components and their subunits and between the various subunits basically. There were highly significant negative correlations between arachin and conarachin ($r = -0.995$), conarachin I ($r = -0.745$), and arachin II ($r = -0.648$) with interaction. There was a highly significant positive correlation between arachin and 35.5 kDa ($r = 0.860$), and there were highly significant negative correlations between arachin and 40.5 kDa ($r = -0.552$), 37.5 kDa ($r = -0.463$), 18 kDa ($r = -0.687$), and 17 kDa ($r = -0.568$). There were highly significant positive correlations between conarachin and 40.5 kDa ($r = 0.555$), 37.5 kDa ($r = 0.467$), 18 kDa ($r = 0.704$), and 17 kDa ($r = 0.543$), and there was a highly significant negative correlation between conarachin and 35.5 kDa. There were highly significant negative correlations between conarachin I and arachin/conarachin ($r = -0.744$) and 35.5 kDa ($r = -0.567$), and there were highly significant positive correlations between conarachin I and 18 kDa ($r = 0.742$), 17 kDa ($r = 0.790$), and 15.5 kDa ($r = 0.580$). There were highly significant negative correlations between conarachin II and arachin/conarachin ($r = -0.648$) and 35.5 kDa ($r = -0.650$); there were highly significant negative correlations between arachin/conarachin and 40.5 kDa ($r = -0.527$), 37.5 kDa ($r = -0.467$), 18 kDa ($r = -0.697$), and 17 kDa ($r = -0.555$); and there was a highly significant positive correlation between arachin/conarachin and 35.5 kDa ($r = 0.848$). There were highly significant positive correlations between 40.5 and 37.5 kDa ($r = 0.535$) and 18 kDa ($r = 0.494$) and there was a highly significant negative correlation between 40.5 and 35.5 kDa ($r = -0.740$). There was a highly significant positive correlation between 37.5 and 18 kDa and there was a highly significant negative correlation between 37.5 kDa and pure kernel rate ($r = -0.461$). There were highly significant negative correlations between 35.5 and 18 kDa ($r = -0.677$) and 17 kDa ($r = -0.477$). The correlation analysis showed that the main components of peanut protein and their subunits interacted with each other. If the content of any one of the components was changed, the content of other components might be affected directly or indirectly. Therefore, the arachin components could be improved by increasing the content of 35.5 kDa subunit so as to improve arachin/conarachin; the contents of conarachin and corresponding subunits could also be increased by reducing the content of 35.5 kDa subunit, which provided a theoretical reference for the selection and breeding of peanut varieties with high arachin or conarachin content.

As for fatty acid composition, there were highly significant correlations ($P < 0.01$) between oleic acid and linoleic acid ($r = -0.41$), UFA and oleic acid ($r = 0.43$) and linoleic acid ($r = 0.46$), MUFA and oleic acid ($r = 0.97$) and linoleic acid ($r = -0.57$). PUFA and oleic acid ($r = -0.66$) and linoleic acid ($r = 0.90$), MUFA and UFA/SFA ($r = 0.40$), and O/L and MUFA ($r = 0.92$), PUFA ($r = -0.93$), oleic acid ($r = 0.84$), and linoleic acid ($r = -0.83$). There was a highly significant negative correlation between oleic acid and linoleic acid, which indicated that the peanuts with high oleic acid content often had low linoleic acid content. Jiang and Duan (1993) found that there was a highly significant negative correlation between oleic acid and linoleic acid in multigrain type, pearl type, Longsheng type, common type, and intermediate type peanuts by analyzing the

Table 2.29 Correlation of peanut fatty acid composition

	Linoleic acid	UFA	SFA	UFA/SFA	MUFA	PUFA	O/L
Oleic acid	−0.41**	0.43**	−0.18	0.24	0.97**	−0.66**	0.84**
Linoleic acid		0.46**	0.26	−0.20	−0.57**	0.90**	−0.83**
UFA			−0.4**	0.51**	0.36 *	0.38*	−0.02
SFA				−0.97 **	−0.36*	0.07	−0.28
UFA/SFA					0.40**	−0.02	0.27
MUFA						−0.73**	0.92**
PUFA							−0.93**
O/L							1.00

Note: **Means highly significant correlation ($P < 0.01$); *means significant correlation ($P < 0.05$)

correlation between oleic acid and linoleic acid contents in different botany types of peanuts, and the correlation coefficients were −0.713, −0.942, −0.979, −0.944, and −0.929, respectively, as shown in Table 2.29.

For endogenous antioxidant of peanuts (Table 2.30), there was a significant or highly significant positive correlation ($P < 0.05$ or $P < 0.01$) between α, γ, and δ-V_E and stigmasterol, campesterol, β-sitosterol, and squalene, there was a highly significant positive correlation ($P < 0.01, r = 0.85$) between γ-V_E and δ-V_E, and the correlation coefficient was the highest, followed by that between campesterol and β-sitosterol ($P < 0.01$, $r = 0.82$), which indicated that the content of one of the unsaponifiable components in peanuts, such as V_E, phytosterols, and squalene, would increase or decrease with the increasing or decreasing of the content of the other component and provide a basis for the selection of peanut varieties with high bioactive substances.

5.3 *Correlation in Processing Quality*

The research and analysis of processing quality indicators of different peanut varieties (Table 2.31) showed that the correlation coefficient among various indicators was low, which remained to be further researched.

5.4 *Correlation Among Sensory Quality, Physicochemical and Nutritional Quality, and Processing Quality*

Some research showed that there were correlations among the sensory quality, physicochemical quality, processing characteristics, and other quality properties of peanut varieties or raw materials. There was a positive correlation ($r = 0.661$) between crude fat and fruit shape (the order is hockey shape<hump shape<bead shape<common shape<bee waist shape<gourd shape<cocoon shape<axe shape). The same trend was also shown in soybeans, which meant that the varieties with

Table 2.30 Correlation among VE, phytosterol, and squalene in peanut

	γ	δ	Stigmasterol	Campesterol	β-sitosterol	Squalene
α	0.36*	0.32*	0.48**	0.64**	0.59**	0.67**
γ		0.85**	0.37*	0.76**	0.63**	0.67**
δ			0.44**	0.78**	0.60**	0.58**
Stigmasterol				0.64**	0.54**	0.52**
Campesterol					0.82**	0.73**
β-Sitosterol						0.67**
Squalene						1.00

Note: **Means highly significant correlation; *means significant correlation

Table 2.31 Analysis of correlation in processing quality of peanut

	Arachin/ conarachin	Protein extraction rate	Oil yield	Kernel yield	Ratio of oleic acid and linoleic acid
Arachin/conarachin	1.000				
Protein extraction rate	0.185	1.000			
Oil yield	−0.084	−0.146	1.000		
Kernel yield	0.267	0.032	0.239	1.000	
Ratio of oleic acid and linoleic acid	−0.091	−0.108	−0.132	0.088	1.000

small soybean seeds generally had high fat content, therefore, pay attention to the selection and breeding of small-grain varieties when selecting and cultivating the varieties with high fat content (Zhou et al. 2008). There was a negative correlation between crude protein and fruit shape (the order was hockey shape<hump shape<bead shape<common shape<bee waist shape<gourd shape<cocoon shape<axe shape), and there was a positive correlation ($r = 0.747$) between grain shape and crude fat (the order of grain shape was cylindrical shape>peach shape>conical shape>oval shape); there was a negative correlation ($r = -0.2719$) between grain shape (the order was oval shape<triangle shape<peach shape<conical shape<cylindrical shape) and crude protein, and there was a positive correlation ($r = 0.516$) between crude protein and 37.5 kDa; the research results of this team were consistent with those of Mozingo et al. (1988). There was a positive correlation ($r = 0.514$) between crude fiber and protein extraction rate; the high crude fiber content might be more conducive to the dissolution of protein so that the protein extraction rate would be increased. There was a significant positive correlation between 23.5 kDa and protein extraction rate ($r = 0.456$). The above analysis showed that there were different degrees of correlations among the sensory quality, physicochemical and nutritional quality, and processing quality of peanut; an indicator might cause or restrict the change of several indicators, and thus the change trend of one indicator might affect the change trend of other indicators. The research in this aspect will be described in the following chapters.

References

Andersen PC, Hill K, Gorbet DW, et al. Fatty acid and amino acid profiles of selected peanut cultivars and breeding lines. J Food Compos Anal. 1998;11(2):100–11.

Arrese EL, Sorgentini DA, Wagner JR, et al. Electrophoretic, solubility, and functional properties of commercial soy protein isolate. J Agric Food Chem. 1991;39(6):1033–6.

Basha SM, Cherry JP. Composition, solubility and electrophoretic properties of proteins isolated from Florunner (*Arachis hypogaea* L.) peanut seeds. J Agric Food Chem. 1976;24:359.

Cheng CL, Wang ZY, Shi YG, et al. Effect of soy protein subunits composition and ratio of 7S to 11 S on the quality and productivity of tofu. China Oils Fats. 2006;31(4):16–9.

Chu XL, Yuan HF, Lu WZ. Progress and application of spectral data pretreatment and wavelength selection methods in NIR analytical technique. Prog Chem. 2004;16(4):528–42.

Cobb WY, Johnson BR. Physicochemical properties of peanuts. In: Peanut: culture and uses. A Symposium. American Peanut Research and Education Society Inc., Stillwater, OK, USA; 1973. p. 209–56.

Dawson RMC, Freinkel N, Jungalwala FB, et al. The enzymic formation of myoinositol 1:2- cyclic phosphate from phosphatidylinositol. Biochem J. 1971;22:605–7.

Dean LL, Hendrix KW, Holbrook CC, et al. Content of some nutrients in the core of the core of the peanut germplasm collection. Peanut Sci. 2009;36:104–20.

Ding JP. Genetic research of oleic/linoleic (O/L) ratio in peanut. J Henan Normal Univ (Natur Sci). 2011;39(4):127–9.

Feng ZY. The GC analysis methods of phytosterols in common foods and research on its application. Nanchang: Nanchang University; 2006.

Gornik HL, Creager MA. Arginine and endothelial and vascular health. J Nutr. 2004;134(10): 2880S–7S.

Guo HH, Yang LP, Wan SB, et al. Studies on peanut quality regional planning in China—guiding thoughts, aims and principle. J Agric Sci Technol. 2010;12(2):28–32.

Hariprasanna K, Lal C, Radhakrishnan T. Analysis of diallel cross for some physical-quality traits in peanut (*Arachis hypogaea* L.). Euphytica. 2008;160:49–57.

Hourant P, Baeten V, Morales MT, et al. Oil and fat classification by selected bands of near infrared spectroscopy. Appl Spectrosc. 2000;54(8):1168–74.

Huang HN, Liu W, Zhang WZ. Application study on stripe picture of natural vitamin E in vegetable oil by HPLC. Strait J Prev Med. 2001;7(3):10–3.

Jiang HF, Duan NX. Correlation analysis between oil and protein content of peanut germplasm resources. J Peanut Sci. 1992;3:13–4.

Jiang HF, Duan NX. Peanut oil quality and correlation analysis between oil, oleic acid and linoleic acid content. J Peanut Sci. 1993;2:5–6.

Jonnala RS, Dunford NT, Dashiell KE. New high-oleic peanut cultivars grown in the southwestern United States. J Am Oil Chem Soc. 2005;82(2):125–8.

Kan JQ. Food chemistry. 1st ed. Beijing: China Agricultural University Press; 2002. p. 112–23.

Li QD, Li Y, Yi XH, et al. The composition and content of fatty acid of 28 almond varieties from Xinjiang. J Mt Agric Biol. 2004;23(4):326–9.

Li B, Li M, Yang X, et al. The physiological function of phytosterols and its application in animal production. Feed Rev. 2011;1:42–4.

Liu GM, Liang ZP. Identification of main quality characteristics of peanut germplasm resources in China. Chinese J Oil Crop Sci. 1993;1:18–21.

Liu C, Wang XS, Ma H. Evaluations on functional properties of soybean protein isolates prepared from cultivars with seed storage protein subunits variation. China Oils Fats. 2008;33(8):31–6.

Luo H, Zhou GY, Fang HB, et al. Studies on biochemical characters relative to direct edible peanut. J Peanut Sci. 2004;4:1–4.

Ma YH, Zhao Z, Li KY, et al. Oil content and composition of almond from different producing area. J Chinese Cereals Oils Assoc. 2009;24(11):70–3.

Misra JB. A mathematical approach to comprehensive evaluation of quality in groundnut. J Food Compos Anal. 2004;17:69–79.
Moriguti JC, Ferriolli E, Donadi EA, et al. Effects of arginine supplementation on the humoral and innate immune response of older people. Eur J Clin Nutr. 2005;59:1362–6.
Mozingo RW, Coffelt TA, Wynne JC. Market grade effects on fatty acid composition of five peanut cultivars. Agron J. 1988;80:73–5.
Mujoo R, Trinh DT, Ng PKW. Characterization of storage proteins in different soybean varieties and their relationship to tofu yield and texture. Food Chem. 2003;82:265–73.
Nelson RG. Proximate, fatty acid and sterol composition of aboriginal peanut (*Arachis hypogaea* L) seeds from Bolivia. J Sci Food Agric. 1997;73:349–56.
Nelson RG, Carlos AG. Lipid, protein, and ash contents, and fatty acid and sterol composition of peanut (*Arachis hypogaea* L.) seeds from Ecuador. Peanut Sci. 1995;22:84–9.
Özcan MM. Some nutritional characteristics of kernel and oil of peanut (*Arachis hypogaea* L.). J Oleo Sci. 2010;59(1):1–5.
Prakash V, Rao MS. Physicochemical properties of oilseed proteins. CRC Crit Rev Biochem. 1986;20(3):286–301.
Qiao JC, Zhang TC, Jiang Y, et al. Anti-oxidation study for thraustochytrids in vivo. Food Res Dev. 2011;32(9):175–8.
Ramakanth SJ, Nurhan TD, Kenton ED. Tocopherol, phytosterol and phospholipid compositions of new high oleic peanut cultivars. J Food Compos Anal. 2006;19:601–5.
Saio K. Food processing characteristics of soybean 11S and 7S protein part I. Effect of difference of protein component among soybean varieties on formation of tofu-curd. Agric Biol Chem. 1969;33:1301–8.
Salleh NM, Koji T, et al. Gelling properties of soybean β-conglycinin having different subunit compositions. Biosci Biotechnol Biochem. 2004;68(5):1091–6.
Serudo RL, Oliveira LC, Rocha JC, et al. Reduction capability of soil humic substances from the Rio Negro basin, Brazil, towards Hg(II) studied by a multimethod approach and principal component analysis (PCA). Geoderma. 2007;138(3–4):229–36.
Shin EC, Pegg RB, Phillips RD, et al. Commercial peanut (*Arachis hypogaea* L.) cultivars in the United States: phytosterol composition. J Agric Food Chem. 2010a;58:9137–46.
Shin EC, Pegg RB, Phillips RD, Eitenmiller RR. Commercial runner peanut cultivars in the USA: fatty acid composition. Eur J Lipid Sci Technol. 2010b;112:195–207.
Shokarii EH, Esen A, Mozingo RW. Immunological characterization of a 36 kD polypeptide in peanuts (*Arachis hypogaea* L.). Peanut Sci. 1991;18(1):11–5.
Song P, Zhou RB, Wei AC. Review of the effect of soybean protein subunits on its functional properties. Sci Technol Cereals Oils Foods. 2010;18(5):19–22.
Song LH, Liu LF, Liu HY, et al. Establishment of model of seeds protein content in peanut using near infrared spectroscopy. Chinese Agric Sci Bull. 2011;27(15):85–89.20.
Tay SL. Physicochemical properties of 7S and 11S protein mixtures coagulated by glucono-δ-lactone. J Food Sci. 2004;69(4):139–43.
Teng Z, Liu Q, Guo DL. The content statistics and classify analyse for major nutrition natters of peanut. Food Res Dev. 2003;24(40):84–5.
Utsumi S, Kinsella JE. Forces involved in soy protein gelation: effects of various reagents on the formation, hardness and solubility of heat-induced gels made from 7S, 11S and soy isolate. J Food Sci. 1985a;50:1278–82.
Utsumi S, Kinsella JE. Structure function relationships in food proteins: subunit interactions in heat-induced gelation of 7S, 11S, and soy isolate proteins. J Agric Food Chem. 1985b;33(2): 297–303.
Wan SB. Science of peanut quality. 2nd ed. Beijing: China Agricultural Science and Technology Press; 2007. p. 111–2.
Wang ML, Chen CY, Davis J, et al. Assessment of oil content and fatty acid composition variability in different peanut subspecies and botanical varieties. Plant Genet Resour: Charact Utilization. 2009;8(1):71–3.

Wang ML, Barkley NA, Chinnan M, et al. Oil content and fatty acid composition variability in wild peanut species. Plant Genet Resour: Charact Utilization. 2010;8(3):232–4.

Wang CT, Tang YY, Wang XZ, et al. Evaluation of groundnut genotypes from China for quality traits. J SAT Agric Res. 2011;9:1–5.

Williams P, Norris K, editors. Near-infrared technology in the agricultural and food industries. 2nd ed. St. Paul: AACC; 2001.

Wu LR, Guan LM, Sun XL, et al. Variety selection and high-yield cultivative of export special-purpose peanut. Seed Sci Technol. 2008;5:61–2.

Yamada T. Dissociation-Association behavior of Arachin between dimeric and monomeric forms. Agric Biol Chem. 1979;43(12):2549–56.

Yang XQ, Chen Z, Zhao MM. Separation and characterization of peanut proteins. J Chin Cereals Oils Assoc. 2001;16(5):25–8.

Yang BT, Chen FX, Mo WL, et al. Research on vitamin E contents of edible oil in China. Grain Oils Process. 2009a;9:52–5.

Yang QL, Zhang CS, Cao YL, et al. The correlation and path analysis between protein content and amino acids composition in peanut seed. Acta Agric Boreali-Sinica. 2009b;24:72–4.

Young CT, Mason ME. Free arginine content of peanuts (*Arachis hypogaea* L.) as a measure of seed maturity. J Food Sci. 1972;37:722–5.

Zhang JS. The effects of fatty acid compositions and endogenous antioxidants of peanut varieties on oil stabiliy. Beijing: Chinese Academy of Agricultural Sciences; 2012.

Zhang HR, Wang Q. Study on correlativity of fatty acid and oxidation index during initial autoxidation stages. Food Sci Technol. 2010;35(3):134–7.

Zhang JS, Wang Q, Liu HZ, et al. Comparative analysis of tocopherol and phytosterol composition of peanut cultivars from different regions. Food Sci. 2012;33(22):191–5.

Zhou EY, Liu LJ, Zu W, et al. Study on relationship between agronomic traits and quality traits in spring soybean. J Northeast Agric Univ. 2008;39(2):145–9.

Chapter 3
Quality Characteristics of Peanut Products

Peanut processing and consumption patterns vary in different countries. In the United States, 57% of peanuts are used for peanut butter, about 23% for roasted peanut, 19% for desserts and ingredients, and only 1% for oil; in China, 53% of peanuts are used for oil extraction, 40% for food, 3% for exports, and 4% for seed reservation. At present, the main peanut-processed products in China include peanut oil, peanut protein (powder), peanut-texturized protein, peanut beverage, peanut butter, and peanut candy. The annual output of peanut oil is about 2.5 million tons, and its annual consumption accounts for 8% of the total edible oil; meanwhile, the total content of unsaturated fatty acid is more than 85%, which is very similar to the fatty acid composition of olive oil. With the efficacy of reducing the incidence of cardiovascular disease, it is known as "Chinese olive oil", so peanut oil has been the leading domestic peanut-processed products and the research focus in the field of peanut processing. At the same time, in recent years, with the promotion of peanut oil cold pressing technology, the deep development and utilization of peanut protein have aroused domestic attention, especially the preparation, structure, and functional evaluation of peanut protein, and the influence of different varieties of protein components and subunit contents on the gel property, solubility, emulsibility, and other functional properties has increasingly become the research focus in this field. Based on the research of the processing characteristics of different peanut varieties, the team systematically studied the functional properties of peanut protein such as preparation, gelation, and solubility, as well as the composition and oxidation stability of peanut oil, and analyzed the processing characteristics of peanut oil, peanut protein, and other main peanut-processed products, so as to provide China's peanut processing enterprises with technical support to produce high-quality peanut-processed products.

Q. Wang, *Peanut Processing Characteristics and Quality Evaluation*,
DOI 10.1007/978-981-10-6175-2_3

1 Quality Characteristics of Separated Protein from Peanut

This team prepared separated protein from peanut using 66 different peanut varieties and the same alkali-soluble acid deposition processing technology, as well as conducted assay determination for 12 quality indicators (Table 3.1) for preparing the separated protein from peanut, including sensory quality, physicochemical and nutritional quality, and processing quality. The results are shown in Table 3.2.

1.1 Edible Quality

Sensory evaluation results indicated that the color of peanut protein is milky white, the shape is powder, and the difference between the peanut varieties is not significant. Therefore, the sensory quality of peanut protein will not be analyzed in detail.

1.2 Physicochemical and Nutritional Quality

According to the figures shown in Table 3.2, the coefficients of variation of crude fat and crude fiber were 63.56% and 77.57%, respectively, which indicated that the crude fat and crude fiber of different peanut protein isolate were different. It was found by comparing mean and median in different varieties that the differences of crude fat and crude fiber were 10.32% and 17.86%, respectively; the median of other qualities was close to their means, which indicated that the outliers of these data were few. The coefficient of variation of protein purity was small (2.10%), with a variation range of 85.64–94.81%, which was consistent with the research results (88.69–94.22%) of Kim et al. (1992).

1.3 Processing Quality

The analysis results of water holding capacity, oil holding capacity, hardness, elasticity, cohesion, solubility, and other main processing quality characteristics for separated protein from peanut are shown in Table 3.3. It can be seen that the coefficient of variation of oil holding capacity is 8.89%, which is small, indicating that their dispersion degree is insignificant; the variation range of peanut protein solubility is 57.63–93.44%. Compared with other literatures, the protein solubility of the peanut varieties selected in this research is better than that in other researches. For example, Berardi and Cherry (1981) found that the variation range of solubility of separated protein from peanut was 15.1–55.6%; Madhavi et al. (1989) found that

Table 3.1 Quality characteristics of separated protein from peanut

Sensory quality		Physicochemical and nutritional quality				Processing quality					
Color and luster	Shape	Crude fat	Purity of protein	Ash content	Crude fiber	Gelation					Solubility
						Hardness	Elasticity	Cohesion	Water holding capacity	Oil holding capacity	

Table 3.2 Physicochemical and nutritional quality analysis of separated protein from peanut

	Range of variation/%	Mean/%	Standard deviation	Upper quartile/%	Median/%	Lower quartile/%	Coefficient of variation/%	Data variation/%
Ash content	0.74–2.35	1.35	0.32	1.11	1.32	1.50	24.09	2.22
Crude fat	0.54–8.84	2.52	1.60	1.35	2.26	3.26	63.56	10.32
Crude fiber	0.00–1.02	0.28	0.22	0.14	0.23	0.34	77.57	17.86
Purity of protein	85.64–94.81	90.57	1.90	89.44	90.51	91.89	2.10	0.07

Table 3.3 Analysis of processing quality of separated protein from peanut

	Variation range/%	Mean/%	Standard deviation	Upper quartile/%	Median/%	Lower quartile/%	Coefficient of variation/%	Data variation/%
Water holding capacity	0.74–1.38	1.07	0.18	0.94	1.08	1.23	16.90	0.93
Oil holding capacity	1.04–1.71	1.34	0.12	1.26	1.33	1.42	8.89	0.75
Hardness	0.37–4.26	1.60	0.86	0.91	1.60	2.09	53.37	0.00
Elasticity	0.53–0.97	0.75	0.11	0.68	0.75	0.82	15.09	0.00
Cohesion	0.28–0.59	0.39	0.06	0.35	0.39	0.43	15.04	0.00
Solubility	57.63–93.44	79.33	9.43	81.91	80.12	76.76	11.89	0.97

the solubility of separated protein from peanut was 33.073%. The variation range of water holding capacity of peanut protein is 0.74–1.38%, which is lower than that in the researches of Berardi and Cherry (1981) and Ihekoronye (1986). The results show that the variation range of water holding capacity of separated protein from peanut is 1.30–1.60% water/g separated protein.

1.4 Correlation Between Quality Characteristics

It was found through correlation analysis (Table 2 of Appendix 3) that the solubility of protein showed a significant negative correlation with the hardness ($r = -0.687$) and cohesion ($r = -0.588$) of protein solubility and gelation, indicating that the varieties with good protein solubility may have poor gelation. This result is similar to that of Li (2009), and these factors may be related to the composition, structure, and subunit content of protein.

1.5 Evaluation Method of Gelation

The gelation of peanut protein is often evaluated by two aspects of texture and other indicators (water holding capacity, oil holding capacity). Texture property is the main indicator to evaluate the gelation of proteins. The greater the hardness, elasticity, and cohesion are, the better the texture property is. At present, there is no uniform method for evaluating the gelation of peanut protein. The evaluation may be conducted by analyzing the hardness, or the three texture indicators or texture indicators, and water holding capacity and oil holding capacity. This book has conducted in-depth analysis on the relationship between the above indicators to determine the evaluation method of gelation.

1.5.1 Evaluation Indicators

Texture is related to brittle, crisp, hard, slippery, sticky, and other mechanical sensory properties of gel, and it may indicate the state of gel from the appearance. Szczesniak (1963) divided the texture property into three main categories: mechanical, geometric, and others (fat and moisture). The evaluation of texture analyzer (TPA) on food is based on the three categories above, which are actually three stages at which texture properties are perceived, namely, first bite (initial stage), chewing (second stage), and remaining (third stage). TPA consists of analog test equipment, which may produce a number of instrument parameters for the purpose of simulating the parameters produced during the human chewing process (Szczesniak 2002). Szczesniak (1963) studied the relationship between sensory results and texture characteristics. The results showed that there was a significant

Table 3.4 Analysis of correlation between texture indicator and other indicators

	Water holding capacity	Oil holding capacity
Water holding capacity	1.000	
Oil holding capacity	−0.360	1.000
Hardness	0.120	−0.080
Elasticity	−0.170	−0.070
Cohesion	0.390	−0.108

positive correlation between the viscosity of sensory evaluators and that of TPA ($r = 0.89$, $P = 0.054$), and there was a significant positive correlation between the cohesion of sensory evaluators and that of TPA ($r = 0.89$, $P = 0.045$), so TPA data may be very effective in evaluating the gelation of food (Dubost 2001).

The correlation between texture indicators and water holding capacity and oil holding capacity was analyzed (Table 3.4). It was shown from the table that the correlation value among water holding capacity, oil holding capacity, and three texture indicators was very small, while texture characteristic was the important indicator to evaluate the gelation of protein, so the gelation of protein was evaluated by analysis of the three texture indicators.

1.5.2 Evaluation Equation

Texture indicator is the main parameter for the evaluation of gelation. The greater the three indicators are, the better the values are. In order to get a unified evaluation method, the three indicators should be normalized into an indicator to reflect the problems.

1.5.2.1 Outlier of Hardness, Elasticity, and Cohesion

Boxplot method was used to analyze the outlier of hardness, elasticity, and cohesion of 66 peanut varieties in Table 3.1. It was found from Fig. 3.1 that there was an outlier in hardness (Huayu 19), one outlier in cohesion (Yueyou 40), while no outlier in elasticity. Based on the removal of these two indicator outliers, the original data became 64 varieties for follow-up analysis.

1.5.2.2 Normalization Method: Adopt Formula

Define m of indicators to be $(x_{i,1}, x_{i,2}, x_{i,3} n x_{i,m})$ $i = 1, 2, \ldots, n$ $(n = 40)$, respectively, and obtain matrix Z:

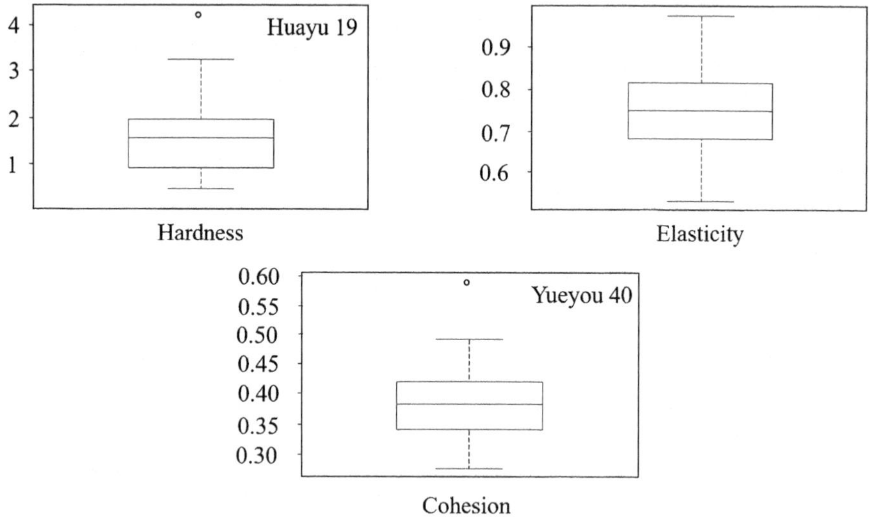

Fig. 3.1 Analysis of outlier of texture indicator

$$Z = \begin{pmatrix} x_{1,1} & x_{1,2} & x_{1,3} \\ x_{2,1} & x_{2,2} & x_{2,3} \\ n & n & n \\ x_{n,1} & x_{n,2} & x_{n,3} \end{pmatrix} \tag{3.1}$$

Standardize matrix Z to obtain matrix Z:

$$Z = \begin{pmatrix} x_{1,1} & x_{1,2} & x_{1,3} \\ x_{2,1} & x_{2,2} & x_{2,3} \\ n & n & n \\ x_{n,1} & x_{n,2} & x_{n,3} \end{pmatrix} \tag{3.2}$$

where $x_{i,j} = \frac{x_{i,j} - \bar{x}_j}{s_j}, \bar{x}_j = \frac{1}{n}\sum_{k=1}^{n} x_{k,j}, s_j = \sqrt{\frac{1}{n-1}\sum_{k=1}^{n}\left(x_{k,j} - \bar{x}_j\right)^2}, i = 1, 2, n, n,$ $j = 1, 2, 3.$

On the basis of Z, two matrixes were established, respectively, **S** and **W**, where

$$S = \begin{pmatrix} s_{1,1} & s_{1,2} & n & s_{1,n} \\ s_{2,1} & s_{2,2} & n & s_{2,n} \\ n & n & n & n \\ s_{n,1} & s_{n,2} & n & s_{n,n} \end{pmatrix} \tag{3.3}$$

$$s_{ij} = e^{-\sqrt{(x_{i1}-x_{j1})^2+(x_{i2}-x_{j2})^2+(x_{i3}-x_{j3})^2}}, i = 1, 2, n, n, j = 1, 2, 3,$$

and

$$W = \begin{pmatrix} \sum_{i=1}^{n} s_{i,1} & 0 & n & 0 \\ 0 & \sum_{i=1}^{n} s_{i,2} & n & 0 \\ n & n & n & n \\ 0 & 0 & n & \sum_{i=1}^{n} s_{i,n} \end{pmatrix} \quad (3.4)$$

$P = W^{-1}S$, where W^{-1} is the inverse matrix of W.

Transform the three indicators for evaluating gelation texture into one indicator using Formulas 3.1–3.4, that is, transform the three-dimensional data of gelation into one-dimensional data, and define the three-dimensional data as the comprehensive value of gelation of peanut protein.

1.5.2.3 Correlation Between Comprehensive Value of Gelation and Original Indicator

Analyze the correlation between the comprehensive value of gelation and three indicators of gelation texture (Table 3.5).

It was found through correlation analysis that the maximum correlation coefficient between the comprehensive value of gelation and hardness was 0.87, and the correlation coefficient with elasticity and cohesion was relatively small, but all of them were positively correlated. The results were in good agreement with the requirements that the greater the gelation hardness, elasticity, and cohesion were, the better they were, indicating that the gelation obtained by this method is accurate and reliable.

1.5.2.4 Establishment of Evaluation Equation of Comprehensive Value of Gelation

Regression analysis was made for hardness, elasticity, and cohesion based on the normalized data, and the equation obtained was

$$\text{Gelation} = 0.0268 + 0.1618 \times \text{hardness} + 0.3781 \times \text{elasticity} + 1.1573 \times \text{cohesion} \quad (3.5)$$

Table 3.5 Analysis of correlation between the comprehensive value of gelation and other texture indicators

	Gelation		Gelation
Gelation	1.00	Elasticity	0.41
Hardness	0.87	Cohesion	0.47

Table 3.6 Quality characteristics of peanut oil

	Evaluation indicators
Sensory quality	Color, taste, and transparency
Physicochemical and nutritional quality	Moisture and volatile matter, specific gravity, refractive index, unsaponifiable matter
Processing quality	Induction time, peroxide value, acid value, iodine value, saponification value

2 Peanut Oil Quality Characteristics

There are rich peanut resources in China, and the difference in the quality of different peanut varieties affects the quality of peanut oil. We have analyzed and determined the 12 indicators (Table 3.6) related to the sensory quality, physicochemical and nutritional quality, and processing quality of 45 different peanut varieties in China's main production areas.

2.1 Edible Quality

The determination (color was measured by Lovibond test method, with reference to GB/T5525-85) and statistical analysis of sensory quality of 45 varieties of peanut oil are shown in Table 3.7. It can be seen that different varieties of peanut oil had the inherent smell and taste of peanut oil, without significant difference. However, the 45 varieties all processed were yellowish and transparent. Of them, the variation range of red value is 0.40–1.70, and the variation range of yellow value is 2.80–13.00; meanwhile, the color difference between different varieties was large, and the variation coefficients of red value and yellow value were, respectively, 39.92% and 43.15%.

2.2 Physicochemical and Nutritional Quality

The determination and statistical analysis of water, volatile matter, and unsaponifiable matter content for the peanut oil processed from 45 varieties of peanut are shown in Table 3.8. It can be seen that the variation range of water and

Table 3.7 Color of different varieties of peanut oil

No.	Name	Color		No.	Name	Color	
		Red	Yellow			Red	Yellow
1	Zhonghua 8	0.65 ± 0.07	11.00 ± 0.00	26	Black peanut	0.55 ± 0.07	5.10 ± 0.28
2	Sanhua 7	0.40 ± 0.00	3.50 ± 0.00	27	034–256-1	0.70 ± 0.00	6.45 ± 0.07
3	Silihong	0.60 ± 0.00	2.80 ± 0.00	28	Jihua 9814	0.80 ± 0.00	10.50 ± 0.78
4	Luhua 11	0.70 ± 0.00	5.50 ± 0.00	29	Yuhua 15	0.70 ± 0.00	5.50 ± 0.00
5	Bianhua 3	0.90 ± 0.00	9.25 ± 0.21	30	Yuhua 9326	0.60 ± 0.07	4.40 ± 0.07
6	Haihua 1	0.65 ± 0.07	4.40 ± 0.00	31	Yuhua 9327	0.90 ± 0.00	6.10 ± 0.07
7	Shuangji 2	0.40 ± 0.00	5.35 ± 0.07	32	Kainong 30	0.50 ± 0.00	9.90 ± 0.00
8	Shanhua 9	0.85 ± 0.07	3.35 ± 0.07	33	Kainong 37	0.80 ± 0.00	4.00 ± 0.00
9	Fenghua 5	0.45 ± 0.07	4.30 ± 0.28	34	Yuanza 9102	0.50 ± 0.00	5.40 ± 0.00
10	White peanut	0.60 ± 0.14	3.95 ± 0.07	35	Heyou 11	1.05 ± 0.35	4.35 ± 0.49
11	Fenghua 1	0.60 ± 0.00	4.40 ± 0.00	36	Guihua 771	1.20 ± 0.00	8.05 ± 0.07
12	Fenghua 3	0.70 ± 0.00	6.50 ± 0.00	37	Yueyou 86	1.25 ± 0.78	11.45 ± 2.19
13	Fenghua 4	1.70 ± 0.00	13.00 ± 0.00	38	Shanyou 250	1.25 ± 0.07	6.90 ± 0.71
14	Xuhua 5	0.70 ± 0.00	5.40 ± 0.00	39	Zhanhua 82	0.90 ± 0.00	12.00 ± 0.00
15	Yuanhua 8	0.80 ± 0.00	6.50 ± 0.00	40	Pearl red	1.15 ± 0.21	11.00 ± 0.00
16	Xuhua 13	0.65 ± 0.07	7.35 ± 0.07	41	Longhua 243	1.00 ± 0.00	6.35 ± 0.21
17	Xuhua 14	0.70 ± 0.00	6.40 ± 0.00	42	Yueyou 14	1.00 ± 0.14	5.70 ± 0.71
18	Huayu 19	0.55 ± 0.07	3.35 ± 0.07	43	Minhua 9	0.85 ± 0.35	5.90 ± 0.95
19	Huayu 20	0.45 ± 0.07	3.25 ± 0.07	44	Yueyou 45	1.55 ± 0.07	6.85 ± 0.21
20	Huayu 22	0.60 ± 0.00	5.70 ± 0.00	45	Yueyou 40	1.70 ± 0.00	13.00 ± 0.00
21	Huayu 23	0.60 ± 0.00	6.20 ± 0.00		Average value	0.79	6.65
22	Huayu 28	0.60 ± 0.00	13.00 ± 0.00		Maximum value	1.70	13.00
23	Huayu 31	0.60 ± 0.00	5.35 ± 0.07		Minimum value	0.40	2.80
24	Baisha 1016	0.55 ± 0.07	5.30 ± 0.14		Variation coefficient %	39.92	43.15
25	Colorful peanut	0.80 ± 0.00	5.20 ± 0.00				

Table 3.8 Physicochemical and nutritional quality indicators of different varieties of peanut oil

No.	Name	Water and volatile matter %	Unsaponifiable matter g/kg	No.	Name	Water and volatile matter %	Unsaponifiable matter g/kg
1	Zhonghua 8	0.07 ± 0.00	4.24 ± 0.49	26	Black peanut	0.08 ± 0.01	17.21 ± 2.03
2	Shanhua 7	0.05 ± 0.01	19.76 ± 1.23	27	034–256-1	0.06 ± 0.02	9.47 ± 1.43
3	Silihong	0.08 ± 0.00	12.10 ± 1.22	28	Jihua 9814	0.11 ± 0.01	11.76 ± 0.76
4	Luhua 11	0.11 ± 0.02	9.74 ± 1.54	29	Yuhua 15	0.09 ± 0.01	8.22 ± 0.83
5	Bianhua 3	0.09 ± 0.01	13.23 ± 0.98	30	Yuhua 9326	0.08 ± 0.01	10.66 ± 1.28
6	Haihua 1	0.09 ± 0.00	10.24 ± 0.67	31	Yuhua 9327	0.07 ± 0.00	12.23 ± 3.02
7	Shuangji 2	0.05 ± 0.04	9.67 ± 1.29	32	Kainong 30	0.05 ± 0.04	10.02 ± 1.79
8	Shanhua 9	0.15 ± 0.01	11.70 ± 1.12	33	Kainong 37	0.11 ± 0.13	17.32 ± 2.83
9	Fenghua 5	0.20 ± 0.00	17.91 ± 2.00	34	Yuanza 9102	0.05 ± 0.05	18.65 ± 1.91
10	White peanut	0.08 ± 0.04	10.95 ± 0.97	35	Heyou 11	0.15 ± 0.03	
11	Fenghua 1号	0.04 ± 0.00	8.95 ± 1.72	36	Guihua 771	0.22 ± 0.02	
12	Fenghua 3	0.08 ± 0.00	6.08 ± 1.02	37	Yueyou 86	0.20 ± 0.01	
13	Fenghua 4	0.08 ± 0.00	9.14 ± 2.00	38	Shanyou 250	0.16 ± 0.04	
14	Xuhua 5	0.07 ± 0.00	7.70 ± 1.23	39	Zhanhua 82	0.18 ± 0.01	
15	Yuanhua 8	0.09 ± 0.00	9.00 ± 2.01	40	Pearl red	0.16 ± 0.01	
16	Xuhua 13	0.05 ± 0.01	11.28 ± 1.78	41	Longhua 243	0.20 ± 0.03	
17	Xuhua 14	0.06 ± 0.02	11.57 ± 1.02	42	Yueyou 14	0.16 ± 0.01	
18	Huayu 19	0.07 ± 0.00	11.13 ± 1.09	43	Minhua 9	0.15 ± 0.01	
19	Huayu 20	0.05 ± 0.03	10.66 ± 1.46	44	Yueyou 45	0.18 ± 0.01	
20	Huayu 22	0.04 ± 0.01	8.01 ± 0.29	45	Yueyou 40	0.16 ± 0.01	
21	Huayu 23	0.08 ± 0.01	6.34 ± 0.54		Average value	0.10	11.39
22	Huayu 28	0.07 ± 0.01	9.70 ± 1.43		Maximum value	0.22	19.76
23	Huayu 31	0.10 ± 0.00	13.61 ± 0.98		Minimum value	0.04	4.24
24	Baisha 1016	0.06 ± 0.01	18.83 ± 1.71		Variation coefficient %	49.74	33.31
25	Colorful peanut	0.08 ± 0.00	10.05 ± 1.98				

volatile matter was 0.04–0.22%, and the difference between varieties was large, and the variation coefficient was 49.74%. The variation range of unsaponifiable matter was 4.24–19.76 g/kg, and the variation coefficient was 33.31%, which was large.

2.3 *Processing Quality*

The analysis results of five processing quality indicators (including peroxide value, acid value) for 45 varieties of peanut oil are shown in Table 3.9. It can be seen that the variation ranges of peroxide value, acid value, induction time, iodine value, and saponification value of 45 varieties were 0.31–4.95 mmol/kg, 0.22–2.57 mgKOH/g, 3.59–5.76 h, 88.13–116.14 g/100 g, and 148.47–217.13 mg/g, respectively. The variation coefficients of indicators were large, the variation coefficient of peroxide value was 56.04%, and the variation coefficient of acid value was 96.78%, indicating that there was large difference in the quality of peanut oil processed from different varieties of peanut.

We determined the oxidation induction time of different varieties of peanut oil at 120 °C using Rancimat oil oxidation stabilizer and comparatively studied the oxidation stability of different varieties of peanut oil. The oxidation induction curve of peanut oil is shown in Fig. 3.2. Carboxylic acid and other volatile substances were generated after oil oxidation, Rancimat method was used to automatically draw the change curve of conductivity with time by testing the change in conductivity caused due to the abovementioned volatile matters, and thus the induction time under accelerated oxidation conditions was calculated. Induction time was used to represent the oxidation stability of oil; the longer the induction time was, the better the oil oxidation stability was; the shorter the induction time was, the worse the oil oxidation stability was. It was found from Table 3.9 that there was a significant difference ($P < 0.01$) in the oxidation induction time between different varieties of peanut oil tested, the average induction time was 4.53 h, the variation amplitude was 2.17 h, and the variation coefficient was 12.04%. Among them, the peanut oil variety with the longest induction time was Shanhua 7, being 5.76 h; the peanut oil variety with the shortest induction time was Yuhua 15, being 3.59 h. According to Worthington et al. (1972), there was difference in the oxidation stability of oil processed from 82 different genotypes of peanut, which was similar to the results of this book. The length of the induction time reflected the stability of oil oxidation (Josep 1993), and the above analysis showed that among 45 peanut oil varieties, the oil oxidation stability of Shanhua 7 was the best.

2.4 *Correlation Between Quality Characteristics*

We determined the indicators of sensory quality, physicochemical and nutritional quality, and processing quality of 45 varieties of peanut oil and conducted

Table 3.9 Processing quality indicators of different varieties of peanut oil

No.	Name	Peroxide value mmol/kg	Acid value mgKOH/g	Induction value h	Iodine value g/100 g	Saponification value mg/g
1	Zhonghua 8	2.72 ± 0.11	0.62 ± 0.41	4.19 ± 0.06	88.14 ± 0.03	177.33 ± 0.11
2	Shanhua 7	2.24 ± 0.55	0.27 ± 0.00	5.76 ± 0.07	96.88 ± 0.12	176.18 ± 0.10
3	Silihong	2.70 ± 0.01	0.36 ± 0.00	5.19 ± 0.16	97.63 ± 0.01	170.65 ± 0.06
4	Luhua 11	3.28 ± 0.05	0.33 ± 0.02	4.67 ± 0.10	93.16 ± 0.03	175.64 ± 0.00
5	Bianhua 3	2.20 ± 0.01	0.64 ± 0.03	4.61 ± 0.10	93.22 ± 0.07	148.48 ± 0.09
6	Haihua 1	3.08 ± 0.02	0.27 ± 0.03	5.23 ± 0.17	90.18 ± 0.21	177.34 ± 0.01
7	Shuangji 2	2.57 ± 0.49	0.31 ± 0.00	4.45 ± 0.09	92.63 ± 0.02	170.98 ± 0.07
8	Shanhua 9	2.97 ± 0.18	0.28 ± 0.00	4.39 ± 0.14	98.07 ± 0.19	174.85 ± 0.12
9	Fenghua 5	2.37 ± 0.02	0.36 ± 0.01	3.74 ± 0.27	93.15 ± 0.12	176.73 ± 0.19
10	White peanut	2.13 ± 1.21	2.57 ± 0.01	4.62 ± 0.16	97.25 ± 0.08	175.49 ± 0.13
11	Fenghua 1	3.07 ± 0.44	0.38 ± 0.01	3.91 ± 0.24	96.25 ± 0.12	175.49 ± 0.08
12	Fenghua 3	2.45 ± 0.00	0.42 ± 0.01	3.91 ± 0.03	91.79 ± 0.07	177.18 ± 0.01
13	Fenghua 4	1.04 ± 0.03	0.32 ± 0.01	4.37 ± 0.09	88.69 ± 0.06	177.18 ± 0.23
14	Xuhua 5	2.73 ± 0.00	0.39 ± 0.06	3.81 ± 0.06	93.04 ± 0.13	166.79 ± 0.09
15	Yuanhua 8	3.72 ± 0.06	0.36 ± 0.04	3.73 ± 0.03	100.92 ± 0.11	186.27 ± 0.00
16	Xuhua 13	3.41 ± 0.44	0.37 ± 0.02	3.82 ± 0.23	101.70 ± 0.13	188.28 ± 0.01
17	Xuhua 14	3.77 ± 0.13	1.05 ± 0.01	4.27 ± 0.10	101.85 ± 0.05	186.27 ± 0.12
18	Huayu 19	3.23 ± 0.03	0.41 ± 0.02	4.71 ± 0.09	98.64 ± 0.10	183.63 ± 0.10
19	Huayu 20	2.04 ± 0.20	0.39 ± 0.00	4.63 ± 0.04	98.27 ± 0.03	181.99 ± 0.02
20	Huayu 22	3.23 ± 0.06	0.41 ± 0.01	4.79 ± 0.13	94.16 ± 0.00	186.54 ± 0.04
21	Huayu 23	2.19 ± 0.04	0.67 ± 0.03	5.63 ± 0.11	99.15 ± 0.12	183.27 ± 0.05
22	Huayu 28	2.08 ± 0.03	0.54 ± 0.01	3.94 ± 0.07	92.59 ± 0.04	181.52 ± 0.07
23	Huayu 31	1.41 ± 0.22	0.37 ± 0.01	4.44 ± 0.14	100.53 ± 0.05	185.64 ± 0.19
24	Baisha 1016	2.27 ± 0.06	0.36 ± 0.01	4.36 ± 0.07	108.98 ± 0.06	187.58 ± 0.23
25	Colorful peanut	0.81 ± 0.01	1.20 ± 0.00	4.97 ± 0.26	103.69 ± 0.11	181.14 ± 0.21
26	Black peanut	2.89 ± 0.25	0.53 ± 0.04	4.81 ± 0.06	103.96 ± 0.13	185.92 ± 0.08

27	034-256-1	3.13 ± 0.34	0.38 ± 0.00	5.23 ± 0.17	96.16 ± 0.07	203.70 ± 0.10
28	Jihua 9814	1.69 ± 0.07	0.48 ± 0.06	4.63 ± 0.07	108.62 ± 0.12	190.04 ± 0.05
29	Yuhua 15	2.33 ± 0.01	0.23 ± 0.00	3.59 ± 0.01	107.35 ± 0.04	217.14 ± 0.09
30	Yuhua 9326	2.28 ± 0.23	0.36 ± 0.01	4.04 ± 0.18	116.15 ± 0.03	179.49 ± 0.12
31	Yuhua 9327	1.61 ± 0.01	3.02 ± 0.01	4.31 ± 0.01	98.69 ± 0.08	203.20 ± 0.21
32	Kainong 30	3.43 ± 0.04	0.43 ± 0.00	4.49 ± 0.42	107.08 ± 0.05	188.68 ± 0.00
33	Kainong 37	3.20 ± 0.00	0.56 ± 0.00	4.49 ± 0.42	113.40 ± 0.12	193.60 ± 0.01
34	Yuanza 9102	4.95 ± 0.06	0.24 ± 0.00	3.70 ± 0.03	103.26 ± 0.06	186.93 ± 0.00
35	Heyou 11	0.31 ± 0.00	0.22 ± 0.00	4.31 ± 0.01		
36	Guihua 771	0.50 ± 0.02	0.36 ± 0.09	4.19 ± 0.03		
37	Yueyou 86	0.50 ± 0.004	0.37 ± 0. 03	4.59 ± 0.06		
38	Shanyou 250	0.52 ± 0.11	0.59 ± 0.18	5.43 ± 0.01		
39	Zhanhua 82	0.50 ± 0.008	0.22 ± 0.00	4.16 ± 0.07		
40	Pearl Red	0.48 ± 0.04	1.98 ± 0.00	5.12 ± 0.04		
41	Longhua 243	0.52 ± 0.03	0.34 ± 0.12	5.63 ± 0.09		
42	Yueyou 14	0.46 ± 0.11	0.81 ± 0.00	4.63 ± 0.04		
43	Minhua 9	0.45 ± 0.00	0.62 ± 0.00	5.02 ± 0.04		
44	Yueyou 45	0.51 ± 0.07	0.53 ± 0.09	4.79 ± 0.21		
45	Yueyou 40	0.50 ± 0.006	0.57 ± 0.09	4.58 ± 0.04		
	Average value	2.09	0.59	4.53	98.96	182.37
	Maximum value	4.95	2.57	5.76	116.14	217.13
	Minimum value	0.31	0.22	3.59	88.13	148.47
	Variation coefficient %	56.04	96.78	12.04	6.90	6.42

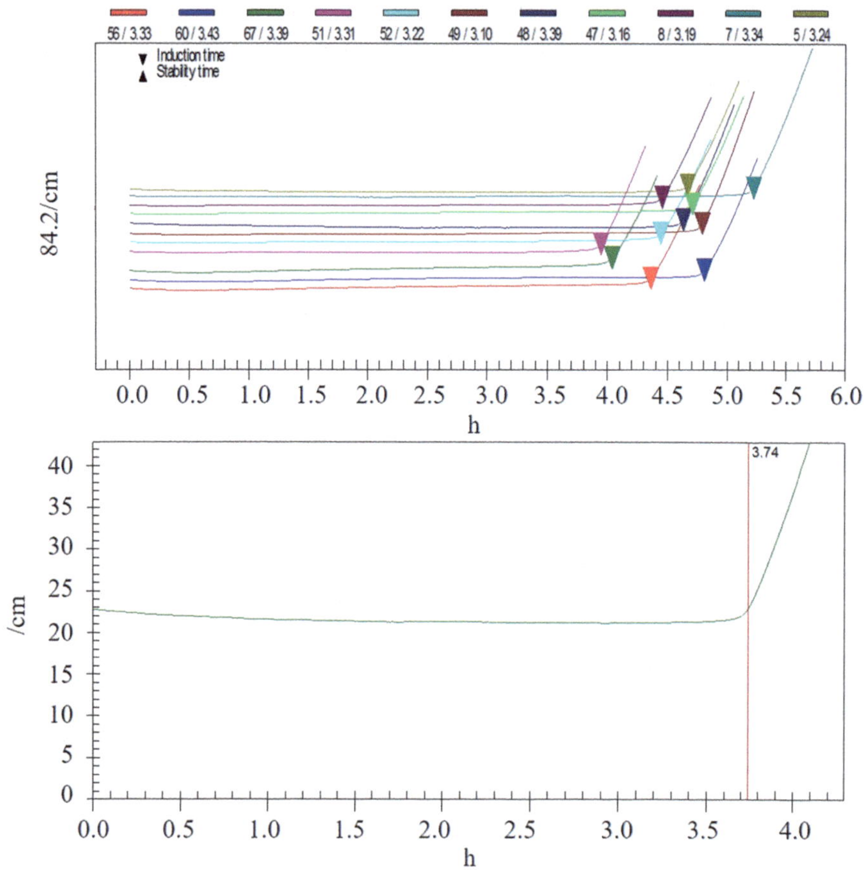

Fig. 3.2 Oxidation induction curve of peanut oil at 120 °C

correlation analysis. The results are shown in Table 3.10. It can be seen that except that the correlation coefficient between yellow and red was 0.493, and the correlation coefficient between saponification value and iodine value was 0.477, the correction coefficient between other indicators was small, indicating that the relationship between indicators was small, so it could be considered that the indicators were basically independent of each other, and they could be used as independent indicators to evaluate peanut oil.

Table 3.10 Correlation analysis between the quality indicators of peanut oil

	Red	Yellow	Water and volatile matter	Unsaponifiable matter	Induction time	Peroxide value	Acid value	Iodine value	Saponification value
Red	1.000								
Yellow	0.493**	1.000							
Water and volatile matter	0.154	−0.024	1.000						
Unsaponifiable matter	−0.229	−0.336	0.076	1.000					
Induction time	−0.066	−0.245	−0.137	0.105	1.000				
Peroxide value	−0.096	−0.034	0.346*	−0.121	−0.186	1.000			
Acid value	−0.049	−0.003	0.341*	0.065	0.070	0.071	1.000		
Iodine value	−0.184	−0.228	−0.119	0.386*	−0.123	−0.239	−0.109	1.000	
Saponification value	−0.014	−0.012	−0.13	0.022	−0.136	−0.383*	0.300	0.477**	1.000

Authors: Wang Li, Wang Qiang, Zhang Jianshu, Du Yin, Lin Weijing

Note: (*) $P < 0.05$, (**) $P < 0.01$ showed that there was a significant or very significant difference between the data, respectively

Table 3.11 Sensory evaluation analysis of 26 varieties of peanut butter

Sample No	Average value	Standard deviation	Variation amplitude	Variation coefficient %
Color (9 scores)	6.74	1.07	8.40–4.29	15.84
Aroma (9 scores)	6.88	0.67	8.20–5.80	9.83
Texture state (9 scores)	6.83	0.36	7.40–6.00	5.47
Taste (9 scores)	6.79	0.58	7.80–5.29	8.66
Spreading (9 scores)	7.23	0.43	8.00–6.29	6.00
Overall acceptability (9 scores)	7.07	0.69	8.20–5.57	9.92

3 Quality Characteristics of Peanut Butter

3.1 Edible Quality

There was a certain difference in the sensory quality of 26 varieties of peanut butter. The results showed (Table 3.11) that there were different degrees of variation in the color, aroma, and taste of peanut butter processed from different varieties of peanut. According to sensory evaluators, the peanut butter processed from Huayu 22 (Shandong) variety was the best; it was yellow brown and shiny and could cause appetite, with pure braked peanut aroma, good tissue state, uniform butter, delicate taste, excellent viscosity, and smooth spreading; the overall acceptability score of its sensory score was 8.2, followed by Luhua 17, osmanthus 17, P905, Fuhua 12, and so on. We determined the hardness and viscosity of different varieties of peanut butter using texture analyzer. The result showed that (Table 3.12) there was a large difference in the hardness and viscosity of different varieties of peanut butter, with change ranges of 75.26-233.16 and 27.33-195.44, respectively, and the variation coefficients were 41.03% and 59.31%. Among them, the variety with the largest difference in hardness and viscosity was Jihuatian 1 (Hebei), and the variety with the smallest difference was Guihua 22 (Guangxi). Some studies have shown that the taste of sensory score and spreading have a certain degree of correlation with the viscosity determined by texture analyzer. Among them, the viscoelasticity determined by the instrument is able to predict the taste of sensory score, namely, the viscosity at the first mouth, the sample with smooth taste has low hardness and viscosity in the instrument (Abegaz and Kerr 2006), which is consistent with the results of this study. We determined the color of different varieties of peanut butter using color difference meter. Its significant difference was small; the variation coefficients of L^*, C^*, and H were smaller than 10, indicating that the baking degree of different varieties of peanut was similar to each other. Riveros et al. (2010) reported that when the color of peanut butter (L^* value) was 50 ± 1, its color was good, which had a certain influence on product appearance and overall acceptability by consumers. There was a small difference between this experiment and its study, and the value of Silihong was closest to it.

Table 3.12 Analysis of texture and color of 26 varieties of peanut butter

Peanut butter	Average value	Standard deviation	Variation amplitude	Variation coefficient %
Hardness (g)	114.40	46.90	72.96–233.16	41.03
Viscosity (g)	73.16	43.16	72.96–233.16	59.31
*L**	57.36	2.81	51.36–61.99	4.98
*C**	29.25	2.34	23.79–35.50	8.74
H	80.38	3.21	72.11–86.24	4.11

Note: *L**: brightness (black =0°, white = 100°), *C**: saturation, *H*: chromaticity (0° = red, 90° = yellow)

3.2 Analysis of Physicochemical and Nutritional Quality

The nutritional contents of peanut butter play a vital role in the quality of the final product. The physicochemical and nutritional quality analysis of 26 varieties of peanut butter is shown in Table 3.13. It can be seen from the table that the average value of crude protein content in peanut butter was 23.64% and the average value of crude fat content was 47.76%; the difference between the crude protein and crude fat contents in peanut was small, indicating that the protein and fat content loss during the processing of peanut butter was small; the water content range is 0.55–1.05%, which met the peanut butter water content limit ($\leq$1.5 g) specified in agricultural industry standards; although the variation range of other components was not clearly defined, the difference was less than that in peanut. Navnitkumar and Chun (2002) analyzed the quality of seven varieties of peanut butter in India. The results showed that the difference in the fat and protein content of the butters processed from different varieties of peanut was small, and the research of Navnitkumar and Chun (2002) showed that during the process of making peanut butter, the change in its physicochemical and nutritional quality indicators was small. It can be seen that the physicochemical and nutritional quality loss was not significantly different during the process of making peanut into butter, and the difference in the physicochemical and nutritional quality among different peanut varieties was small. In peanut fatty acid composition, the component with the highest content was oleic acid and linoleic acid, and the research of Özcan and Seven (2003) showed that the difference in the content change of oleic acid and linoleic acid was small during the process of making peanut into butter; the research of Savage and Keenan (1994) and Lopez et al. (2001) showed that the ratio between oleic acid and linoleic acid (*O*/*L*) was an important indicator to measure the storability of peanut raw materials and their products, i.e., the higher the ratio was, the more stable the oil quality was; the physiological functions of oleic acid included lowering cholesterol, regulating blood lipids, and decreasing blood sugar (Wahrbur 2004); oleic acid was also known as “safe fatty acid,” and its content is an important indicator to assess fat quality (Liu et al. 2010). Therefore, the quality of peanut butter was significantly

Table 3.13 Analysis of physicochemical and nutritional quality of 26 varieties of peanut butter

Peanut butter	Average value	Standard deviation	Variation amplitude	Variation coefficient %
Water %	0.78	0.13	0.55–1.05	17.11
Ash %	3.19	0.27	2.58–3.50	8.48
Protein %	23.64	2.36	19.53–28.01	10.14
Total sugar %	13.22	2.91	9.66–21.82	22.31
Fat %	47.76	3.19	42.13–53.30	6.76
Fiber %	7.39	2.81	2.94–12.25	38.14
Oleic acid mg/100 g	21.82	5.74	13.40–36.49	27.91
Linoleic acid mg/100 g	14.28	5.14	1.84–20.52	36.78
Grain size (μm)	46.37	8.81	24.05–68.90	19.27

influenced by oleic acid and linoleic acid. The content ranges of oleic acid and linoleic acid in the peanut varieties selected in this experiment were 13.40–36.49 and 1.84–20.52, respectively, and the variety with highest oleic acid content was P905, which showed that the oleic acid and linoleic acid contents in 26 varieties of peanut were significantly different, and this had a great impact on the quality of the prepared peanut butter. China's agricultural industry standard stipulates that the grain size of peanut butter should be greater than 100 meshes but not clearly specifies its range. The average grain size of peanuts processed from 26 varieties of peanut in this research was 46.37 μ, and there was a certain difference in the grain sizes of different varieties. Among them, the two varieties of peanut butter of Guihuahong 95 and Luhua 18 had a significant difference.

3.3 *Stability Analysis*

The peroxide value is usually used to measure the oxidation degree at the initial oxidation of fat. The lower the peroxide value is, the better the initial stability of peanut butter is, and the less the oxidative rancidity and other similar phenomena is. The initial stability of 26 varieties of peanut butter is shown in Fig. 3.3. The variation range of peroxide value was 0.004–0.259 g/100 g, the value of Fuhua 18 (Liaoning) was the highest, and the value of 13–2 (Henan) was the lowest, showing that the stability of 13–2 was the best and the stability of Fuhua 18 was the worst among the 26 varieties of peanut butter when they were fresh. According to China's agricultural industry standards, the peroxide value of peanut butter (NY/T 958-2006 peanut butter) should be lower than 0.25 g/100 g, so it can be seen that Fuhua 18 peanut butter had exceeded the range specified in national standard when it was fresh. Riveros et al. (2010) mainly measured the stability of high oleic acid

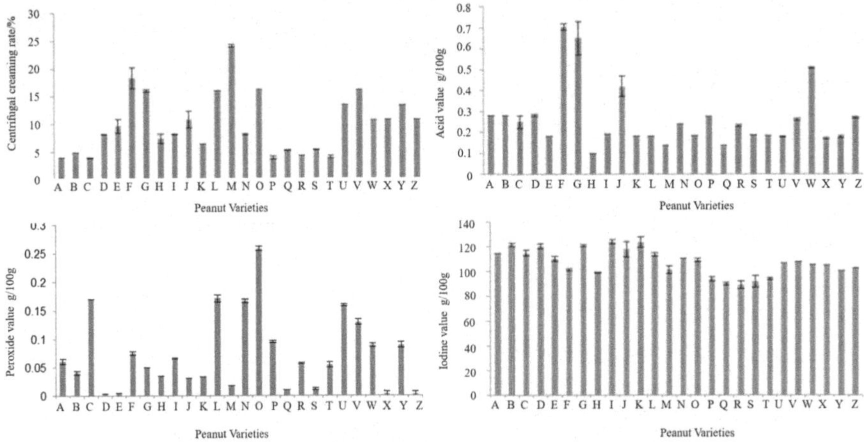

A: Huayu 22 (Shandong), B: Huayu 22 (Xinjiang), C: Weihua 8, D: 12-2 butter, E: 13-3 butter, F: Guihua 22, G: Guihuahong 95, H: P905, I: Jihuatian 1, J: Luhua 17, K: Fenghua 1, L: Guihua 17, M: Gaoyousuan, N: Xiaobaisha, O: Fuhua 18, P: Fuhua 12, Q: Kai 17-15, R: Luhua 18, S: Yuhua 9414, T: Shangyan 9658, U: Huayu 20, V: Fuhua 16, W: Silihong, X: Meihong, Y:Shanhua 9, Z: Fuhua 19

Fig. 3.3 Twenty-six analysis of initial stability of 26 varieties of peanut butter

and ordinary oleic acid using the change in peroxide value. The results showed that the stability of butter processed from high oleic acid peanut was good, and it was the variety with the lowest peroxide value content. Therefore, it can be seen from the figures that the initial oxidation stability of high oleic peanut butter is relatively good. The variation range of acid value was 0.09–0.705 g [KOH]/100 g, and the variety with the lowest acid value was P905 (Shandong). The greater the acid value was, the greater the degree of oxidative deterioration was, showing that the initial acid values of different varieties of peanut butter were significantly different. The variation range of iodine value was 99.22–123.997 g/100 g, and the vibration coefficient of iodine value was 70.15%, showing that the difference in iodine value was large among different varieties. The results obtained by Özcan and Seven (2003) after researching the iodine value of different varieties of peanut butter were consistent with the research results of this experiment. The variation range of centrifugal creaming rate was 4.0–18.38%, and the variation coefficient was 50.00%, showing that the difference in centrifugal creaming rate among different varieties was large. The lower the centrifugal creaming rate was, the smaller the degree of fat separation of peanut butter, and the better the stability effect was. Li et al. (2013) and Totlani (2002) found that the fat separation degree of peanut butter was good when the centrifugal creaming rate was low. Hinds et al. (1994) showed that the largest value of fat separation degree of fresh peanut butter was 0.5 ml within 24 h of storage according to the regulations of US Department of Agriculture. According to the variation coefficients (93.64%, 58.56%, and 50.00%) of peroxide value, acid value, and centrifugal creaming rate, the initial stabilities of butters processed from different varieties were significantly different.

Table 3.14 Analysis of quality correlation of 19 indicators of 26 varieties of peanut butter

	Protein	Water	Ash	Fiber	Total sugar	Oleic acid	Linoleic acid	Hardness	Viscosity	*L**	*C**	*H*	POV	AV	IV	Eccentricity	Receptiveness	Grain size
Fat	−0.227	−0.092	0.014	−0.069	−0.145	0.150	−0.075	−0.436*	−0.428*	0.280	0.080	0.112	−0.460*	0.116	−0.380	−0.115	0.020	−0.321
Protein	1	0.027	−0.068	−0.349	−0.602**	0.073	0.072	0.071	−0.142	−0.182	0.255	−0.249	0.226	0.233	−0.209	0.578**	−0.165	0.023
Water		1	−0.105	0.226	0.096	−0.155	0.002	0.099	0.073	−0.079	0.081	−0.070	0.169	−0.072	0.191	−0.024	−0.060	−0.149
Ash			1	0.080	−0.124	0.241	−0.107	−0.081	0.059	−0.215	0.275	−0.166	0.242	−0.266	−0.387	−0.080	0.122	−0.156
Fiber				1	0.191	−0.333	0.273	0.077	0.191	0.006	−0.008	−0.025	−0.030	−0.168	0.274	−0.250	0.239	0.002
Total sugar					1	−0.193	−0.327	0.422*	0.494*	−0.066	−0.324	0.153	−0.170	−0.044	0.478*	−0.323	−0.173	0.163
Oleic acid						1	−0.667**	−0.442*	−0.356	0.050	0.056	0.080	−0.248	−0.545**	−0.537**	−0.040	0.236	−0.287
Linoleic acid							1	0.026	0.031	0.170	0.101	−0.070	0.235	0.326	−0.021	−0.075	0.039	0.211
Hardness								1	0.842**	−0.410*	0.054	−0.320	0.159	0.192	0.377	−0.020	−0.496**	0.296
Viscosity									1	−0.514**	0.255	−0.503**	0.134	−0.162	0.266	−0.151	−0.193	0.195
*L**										1	−0.709**	0.859**	−0.212	0.090	−0.287	−0.183	0.068	−0.187
*C**											1	−0.912**	0.084	−0.125	−0.168	0.224	0.134	0.195
H												1	−0.116	0.106	−0.042	−0.237	−0.005	−0.231
POV													1	−0.081	0.079	0.182	0.093	0.163
AV														1	0.214	0.269	−0.428*	0.189
IV															1	0.102	−0.281	0.413*
Eccentricity																1	−0.268	0.350
Receptiveness																	1	−0.368

Note: **represent highly significant correlation ($P < 0.01$); *represent significant correlation ($P < 0.05$)

3.4 Analysis of Quality Correlation of Peanut Butter

In the processing industry of peanut, edible quality, physicochemical and nutritional quality, and stability have always been the focus of attention. The results of correlation analysis are shown in Table 3.14: the hardness and viscosity of peanut butter were significantly positively correlated ($r = 0.842$ **), indicating that the viscosity of peanut butter increased with the increase of hardness, which was consistent with the research results of Navnitkumar and Chun (2002) and Abegaz (2003). L^* value, C^*, and value, H value were significantly correlated ($r = -0.709$ **, $r = 0.0.859$ **), indicating that the more bright the peanut was, the more brown yellow its color was and the lower its color angle saturation was. There was a significant negative correlation between oleic acid and linoleic acid ($r = -0.667$ *). It was found by Özcan and Seven (2003) that after making peanut into butter, the variation difference of oleic acid and linoleic acid content was small, O/L value was not changed, indicating that the higher the oleic acid content of peanut was, the lower the linoleic acid content of peanut was, which was consistent with the research results of Zhang (2012) and Shin et al. (2010)

References

Abegaz EG. The role of moisture and common ingredients in quality changes in stored model peanut confections. Athens: The University of Georgia; 2003.

Abegaz EG, Kerr WL. Effect of moisture, sugar and tertiary butylhydroquinone on color, texture and microstructure of peanut paste. J Food Qual. 2006;29(6):643–57.

Berardi LC, Cherry JP. Functional properties of co-precipitated protein isolates from cottonseed, soybean and peanut flours. Food Sci. 1981;14:283.

Dubost NJ. Development and optimization of peanut soy spread [D]. Georgia: University of Georgia; 2001.

Hinds MJ, Chinnan MS, Beuchat LR. Unhydrogenated palm oil as a stabilizer for peanut butter. J Food Sci. 1994;59(4):816–20.

Ihekoronye AI. Functional properties of meal products of the Nigerian 'Red Skin' groundnut (Arachis hypogaea L.). J Sci Food Agric. 1986;37(10):1035–41.

Josep SB. Oil content, stability, and fatty acid composition of the main varieties of Catalonian hazelnuts (*Corylus avellana* L.). Food Chem. 1993;48:237–41.

Kim N, Kim YJ, Nam YJ. Characteristics and functional properties of protein isolates from various peanut (*Arachis hypogaea* L.) cultivars. J Food Sci. 1992;57(2):406–10.

Li JJ. Screening and applicability of special soybean varieties for processing in China [D]. Shijiazhuang: Hebei Agricultural University; 2009.

Li LF, Xu YL, Wang JL, et al. Effect of wheat fiber and compound emulsifier on the stability of peanut butter. Sci Technol Food Ind. 2013;34(22):232–7.

Liu SC, Deng CJ, Zhong YQ, et al. Optimizing extraction and studying physicochemical property of papaya seed oil[J]. J Chin Cereals Oils Assoc. 2010;25(8):42–6.

López Y, Smith OD, Senseman SA, Rooney WL, López Y, Smith OD. Genetic factors influencing high oleic acid content in spanish market-type peanut cultivars. Cropence. 2001;41(1):51–6.

Madhavi DL, Subramanyam L, Chand N, et al. Effect of solvent extraction on removal of nutty flavor and functional properties of groundnut flour. J Food Qual. 1989;11:375–85.

Navnitkumar, Chun Ji-Yeon. Vitamin E content and stability in peanuts and peanut products during processing and storage. Athens: The University of Georgia; 2002.
Özcan M, Seven S. Physıcal and chemıcal analysıs and fatty acıd composıtıon of peanut, peanut oıl and peanut butter from ÇOM and NC-7 cultıvars. Grasas Aceites. 2003;54(1):12–8.
Riveros CG, Mestrallet MG, Gayol MF. Effect of storage on chemical and sensory profiles of peanut pastes prepared with high-oleic and normal peanuts. J Sci Food Agric. 2010;90:2694–9.
Savage GP, Keenan JI. The composition and nutritive value of groundnut kernels. In: Smartt J, editor. The groundnut crop: a scientific basis for improvement. London: Chapman & Hall; 1994. p. 173–213.
Szczesniak AS. Classification of textural characteristics. J Food Sci. 1963;28(4):385–9.
Szczesniak AS. Texture is a sensory property. Food Qual Prefer. 2002;13(4):215–25.
Totlani VM. Investigation of centrifugal and rheological techniques to predict stability of peanut butter. Athens: University of Georgia; 2002.
Wahrbur GU. What are the health effects of fat. Eur J Nutr. 2004;43(1):6–11.
Worthington RE, Hammons RO, Allison JR. Varietal differences and seasonal effects on fatty acid composition and stability of oil from 82 Peanut genotypes [J]. J Agric Food Chem. 1972;20:727–30.
Zhang JS. Effects of fatty acid composition and endogenous antioxidants on the stability of different peanut oils[D]. Beijing: Chinese Academy of Agricultural Sciences; 2012.

Chapter 4
Relationship Between Raw Material Quality and Product Quality of Peanut

There is a certain relationship between the raw material quality and product quality of peanut. Grosso et al. (2000) analyzed the quality characteristics of 17 wild peanut varieties and their peanut oil in Argentina and found that there were significant differences in the quality of different varieties and peanut oil prepared; the variation ranges of protein, fat, oleic acid/linoleic acid, and iodine value were 25.0–30.1%, 45.7–51.8%, 0.65–1.38%, and 99.2–110.4% separately; and the quality characteristics of these wild varieties and stability of peanut oil were worse than those of cultivars. The research conducted by Shin et al. (2010) showed that the fatty acid composition of peanut affected the oxidation stability of peanut oil, and the higher the oleic acid/linoleic acid ratio, the higher the oxidation stability of oil (Worthington et al. 1972; Brown 1975; Savage 1994; Lopez et al. 2001). Grosso et al. (1994) and Özcan and Seven (2003) analyzed the impacts of physiochemical characteristics and fatty acid composition and contents of different peanut varieties on the stability of peanut butter. Misra (2004) assigned the quality indicators of peanuts and established mathematical models to evaluate the comprehensive quality of different peanut varieties so as to provide a basis for the special peanut varieties for candies. At present, there is less systematic research on the relationship between the raw material quality characteristics and product quality characteristics of peanut. Wang (2012) determined the raw material quality indicator closely related to product quality by analyzing the relationship between raw material quality and product quality of peanut and established the relationship model between them by using the supervised principal component regression analysis. The gelation and solubility of protein, oxidation stability of peanut oil, and quality of export peanut of unknown peanut varieties can be predicted, and thus the special varieties suitable for processing can be selected through the model. It is of great significance to research the intrinsic relationship between peanut varieties or raw material quality characteristics and quality characteristics of processed products, establish the correlation model, and scientifically evaluate and forecast the processing purposes of different peanut varieties or raw materials for full and

Q. Wang, *Peanut Processing Characteristics and Quality Evaluation*,
DOI 10.1007/978-981-10-6175-2_4

reasonable utilization of peanut variety resources and improvement of processed product quality.

1 Relationship Between Raw Material Quality and High-Gelation Protein Quality of Peanut

Gelation is one of the important functional properties of peanut protein. The gelation of protein is influenced by external factors and intrinsic factors. The protein concentration, heating time, heating temperature, ion strength and species, pH value, and aging time were the external factors that affected gelation (Saio and Watanabe 1978); the protein content, structure, and amino acid composition were the intrinsic factors. The research conducted by Young et al. (1974) showed that there were great differences in the protein contents of peanuts in different regions and different varieties of peanuts. Prakash and Narasinga (1986) and Yang et al. (1998) found that the conarachin had more sulfur-containing amino acids than arachin. The sulfur-containing amino acid is the main influencing factor of gel formation, so the varieties with high conarachin contents have better gelation of protein. There are a number of peanut quality indicators, and how each quality affects the gelation of peanut protein is one of the important problems to be solved currently. The supervised principal component regression analysis (Bair et al. 2006) has been widely used in the modern agricultural science and related disciplines. Liu Xuhua et al. (2009), Roberts and Martin (2006), and Bair and Tibshirani (2004) used the supervised principal component regression analysis for the selection of near-infrared spectrum wavelengths, selection of environmental pollution indicators, and research of influence factors of disease, which played a role in the "less and fine" problem response. Wang (2012) used the supervised principal component regression to analyze the relationship between peanut quality and protein gelation, which provided a basis for the evaluation and prediction of protein gelation of certain peanut variety.

2 Theoretical Analysis Model of Quality Evaluation of Peanut with High-Gelation Protein

The 64 peanut varieties in Sect. 1.5.2.1, Chap. 3, were taken as the research objects, the model was established by using the supervised principal component regression analysis, 43 varieties were randomly selected from 64 varieties as the model set, and the remaining 21 varieties were considered as a validation model set.

2.1 *Quality Evaluation Indicators of Gel-Type Protein*

Correlation analysis (Table 4.1) was carried out for the comprehensive value of gelation of peanut protein and other indicators of peanut protein. Results showed that there was no significant relationship between comprehensive value of gelation and other physicochemical indicators of protein at 0.05 level; therefore, the impacts of these four indicators on gelation of peanut protein were not considered, and the gelation was considered as the quality evaluation indicator of gel-type peanut protein.

2.2 *Establishment of Model*

2.2.1 Correlation Between Peanut Quality and Gelation

After the correlation analysis was carried out for the peanut quality indicators and gelation of 64 varieties in Sect. 1.5.2.1, Chap. 3 (Table 4.2), it was found that there were highly significant positive correlations between cystine, leucine, arginine, and other indicators and gelation, and these indicators might be important indicators that affected the gelation of peanut protein.

2.2.2 Supervised Principal Component Regression Analysis Modeling

The independent variables that had a strong correlation with the corresponding variables were used for the supervised principal component regression analysis; the regression independent variable was selected based on the correlation coefficients of response variables and their respective variables, and after that, the principal component regression analysis was carried out for the selected regression independent variables.

The independent variables were selected with correlation coefficients exceeding a certain threshold and then carried out for:

1. Analysis of independent variables

   The outliers (Shanhua 7, Bianhua 3, and other nine varieties) of peanut quality were removed by using the boxplots, so the remaining 34 varieties were analyzed.

2. Quality evaluation indicators of peanuts suitable for gel-type protein processing

   After the regression coefficient significance analysis was carried out for the single peanut quality indicator and comprehensive value of protein gelation in accordance with the idea of regression analysis, it was found that there were significant correlations between 12 indicators and gelation at 0.05 level (Table 4.3).

Table 4.1 Correlation between gelation and other indicators

Indicator	Correlation coefficient	*P* value	Indicator	Correlation coefficient	*P* value
Ash	0.300	0.045	Crude fat	0.012	0.940
Crude fiber	0.090	0.550	Protein purity	−0.210	0.190

Table 4.2 Correlation between various quality indicators of peanut and protein gelation

Indicator	Correlation coefficient	Indicator	Correlation coefficient	Indicator	Correlation coefficient
Fruit shape	−0.593[a]	Cysteine	0.708[b]	Conarachin II	0.014
Red skin	−0.221	Valine	−0.108	Globulin/myoglobulin	−0.356[a]
Grain shape	−0.264	Methionine	−0.007	40.5 kDa	0.366[a]
Hundred fruit weight	0.256	Isoleucine	−0.300	37.5 kDa	−0.187
Hundred kernel weight	0.283	Leucine	0.606[b]	35.5 kDa	−0.148
Water content	−0.019	Tyrosine	−0.283	23.5 kDa	−0.744[b]
Crude fat	−0.211	Phenylalanine	0.063	18 kDa	0.263
Crude protein	0.808[a]	Lysine	0.094	17 kDa	0.274
Total sugar	0.198	Histidine	−0.220	15.5 kDa	0.315[a]
Ash	−0.103	Tryptophan	−0.020	Protein extraction rate	0.028
Crude fiber	−0.694[a]	Arginine	0.637[b]	Pure kernel rate	−0.174
Total amino acid	0.715[a]	Arachin	−0.370[a]	Serine	−0.259
Asparaginic acid	−0.323[a]	Conarachin	0.381[a]	Glutamic acid	−0.253
Threonine	−0.260	Conarachin I	0.392[a]	Proline	−0.403[b]
		Alanine	−0.205	Glycine	−0.321[a]

[a] Significant correlation
[b] Extremely significant correlation

3. Correlation in peanut quality

Correlation analysis (Table 4.4) was carried out for the 12 selected quality indicators. One of the indicators with the correlation coefficients significantly higher than 0.800 at 0.01 level was expressed by another indicator, so the indicators retained were fruit shape, crude protein, crude fiber, glycine, cystine, arginine, leucine, conarachin I, arachin/conarachin, and 23.5 kDa.

Table 4.3 Regression significance indicators of various peanut quality indicators and protein gelation

No.	Indicator	*P* value	No.	Indicator	*P* value	No.	Indicator	*P* value
1	Fruit shape	0.011	5	Cystine	0.037	9	Myoglobulin	0.033
2	Crude protein	0.018	6	Leucine	0.009	10	Myoglobulin I	0.003
3	Crude fiber	0.010	7	Tyrosine	0.030	11	Globulin/ myoglobulin	0.030
4	Glycine	0.050	8	Arginine	0.005	12	23.5 kDa	0.001

4. Principal component analysis

The principal component analysis and dimensionality reduction (Table 4.5) were carried out for the ten selected indicators. Through the principal component analysis, it was found that the cumulative contribution rate of the first six principal components was 93.4%, more than 85% (Jolliffe 2002). Therefore, the first six principal components could express the information of original principal component, and the original ten indicators were transformed into six new indicators, which played a role in dimensionality reduction. The eigenvectors (not listed here) of the first six principal components were output by SAS, the linear relations between the first six principal components and various indicators were obtained, and then the scores of various principal components were calculated. The normalized data was substituted into the relations, and the scores of the principal components of all evaluation objects could be obtained (Table 4.6).

5. Establishment of regression equation

The dependent variable for establishment of regression equation should conform to the normal distribution. So Box-Cox transformation (Fitzmaurice et al. 2007) between the principal components and gelation was carried out to make the dependent variable conform to the normal distribution. Then the regression analysis was conducted again, and $\lambda = 0.25$ was obtained and put into Formula 4.1,

Box-Cox transformation of dependent variable Y

$$y = \begin{cases} \dfrac{(x)^{\lambda} - 1}{\lambda} & \lambda \neq 0 \\ \ln(x) & \lambda = 0 \end{cases} \tag{4.1}$$

The transformation for factors that did not conform to normal distribution was conducted by this method.

Formula 4.2 shown below was obtained:

$$y_1 = 4(\sqrt[4]{y} - 1) \tag{4.2}$$

y_1 – Value of protein gelation after changing into normal distribution
y – Original value of protein gelation

Table 4.4 Correlation analysis among various peanut quality indicators

	Fruit shape	Crude protein	Crude fiber	Glycine	Cystine	Leucine	Tyrosine	Arginine	Myoglobulin	Myoglobulin I	Globulin/ myoglobulin	23.5 kDa
Fruit shape	1.000											
Crude protein	0.347[a]	1.000										
Crude fiber	0.241	0.293	1.000									
Glycine	0.166	0.684[b]	0.174	1.000								
Cystine	0.035	−0.117	−0.304	−0.087	1.000							
Leucine	0.029	0.699[b]	0.252	0.580[b]	−0.398[b]	1.000						
Tyrosine	−0.084	0.508[b]	0.175	0.397[a]	−0.404[b]	0.844[b]	1.000					
Arginine	0.236	0.732[b]	0.409[b]	0.635[b]	−0.401[b]	0.649[b]	0.399[b]	1.000				
Myoglobulin	0.125	0.022	−0.065	0.013	0.433[b]	−0.271	−0.340[a]	−0.123	1.000			
Myoglobulin I	0.022	−0.094	−0.206	−0.146	0.660[b]	−0.262	−0.252	−0.380[a]	0.777[b]	1.000		
Globulin/ myoglobulin	−0.138	−0.013	0.080	0.018	−0.430[b]	0.290	0.354[a]	0.126	−0.996[b]	−0.769[b]	1.000	
23.5 kDa	0.133	0.188	0.157	0.302	−0.526[b]	0.198	0.171	0.272	−0.271	−0.457[b]	0.282	1.000

[a] Significant correlation
[b] Extremely significant correlation

Table 4.5 Eigenvalues of correlation matrixes

	Characteristic root	Variance contribution rate/%	Accumulative variance contribution rate/%		Characteristic root	Variance contribution rate/%	Accumulative variance contribution rate/%
1	4.5573	0.4557	0.4557	6	0.6146	0.0615	0.9355
2	1.4941	0.1494	0.6051	7	0.2241	0.0224	0.9579
3	1.1379	0.1138	0.7189	8	0.2031	0.0203	0.9782
4	0.9112	0.0911	0.8100	9	0.1418	0.0142	0.9924
5	0.6397	0.064	0.8740	10	0.0762	0.0076	1.0000

Table 4.6 Scores of principal components

No.	First principal component	Second principal component	Third principal component	Fourth principal component	Fifth principal component	Sixth principal component
1	0.4641	−0.3870	1.3948	−0.2171	0.9044	−0.3155
2	0.7874	−0.0702	2.1271	−0.2155	0.7018	−1.1840
3	−4.2626	−0.5446	0.5535	−0.1516	−0.3927	0.7502
4	−3.5935	−1.0486	−0.0773	0.1140	0.4455	0.3378
5	−3.7253	1.2811	0.7876	−0.8349	−1.5736	0.7233
6	−1.2750	−1.8653	1.9659	0.5216	−1.1253	0.6544
7	−4.4915	0.8357	−0.7157	0.0036	1.1619	0.4854
8	1.3876	0.6982	0.2454	1.0741	1.0068	0.3307
9	0.5481	0.4059	−0.3432	0.6747	1.2072	−0.2061
10	−0.1961	1.2793	−0.8121	0.4543	−0.2816	−0.2233
11	−0.5267	−0.4612	1.1758	1.0341	0.4232	−0.1939
12	−0.7802	1.1919	−1.2838	1.8728	−0.9229	−0.0288
13	−0.5886	0.1334	0.9913	0.1349	−0.2061	−0.1416
14	0.6037	2.4966	−0.1545	−0.1808	−0.2540	−0.3469
15	0.7456	−0.1347	−0.6517	1.6144	0.9535	0.9182
16	−0.6850	−1.2218	0.9205	−1.3482	0.8575	0.4393
17	−0.1302	−1.4084	0.4132	−0.9613	1.0766	−0.7440
18	2.8711	−1.3535	0.0550	1.0268	0.3246	0.8829
19	2.1061	−1.1659	0.1342	0.4363	0.1378	1.2214
20	2.6249	0.2007	−0.0655	0.3278	1.0297	0.1288
21	−0.0965	0.0316	−0.2373	−0.2867	−0.2826	−0.1183
22	0.9782	−2.2987	−1.2513	0.5018	−0.7548	0.4037
23	0.6121	1.4390	−0.5815	−1.3894	−0.0004	−0.7921
24	1.9155	−0.4486	−0.4687	0.5073	−0.4099	−0.6846
25	−0.1796	−0.7250	−1.5802	0.3584	0.6398	−0.3477
26	1.6786	−0.2999	−0.0527	0.7632	−1.1519	−1.0118
27	1.0199	−0.9869	0.1179	0.1124	−0.0577	−1.3018
28	−1.4437	−1.9255	−0.8564	−0.0340	−0.0250	−0.7871
29	−0.1355	−0.5718	−2.0695	−0.7734	0.7561	0.6745
30	−0.2469	−1.6777	−0.7874	−0.6395	−0.0632	−0.4344
31	1.4801	−1.1690	−0.3625	−0.2094	−0.8406	−0.0509
32	−0.0196	−0.5177	−1.5923	−0.2227	−0.2798	−0.0800
33	−0.1356	−0.5718	−2.0695	−0.7734	0.7561	0.6745
34	1.9155	−0.4486	−0.4687	0.5073	−0.4099	−0.6846

The regression analysis showed that the regression coefficients of principal components and gelation were significant at 0.05 level (Table 4.7). Therefore, the relationship between the principal components and gelation was established, the determination coefficient of this model was $R^2 = 0.875$, and then the relationships between various indicators and gelation were established. The results are shown in Formula 4.3.

Table 4.7 Regression coefficient significance between gelation and principal components (theoretical analysis)

Variable	Coefficient	*P* value	Variable	Coefficient	*P* value
Intercept	−0.0115514	0.6512	PC4	−0.0148578	0.04590
PC1	−0.0453256	0.0036	PC5	−0.0307763	0.03314
PC2	−0.0251736	0.02504	PC6	−0.0371545	0.03499
PC3	0.0326287	0.01803			

$$\text{Gelation} = \left\{ \begin{array}{l} 1.20143 - 0.00748 \times \text{fruit shape} + 0.00154 \times \text{crude protein} - 0.00632 \\ \times \text{crude fibre} - 0.05178 \times \text{glycine} + 0.010843 \times \text{cystine} - 0.02178 \\ \times \text{leucine} + 0.004128 \times \text{arginine} + 0.001321 \times \text{conarachin I} - 0.03676 \\ \times \text{globulin/myoglobulin} - 0.00382 \times 23.5\,\text{kDa} \end{array} \right\} \tag{4.3}$$

2.2.3 Validation of Model

In order to inspect the accuracy and promotion applicability of the established model, the remaining 21 varieties were used for verification. Through the analysis of outlier, it was found that the gelation of Huaguanwang was an outlier, so it was removed. The ten indicators (fruit shape, crude protein, crude fiber, glycine, cystine, arginine, leucine, conarachin I, globulin/myoglobulin, and 23.5 kDa) of 20 peanut varieties were put into Formula 4.3 to calculate the gelation of 20 varieties, the regression analysis was conducted for the model calculation results, and comprehensive value of gelation and the correlation coefficient was 0.937 (Fig. 4.1).

2.3 *Practical Application Model of Quality Evaluation of Peanut with High-Gelation Protein*

The quality evaluation model suitable for processing of peanuts with gel-type protein established in 1.1 was better, and it could predict the protein gelation degrees of new peanut varieties, but there were ten indicators in this model, which were difficult to apply in the industry. Therefore, this model was optimized so that few indicators were used to reflect more information.

The evaluation method of gelation of peanut protein is shown in Sect. 2.1, and the correlation analysis between peanut quality and gelation is shown in Sect. 2.3.1.1. The analysis of independent variable is shown in Sect. 2.3.1.4 (1).

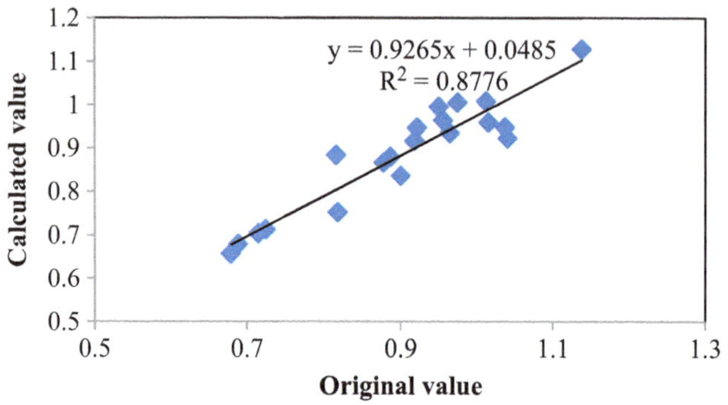

Fig. 4.1 Fitting chart of original values of gelation and calculated values of theoretical analysis model

2.3.1 Establishment of Model

2.3.1.1 Selection of Peanut Quality Evaluation Indicator

After the regression coefficient significance analysis was carried out for the single peanut quality indicator and protein gelation in accordance with the idea of regression analysis, it was found that there were significant correlations between seven indicators and gelation at 0.01 level (Table 4.8).

2.3.1.2 Correlation Analysis of Peanut Quality

Correlation analysis (Table 4.9) was carried out for the seven selected quality indicators. One of the indicators with the correlation coefficients significantly higher than 0.600 at 0.01 level was expressed by another indicator, so the indicators retained were fruit shape, crude protein, arginine, conarachin I, and 23.5 kDa protein subunit.

2.3.1.3 Principal Component Analysis

The principal component analysis (Table 4.10) was carried out for the five selected indicators. Through the principal component analysis, it was found that the cumulative contribution rate of the first three principal components was 85.19%. Therefore, the first three principal components could express the information of original principal component, and the original five indicators were transformed into three new indicators, which played a role in dimensionality reduction. The eigenvectors (not listed here) of the first six principal components were output by SAS, the linear relations between the first three principal components and various indicators were

Table 4.8 Regression significance indicators of peanut quality and gelation

No.	Indicator	*P* value	No.	Indicator	*P* value
1	Fruit shape	0.011	5	Arginine	0.005
2	Crude protein	0.018	6	Conarachin I	0.003
3	Crude fiber	0.01	7	23.5 kDa	0.001
4	Leucine	0.009			

obtained, and then the scores of various principal components were calculated by using the relations between the principal components and independent variables.

2.3.1.4 Supervised Principal Component Regression Analysis Modeling

The dependent variable for establishment of regression equation should conform to the normal distribution. The gelation data did not conform to the normal distribution, so the Box-Cox transformation between the principal components and gelation was carried out to make the dependent variable conform to the normal distribution. Then the regression analysis was conducted again, $\lambda = 0.00$ was obtained and put into Formula 4.1, and Formula 4.4 was obtained:

$$y_1 = \log\ y \tag{4.4}$$

$y1$ – Value of protein gelation after changing into normal distribution, y – Original value of protein gelation

The regression analysis found that the regression coefficients of principal components and gelation were significant at 0.05 level (Table 4.11). Therefore, the relationship between the principal components and gelation was established, the determination coefficient of this model was $R^2 = 0.768$, and then the relationships between various indicators and gelation were established. The results are shown in Formula 4.5.

$$\text{Gelation} = e^{(1.571-0.02474*\text{fruit shape}+0.007009*\text{crude protein}-0.04351*\text{arginine}-0.005*\text{conarachin I}-0.06057*23.5\text{kDa})} \tag{4.5}$$

2.3.2 Verification of Model

In order to inspect the accuracy and promotion applicability of the established model, the remaining 21 varieties were used for verification. Through the analysis of outlier, it was found that the gelation of Huaguanwang was an outlier, so it was removed. The five indicators (fruit shape, crude protein, arginine, conarachin I, and

Table 4.9 Correlation of peanut quality

	Fruit shape	Crude protein	Crude fiber	Leucine	Arginine	Myoglobulin I	23.5 kDa
Fruit shape	1.000						
Crude protein	0.347[a]	1.000					
Crude fiber	0.241	0.293	1.000				
Leucine	0.029	0.699[b]	0.252	1.000			
Arginine	0.236	0.732[b]	0.409[b]	0.649[b]	1.000		
Myoglobulin I	0.022	−0.094	−0.206	−0.262	−0.380[a]	1.000	
23.5 kDa	0.133	0.188	0.157	0.198	0.272	−0.457[b]	1.000

[a] Significant correlation
[b] Extremely significant correlation

Table 4.10 Eigenvalues of correlation matrixes

Number of principal components	Variance contribution rate/%	Accumulative variance contribution rate/%	Number of principal components	Variance contribution rate/%	Accumulative variance contribution rate/%
1	0.4281	0.4281	4	0.0936	0.9455
2	0.2580	0.6861	5	0.0545	1.0000
3	0.1658	0.8519			

Table 4.11 Regression coefficient significance between gelation and principal components (practical application)

Variable	Coefficient	*F* value	*P* value
Intercept	−1.5192579	7.05	0.01
PC1	−0.0571799	12.92	0.0010
PC2	0.0116931	0.10	0.7501
PC3	−0.0812101	5.64	0.0233

23.5 kDa) of 20 peanut varieties were put into Formula 4.5 to calculate the gelation of 20 varieties, the regression analysis was conducted for the model calculation results, and comprehensive value of gelation and the correlation coefficient was 0.718 (Fig. 4.2).

This model was adopted to predict the protein gelation of 86 peanut varieties in China; the protein gelation of 152 varieties (predicted varieties and modeled varieties) is shown in Table 3, Appendix 3.

3 Relationship Between Raw Material Quality and High-Solubility Protein Quality of Peanut

The protein solubility refers to the solution performance of protein in aqueous solution or salt solution, and it is closely related to the protein composition and structure. The protein solubility directly affects the quality of its processed product simultaneously. This book analyzed the relationship between the quality characteristics of 66 peanut varieties and protein solubility and established the quality evaluation model of peanut with high-solubility protein suitable for beverage processing by using the supervised principal component regression analysis method. Through this model, the protein solubility degrees of new peanut varieties can be predicted to find the special varieties for processing in order to improve the product quality and increase enterprise efficiency.

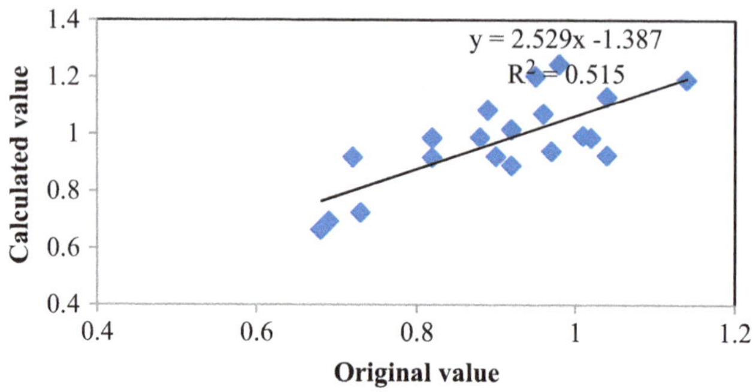

Fig. 4.2 Fitting chart of original values of gelation and calculated values of practical application model

3.1 Theoretical Analysis Model of Quality Evaluation of Peanut with High-Solubility Protein

The 66 peanut varieties in Table 2.1, Chap. 2, were taken as the research objects, the model was established by using the supervised principal component regression analysis, 45 varieties were randomly selected from 66 varieties as the model set, and the remaining 21 varieties were considered as a validation model set.

3.1.1 Quality Evaluation Indicators of Soluble Protein

Correlation analysis (Table 4.12) was carried out between the solubility of peanut protein and other indicators. It could be seen from the table that the correlation between the protein solubility and other protein components, such as ash, crude fat, crude fiber, and protein purity, was not significant; therefore, other components of protein were not analyzed, and only the protein solubility was considered as the analysis indicator of soluble protein.

3.1.2 Establishment of Model

3.1.2.1 Outlier

Through the analysis of solubility (dependent variable) and peanut quality (independent variable) by using the boxplots, it was found that there was no outlier in solubility; there were 11 outliers in peanut quality (Zhonghua 8, Shanhua 7, etc.), so the remaining 34 varieties were analyzed.

Table 4.12 Correlation between solubility and protein quality

	Ash	Crude fat	Crude fiber	Protein purity	Solubility
Ash	1.000				
Crude fat	0.043	1.000			
Crude fiber	0.024	0.016	1.000		
Protein purity	−0.307*	−0.309*	−0.006	1.000	
Solubility	−0.196	0.111	−0.126	0.218	1.000

3.1.2.2 Supervised Principal Component Regression Analysis Modeling

1. Regression significance between independent variable and dependent variable

The regression significance analysis (Table 4.13) was carried out for the protein solubility and peanut quality, and it was found that there were significant correlations between 15 quality indicators and solubility at 0.05 level.

2. Correlation among indicators

The correlation among 15 indicators of 34 varieties was analyzed (Table 4.14), and it was found that there was a negative correlation between crude fat and methionine ($r = -0.508$), a positive correlation between crude protein and arginine ($r = 0.500$), a negative correlation between cystine and arginine ($r = -0.505$), a positive correlation between cystine and conarachin I ($r = 0.668$), a negative correlation between cystine and 18 kDa protein subunit ($r = 0.589$), a positive correlation between cystine and 17 kDa protein subunit ($r = 0.514$), a positive correlation between methionine and 17 kDa protein subunit ($r = 0.533$), a positive correlation between conarachin I and 18 kDa protein subunit ($r = 0.781$), a positive correlation between conarachin I and 18 kDa protein subunit ($r = 0.766$), a positive correlation between 40.5 kDa protein subunit and 37.5 kDa protein subunit ($r = 0.500$), a positive correlation between 37.5 kDa protein subunit and 18 kDa protein subunit ($r = 0.506$), a positive correlation between 18 kDa protein subunit and 17 kDa protein subunit ($r = 0.570$), and a positive correlation between protein extraction rate and cystine ($r = 0.403$). Therefore, one indicator was used to reflect the information of another indicator, the four indicators (18 kDa, 17 kDa, and 40.5 kDa protein subunits and methionine) were deleted, and the follow-up analysis was carried out for the remaining 11 indicators.

3. Principal component analysis

The principal component analysis was carried out for original data of 11 indicators after standardization, it was found that the cumulative contribution rate of the first six principal components was 89.43% (Table 4.15) and greater than 85%, and the first six principal components could express all the information of original data completely; therefore, the first six principal components were selected for analysis.

The linear relations between various principal components and independent variables were established (Formula 4.5):

Table 4.13 Regression coefficient significance between solubility and peanut quality (theoretical analysis)

No.	Indicator	Regression coefficient	No.	Indicator	Regression coefficient	No.	Indicator	Regression coefficient
1	Crude fat	0.001	6	Arginine	0.044	11	18 kDa	0.002
2	Crude protein	0.007	7	Myoglobulin I	0.007	12	17 kDa	0.023
3	Total sugar	0.020	8	40.5 kDa	0.023	13	15.5 kDa	0.001
4	Cystine	0.002	9	37.5 kDa	0.035	14	Protein extraction rate	0.010
5	Methionine	0.004	10	23.5 kDa	0.025	15	Pure kernel rate	0.030

Table 4.14 Correlation coefficient among various quality indicators

	Crude fat	Crude protein	Total sugar	Cystine	Methionine	Arginine	Myoglobulin I	40.5 kDa	37.5 kDa	23.5 kDa	18 kDa	17 kDa	15.5 kDa	Protein extraction rate	Pure kernel rate
Crude fat	1.000														
Crude protein	−0.415[a]	1.000													
Total sugar	−0.195	−0.188	1.000												
Cystine	−0.379[a]	−0.099	0.367[a]	1.000											
Methionine	−0.508[b]	0.499[b]	0.256	0.472[b]	1.000										
Arginine	0.249	0.500[b]	−0.463[b]	−0.505[b]	−0.236	1.000									
Myoglobulin I	−0.360[a]	−0.049	0.256	0.668[b]	0.343[a]	−0.393[a]	1.000								
40.5 kDa	−0.397[a]	0.092	0.186	0.302	0.149	−0.039	0.285	1.000							
37.5 kDa	−0.352[a]	0.321	−0.199	0.093	0.345[a]	0.158	0.272	0.500[b]	1.000						
23.5 kDa	0.097	0.367[a]	−0.372[a]	−0.410[a]	−0.063	0.375[a]	−0.372[a]	−0.244	−0.143	1.000					
18 kDa	−0.388[a]	0.092	−0.031	0.589[b]	0.381[a]	−0.303	0.781[b]	0.455[b]	0.506[b]	−0.228	1.000				
17 kDa	−0.363[a]	0.102	0.137	0.514[b]	0.533[b]	−0.374[a]	0.766[b]	0.092	0.330	−0.226	0.570	1.000			
15.5 kDa	0.122	−0.313	0.405[a]	0.094	−0.322	−0.034	0.267	−0.072	−0.404[a]	−0.235	−0.229	−0.189	1.000		
Protein extraction rate	−0.114	0.000	0.193	0.403[a]	0.159	−0.067	0.167	0.287	−0.085	−0.237	0.195	0.152	−0.059	1.000	
Pure kernel rate	0.406[a]	−0.241	0.097	−0.057	−0.435[a]	0.271	−0.140	0.002	−0.366	0.135	−0.318	−0.377	0.490	0.162	1.000

[a] Significant correlation
[b] Extremely significant correlation

Table 4.15 Eigenvalues of correlation matrixes

Number of principal components	Variance contribution rate/%	Accumulative contribution rate/%	Number of principal components	Variance contribution rate/%	Accumulative contribution rate/%
1	0.3067	0.3067	6	0.0509	0.8943
2	0.2483	0.5550	7	0.0413	0.9356
3	0.1210	0.6760	8	0.0371	0.9727
4	0.0955	0.7715	9	0.0163	0.9889
5	0.0719	0.8435	10	0.0111	0.9960
			11	0.0071	1.0000

$$
\begin{aligned}
\mathrm{PC1} = & -0.080281^{*} \times 1 - 1.746117^{*} \times 2 + 0.63678^{*} \times 3 + 2.326015^{*} \\
& \times 4 - 1.72047^{*} \times 5 + 0.177562^{*} \times 6 + 0.009678^{*} \times 7 - 0.328639^{*} \\
& \times 8 + 0.164427^{*} \times 9 - 0.015374^{*} \times 10 + 51.9652 \\
\mathrm{PC2} = & \, 0.148956^{*} \times 1 - 0.251102^{*} \times 2 + 0.064918^{*} \times 3 - 0.570821^{*} \\
& \times 4 - 0.097447^{*} \times 5 - 0.063034^{*} \times 6 - 00.303372^{*} \times 7 - 0.026785^{*} \\
& \times 8 + 0.347562^{*} \times 9 + 0.0855^{*} \times 10 - 3.03016 \\
\mathrm{PC3} = & -0.093136^{*} \times 1 + 0.297978^{*} \times 2 + 0.064391^{*} \times 3 + 0.740778^{*} \\
& \times 4 + 1.680676^{*} \times 5 + 0.111877^{*} \times 6 + 0.034164^{*} \times 7 + 0.14043^{*} \\
& \times 8 + 0.377383^{*} \times 9 + 0.077471^{*} \times 10 - 22.843 \\
\mathrm{PC4} = & \, 0.126717^{*} \times 1 - 0.144657^{*} \times 2 - 0.152142^{*} \times 3 + 0.189509^{*} \\
& \times 4 + 1.227407^{*} \times 5 + 0.117329^{*} \times 6 + 0.341119^{*} \times 7 - 0.364707^{*} \\
& \times 8 + 0.062603^{*} \times 9 + 0.030766^{*} \times 10 - 7.49824 \\
\mathrm{PC5} = & \, 0.03432^{*} \times 1 - 0.072938^{*} \times 2 - 0.187236^{*} \times 3 + 2.303728^{*} \\
& \times 4 - 0.774976^{*} \times 5 + 0.149748^{*} \times 6 - 0.113836^{*} \times 7 + 0.447701^{*} \\
& \times 8 - 0.153098^{*} \times 9 + 0.040257^{*} \times 10 - 11.3983 \\
\mathrm{PC6} = & \, 0.041337^{*} \times 1 + 0.025688^{*} \times 2 + 0.151766^{*} \times 3 + 1.957545^{*} \\
& \times 4 + 0.135431^{*} \times 5 - 0.142333^{*} \times 6 + 0.151649^{*} \times 7 - 0.042339^{*} \\
& \times 8 - 0.443624^{*} \times 9 + 0.092585^{*} \times 10 - 6.76348
\end{aligned}
\tag{4.6}
$$

(X_1–X_{11}, respectively, represent crude fat, crude protein, total sugar, cystine, arginine, conarachin I content, 37.5 kDa, 23.5 kDa, 15.5 kDa, protein extraction rate, and pure kernel rate)

The scores of various principal components were calculated through Formula 4.6 (Table 4.16).

4. Establishment of the model

The dependent variable for establishing regression equation should conform to the normal distribution. The solubility did not conform to the normal distribution, so the Box-Cox transformation between the principal components and solubility was carried out to make the dependent variable conform to the normal distribution.

Table 4.16 Scores of principal components

	Principal component 1	Principal component 2	Principal component 3	Principal component 4	Principal component 5	Principal component 6
1	9.128457	0.586142	1.263228	0.488481	0.998709	−0.64193
2	4.055325	−0.03317	−0.40984	−0.41455	1.00594	0.967
3	12.50248	2.632961	0.661499	0.438696	0.760556	0.630236
4	−5.58362	−2.9336	0.402749	−0.80557	0.723483	−0.45169
5	−6.18265	−2.97037	−1.08303	0.396819	0.701952	0.427791
6	−2.98548	−2.3555	−0.49942	−0.1417	−1.00616	−0.61672
7	1.286049	−1.7547	0.067458	−0.13806	−0.17406	0.914712
8	2.36193	−1.05599	−0.42174	−0.11178	−0.24326	0.988968
9	−0.73141	−2.38592	−0.43925	−0.06557	1.751362	−0.39727
10	1.405179	−2.40412	−1.11371	−0.4439	−0.54349	0.942261
11	−3.70295	−2.79844	2.235445	0.246221	−0.66899	−1.06083
12	0.934328	0.954487	0.685058	−2.55808	0.360089	−0.6314
13	2.334796	−0.9052	−0.94097	1.815781	−1.09002	−0.20723
14	4.658445	0.196305	−0.35942	0.0336	−2.21994	−0.27171
15	−3.71362	0.694966	0.683042	−0.78556	−0.01451	0.515567
16	−3.63587	−0.94202	0.31259	0.969273	1.54273	−0.05674
17	−0.84592	1.223292	−1.3393	0.620929	1.011862	0.657335
18	−0.70769	1.602587	0.579778	−1.16765	−0.07982	1.592447
19	−6.30716	−0.60286	0.363321	−1.26953	0.200859	−0.25295
20	−2.28518	0.764689	1.774076	−0.18861	−1.43618	1.12523
21	−1.65067	0.093086	−0.67158	−0.33932	−0.23503	0.588474
22	2.642411	2.129579	0.709512	−0.38825	−0.36323	−0.45665
23	−3.54899	0.823931	0.708936	0.249854	0.037391	−0.30888
24	3.134838	1.101168	−0.05376	−1.38574	−0.0897	0.165431
25	−2.02424	−0.4229	1.189127	2.782385	−0.75203	0.810314
26	−6.08692	0.185245	1.47213	−0.06722	0.33186	−0.08137
27	1.965937	1.381517	−0.24452	0.455975	−0.01432	−0.80016
28	−4.44853	0.856418	1.719864	0.475489	0.452442	−0.76549
29	−3.64881	0.998784	0.199703	0.864404	−0.26771	−0.7761
30	4.096093	1.63986	−1.77585	0.09324	−0.06024	−1.11503
31	2.29413	0.765537	−2.58132	−0.28207	0.179869	−0.36242
32	3.917485	2.513473	−1.4321	1.418339	0.259103	0.039574
33	3.247681	0.037023	−1.4404	−1.2357	−1.43998	−0.70713
34	−1.879	0.38483	−0.22032	0.444434	0.377673	−0.40486

Through the regression analysis, $\lambda = 1$ was obtained and put into Formula 4.1, and Formula 4.7 was obtained:

$$y_1 = y - 1 \tag{4.7}$$

Table 4.17 Regression coefficient significance between solubility and principal component (theoretical analysis)

Variable	Coefficient	*P* value
Intercept	1.3172222	4.135×10^{-20}
PC1	0.1278960	0.0002
PC2	−0.1373244	0.0093

After the regression analysis was carried out for y_1 and various principal components, the relationship between various indicators and solubility was obtained. The regression analysis showed that the regression coefficients of solubility of the first and fifth principal components were significant at 0.05 level (Table 4.17). The relations between these two principal components and solubility were established and then Formula 4.8 was obtained.

$$\begin{aligned} \text{Solubility} = {} & 0.770362 \times \text{crude fat} - 0.60393 \times \text{crude protein} \\ & -0.91626 \times \text{total sugar} - 8.32449 \times \text{cystine} \\ & +3.214817 \times \text{arginine} - 0.21846 \times \text{conarachin I} \\ & -1.16885 \times 37.5\ \text{kDa} + 1.81934 \times 23.5\ \text{kDa} + 1.018139 \\ & \times 15.5\ \text{kDa} - 0.44476 \times \text{protein extraction rate} \\ & +0.207081 \times \text{pure kernel rate} + 47.67507 \end{aligned} \tag{4.8}$$

3.1.3 Verification of Model

In order to inspect the accuracy and promotion applicability of the model, the remaining 21 varieties were used for validation. Through the analysis of outlier, it was found that Zhongnong 108 was the outlier of solubility, so the remaining 20 varieties were analyzed. The 11 indicators (crude fat, crude protein, total sugar, cystine, arginine, conarachin I, 37.5 kDa, 23.5 kDa, 15.5 kDa, protein extraction rate, and pure kernel rate) of 20 peanut varieties were put into Formula 4.8 to calculate the solubility of 20 varieties, the regression analysis was conducted for the calculated value and measured value of model, and the correlation coefficient was 0.820 (Fig. 4.3).

3.2 Practical Application Model of Quality Evaluation of Peanut with High-Solubility Protein

The quality evaluation model suitable for processing of peanuts with soluble protein was established, and it could predict the protein solubility degrees of unknown peanut varieties, but there were 11 indicators in this model, which were

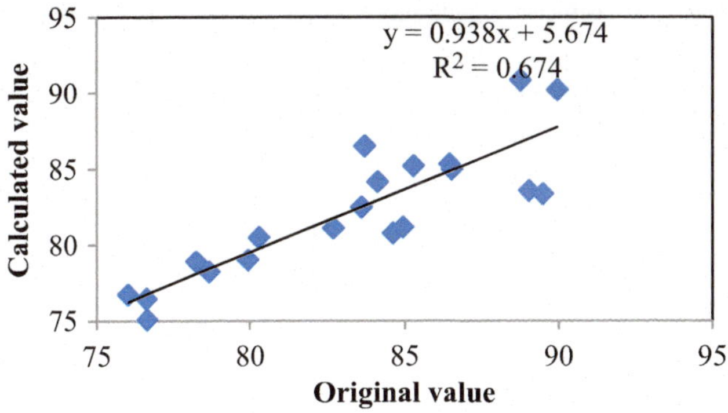

Fig. 4.3 Diagram of relationship between original values of solubility and calculated values of theoretical analysis model

not avail to apply in the actual production. Therefore, this model was optimized so that few indicators were used to reflect more information.

3.2.1 Establishment of Model

3.2.1.1 Outlier

See Sect. 3.1.2.1.

3.2.1.2 Regression Significance Between Independent Variable and Dependent Variable

Regression significance analysis was carried out for the protein solubility and peanut quality, and it was found that there were significant correlations between eight quality indicators and solubility at 0.01 level (Table 4.18).

3.2.1.3 Correlation in Peanut Quality

Through the correlation analysis of eight indicators which had significant relationships with protein solubility values at 0.01 level, it was found that (Table 4.19) there were significant correlations between crude fat and crude protein, crude fat and methionine, crude protein and methionine, cystine and myoglobulin I, 18 kDa protein subunit and protein extraction rate, and myoglobulin I and 18 kDa protein subunit, so the four indicators (crude protein, cystine, conarachin I, and 15.5 kDa protein subunit) were used for follow-up analysis.

Table 4.18 Regression coefficient significance between solubility and peanut quality (practical application)

No.	Indicator	Regression coefficient	No.	Indicator	Regression coefficient
1	Crude fat	0.001	5	Myoglobulin I	0.007
2	Crude protein	0.007	6	18 kDa	0.002
3	Cystine	0.002	7	15.5 kDa	0.001
4	Methionine	0.004	8	Protein extraction rate	0.010

3.2.1.4 Supervised Principal Component Regression Analysis Modeling

The dependent variable for the establishment of regression equation should conform to the normal distribution. The solubility did not conform to the normal distribution, so the Box-Cox transformation between various quality indicators and solubility was carried out to make the dependent variable conform to the normal distribution. Through the regression analysis, $\lambda = 1.25$ was obtained and put into Formula 4.1, and Formula 4.9 was obtained:

$$Y_1 = \frac{4}{5}\left(y^{\frac{5}{4}} - 1\right) \tag{4.9}$$

The regression analysis showed that the regression coefficients between various indicators and solubility were significant at 0.05 level. The regression equations between various indicators and solubility were established and then Formula 4.10 was obtained:

$$\text{Solubility} = \begin{pmatrix} -1.49017^{*}\text{crude protein} - 3.3775^{*}\text{cystine} - 0.39096^{*}\text{conarachin I} \\ +6.016274^{*}15.5\ \text{kDa} + 266.7366 \end{pmatrix}^{\frac{4}{5}} \tag{4.10}$$

3.2.2 Verification of Model

In order to inspect the accuracy and promotion applicability of the model, the remaining 21 varieties were used for verification. Through the analysis of outlier, it was found that Zhongnong 108 was the outlier of solubility, so the remaining 20 varieties were analyzed. The four indicators (crude protein, cystine, conarachin I, and 15.5 kDa) in 20 peanut varieties were put into Formula 4.10 to calculate the solubility of 20 varieties, the regression analysis was conducted for the calculated value and measured value of model, and the correlation coefficient was 0.699 (Fig. 4.4). Therefore, this method was used as the quality evaluation model suitable for processing of peanuts with soluble protein.

Table 4.19 Correlation of peanut quality

	Crude fat	Crude protein	Cystine	Methionine	Myoglobulin I	18 kDa	15.5 kDa	Protein extraction rate
Crude fat	1.000							
Crude protein	−0.415[a]	1.000						
Cystine	−0.379[a]	−0.099	1.000					
Methionine	−0.508[b]	0.499[b]	0.472[b]	1.000				
Myoglobulin I	−0.360[a]	−0.049	0.668[b]	0.343[a]	1.000			
18 kDa	−0.388[a]	0.092	0.589[b]	0.381[a]	0.781[b]	1.000		
15.5 kDa	0.122	−0.313	0.094	−0.322	0.267	−0.229	1.000	
Protein extraction rate	−0.114	0.131	0.403[a]	0.159	0.167	0.195	−0.059	1.000

[a] Significant correlation
[b] Extremely significant correlation

Fig. 4.4 Diagram of relationship between original values of solubility and calculated values of practical application model

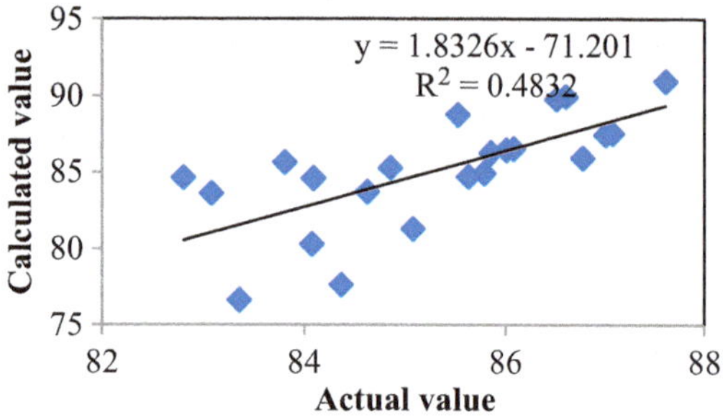

This model was adopted to predict the protein solubility of 86 peanut varieties in China. The protein solubility of 152 varieties (predicted varieties and modeled varieties) is shown in Table 4, Appendix 3.

4 Relationship Between Peanut Raw Material Quality and Peanut Oil Quality

With rich nutrition and fragrant and pure smell, the peanut oil is a good kind of cooking oil and one of the main edible oils in China. With the improvement of people's living standards, people gradually focus on the quality of peanut oil, and the research on the relationship between peanut raw material quality and peanut oil quality has also attracted the attention of peer experts. There have been some reports about the quantitative relation between fatty acid ratio mode and oil oxidation stability. Worthington et al. (1972) determined the fatty acid composition and oil stability of 82 different genotype peanut varieties in the United States and established the relations between the peanut oil stability and fatty acid composition. Yan (1998) studied the quantitative relationship between oxidative stability index (OSI) and fatty acid composition (FAC): $\mathrm{OSI}(h) = 7.5123 + \%C16:0(0.2733) + \%C18:0(0.0797)\%C18:1(0.0159) + \%C18:2 \times (-0.1141) + \%C18:3 \times (-0.3962)$ where%C16:0–C16:0 percentage content in oil. The OSI predicted values of soybean oil, corn oil, sunflower seed oil, rapeseed oil, and olive oil were predicted by this quantitative relationship. Results showed that the error between the predicted value and experimental value was less than 10%. Li Hongyan (2009) established a model through the Matlab mathematical tools to discuss the multiple quadratic mathematical model between the POV value of natural edible vegetable oil and fatty acid composition. The verification experiment showed that the simulation effect was good, and it could essentially reflect the effects of fatty acid composition on oil oxidation and predict the stability of vegetable oil through the change of fatty acid composition. Due to the diversification of peanut varieties, there were great differences in the quality of different varieties of peanuts, and the quality of peanut oil prepared with these peanut varieties as raw materials was different. At present, there has been less research on the effects of quality of different varieties of peanut raw materials on their oil quality. Only Zhang (2012) and Zhang et al. (2012) studied on the relationship between the raw material quality and peanut oil quality of 45 peanut varieties and established the relationship model between peanut raw material quality and peanut oil quality. The model could help to predict the differences of quality prepared by different varieties of peanuts and provide the raw material guarantee for high-quality peanut oil processing and provide a basis for the selection and breeding of peanut varieties for oil expression simultaneously.

4.1 Correlation Between Fatty Acid Composition and Stability of Peanut Oil

The correlation analysis between peanut fatty acid composition (UFA, SFA, UFA/SFA, MUFA, PUFA, and O/L) and induction time of peanut oil of 45 peanut varieties in Table 2-1-1 was carried out. The results were shown in Table 4.20.

In the correlation between fatty acid composition and induction time, the correlation between linoleic acid, PUFA, *O/L*, and induction time reached the significant or highly significant level. There was a highly significant negative correlation between linoleic acid and induction time ($r = -0.57$, $P < 0.01$), which indicated that the higher the linoleic acid content, the shorter the induction time of peanut oil and the worse the oil oxidation stability, and the research results were consistent with those of Worthington et al. (1972) and Firestone (1993). There was a highly significant negative correlation between PUFA and induction time ($P < 0.01$, $r = -0.50$), which indicated that the oil with high PUFA content had poor oxidative stability and could not be resistant to storage, and the research results were consistent with those of Li et al. (2010). In addition, the double bonds of PUFA (linoleic acid and linolenic acid) in peanuts were all isolated double bonds, and the methylene groups in the isolated double bonds were activated by double bonds on both sides and their nature was active; a hydrogen was removed in the reaction to form radicals and then the radicals resonated with the double bonds on both sides, so the energy required for dehydrogenation was low and it was easy to produce peroxide (ROOH) and thus cause oxidation (Zhou et al. 2002; Zhang and Shi 2009; Zhang and Wang 2010). There was a highly significant positive correlation between O/L and induction time ($P < 0.01$, $r = 0.47$), which indicated that the higher the O/L, the better the oxidative stability of peanut and its products and the longer the shelf life (Jiang and Duan 1993). Worthington et al. (1972) found that there was a positive correlation between *O/L* and stability of peanut oil ($r = 0.492$), which was similar to this research result.

Table 4.20 Correlation between peanut fatty acid composition and oil stability

		Fatty acid ratio mode						Induction time
	Linoleic acid	UFA	SFA	UFA/SFA	MUFA	PUFA	O/L	OSI
Oleic acid	−0.41**	0.43**	−0.18	0.24	0.97**	−0.66**	0.84**	0.22
Linoleic acid		0.46**	0.26	−0.20	−0.57**	0.90**	−0.83**	−0.57**
UFA			−0.4**	0.51**	0.36*	0.38*	−0.02	−0.22
SFA				−0.97**	−0.36*	0.07	−0.28	−0.24
UFA/SFA					0.40**	−0.02	0.27	0.23
MUFA						−0.73**	0.92**	0.34
PUFA							−0.93**	−0.50**
O/L								0.47**

** means highly significant correlation ($P < 0.01$); * means significant correlation ($P < 0.05$)

4.2 Correlation Between Endogenous Antioxidant and Stability of Peanut Oil

In the correlation between endogenous antioxidants of peanut and induction time of peanut oil (Table 4.21), there was a correlation between V_E and induction time but it was not significant; on the one hand, the antioxidant order of α, γ, and δ-V_E in oil was δ-V_E > γ-V_E > α-V_E (Liu and Yao 2006), and the content of α-V_E in peanuts was the highest among the three V_E isomers, so the distribution proportion of V_E in peanuts was not good for prevention of oil oxidation; on the other hand, the V_E content in peanut oil (17.47–42.61 mg/100 g) was significantly lower than that in corn oil (80–120 mg/100 g), and the V_E in corn oil played an antioxidant effect, while the V_E in peanut oil was not sufficient to produce a significant effect on its oxidative stability (Tong et al. 2009). There was also a correlation between stigmasterol, β-sitosterol, campesterol, and induction time, and the correlation was not significant. Li Rui et al. (2009) found that the solubility of phytosterol in oil was small and only 1.13%, and Tong et al. (2002) found that the phytosterol could not improve the oxidative stability of oil at low concentrations. There was a negative correlation between squalene and induction time ($r = -0.17$), but the correlation was not significant, which might be because the squalene content in peanut was less and only 3.77–16.45 mg/100 g, and it had less effect on the oxidative stability of peanut oil. Xu Li et al. (2007) found that the squalene content in olive oil was high and 136–708 mg/100 g, and the virgin olive oil had very superior oxidation stability due to high squalene content, so the squalene content would affect its antioxidant capacity in oil.

4.3 Establishment of Model

In the 45 varieties selected from Table 2-1-1, 34 varieties were randomly selected as the varieties for model establishment, and the other 11 varieties were selected as the varieties for model verification.

4.3.1 Peanut Oil Quality Indicator Transformation

For some peanut oil, the greater the indicator value, the better the quality, while for some other peanut oil, the smaller the indicator value, the better the quality. Therefore, for convenience of follow-up calculation, all the evaluation indicators of 34 varieties of peanut oil were changed so that the greater the indicator value, the better the quality. The results were shown in Table 4.22.

Table 4.21 Correlation between V_E, phytosterol, and squalene contents in peanut

	γ	δ	Stigmasterol	Campesterol	β-Sitosterol	Squalene	Induction time
α	0.36^{*}	0.32^{*}	0.48^{**}	0.64^{**}	0.59^{**}	0.67^{**}	−0.05
γ		0.85^{**}	0.37^{*}	0.76^{**}	0.63^{**}	0.67^{**}	0.004
δ			0.44^{**}	0.78^{**}	0.60^{**}	0.58^{**}	0.04
Stigmasterol				0.64^{**}	0.54^{**}	0.52^{**}	−0.26
Campesterol					0.82^{**}	0.73^{**}	−0.32
β-Sitosterol						0.67^{**}	−0.19
Squalene							−0.17

** means highly significant correlation; * means significant correlation

Table 4.22 Quality evaluation indicators of different varieties of peanut oil

No.	Name	Sensory quality (Color)		Physicochemical and nutritional quality		Processing quality				
		Red	Yellow	Water and volatile matter	Unsaponifiable matter	Induction time	Peroxide value	Acid value	Iodine value	Saponification value
1	Zhonghua 8	−0.65	−11.00	−0.125	4.24	4.19	−3.231	−0.411	−88.13	−177.32
2	Shanhua 7	−0.40	−3.50	−0.045	19.76	5.76	−2.367	−0.358	−96.87	−176.17
3	Silihong	−0.60	−2.80	−0.080	12.10	5.19	−2.972	−0.280	−97.62	−170.64
4	Luhua 11	−0.70	−5.50	−0.105	9.74	4.67	−3.081	−0.273	−93.15	−175.63
5	Bianhua 3	−0.90	−9.25	−0.085	13.23	4.61	−2.201	−0.643	−93.21	−148.47
6	Haihua 1	−0.65	−4.40	−0.090	10.24	5.23	−4.040	−0.389	−90.17	−177.33
7	Shuangji 2	−0.40	−5.35	−0.050	9.670	4.45	−2.242	−0.271	−92.62	−170.97
8	Shanhua 9	−0.85	−3.35	−0.145	11.70	4.39	−2.699	−0.364	−98.06	−174.84
9	Fenghua 5	−0.45	−4.30	−0.200	17.91	3.74	−6.131	−2.574	−93.14	−176.72
10	White peanut	−0.60	−3.95	−0.080	10.95	4.62	−3.281	−0.328	−97.24	−175.48
11	Fenghua 1	−0.60	−4.40	−0.040	8.95	3.91	−4.573	−0.307	−96.24	−175.40
12	Fenghua 3	−0.70	−6.50	−0.080	6.08	3.91	−4.953	−0.235	−91.78	−177.17
13	Fenghua 4	−1.70	−13.00	−0.080	9.14	4.37	−3.203	−0.563	−88.68	−177.17
14	Xuhua 5	−0.70	−5.40	−0.070	7.70	3.81	−3.428	−0.429	−93.03	−166.78
15	Yuanhua 8	−0.80	−6.50	−0.090	9.00	3.73	−3.717	−0.361	−100.91	−186.26
16	Xuhua 13	−0.65	−7.35	−0.050	11.28	3.82	−4.269	−0.362	−101.69	−188.27
17	Xuhua 14	−0.70	−6.40	−0.055	11.57	4.27	−2.326	−0.230	−101.84	−186.26
18	Huayu 19	−0.55	−3.35	−0.070	11.13	4.71	−3.069	−0.381	−98.63	−183.62
19	Huayu 20	−0.45	−3.25	−0.050	10.66	4.63	−3.232	−0.406	−98.26	−181.98
20	Huayu 22	−0.60	−5.70	−0.040	8.01	4.79	−2.192	−0.668	−94.15	−186.53
21	Huayu 23	−0.60	−6.20	−0.075	6.34	5.63	−3.774	−1.048	−99.14	−183.26
22	Huayu 28	−0.60	−13.00	−0.065	9.70	3.94	−2.892	−0.526	−92.50	−181.51

(continued)

Table 4.22 (continued)

No.	Name	Sensory quality (Color)		Physicochemical and nutritional quality		Processing quality				
		Red	Yellow	Water and volatile matter	Unsaponifiable matter	Induction time	Peroxide value	Acid value	Iodine value	Saponification value
23	Huayu 31	−0.60	−5.35	−0.100	13.61	4.44	−2.724	−0.622	−100.53	−185.63
24	Baisha 1016	−0.55	−5.30	−0.055	18.83	4.36	−4.076	−0.540	−108.98	−187.57
25	Colorful peanut	−0.80	−5.20	−0.080	10.05	4.97	−2.276	−0.356	−103.60	−181.13
26	Black peanut	−0.55	−5.10	−0.075	17.21	4.81	−1.686	−0.477	−103.95	−185.91
27	034-256-1	−0.70	−6.45	−0.060	9.47	5.23	−1.613	−3.020	−96.15	−203.69
28	Ji 9814	−0.80	−10.45	−0.105	11.76	4.63	−2.727	−0.388	−108.61	−190.03
29	Yuhua 15	−0.70	−5.50	−0.085	8.22	3.59	−0.806	−1.201	−107.34	−217.13
30	Yuhua 9326	−0.55	−4.35	−0.080	10.66	4.04	−3.411	−0.375	−116.14	−179.48
31	Yuhua 9327	−0.90	−6.05	−0.070	12.23	4.31	−1.413	−0.372	−98.60	−203.19
32	Kainong 30	−0.50	−9.90	−0.045	10.02	4.05	−3.128	−0.377	−107.07	−188.67
33	Kainong 37	−0.80	−4.00	−0.105	17.32	4.49	−2.454	−0.422	−113.39	−193.59
34	Yuanza 9102	−0.50	−5.40	−0.045	18.65	3.70	−1.037	−0.316	−103.25	−186.92

4.3.2 Data Standardization

The quality of 34 varieties of peanut oil was changed so that the greater the indicator value, the better the quality, so as to reduce the effects of unit differences on the data.

4.3.3 Comprehensive Value of Peanut Oil

The normalized data in Sect. 4.3.2 was added equally and recorded as *Y*, all the data had been standardized and was not affected by the unit, and the data could represent the weight of each data, so the *Y* value was the actual value of overall effect of each indicator on peanut oil and *Y* was the comprehensive value of peanut oil quality.

4.3.4 Supervised Principal Component Regression Analysis Modeling

Significance analysis was carried out for the *Y* value in Sect. 4.3.3 and quality indicator of peanut, and the results showed that there were significant correlations between the following three indicators and *Y* at 0.05 level (Table 4.23).

The dependent variable for establishment of regression equation should conform to the normal distribution. The comprehensive value of peanut oil did not conform to the normal distribution, so the Box-Cox transformation between various quality indicators and comprehensive value of peanut oil was carried out to make the dependent variable conform to the normal distribution. Then the regression analysis was conducted again, and the dependent variable was put into Formula 4.1, and $\lambda = 1$ was obtained, namely, Formula 4.11:

$$Y1 = Y - 1 \tag{4.11}$$

The regression equations between Y_1 and various independent variables were established, and it could be known from the regression coefficients of regression equations that *P* value was significant at 0.05 level (Table 4.24). Therefore, each indicator was incorporated into the equation to obtain Formula 4.12.

$$\begin{aligned} Y_1 = \; & 5.9999982 - 1.3946738^{*}\text{standardized data of crude fat} \\ & +0.6473054^{*}\text{standardized data of oleic acid/linoleic acid} \\ & -1.9303204^{*} \end{aligned} \tag{4.12}$$

Change *Y*1 in the formula to *Y* and convert the standardized data to original data:

Table 4.23 Significant correlation between peanut quality and peanut oil quality

No.	Indicator	Significance
1	Crude fat	0.004
2	Oleic acid/linoleic acid	0.034
3	Polyunsaturated fatty acid	0.043

Table 4.24 Table of regression coefficients

Variable	Coefficient	*F* value	*P* value
Coefficient	5.9999982	173.27	0.0001
Crude fat X_1	−1.3946738	8.35	0.0072
Oleic acid/linoleic acid X_2	0.6473054	1.87	0.01822
Polyunsaturated fatty acid X_3	−1.9303204	16.45	0.0003

$$Y = -0.412547 \times \text{crude fat} + 40.560138 \times \text{oleic acid/linoleic acid} - 0.618986 \times \text{polyunsaturated fatty acid} \quad (4.13)$$

4.4 *Verification of Model*

The crude fat, oleic acid/linoleic acid, and unsaturated fatty acid content as well as the quality indicators of the peanut oil of the other 11 peanut varieties were determined. The peanut oil quality of 11 varieties was changed into that the greater the indicator value, the better the quality, and the data was standardized; the standardized data were added equally, and the sum was Y_3 after the addition; in addition, three indicators (crude fat, oleic acid/linoleic acid, and unsaturated fatty acid content) were put into Formula 4.13 for calculation to obtain the calculated value Y_2 and analyze the correlation coefficient of Y_3 and Y_2 and determine the applicability of the equation. The above Y_3 and Y_2 were fitted and the fitting results were shown in Fig. 4.5. The correlation coefficient was $R = 0.70$ which could be used to predict the comprehensive value of peanut oil of certain variety.

5 Relationship Between Peanut Raw Material Quality and Peanut Butter Quality

5.1 *Analysis of Quality Evaluation Indicator of Peanut Butter*

5.1.1 Correlation Analysis of Peanut Butter

Correlation analysis was carried out for 19 indicators of 20 varieties of peanut butter, and the results in Table 4.25 showed that there were significant negative correlations between fat, total sugar, hardness, viscosity, and peroxide value; the higher the fat content, the worse the hardness and viscosity of peanut butter; on the

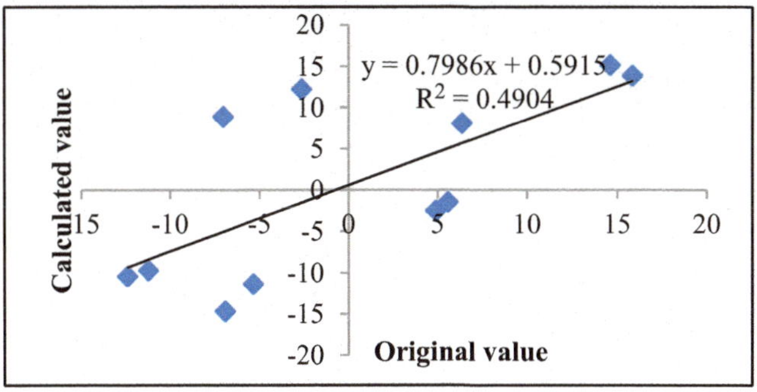

Fig. 4.5 Diagram of relationship between actual values and theoretical values of comprehensive values of peanut oil

contrary, the higher the total sugar content, the better the viscosity and hardness of peanut butter; therefore, the fat and total sugar might be important indicators affecting peanut butter. According to the correlation analysis, it could be seen that, except for water and ash, there were significant differences among all other indicators, so the removal of water, ash, total sugar, linoleic acid, hardness, and H value could play a role in the preliminary dimensionality reduction, which indicated that these indicators were closely related, and one indicator could control or improve the change of another indicator and these indicators were interacted and related. Therefore, according to the correlation analysis, the indicators which had great influences on quality of peanut butter were selected for principal component analysis.

5.1.2 Principal Component Analysis of Peanut Butter

The principal component analysis was carried out for 14 indicators of peanut butter. The results were shown in Table 4.26. Through the principal component analysis, it was found that the cumulative contribution rate of the first six principal components was 84.87%. According to the basic principle that if the cumulative contribution rate of principal component analysis was about 85%, the 14 original indicators were transformed into six new indicators, which played a role in dimensionality reduction. The relationship between various principal components and independent variables was calculated according to Formula 4.2. The results were shown in Table 4.27.

Table 4.25 Correlation analysis of 20 varieties of peanut butter

	Protein	Water	Ash	Fiber	Total sugar	Oleic acid	Linoleic acid	Hardness	Viscosity	L*	C*	H	POV	AV	IV	Eccentricity	Acceptability	Grain size
Fat	−0.280	−0.135	−0.008	−0.102	−0.094	0.205	−0.146	−0.497*	−0.512*	0.439	−0.031	0.343	−0.499*	0.145	−0.368	−0.175	0.078	−0.388
Protein	1	−0.069	−0.066	−0.352	−0.578**	0.186	0.036	−0.041	−0.305	−0.074	0.135	−0.110	0.240	0.203	−0.205	0.629**	−0.155	0.073
Water		1	−0.204	0.304	0.187	−0.073	−0.063	0.100	0.007	0.213	−0.413	0.336	−0.011	−0.171	0.232	−0.158	−0.145	−0.286
Ash			1	0.063	−0.146	0.275	−0.098	−0.092	0.099	−0.010	0.293	−0.082	0.330	−0.478*	−0.439	−0.072	0.200	−0.087
Fiber				1	0.279	−0.424	0.262	0.225	0.304	−0.072	−0.047	−0.045	−0.053	−0.094	0.363	−0.335	0.126	−0.059
Total sugar					1	−0.319	−0.262	0.554*	0.693**	−0.232	−0.206	−0.104	−0.212	−0.040	0.460*	−0.300	−0.238	0.185
Oleic acid						1	−0.664**	−0.421	−0.286	−0.089	0.415	−0.203	−0.245	−0.578**	−0.602**	0.030	0.226	−0.302
Linoleic acid							1	0.014	−0.057	0.302	−0.155	0.234	0.267	0.406	0.040	−0.200	0.074	0.168
Hardness								1	0.841**	−0.427	−0.045	−0.334	0.231	0.126	0.411	−0.011	−0.457*	0.427
Viscosity									1	−0.542*	0.118	−0.481*	0.154	−0.260	0.320	−0.223	−0.130	0.248
L*										1	−0.634**	0.929**	−0.181	0.387	−0.404	−0.158	−0.014	−0.330
C*											1	−0.835**	−0.136	−0.400	−0.146	0.056	0.270	0.193
H												1	−0.031	0.406	−0.197	−0.108	−0.160	−0.326
POV													1	−0.103	0.075	0.127	−0.037	0.077
AV														1	0.225	0.334	−0.425	0.333
IV															1	0.142	−0.324	0.466*
Eccentricity																1	−0.343	0.288
Acceptability																	1	−0.546*

** means highly significant correlation ($P < 0.01$); * means significant correlation ($P < 0.05$)

Table 4.26 Eigenvalues of correlation matrixes

No.	Eigenvalue	Difference	Variance contribution rate	Cumulative contribution rate
1	3.57872638	0.97419057	0.2556	0.2556
2	2.60453580	0.42393454	0.1860	0.4417
3	2.18060127	0.63411825	0.1558	0.5974
4	1.54648301	0.43314414	0.1105	0.7079
5	1.11333887	0.25508178	0.0795	0.7874
6	0.85825709	0.20942362	0.0613	0.8487

5.2 *Supervised Principal Component Regression Analysis Modeling*

5.2.1 Correlation Analysis Between Characteristics of Raw Material and Comprehensive Value of Peanut Butter

Correlation analysis between 26 characteristic indicators (fruit shape, hundred fruit weight, crude fat, crude protein, etc.) of 20 varieties of peanuts and comprehensive values of peanut butter were carried out (Table 4.28). It was found that there were significant correlations between crude fat, total sugar, α-VE, and other indicators and comprehensive values of peanut butter, and these indicators might be important indicators affecting the quality of peanut butter.

5.2.2 Establishment of Model by Principal Component Regression

Twenty varieties of peanut butter were randomly selected from 26 varieties of peanut butter to establish the model by using the principal component regression, and the remaining six varieties were used for model verification.

1. Selection of peanut characteristic indicators suitable for peanut butter processing

   According to the regression analysis idea, the regression coefficient significance analysis was carried out for the single peanut characteristic indicator and comprehensive values of peanut butter. It was found that there were significant correlations between ten indicators and comprehensive values of peanut butter at 0.05 level (Table 4.29).

2. Correlation analysis of peanut quality

   Correlation analysis was carried out for the ten peanut characteristic indicators selected (Table 4.30). Among the indicators with significant correlation at 0.05 level and the correlation coefficient which was greater than 0.05, one indicator was used to indicate another indicator. Therefore, the indicators remained were fat, protein, hundred kernel weight, O/L, and total V_E.

Table 4.27 Scores of principal components

	Y_1	Y_2	Y_3	Y_4	Y_5	Y_6	Y comprehensive value
Huayu 22 (Shandong)	0.544455	1.424542	0.442361	−1.03504	−0.86828	0.675809	0.390095861
Huayu 22 (Xin Jiang)	0.352268	−0.49874	1.926542	0.756663	1.856943	0.14871	0.633656275
Weihua 8	−0.20737	−0.07631	2.570857	−0.7874	1.696326	1.092927	0.528089056
13-2 butter	−0.47109	−0.28842	0.390777	−0.09968	−1.76047	0.832427	−0.25111122
13-3 butter	−0.53983	0.442675	−1.91325	1.508349	−0.25319	0.240797	−0.22672452
Guihua 22	−0.73864	−4.39009	0.730535	−0.35099	0.596101	−0.23562	−1.05735146
Guihua red 95	1.237874	−3.54726	0.792665	2.095584	−1.09101	−0.80466	−0.14656677
P905	−0.99311	2.079225	−1.84619	1.553272	1.049697	−0.42833	0.087298512
Jihuatian 1	5.447304	2.079225	0.983168	1.016169	0.825569	−0.67568	2.43754504
Luhua 17	−0.91509	0.587445	0.775902	1.043931	−1.44365	0.866665	0.058871554
Fenghua 1	0.96947	1.347147	1.314355	0.810797	−1.25529	1.188043	0.902281916
Guihua 17	−0.68713	−0.18973	−1.13966	−1.77407	0.416088	1.630226	−0.53199452
High linoleic acid	−0.19138	−1.0717	−2.79149	0.72717	0.770069	1.425548	−0.53517936
Xiaobaisha	2.725179	−0.00846	−1.09801	−1.91028	−0.07941	−0.48403	0.326196046
Fuhua 18	2.600902	−1.11889	−2.37104	−1.69145	−0.22392	−0.65491	−0.18567801
Fuhua 12	−1.72042	0.235312	1.50203	−1.18223	−0.04242	−0.68034	−0.39786657
Kai 17-15	−2.47119	1.315399	−0.86123	1.216125	1.186078	−0.63211	−0.39027232
Luhua 18	−2.17657	1.050334	0.514745	−1.39153	0.358066	−1.01807	−0.5519948
Yuhua 9414	−1.67924	0.184373	−0.12075	−0.24108	−0.70595	−1.36928	−0.68390956
Shangyan 9658	−1.08639	1.117402	0.197685	−0.26429	−1.03135	−1.11812	−0.2577866

Table 4.28 Correlation analysis between peanut quality and comprehensive value of peanut butter

Indicator	Correlation coefficient	Indicator	Correlation coefficient	Indicator	Correlation coefficient
Fruit shape	−0.415	Water	−0.353	C16:0	0.042
Red skin	0.123	Ash	−0.022	C18:0	0.284
Grain shape	0.447*	Total sugar	−0.623**	C18:1	−0.485*
Hundred fruit weight	0.469*	a-VE	−0.690**	C18:2	0.044
Hundred kernel weight	0.571*	r-VE	−0.055	C18:3	−0.239
Crude protein	0.546*	d-VE	−0.058	C20:0	0.789
Crude fat	0.648**	Total VE	0.590**	C22:0	0.547*
Crude fiber	0.103	C14:0	−0.201	O/L	0.789**

** means highly significant correlation ($P < 0.01$); * means significant correlation ($P < 0.05$)

Table 4.29 Regression significance analysis of peanut characteristics and comprehensive values of peanut butter

No.	Indicator	*P* value	No.	Indicator	*P* value
1	Grain shape	0.048	6	Total sugar	0.001
2	Hundred fruit weight	0.037	7	α-V_E	0.003
3	Hundred kernel weight	0.008	8	Total -V_E	0.001
4	Crude protein	0.002	9	C18:1	0.031
5	Crude fat	0.013	10	*O*/*L*	0.000

** means highly significant correlation ($P < 0.01$); * means significant correlation ($P < 0.05$)

3. Principal component analysis

Principal component analysis and dimensionality reduction were carried out for the five indicators selected (Table 4.31).

Through the principal component analysis, it was found that the cumulative contribution rate of the first four principal components was 91.07% and far greater than 85% (basic principle of principal component analysis). Therefore, the first four principal components could express the information of original principal component, and the original five indicators were transformed into four new indicators, which played a role in dimensionality reduction. The relationship between various principal components and independent variables was calculated according to Formula 4.9. The results were shown in Table 4.32.

4. Establishment of regression equation

According to the regression analysis, it was found that there were significant correlations between the first and second principal components among above four principal components and regression coefficients of comprehensive values of peanut butter at 0.05 level (Table 4.33). Therefore, the relationship between the peanut

Table 4.30 Correlation analysis of peanut characteristics

	Grain shape	Hundred fruit weight	Hundred kernel weight	Crude protein	Crude fat	Total sugar	a-VE	Total V_E	C18:1	*O*/*L*
Grain shape	1	0.438	0.547*	−0.032	0.102	−0.125	0.294	0.349	0.304	0.202
Hundred fruit weight		1	0.857**	0.433	−0.025	0.091	0.090	0.069	0.456*	0.344
Hundred kernel weight			1	0.360	−0.150	0.283	0.157	0.127	0.291	0.171
Crude protein				1	−0.462*	0.329	0.245	0.293	0.091	0.161
Crude fat					1	−0.535*	−0.380	−0.362	0.309	0.081
Total sugar						1	0.391	0.288	−0.359	−0.192
a-VE							1	0.957**	−0.216	−0.133
Total VE								1	−0.256	−0.158
C18:1									1	0.813**
O/L										1

** means highly significant correlation ($P < 0.01$); * means significant correlation ($P < 0.05$)

Table 4.31 Eigenvalues of correlation matrixes

No.	Eigenvalue	Difference	Variance contribution rate	Accumulative contribution rate
1	1.90697711	0.66011170	0.3814	0.3814
2	1.24686541	0.50031247	0.2494	0.6308
3	0.74655294	0.10298350	0.1493	0.7801
4	0.64356944	0.18753433	0.1287	0.9088
5	0.45603511	0.0912		1.0000

Table 4.32 Scores of principal components

	Y_1	Y_2	Y_3	Y_4	*Y* comprehensive value
Huayu 22 (Shandong)	1.422849	−0.07834	0.395133	−0.14673	0.390095861
Huayu 22 (Xin Jiang)	1.922936	−1.30312	0.540816	1.060003	0.633656275
Weihua 8	0.27784	0.813521	0.908521	−1.03018	0.528089056
13-2 butter	0.136496	0.179356	−0.34945	0.395513	−0.25111122
13-3 butter	0.467341	−0.21415	1.368177	0.546669	−0.22672452
Guihua 22	−2.82245	−0.63846	−0.13146	0.188228	−1.05735146
Guihua red 95	−1.14084	−0.85363	−1.1279	−0.1323	−0.14656677
P905	0.572133	2.789738	−1.03004	0.315175	0.087298512
Jihuatian 1	2.56908	−1.3793	−0.29872	0.312479	2.43754504
Luhua 17	0.097623	−0.58133	0.715267	0.815788	0.058871554
Fenghua 1	0.507068	−0.33639	−0.0537	−0.31381	0.902281916
Guihua 17	−2.89005	−0.8567	0.792554	1.330899	−0.53199452
High linoleic acid	0.783234	2.251682	−0.18679	1.183127	−0.53517936
Xiaobaisha	1.433051	−1.1649	−1.15937	−0.54153	0.326196046
Fuhua 18	−0.74522	−0.80518	−1.28909	−0.67719	−0.18567801
Fuhua 12	−0.16792	−0.27991	−0.63708	−1.08269	−0.39786657
Kai 17-15	−1.00661	0.890074	−1.06726	0.853202	−0.39027232
Luhua 18	−0.96895	0.144756	0.394789	−1.06759	−0.5519948
Yuhua 9414	−0.29282	0.688554	0.783395	−1.06225	−0.68390956
Shangyan 9658	−0.15478	0.73378	1.432213	−0.94682	−0.2577866

Table 4.33 Significance of regression coefficient

Variable	Coefficient	*P* value
Intercept	4.639E-7	1.000
PC1	0.385	0.001
PC2	−0.258	0.042

characteristic indicators and comprehensive values of peanut butter was established. The determination coefficient of the model was $R = 0.818$, and then the relationship between the indicators and comprehensive values of peanut butter was established. The results were shown in Formula 4.14.

$$\begin{aligned} Y\ \text{comprehensive value} = {} & 1.371001 - 0.00941 \\ & \times \text{hundred kernel weight} + 0.00852 \times \text{crude protein} \\ & - 0.01506 \times \text{crude fat} + 0.03546 \times \text{total } V_E + 0.044080 \\ & \times O/L \end{aligned} \tag{4.14}$$

5.2.3 Verification of Model

The remaining six varieties were used to validate the accuracy and promotion applicability of the model established. After the five indicators (hundred kernel weight, crude fat, crude protein, total V_E, and O/L) of six peanut varieties were put into the formula, the comprehensive values of peanut butter of six varieties were calculated, and the regression analysis was carried out for the calculated results of model and comprehensive values of peanut butter. The correlation coefficient was $R^2 = 0.788$ (Fig. 4.6).

6 Relationship Between Peanut Raw Material Quality and Export Peanut Quality

Peanut is an important oilseed crop in China and export agricultural product with advantage. Over the years, the peanut and peanut products in China had occupied an important position in the peanut export trade in the world (Yang 2009). The statistics

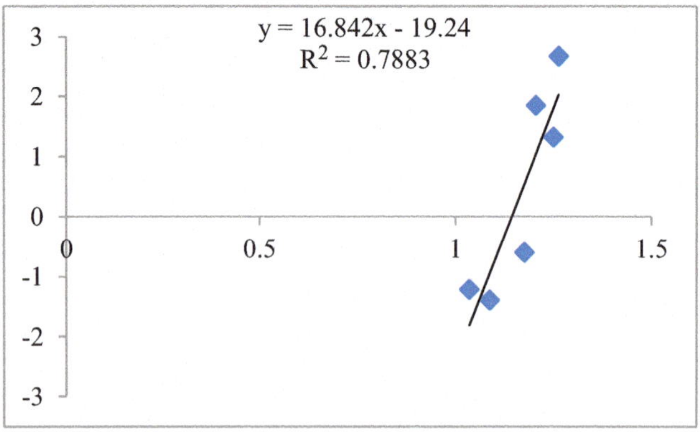

Fig. 4.6 References measured values versus predicted value

of the Food and Agriculture Organization of the United Nations (FAO) showed that the total export volume of unshelled, shelled, and processed peanuts in China reached 29.8%, 33.9%, and 44.6%, respectively, in the export volume of peanut and peanut products of the same kind in the world from 2000 to 2005 (Qu et al. 2008).

Peanut varieties have great impacts on peanut quality. The peanuts imported by the peanut import countries and regions were mainly used for food processing, not oil expression, so they often preferred to import certain varieties of peanuts (Holley and Hammons 1968; Worthington et al. 1972). In addition, with the development of the world economic level, the peanut import countries and regions have increased their requirements on the health, physicochemical, and nutritional indicators of import peanuts; they only are concerned about the size, color, water content, and other indicators of peanuts, but they had higher requirements on the aflatoxin content, sugar content, *O*/*L*, and other indicators now, which restricted the development of peanut export industry in China to a certain extent (Chen and Duan 1994; Jiang and Duan 1998).

Therefore, it was necessary to summarize the health indicators and physicochemical and nutritional indicators involved by the existing export peanut kernels and peeled peanuts, use the principal component analysis and other statistical processing, select the main indicators that affect the quality of export peanut, establish the quality evaluation model of export peanut, improve the quality control level of peanut export enterprises for raw materials, and promote the stable and healthy development of peanut export industry (Chen et al. 2007; Yin et al. 2011; Li et al. 2011).

6.1 Quality Evaluation Indicator System of Export Peanut Kernel

6.1.1 Quality Evaluation Indicator

The quality indicators of export peanut kernel mainly include six indicators: seed coat color, water content, aflatoxin content, oleic acid/linoleic acid ratio, imperfect grain content, and grade. According to the impacts of the indicators on export peanut kernel, a certain score was assigned to each indicator. The sum of scores of all indicators was the total score of the export quality of the variety of peanut. The detailed evaluation indicators and scoring standards were shown in Table 4.34.

6.1.2 Quality Evaluation Method

6.1.2.1 Seed Coat Color

Much attention is paid to the appearance of export peanut. Under normal circumstances, the outer seed coat is pink with bright color, without cracks or black halo

Table 4.34 Quality evaluation standards of export peanut kernel

Indicator	Evaluation standard	Evaluation score
Color (10 points)	Pink and glossy	8–10
	Reddish and light red	4–7
	Other colors and spots on surface	0–3
Water (Wang 2012)	15-water content × 1.5	0–15
Imperfect grain (Yin et al. 2011)	20-imperfect grain × 6.6	0–20
Aflatoxin (Wang 2012)	15-aflatoxin content × 1.5	0–15
Classification (Zhou et al. 2002)	24/28	25
	28/32	20
	34/38	15
	38/42	10
	>42	5
Oleic acid/linoleic acid (Wang 2012)	15× (oleic acid/linoleic acid-1)	15
Comprehensive evaluation (100)	Total score = color + water + imperfect grain + aflatoxin + classification + oleic acid/linoleic acid	100

spots, and the peanuts with orange-yellow inner seed coat are preferred by foreign customers. Therefore, the scores were given to the varieties according to the colors of the outer seed coat and inner seed coat, and the total score was 10 points.

6.1.2.2 Aflatoxin Content

Aflatoxin is an important factor affecting the export peanut kernel. The requirements for aflatoxin content in peanut kernels are different in different countries and regions. The EU required that the total content of aflatoxin should not exceed 4 μg/kg, Argentina required that the content of aflatoxin B1 should not exceed 5 μg/kg, Japan required that the content of aflatoxin B1 should not exceed 10 μg/kg, and Australia required that the total aflatoxin content should not exceed 15 μg/kg (Ding et al. 2011). However, more and more attention has been paid to the harm of aflatoxin to the human body. At present, the aflatoxin content in most of the peanut kernels of export enterprises in China does not exceed 10 μg/kg. Therefore, this book gave 15 points to the indicator (aflatoxin content), the score of aflatoxin content was calculated in accordance with the formula (15-aflatoxin content × 1.5), and the score was 0 when the aflatoxin content in peanut exceeded 10 μg/kg.

6.1.2.3 Oleic Acid/Linoleic Acid Ratio (*O*/*L* Ratio)

Much attention is paid to the *O*/*L* ratio of peanuts by the peanut import countries, and the *O*/*L* ratio of most of the peanut varieties in China is low, so the *O*/*L* ratio has been the main factor which restricts the peanut export in China and international trade competitiveness. This book gave 15 points to this indicator. Under normal conditions, the *O*/*L* ratio of peanut varieties in China is mostly above 1.0 and below 2.0. Therefore, the score value of this ratio was calculated according to the formula $15\times$ (*O*/*L* ratio-1).

6.1.2.4 Imperfect Grain

Imperfect grains include moldy grains, broken grains, insect-damaged grains, grains damaged by rats, etc. Under normal conditions, the peanut import countries require that the percentage of imperfect grains accounting for the total amount in peanuts should be less than 3%. This book gave the indicator 20 points and the corresponding score was calculated in accordance with the formula: 20 – imperfect grain content $\times$ 6.6.

6.1.2.5 Classification

At present, the export peanut kernel can be divided into two classifications: large peanut kernel and small peanut kernel. The export volume of large peanut kernels is great and accounts for more than 80% of total export volume. Therefore, when considering the classification scores of export peanuts, the large peanut kernels are dominated, supplemented by small peanut kernels.

The current large export peanut kernels can be graded as 24/28, 28/32, 34/38, and 38/42. The small peanut kernels can be graded as 25/35, 35/40, 40/50, 50/60, 60/70, and 70/80 (grains/ounce). After the investigation and research conducted on a number of peanut export enterprises in Shandong, it was found that the export volume of 24/28 grade large peanut kernels was the largest, secondarily that of 28/32 grade, thirdly that of 34/38 grade, and finally that of 38/42 grade. The export volume of small peanuts was significantly lower than that of large peanuts. The peanut specification was the main factor affecting the export, so this book gave the indicator 25 points. At the same time, according to the information obtained from investigation and research, certain scores were given to all grades in accordance with the export volume.

6.1.3 Correlation Between Peanut Raw Material Quality and Export Peanut Kernel Quality

The correlation between the quality indicators of peanut raw materials and quality of export peanut kernel was shown in Table 4.35. Water, ratio of oleic acid and

Table 4.35 Correlation analysis between quality indicators of peanut raw materials and export peanut kernels

	Water	Seed coat color	Imperfect grain	Aflatoxin	Classification	Oleic acid/linoleic acid	Total score
Hundred fruit weight	−0.112	−0.006	0.008	−0.039	0.612[b]	0.036	0.503[b]
Hundred kernel weight	−0.155	−0.029	0.048	−0.119	0.676[b]	0.028	0.536[b]
Number of a kilogram of fruits	0.082	0.089	−0.021	0.019	−0.397[b]	0.068	−0.267[a]
Number of a kilogram of kernels	0.102	0.053	−0.031	−0.140	0.259	−0.110	−0.141
Kernel yield	0.113	−0.037	−0.031	0.027	−0.177	−0.023	−0.155
Fat content	0.056	0.377[b]	−0.020	0.119	−0.206	−0.008	−0.048
Protein content	−0.125	−0.157	0.123	−0.106	−0.069	−0.272[a]	−0.214
Protein extraction rate	−0.048	0.076	0.022	0.076	0.075	−0.084	0.058
Nitrogen solubility index	−0.099	0.012	0.008	−0.218	−0.051	0.106	−0.035
Total sugar content	0.008	0.068	0.024	0.187	0.027	0.053	0.089
Polysaccharide content	−0.039	0.017	−0.002	0.192	−0.005	0.301	0.138
Ash	0.116	−0.236	−0.113	−0.169	0.095	−0.323[a]	−0.124
Oleic acid	−0.026	0.104	0.122	0.130	0.023	0.942[b]	0.448[b]
Linoleic acid	0.029	−0.071	−0.134	−0.101	−0.010	−0.954[b]	−0.430[b]
Oleic acid/linoleic acid	−0.047	0.102	0.146	0.090	0.014	0.981[b]	0.451[b]
Water	−1.000[b]	0.246	0.922[b]	0.125	0.175	0.091	0.258[a]
Seed coat color	−0.246	1.000	0.154	0.214	0.007	0.124	0.353[b]
Imperfect grain	0.924[b]	−0.152	−1.000[b]	−0.122	−0.168	−0.156	−0.269[a]
Aflatoxin content	0.125	−0.215	−0.125	−1.000	−0.011	−0.038	−0.210

[a] Significant correlation
[b] Extremely significant correlation

linoleic acid, seed coat color, imperfect grain, and aflatoxin content were not only the quality indicators of peanut raw materials but also the important quality indicators of export peanut kernels, so these five indicators were also included in the peanut quality indicators for correlation analysis.

The results showed that there were highly significant correlations between hundred fruit weight ($r = 0.503$), hundred kernel weight ($r = 0.536$), seed coat color ($r = 0.353$), oleic acid content ($r = 0.448$), linoleic acid content ($r = 0.430$), oleic acid/linoleic acid ratio, and comprehensive scores of export peanuts ($p < 0.01$); there were highly significant correlations between the number of a kilogram of fruits ($r = -0.267$), water content ($r = 0.258$), ratio of imperfect grains ($r = -0.269$*), and comprehensive scores of export peanut kernels ($p < 0.05$). There were highly significant correlations between hundred fruit weight ($r = 0.612$), hundred kernel weight ($r = 0.676$), the number of a kilogram of fruits ($r = -0.397$), and classification scores. There was a highly significant correlation between protein content ($r = -0.272$) and oleic acid/linoleic acid content and a highly significant correlation between water ($r = 0.922$) and imperfect grains; there was a significant correlation between ash ($r = -0.323$) and oleic acid/linoleic acid ratio.

6.1.4 Evaluation Indicator System

The 62 × 14 matrix was built by using the 14 indicators (hundred kernel weight (X_1), kernel yield (X_2), fat content (X_3), protein content (X_4), protein extraction rate (X_5), nitrogen solubility index (X_6), total sugar content (X_7), polysaccharide content (X_8), ash (X_9), oleic acid/linoleic acid (X_{10}), water (X_{11}), seed coat color (X_{12}), ratio of imperfect grain (X_{13}), and aflatoxin content (X_{14})) of 65 varieties of peanuts for principal component analysis. The results were shown in Table 4.36.

It could be seen from Table 4.36 that the information amount formed by the first six principal components accounted for 69.79% of the total information amount, and the original variable information was basically retained. The component matrix of six principal components was further extracted and the component matrix factor load was observed (Table 4.37).

It could be seen from Table 4.37 that the protein content (X_4), polysaccharide content (X_8), and oleic acid/linoleic acid ratio (X_{10}) played decisive roles in the principal component 1; the water (X_{11}) and imperfect grain (X_{13}) played decisive roles in the principal component 2; the hundred kernel weight (X_1) and fat content (X_3) played decisive roles in the principal component 3; the aflatoxin content (X_{14}) played a decisive role in the principal component 4; the kernel yield (X_2) played a decisive role in the principal component 5; and the total sugar content (X_7) played a decisive role in the principal component 6.

Through the principal component analysis and dimensionality reduction processing, 14 indicators of peanut raw materials were reduced to 10 indicators

Table 4.36 Explained total variance

Component	Initial eigenvalue			Extracted sum of squares loading		
	Total	Variance contribution rate%	Accumulative total%	Total	Variance contribution rate%	Accumulative total%
1	2.496	17.832	17.832	2.496	17.832	17.832
2	1.932	13.803	31.635	1.932	13.803	31.635
3	1.795	12.821	44.455	1.795	12.821	44.455
4	1.348	9.627	54.082	1.348	9.627	54.082
5	1.153	8.238	62.320	1.153	8.238	62.320
6	1.046	7.473	69.792	1.046	7.473	69.792
7	0.935	6.679	76.471			
8	0.836	5.974	82.446			
9	0.726	5.185	87.631			
10	0.572	4.089	91.720			
11	0.498	3.559	95.279			
12	0.374	2.671	97.950			
13	0.230	1.640	99.590			
14	0.057	0.410	100.000			

Table 4.37 Component matrix

Indicator		Principal component					
		1	2	3	4	5	6
Hundred kernel weight	X_1	0.397	0.194	0.714	−0.011	−0.111	−0.109
Kernel yield	X_2	0.042	−0.052	−0.407	0.261	0.522	−0.095
Fat content	X_3	0.166	0.079	−0.764	−0.086	0.006	−0.012
Protein content	X_4	−0.643	0.132	0.071	0.258	−0.096	−0.273
Protein extraction rate	X_5	−0.034	−0.034	0.050	−0.104	0.924	0.010
Nitrogen solubility index	X_6	0.046	0.131	−0.166	0.737	0.207	−0.035
Total sugar content	X_7	0.146	0.001	−0.113	−0.085	−0.09	0.877
Polysaccharide content	X_8	0.779	0.075	0.318	−0.111	−0.04	−0.098
Ash	X_9	−0.443	−0.125	0.414	0.072	0.139	0.558
Oleic acid/linoleic acid	X_{10}	0.722	0.036	−0.162	0.210	−0.037	0.022
Water	X_{11}	0.011	0.973	0.085	−0.005	−0.019	−0.047
Seed coat color	X_{12}	0.373	0.343	−0.095	−0.342	0.228	−0.159
Imperfect grain	X_{13}	−0.011	−0.944	0.037	−0.037	0.062	−0.007
Aflatoxin content	X_{14}	−0.108	−0.109	0.254	0.698	−0.206	−0.063

(hundred kernel weight, kernel yield, fat content, protein content, total sugar content, polysaccharide content, oleic acid/linoleic acid, water, imperfect grain, and aflatoxin content), and these ten indicators were taken as the quality evaluation indicators of export peanut kernels.

6.1.5 Quality Evaluation Model

6.1.5.1 Establishment of Model

With the hundred kernel weight, kernel yield, fat content, protein content, total sugar content, polysaccharide content, oleic acid/linoleic acid, water, imperfect grain, and aflatoxin content of peanut as the independent variables, and the total score of export peanut quality as the dependent variable (Y), the multiple linear regression equation was established by using the stepwise regression method (stepwise multiple regression).

The specific steps of stepwise regression method were as follows: calculate the contributions of the independent variables X and dependent variable Y, respectively, first, and select the independent variable with the largest contribution in accordance with the order from large to low to enter into the equation, and then recalculate the contributions of the independent variables X and dependent variable Y and put them into the equation again. Check whether the variables which had been in the equation do not have statistics significance any longer due to the introduction of new variables simultaneously. If they had no statistics significance surely, remove the first independent variable. Repeat the step until no variable in the equation could be removed and no variable outside the equation could be introduced. Finally, retain the hundred kernel weight, oleic acid/linoleic acid ratio, and aflatoxin content. The regression equation shown below was obtained:

$$Y = 26.291 + 0.234\,X1 + 14.609\,X10 - 2.964X14 \qquad (4.15)$$

The effectiveness of quality regression model of export peanut kernel was determined through the determination of determination coefficient (R^2) and F test. Therefore, the summary and variance analysis were carried out for the model. From the summary of the model (Table 4.38), it could be obtained that the square root (R) of the determination coefficient of the above custom regression model was 0.573, which showed that the model was well fitted with the data. The statistics of regression equation was $F = 22.366$, and the significance test result Sig. = 0.000 was less than 0.05, which indicated that the linear relation of multiple regression equation was significant. It could be seen that the comprehensive score of export peanut kernel quality predicted by the multiple linear regression model established in this book was effective (Table 4.39).

6.1.5.2 Verification of Model

Ten peanut varieties were selected outside the 65 peanut varieties used in the quality regression model of export peanut kernel, and they were Longhua 243, Yueyou 45, Yueyou 52, Yueyou 86, Zhanhua 82 (Oil), Heyou 11, Yuhua 15, Yuhua 9326, Guihua 771 and Yuanza 9307, respectively. The quality analysis

Table 4.38 Summary of quality model of export peanut kernel

Model	R	R^2	Adjustment R^2	Estimated standard error
1	0.757	0.573	0.547	5.59971

Table 4.39 Table of variance analysis

Model	Sum of squares	Degree of freedom	Average variance	F	Sig.
Regression	2103.959	3	701.320	22.366	0.000
Residual	1567.836	50	31.357		
Total	3671.795	53			

Table 4.40 Property analysis of ten peanut varieties

			O/L (X_{10})		
No.	Peanut variety	Hundred kernel weight/g (X_2)	O/L (X_{10})	Aflatoxin content μg/kg (X_{14})	Predicted score
1	Longhua 243	68.10	1.29	0.25	60.29
2	Yueyou 45	63.70	0.84	0.49	51.96
3	Yueyou 52	76.50	1.04	0.05	59.24
4	Yueyou 86	63.55	0.86	1.09	50.44
5	Zhanhua 82 (oil)	50.55	1.03	0.54	51.60
6	Heyou 11	65.85	1.02	1.51	52.13
7	Yuhua 15	99.30	1.19	1.95	61.19
8	Yuhua 9326	88.00	1.10	1.47	58.61
9	Guihua 771	65.20	0.96	0.16	55.11
10	Yuanza 9307	74.90	0.99	1.55	53.70

results of these ten kinds of peanut kernels are shown in Table 4.40. The determination results of three qualities of ten kinds of peanuts were put into Formula 4.15, and the predicted values were compared with the experimental values for verification. The results showed that the correlation coefficient between comprehensive score of export peanut kernel (Table 4.41) and predicted value of model was 0.871, and the significance was 0.001, which indicated that they were highly significantly correlated at 0.01 level (Table 4.42).

There were certain errors between the predicted values and actual comprehensive scores of ten varieties. The relative error of Yuanza 9307 reached 3.92%, and the errors of other varieties were all less than 3%, which indicated that the quality regression prediction model of export peanut kernel could effectively predict the quality of unknown varieties of export peanut kernels.

Table 4.41 Comprehensive scores of quality of ten kinds of peanut kernels

Peanut Variety	Seed coat color	Score of seed coat	Water/ %	Score of water	Imperfect grain	Score of imperfect grain	Aflatoxin	Score of aflatoxin	Grain/ ounce	Classification score	O/L	Score of *O/L*	Total score
Longhua 243	Light red	5	4.06	8.91	1.35	11.10	0.25	19.63	41.63	10.00	1.29	4.30	58.94
Yueyou 45	Light red	5	4.05	8.93	1.33	11.22	0.49	19.27	44.51	5.00	0.84	0.00	49.41
Yueyou 52	Light red	5	4.01	8.99	1.31	11.38	0.05	19.92	37.06	15.00	1.04	0.60	60.88
Yueyou 86	Light red	5	4.32	8.52	1.23	11.90	1.09	18.36	44.61	5.00	0.86	0.00	48.79
Zhanhua 82 (oil)	Light red	5	4.22	8.67	1.24	11.79	0.54	19.19	56.08	5.00	1.03	0.48	50.13
Heyou 11	Light red	5	4.08	8.88	1.34	11.14	1.51	17.74	43.05	5.00	1.02	0.30	48.06
Yuhua 15	Pink	8	4.07	8.90	1.33	11.25	1.95	17.07	28.55	10.00	1.19	2.91	58.12
Yuhua 9326	Pink	8	2.03	11.96	1.73	8.58	1.47	17.80	32.22	10.00	1.10	1.52	57.85
Guihua 771	Pink	8	2.00	12.00	1.78	8.26	0.16	19.76	43.48	5.00	0.96	0.00	53.02
Yuanza 9307	Light red	6	3.20	10.20	1.46	10.37	1.55	17.67	37.85	15.00	0.99	0.00	59.24

Table 4.42 Verification analysis of quality prediction model of export peanut kernel

Variety	Predicted value	Comprehensive value	Relative error (%)	Correlation coefficient	Significance level
Longhua 243	60.29	58.94	0.95	0.839	0.002
Yueyou 45	51.96	49.41	1.81		
Yueyou 52	59.24	60.88	1.16		
Yueyou 86	50.44	48.79	1.17		
Zhanhua 82 (oil)	51.60	50.13	1.04		
Heyou 11	52.13	48.06	2.88		
Yuhua 15	61.19	58.12	2.17		
Yuhua 9326	58.61	57.85	0.54		
Guihua 771	55.11	53.02	1.48		
Yuanza 9307	53.70	59.24	3.92		

6.2 Quality Evaluation Indicator System of Peeled Export Peanut Kernel

6.2.1 Quality Evaluation Indicator

The quality of peeled export peanut kernel mainly includes seven indicators: water content, aflatoxin content, oleic acid/linoleic acid ratio, imperfect grain content, red skin residual rate, half grain content, and grade. According to the impacts of the indicators on export peanut kernel, a certain score was assigned to each indicator. The sum of scores of all indicators was the total score of the export quality of the variety of peanut. The detailed evaluation indicators and scoring standards were shown in Table 4.43.

6.2.2 Quality Evaluation Method

6.2.2.1 Water

The water content of peanut kernel was determined by using the method of weight change determination by conventional heating, and 15 points were given to this indicator.

6.2.2.2 Imperfect Grain

Imperfect grains included moldy grains, broken grains, insect-damaged grains, grains damaged by rats, etc. Under normal conditions, the peanut import countries required that the percentage of imperfect grains accounting for the total amount in

Table 4.43 Quality evaluation standard of peeled export peanut kernel

Indicator	Sensory evaluation standard	Evaluation score
Water (Wang 2012)	15-water content × 3	0–15
Imperfect grain (Wang 2012)	15- imperfect grain × 5	0–15
Half grain	10-proportion of half grain × 0.5	0–10
Aflatoxin (Yin et al. 2011)	20- aflatoxin content × 4	0–20
Classification (Wang 2012)	25/29	15
	29/33	11
	35/39	13
	39/43	9
	41/46	9
	46/51	8
	51/61	5
Oleic acid/linoleic acid	10× (oleic acid/linoleic acid-1)	15
Red skin residual rate (Wang 2012)	15-residual seed coat rate *7.5	15
Comprehensive evaluation (100)		100

peanuts should be less than 3%. This book gave this indicator 15 points and the corresponding score was calculated in accordance with the formula: 15-imperfect grain content × 5.

6.2.2.3 Half Grain

The indicator was determined according to the counting method, and 10 points were given to this indicator because the requirements of importer for this indicator were not high.

6.2.2.4 Aflatoxin

The aflatoxin in peanut is generally concentrated on the red skin. After the peanut is peeled, its aflatoxin content will be significantly reduced. Therefore, the peanut import countries and regions have relatively high requirements for the aflatoxin content in peeled peanut kernels, and the content should not be higher than 5 μg/kg. Therefore, this book gave 20 points to the indicator (aflatoxin content), the score of aflatoxin content was calculated in accordance with the formula (20-aflatoxin content × 4), and the score was 0 when the aflatoxin content in peanut exceeded 5 μg/kg.

6.2.2.5 Classification

At present, the peeled export peanut kernel can be divided into two classifications: large peanut kernel and small peanut kernel. The export volume of large peanut kernels is great and accounts for more than 80% of total export volume. Therefore, when considering the classification scores of export peanuts, the large peeled peanut kernels are dominated, supplemented by small peeled peanut kernels.

The current large and small peeled export peanut kernels can be graded as 25/29, 29/33, 35/39, 39/43, 41/46, 46/51 and 51/61 (grains/ounce). After the investigation and research conducted on a number of peanut export enterprises in Shandong, it was found that the export volume of 25/29 grade large peanut kernels was the largest, secondarily that of 35/39 grade, thirdly that of 29/33 grade, and that of other grades is less. Therefore, this book gave this indicator 15 points. At the same time, according to the information obtained from investigation and research, certain scores were given to all grades in accordance with the export volume.

6.2.2.6 Oleic Acid/Linoleic Acid Ratio

Much attention is paid to the oleic acid/linoleic acid ratio by the peanut import countries and the oleic acid/linoleic acid ratio of most of the peanut varieties in China is low, so the oleic acid/linoleic acid ratio has been the main factor which restricts the peanut export in China and international trade competitiveness. This book gave 10 points to this indicator. Under normal conditions, the oleic acid/linoleic acid ratio of peanut varieties in China is mostly above 1.0 and below 2.0. Therefore, the score value of this ratio was calculated according to the formula 10× (oleic acid/linoleic acid-1).

6.2.2.7 Red Skin Residual Rate

Under normal conditions, the importers require that the indicator should not exceed 2%. Therefore, this book gave 15 points to this indicator, the score was calculated in accordance with the formula 15-residual seed coat rate × 7.5, and the score was determined as 0 if this indicator exceeded 2%.

6.2.2.8 Total Score of Peeled Export Peanut Kernel Quality

The sum of the scores of the above seven indicators was the total score of this variety which could be used for correlation analysis.

6.2.3 Correlation Between Peanut Raw Material Quality and Peeled Peanut Kernel Quality

The correlation analysis results (Table 4.44) of peanut raw material quality and peeled peanut kernel quality showed that there were significant positive correlations between hundred fruit weight ($r = 0.350$) and hundred kernel weight ($r = 0.396$) and the total score of peeled peanut kernel quality; there were negative correlations between the number of a kilogram of fruits ($r = -0.387$), proportion of imperfect

Table 4.44 Correlation between peanut raw material quality and peeled peanut kernel quality

Indicator	Water content	Proportion of imperfect grain	Proportion of half grain	Aflatoxin content	Classification score	Oleic acid/linoleic acid ratio	Red skin residual rate	Total score
Hundred fruit weight	0.016	−0.049	−0.043	0.114	0.823[b]	0.086	0.125	0.350[b]
Hundred kernel weight	0.014	−0.046	−0.058	0.095	0.891[b]	0.097	0.123	0.396[b]
Number of a kilogram of fruits	0.050	0.139	0.005	−0.005	−0.613[b]	0.038	0.027	−0.387[b]
Number of a kilogram of kernels	0.143	−0.117	−0.165	−0.038	−0.249	−0.110	0.163	−0.122
Kernel yield	−0.072	0.105	0.056	−0.138	−0.146	−0.091	−0.092	0.015
Fat content	0.053	0.003	−0.014	−0.139	−0.176	−0.021	0.079	−0.097
Protein content	0.097	−0.084	−0.040	−0.114	−0.106	−0.129	−0.085	−0.051
Protein extraction rate	0.010	−0.155	−0.041	−0.171	−0.014	−0.049	−0.158	0.185
Nitrogen solubility index	−0.088	−0.047	0.065	0.018	−0.024	0.198	−0.123	0.125
Total sugar content	0.076	0.033	−0.046	0.153	−0.122	0.045	0.090	−0.196
Polysaccharide content	0.054	0.303[a]	−0.064	−0.105	0.047	0.301[a]	0.022	0.201
Ash	−0.009	0.001	0.046	0.071	−0.055	−0.203	−0.162	−0.083
Oleic acid	0.127	0.067	−0.155	0.130	0.022	0.981[b]	0.149	0.214
Linoleic acid	−0.097	0.011	0.123	−0.197	0.013	−0.986[b]	−0.127	−0.207
Oleic acid/linoleic acid	0.105	0.041	−0.138	0.161	0.017	1.000[b]	0.142	0.229
Water	0.471[b]	−0.135	−0.469[b]	−0.220	0.104	0.049	0.428[b]	−0.095
Imperfect grain	−0.015	−1.000[b]	−0.017	0.050	−0.037	0.036	−0.107	−0.325[a]
Aflatoxin	0.197	0.049	−0.241	−1.000[b]	0.132	0.155	0.064	−0.257[a]

[a] Significant correlation
[b] Extremely significant correlation

grain ($r = -0.325$), and aflatoxin content ($r = -0.257$) and the total score of export peanut kernel quality.

There were significant correlations between the hundred fruit weight ($r = 0.823$), hundred kernel weight ($r = 0.891$), and the number of a kilogram of fruits ($r = -0.613$) of peanut raw materials and the classification score of peeled peanut kernel, respectively. There were highly significant correlations between the oleic acid content ($r = 0.981$) and linoleic acid content ($r = -0.986$) of peanut raw materials and the oleic acid/linoleic acid ratio of peeled peanut kernel. There were highly significant correlations between the water content of peanut raw materials and water content ($r = 0.471$), proportion of half grain ($r = -0.469$), and red skin residual rate ($r = 0.428$) of peeled peanut kernel, respectively.

The raw materials of export peanut kernels and peeled peanut kernels were the same, so the method was the same as that in Sect. 6.1.4 during the dimensionality reduction, and the results obtained were consistent. Therefore, 14 indicators of peanut raw materials were reduced to 10 indicators (hundred kernel weight, kernel yield, fat content, protein content, total sugar content, polysaccharide content, oleic acid/linoleic acid, water, imperfect grain and aflatoxin content). These ten indicators were taken as the processing characteristic and quality evaluation indicators of peeled peanut kernels.

6.2.4 Establishment of Model

With the hundred kernel weight (X_1), kernel yield (X_2), fat content (X_3), protein content (X_4), total sugar content (X_5), polysaccharide content (X_6), oleic acid/linoleic acid ratio (X_7), water (X_8), imperfect grain (X_9), and aflatoxin content (X_{10}) of peanut raw materials as the independent variables, and the total score of peeled peanut kernel quality as the dependent variable (Y), the multiple linear regression equation was established by using the stepwise regression method. The results showed that the hundred kernel weight (X_1), oleic acid/linoleic acid ratio (X_7), and aflatoxin content (X_{10}) were included (Table 4.45) to form the final regression equation:

$$Y = 30.839 + 0.091X_1 + 5.648X_7 - 1.909X_{10} \tag{4.16}$$

From the table of regression coefficients (Table 4.45), it could be seen that the significant levels of hundred kernel weight, oleic acid/linoleic acid ratio and aflatoxin content were less than 0.05, which indicated that the independent variables and dependent variables had significant impact effects and there was a significant linear regression relationship.

The effectiveness of quality regression model of peeled peanut kernel was determined through the F test. Therefore, the variance analysis was carried out for the model. From the variance analysis (Table 4.46), it could be obtained that the statistics of regression equation was $F = 5.967$ and the significance test result Sig. = 0.001, which indicated that the linear relation of multiple regression equation was significant. It could be seen that the comprehensive score of peeled peanut

Table 4.45 Table of quality regression model coefficient of peeled peanut kernel

Model	Nonstandard coefficient		Standard coefficient		
	Regression coefficient	Standard error	Trial	*t*	Sig.
Constant	30.839	4.110		7.503	0.000
Hundred kernel weight	0.091	0.032	0.341	2.796	0.007
Oleic acid/linoleic acid ratio	5.648	2.584	0.266	2.186	0.034
Aflatoxin content	−1.909	0.921	0.253	2.073	0.043

Table 4.46 Table of variance analysis

Model	Sum of squares	Degree of freedom	Average variance	*F*	Sig.
Regression	401.317	3	133.772	5.967	0.001
Residual	1120.925	50	22.419		
Total	1522.242	53			

Table 4.47 Predicted scores of ten peanut varieties

Peanut variety	Hundred kernel weight/g	*O*/*L*	Aflatoxin content μg/kg	Predicted score
Longhua 243	68.10	1.29	0.25	43.83
Yueyou 45	63.70	0.84	0.49	40.42
Yueyou 52	76.50	1.04	0.05	43.58
Yueyou 86	63.55	0.86	1.09	39.38
Zhanhua 82 (oil)	50.55	1.03	0.54	40.24
Heyou 11	65.85	1.02	1.51	39.71
Yuhua 15	99.30	1.19	1.95	42.90
Yuhua 9326	88.00	1.10	1.47	42.26
Guihua 771	65.20	0.96	0.16	41.89
Yuanza 9307	74.90	0.99	1.55	40.29

kernel quality predicted by the multiple linear regression model established in this book was effective.

6.2.5 Verification of Model

Ten peanut varieties were selected outside the 62 peanut varieties used in the quality regression model of peeled export peanut kernel, and they were Longhua 243, Yueyou 45, Yueyou 52, Yueyou 86, Zhanhua 82 (oil), Heyou 11, Yuhua 15, Yuhua 9326, Guihua 771, and Yuanza 9307, respectively. The quality analysis results of these ten peanut varieties were shown in Table 4.47. The determination results of three qualities of ten kinds of peanuts were put into Formula 4.16. The predicted values obtained were compared with the experimental values for verification (Table 4.48). The results showed that the relative error between the

Table 4.48 Actual scores of ten peanut varieties

Peanut variety	Water/%	Score of water	Proportion of imperfect grain/%	Score of imperfect grain	Proportion of half grain/%	Score of half grain	Aflatoxin content μg/kg	Score of aflatoxin content	Classification	Classification score	*O*/*L*	Score of *O*/*L*	Red skin residual rate/%	Red skin residual rate score	Total score
Longhua 243	4.28	2.16	0.88	10.60	13.35	3.33	0.25	10.68	41.85	9	1.29	9.71	1.59	3.075	56.87
Yueyou 45	4.23	2.31	0.97	10.15	13.55	3.23	0.49	7.68	44.74	9	0.84	10.16	1.51	3.675	56.56
Yueyou 52	4.29	2.13	0.73	11.35	13.65	3.18	0.05	8.08	37.25	13	1.04	9.96	1.62	2.85	62.27
Yueyou 86	4.55	1.35	1.03	9.85	12.95	3.53	1.09	7.44	44.85	9	0.86	10.14	1.74	1.95	51.46
Zhanhua 82 (oil)	4.48	1.56	1.12	9.40	12.78	3.61	0.54	9.36	56.38	5	1.03	9.97	1.69	2.325	49.70
Heyou 11	4.19	2.43	0.99	10.05	13.98	3.01	1.51	10.2	43.28	9	1.02	9.98	1.42	4.35	52.78
Yuhua 15	4.15	2.55	1.07	9.65	14.23	2.89	1.95	6.44	28.70	15	1.19	9.81	1.43	4.275	56.37
Yuhua 9326	4.13	2.61	0.95	10.25	13.87	3.07	1.47	8.36	32.39	11	1.10	9.90	1.16	6.3	57.24
Guihua 771	4.10	2.70	0.88	10.60	13.69	3.16	0.16	8.68	43.71	9	0.96	10.04	1.61	2.925	57.78
Yuanza 9307	4.14	2.58	1.05	9.75	13.89	3.06	1.55	8.2	38.05	13	0.99	10.01	1.43	4.275	56.47

Table 4.49 Verification of quality prediction model of peeled peanut kernel

Variety	Predicted value	Comprehensive score	Relative error (%)	Correlation coefficient	Significance level
Longhua 243	43.83	56.87	9.22	0.741	0.014
Yueyou 45	40.42	56.56	11.41		
Yueyou 52	43.58	62.27	13.21		
Yueyou 86	39.38	51.46	8.54		
Zhanhua 82 (oil)	40.24	49.70	6.69		
Heyou 11	39.71	52.78	9.24		
Yuhua 15	42.90	56.37	9.52		
Yuhua 9326	42.26	57.24	10.59		
Guihua 771	41.89	57.78	11.23		
Yuanza 9307	40.29	56.47	11.44		

comprehensive score of peeled export peanut kernel and predicted value of model was about 10%, the correlation coefficient was 0.741, and the significance was 0.014, which indicated that they were highly significantly correlated at 0.01 level (Table 4.49).

References

Bair E, Tibshirani R. Semi-supervised methods to predict patient survival from gene expression data [J]. PLoS Biol. 2004;2(4):0511–22.

Bair E, Hastie T, Paul D, et al. Prediction by supervised principal components [J]. J Am Stat Assoc. 2006;101(473):119–37.

Brown DF. Effect of variety growing location and their interaction on the fatty acid composition of peanuts[J]. J Food Sci. 1975;40:1055–60.

Chen YS, Duan NX. Study on main quality shape of peanut germplasm resources in Fujian [J]. J Peanut Sci. 1994;4:9–11.

Chen SL, Li YR, Cheng ZS, et al. Principal component analysis and cluster analysis on biological characters of the peanut cultivars [J]. J Peanut Sci. 2007;36(2):28–34.

Ding XX, Li PW, Zhou HY, et al. Effect of aflatoxin maximum limit standards on consumer dietary risk and peanut industry in China [J]. Chin J Oil Crop Sci. 2011;33(2):180–4.

Firestone D. Worldwide regulation of frying fats and oils [J]. Inform. 1993;4:1366–71.

Fitzmaurice GM, Lipsitz SR, Parzen M. Approximate median regression via the box-cox transformation [J]. Am Stat. 2007;61(3):233–8.

Grosso NR, Lamarque A, Maestri DM, et al. Fatty acid variation of runner peanut (Arachis hypogaea L.) among geographic localities from córdoba (Argentina) [J]. J Am Oil Chem Soc. 1994;71(5):541–2.

Grosso NR, Nepote V, Guzmán CA. Chemical composition of some wild peanut species (Arachis hypogaea L.) seeds [J]. J Agric Food Chem. 2000;48:806–9.

Holley KT, Hammons RO. Strain and seasonal effects on peanut characteristics [J]. Univ Ga Coll Agric Exp Sta Res Bull. 1968;32:1–27.

Jiang HF, Duan NX. Correlation analysis of peanut grease quality and oil content, oleic acid and linoleic acid content [J]. J Peanut Sci. 1993;2:5–6.

Jiang HF, Duan NX. Ren Comprehensive evaluation of peanut germplasm resources [J]. Chin J Oil Crop Sci. 1998;20(3):31–5.

Jolliffe IT. Principal component analysis[M]. 2nd ed. Softcover of orig. ed. New York: Springer; 2002.

Li HY. Study on the steady state of polyunsaturated fatty acids in infant formula [D]. Nanchang: Nanchang University; 2009.

Li R, Zhao SL, Xia QY, et al. Comparison of fat solubility and antioxidation of conjugated Linoleyl B-Sitosterol and B-Sitosterol [J]. J Chin Cereals Oils Assoc. 2009;24(5):91–4.

Li HY, Zheng ZY, Li J, et al. Study on oxidative stability of plant oils with different fatty acid composition [J]. Sci Technol Food Ind. 2010;31(1):173–82.

Li QH, Huang JT, Chen HL, et al. Principal component and genetic distance determine in 27 peanut germplasm resources [J]. J Plant Genet Resour. 2011;12(4):519–24.

Liu Q, Yao H.Y. Antioxidants in oilseeds [J]. China Food Addit, 2006(2): 95–99.

Liu XH, Xu XZ, He XK, et al. Study on the application of supervised principal component regression procedure to near-infrared spectroscopy quantitative analysis [J]. Spectrosc Spectr Anal. 2009;29(11):2959–61.

Liu C, Gong AN, Liu L, et al. Reviewed on the nutrient and processed products of oat[J]. Aem Prod Proc. 2015;4:67–70.

Lopez Y, Smith OD, Senseman SA, et al. Genetic factors influencing high oleic acid content in Spanish market-type peanut cultivars [J]. Crop Sci. 2001;41:51–6.

Misra JB. A mathematical approach to comprehensive evaluation of quality in groundnut [J]. J Food Compos Anal. 2004;17:69–79.

Özcan M, Seven S. Physical and chemical analysis and fatty acid composition of peanut, peanut oil and peanut butter from ÇOM and NC-7 cultivars [J]. Grasas Aceties. 2003;54:12–8.

Prakash V, Narasinga RMS. Physicochemical properties of oilseed proteins[J]. Crit Rev Biochem Mol Biol. 1986;20(3):286–98.

Qu BX, Luo QY, Zhang Q, et al. Development of Chinese peanut industry and the safety of vegetable oil supply strategies [J]. Food Nutr China. 2008;11:13–5.

Roberts S, Martin M. Using supervised principal components analysis to assess multiple pollutant effects[J]. Environ Health Perspect. 2006;114(12):1877–82.

Saio K, Watanabe T. Differences in functional properties of 7S and 11S soybean proteins[J]. J Texture Stud. 1978;9:135–57.

Savage GP, Keenan JI. The composition and nutritive value of groundnut kernels. In: Smartt J, editor. The groundnut crop: a scientific basis for improvement. London: Chapman & Hall; 1994. p. 173–213.

Shin EC, Pegg RB, Phillips RD, et al. Commercial runner peanut cultivars in the USA: fatty acid composition[J]. Eur J Lipid Sci Technol. 2010;112:195–207.

Tong W, Kevin BH, Robert M. Antioxidant activity of phytosterols, oryzanol, and other phytosterol conjugates[J]. J Am Oil Chem Soc. 2002;79(12):1201–6.

Tong YW, Chen FX, Yang BT. Study on oxidative stability of different edible vegetable oils [J]. China Oils Fats. 2009;34(2):31–34.

Wang L. Study on processing characteristics and quality evaluation of peanut protein [D]. Beijing: Chinese Academy of Agricultural Sciences; 2012.

Warthington RE, Hammons RO, Allison JR. Varietal differences and seasonal effects on fatty acid composition and stability of oil from 82 peanut genotypes[J]. J Agric Food Chem. 1972;20:727–30.

Xu L, Wang R, Cao XB, et al. Study on the variation of anti-oxidates in olive oil [J]. J Henan Univ Technol. 2007;28(2):38–41.

Yan HC. A study on vegetable oil blends [J]. Food Chem. 1998;62(2):191–5.

Yang J. Development of peanut industry in China and its countermeasures [J]. Food Nutr China. 2009;1:17–9.

Yang XQ, Zhang SH, Li Y. Extraction and isolation of peanut 2S protein and its partial properties [J]. J South China Univ Technol (Nat Sci Ed). 1998;26(4):1–5.

Yin DM, Zhang XG, Wang Y, et al. Principal component analysis and comprehensive evaluation on quality traits of peanut parents [J]. J Plant Genet Resour. 2011;12(4):507–12.

Young CT, Waller GR, Matlock RS, et al. Some environmental factors affecting free amino acid composition in six varieties of peanuts[J]. J Am Oil Chem Soc. 1974;51(6):265–8.

Zhang JS. Effects of fatty acid composition and endogenous antioxidants on the stability of different peanut oils [D]. Beijing: Chinese Academy of Agricultural Sciences; 2012.

Zhang WH, Shi B. Theoretical study on the relationship between structure and autoxidation of unsaturated fatty acids [J]. Leather Sci Eng. 2009;19(4):5–9.

Zhang HR, Wang Q. Study on correlation between fatty acids and oxidative indexes in early stage of fatty oxidation [J]. Food Sci Technol. 2010;35(3):134–7.

Zhang JS, Wang Q, Liu HZ, et al. Comparative analysis of tocopherol and Phytosterol composition of peanut cultivars from different regions [J]. Food Sci. 2012;33(22):191–5.

Zhou HL, Chen JL, Zhang XS, et al. Study of Oxidating mechanism of unsaturated fatty acid [J]. West Leather. 2002;4:23–7.

Chapter 5
Peanut Processing Suitability Evaluation Standards

Developed countries have attached great importance to the research and formulation of peanut processing suitability evaluation standards. According to the consumption habits of peanuts and their processed products, the United States and India pay high attention to the selection of peanut varieties suitable for candy processing and research of suitability evaluation indicators. Misra (2004) first proposed the evaluation indicators of peanut varieties suitable for candy processing: slender varieties with taper ends, an average grain weight of about 0.55 g, pink seed kernel or light brown seed coat, high protein content (more than 25%), low fat content (less than 45%), and high sucrose content (more than 5%). The edible peanuts in China with protein content of more than 26%, 23%–26%, and less than 23% which have conventional sensory quality are divided into Class I, Class II, and Class III (edible peanut-NY/T1067-2006); the peanuts for oil with oil content of more than 51.0%, 48.0%–51.0%, and less than 48.0% which have conventional sensory quality are divided into Class I, Class II, and Class III (Peanuts for oil-NY/T1068-2006). There are no relevant provisions on the evaluation standards which affect the peanut protein function characteristic, stability of peanut oil and quality indicators of export peanut. Wang et al. (2012) classified the protein gelation and solubility of different peanut varieties and quality characteristics of peanut oil and export peanut by using K-means cluster analysis (Ramesh et al. 2013; Kanungo et al. 2002). Based on the evaluation model of various kinds of peanut products, the quality evaluation standards of peanuts suitable for processing of gel-type and soluble protein, peanut oil, and export peanut were established, the processing suitability of different varieties of peanuts was classified, and the results were validated to select the special varieties suitable for processing of gel-type protein and soluble protein, peanut oil, and export peanut so as to provide the basis for processing and utilization of peanut varieties.

Q. Wang, *Peanut Processing Characteristics and Quality Evaluation*,
DOI 10.1007/978-981-10-6175-2_5

1 Quality Evaluation Standards of Peanuts Suitable for High-Gelation Protein Processing

Gelation is one of the main functional properties of protein, and it is also the functional property with the most market development prospect in practical application. Among a number of soybean protein products, the gel-type soybean protein is the soybean protein product with the maximum yield in the domestic market and is widely used as an additive in the ham sausage production. The research and development of gelation of peanut protein has just begun relative to soybean protein. The gelation property of peanut protein is closely related to the composition, structure, and subunit content of protein which are determined by varieties. Different peanut varieties have different protein composition and structures and different protein function properties. In recent years, the research on quality evaluation technologies, methods, and standards of peanuts suitable for gel-type protein processing and the protein composition, subunit content, and protein function properties of different peanut varieties have become the focus of research in this field. This section focused on the introduction of quality evaluation standards of peanuts suitable for gel-type protein processing, including theoretical analysis evaluation standards and practical application evaluation standards, so as to provide the basis for the selection of peanut varieties or raw materials for gel-type peanut protein.

1.1 Theoretical Analysis Evaluation Standards

1.1.1 Gelation Classification Standards

Cluster analysis was carried out for the gelation of the peanut varieties used for model establishment and validation in Sect. 1, Chap. 4, the classification number was three, and the categories were determined according to the Formulas 5.1 and 5.2.

The objective function used the square error criterion, i.e.:

$$E = \sum \sum |P - m_i|^2 \tag{5.1}$$

where E is the sum of square errors of various cluster objects, P is the cluster object, and m_i is the mean of various cluster objects of C_i, i.e.:

$$m_i = \frac{\sum p \in C_i P}{|C_i|} \tag{5.2}$$

Table 5.1 Protein gelation suitability

Classification	Standard	Number of samples	Sample name
Suitable	≥1.08	14	Luhua 11, Shuangji 2, Bianhua 3, Fenghua 1, Kainong 30, Fenghua 3, Shanhua 7, Minhua 9, Zhanhua 82, Yuhua 15, Haihua 1, Honghua 1, Ji 9814, Yueyou 14
Basically suitable	0.85–1.08	34	034-256-1, Shanhua 9, Huayu 16, Luhua 9, Kainong 37, Lufeng 2, Luhua 14, Zhonghua 8, Guihua 771, Shanyou 250, Xuhua 13, Qinglan 8, Silihong, Fenghua 5, Huayu 23, Zhengnong 7, Zhongnong 108, Huayu 8, Yuanhua 8, Hongguan, Hua 17, Pearl Red, Huayu 20, Haiyu 6, Yuhua 9327, Longhua 243, Heyou 11, Black Peanut, Zhonghua 15, Yueyou 86, White Peanut, Yuhua 9326, Quanhua 551, Xuhua 5
Unsuitable	≤0.85	13	Yuanza 9307, Xuhua 15, Baisha 1016, Yuanza 9102, Yueyou 45, Fenghua 4, Colorful Peanut, Huayu 28, Huayu 31, Fenghua 6, Luhua 15, Zhonghua 4, Xianghua 509–77

where $|c_i|$ refers to the number of cluster objects of C_i, the computation complexity of K-means cluster method is $O(knt)$, k is the number of clusters, n is the number of cluster object samples, and t is the number of iterations.

From the analysis results in Chap. 4, it was known that the larger the comprehensive value of gelation, the better; the peanut varieties were divided into three categories (suitable, basically suitable, and unsuitable) (Table 5.1) as the basis for quality evaluation standards of peanuts suitable for gel-type protein processing.

1.1.2 Weight of Evaluation Indicator

1.1.2.1 Weight of Indicators Calculated by Using Regression Coefficient

Formula 5.3 was the normalized formula for the data in Formula 4.3, so the results could reflect the relationship between the indicators better.

$$\begin{aligned} Y_1 = {} & 1.125720 - 0.029940 \times \text{fruit shape} - 0.006180 \times \text{crude protein} \\ & - 0.025260 \times \text{crude fibre} - 0.20712 \times \text{glycine} + 0.043372 \times \text{cystine} \\ & - 0.087110 \times \text{leucine} + 0.016510 \times \text{arginine} + 0.005285 \times \text{conarachin I} \\ & - 0.147030 \times \text{arachin/conarachin} - 0.015300 \times 23.5\ \text{kDa} \end{aligned} \tag{5.3}$$

Y1 is the value of protein gelation after being transformed into that in compliance with normal distribution.

Formula 5.3 was obtained by the supervised principal component analysis, ten peanut quality indicators were involved, and these indicators were not only of theoretical significance but also of guiding significance in practical research.

This research showed that there was a negative correlation between the fruit shape (in the order of hockey shape < hump shape < bead shape < common shape < bee waist shape < gourd shape < cocoon shape < axe shape) and crude protein content and protein gelation (Table 5.5, Appendix 3), which indicated that the change in fruit shape would lead to the change in crude protein content, while the fruit shape also had negative impacts on the gelation. Su Qiuqin (2006) showed that the fruit shape had significant effects on the quality of peanut, for example, if the pod type was in the order of common shape, hockey shape, gourd shape, cocoon shape, and bead shape, the crude fat and linoleic acid contents in peanuts were increased, while the crude protein and oleic acid contents were significantly reduced. Among them, the crude fat content in bead-shaped peanut was the highest, and the crude protein, oleic acid, and linoleic acid contents in common-shaped peanut were the lowest, which indicated that the fruit shape of peanut was closely related to the quality of peanut. Another researcher found that the appearance quality of rice and soybean was closely related to their processing quality. There was a highly significant negative correlation between the grain length and aspect ratio of rice and the taste value of freeze-dried instant rice after re-watering, which indicated that the longer the rice grain, the larger the aspect ratio, the worse the taste of freeze-dried rice (Pu et al. 2007). There was a moderate positive correlation between the grain shape and crude fat content of soybean. From the mean of fat contents of various grain shapes, it could be seen that the fat contents of oblate elliptic, long elliptic, and elliptic to round grains increased gradually, indicating that the soybean grain shape was closely related to the soybean quality (Liang 1982). Another researcher found that the amino acid was closely related to the processing quality, for example, cystine contains disulfide bonds, and disulfide bond was an important factor affecting the formation of gel. Wang and Zhang (2010) found that the gel ability of soybean 11S globulin was stronger than that of 7S protein, because each subunit in 11S consisted of an acidic polypeptide chain and an alkaline polypeptide chain and they were bound by disulfide bonds to make the protein gel hard and brittle. Leucine was a hydrophobic amino acid, and the higher the hydrophobic amino acid content, the better the gelation (Relkin 1998).

The peanut protein composition and subunit content were closely related to the protein processing characteristic. This research showed that conarachin I had positive correlated effects on the gelation of peanut protein (Table 5.5, Appendix 3). Yang Xiaoquan et al. (1998) found that the methionine and cysteine sulfur-containing amino acid in conarachin I was significantly higher than that in defatted protein powder. The methionine content was 2.93 g/100 g protein which was twice higher than that in defatted protein powder; the cysteine content was 8.10 g/100 g protein which was eight times higher than that in defatted powder protein. Murphy et al. (1997) found that the 11S and 7S component contents of soybean, 11S/7S value, A_3 subunit and $A_{1a}A_{1b}A_2$ subunit of 11S component, and α', α, and β subunits of 7S component were correlated with the hardness, rupture, coherence, chewiness, springiness, viscidity, and fragility of tofu. Shi Yanguo et al. (2005) researched the effects of different soybean varieties on the gelation of soybean, and they found that, among the subunit components of soybean protein, 7S/11S was significantly negatively correlated with the hardness and was not correlated with the springiness. Wang Xiansheng et al. (2006) found that A_3 and A_4 subunits had significant effects on the texture characteristics of soybean protein isolate. It could be seen that the differences in protein composition caused by variety difference had significant effects on the gel property of soybean protein.

The other indicators (such as crude fiber and glycine) were closely related to the gelation, and it was to be further researched why these indicators had effects on gelation. Similar to the research results of many other varieties, the results showed that there was no theoretically correlated quality, but the correlation in the test data was relatively significant. For example, there was a positive correlation between the thousand grain weight and ash and the quality of steamed bun, and the unit weight had positive effects on the springiness of steamed bun and had negative effects on the appearance property. There were significant or highly significant positive correlations between the protein content and wet gluten content and the weight, volume, specific volume, and color of steamed bun; there was a positive correlation between the sedimentation value and the volume, specific volume, and springiness of steamed bun, and there was a significant correlation between the wet gluten content and organization structure of steamed bun (Li Zhixi et al. 1998; Zhang Guoquan et al. 2000; Shi Junling et al. 2001a, b).

It could be seen from the Formula 5.3 that the coefficients in the front of various indicators were positive or negative; in order to avoid the effect of negative on the calculation of weight, all the coefficients were increased by 1 for normalization processing so as to calculate the weight of indicators (Table 5.2).

1.1.2.2 Weight Calculated by Correlation Coefficient

The weight of indicator is the main basis for the importance of indicators. Therefore, in order to further investigate the accuracy of weight of various indicators determined by the regression coefficients, the weight of various indicators was further determined by using the correlation coefficients so as to compare with the weight determined by the regression coefficients. Correlation analysis was carried

Table 5.2 Weight determined by regression coefficient

No.	Indicator	Coefficient	Weight	No.	Indicator	Coefficient	Weight
1	Cystine	1.043	11	6	Crude fiber	0.975	10
2	Arginine	1.017	11	7	Fruit shape	0.970	10
3	Conarachin I	1.005	11	8	Leucine	0.913	10
4	Crude protein	0.994	10	9	Globulin/ conarachin	0.853	9
5	23.5 kDa	0.985	10	10	Glycine	0.793	8

out for the ten quality indicators determined in the quality evaluation model of peanuts suitable for gel-type protein processing (Table 5.3), and the normalization processing was carried out for the correlation coefficients between the processing quality evaluation indicators of peanut to determine the weight (Table 5.4).

After the weight values of various indicators were calculated by using the above two methods, it was found that the weight values of various indicators were close to each other, indicating that the various indicators selected were very important. As the above model was established and validated using the formula, the following weight calculated by using the formula was analyzed.

1.1.3 Evaluation Indicator Standards

1.1.3.1 Class Standards of Evaluation Indicators

K-means cluster analysis was carried out for the ten quality indicators of various peanut varieties, and each indicator was divided into three categories according to the Formulas 5.1 and 5.2, namely, Class I (suitable), Class II (basically suitable), and Class III (unsuitable); the weight of various indicators determined in Sect. 1.1.2.1 was considered as the highest score (Table 5.5), namely, Class I and so on, and the corresponding scores of various classes of indicators were given.

1.1.3.2 Classification of Final Scores of Various Variety Resources

The sums of scores of various character indicators were considered as the final scores of various variety resources, and the final scores of various varieties were divided into three categories according to the Formulas 5.1 and 5.2, namely, Class I (suitable), Class II (basically suitable), and Class III (unsuitable) (Table 5.6).

The results of Tables 5.6 and 5.1 were compared, and their matching degrees were 92% for suitable, 82% for basically suitable, and 64% for unsuitable. The evaluation results were good and suitable as the quality evaluation standards of peanuts suitable for gel-type protein processing.

1.2 Practical Application Evaluation Standards

1.2.1 Weight of Evaluation Indicator

The weight of various evaluation indicators was calculated by using the regression coefficients and correlation coefficients, and the method was the same as that in Sects. 1.1.2.1 and 1.1.2.2.

Table 5.3 Summary table of correlation coefficients

	Fruit shape	Crude protein	Crude fiber	Glycine	Cystine	Leucine	Arginine	Conarachin I	Globulin/ conarachin	23.5 kDa	Σ
Fruit shape	1.000	0.257	0.127	0.052	0.090	0.035	0.199	0.130	0.204	0.004	2.098
Crude protein	0.257	1.000	0.303*	0.696**	0.128	0.710**	0.725**	0.102	0.017	0.225	1.729
Crude fiber	0.127	0.303*	1.000	0.206	0.321*	0.272	0.406**	0.245	0.129	0.208	2.187
Glycine	0.052	0.696**	0.206	1.000	0.124	0.595**	0.624**	0.195	0.083	0.357*	1.660
Cystine	0.090	0.128	0.321*	0.124	1.000	0.396**	0.398**	0.671**	0.463**	0.544**	1.342
Leucine	0.035	0.710**	0.272	0.595**	0.396**	1.000	0.645**	0.255	0.287	0.233	2.082
Arginine	0.199	0.725**	0.406**	0.624**	0.398**	0.645**	1.000	0.354*	0.127	0.270	1.596
Conarachin I	0.130	0.102	0.245	0.195	0.671**	0.255	0.354*	1.000	0.804**	0.515**	1.927
Globulin/ conarachin	0.204	0.017	0.129	0.083	0.463**	0.287	0.127	0.804**	1.000	0.370*	1.847
23.5 kDa	0.004	0.225	0.208	0.357*	0.544**	0.233	0.270	0.515**	0.370*	1.000	1.940
Σ	2.098	1.729	2.187	1.66	1.342	2.082	1.596	1.927	1.847	1.940	18.408
Weight	11.397	9.393	11.881	9.018	7.290	11.31	8.670	10.468	10.034	10.538	100

Table 5.4 Weight calculated by correlation coefficient

No.	Indicator	Weight	No.	Indicator	Weight
1	Crude fiber	12	6	Globulin/conarachin	10
2	Fruit shape	11	7	Crude protein	9
3	Leucine	11	8	Glycine	9
4	23.5 kDa	11	9	Arginine	9
5	Conarachin I	10	10	Cystine	7

Table 5.5 Class standards of evaluation indicators (gel type—theoretical analysis)

Indicator		Class I	Class II	Class III
Fruit shape−	Classification value	≤3.03	3.03∼6.71	≥6.71
	Score	12	9	6
Crude protein+	Classification value	≥27.70	24.35∼27.70	≤24.35
	Score	12	9	6
Crude fiber−	Classification value	≤2.53	2.53∼5.01	≥5.01
	Score	11	8	5
Glycine−	Classification value	≤1.34	1.34∼1.89	≥1.89
	Score	2	1.5	1
Cystine+	Classification value	≥0.89	0.51∼0.89	≤0.51
	Score	13	10	7
Leucine−	Classification value	≤1.60	1.60∼2.14	≥2.14
	Score	9	6	3
Arginine+	Classification value	≥3.98	2.95∼3.98	≤2.95
	Score	12	9	6
Conarachin I+	Classification value	≥29.35	23.8∼29.35	≤23.8
	Score	12	9	6
Globulin/conarachin-	Classification value	≤1.08	1.08∼1.42	≥1.42
	Score	7	4	1
23.5 kDa−	Classification value	≤20.83	20.83∼23.72	≥23.72
	Score	12	9	6

1.2.2 Evaluation Indicator Standards

1.2.2.1 Class Standards of Evaluation Indicators

K-means cluster analysis was carried out for the five quality indicators of various peanut varieties, and each indicator was divided into three categories according to the Formulas 5.1 and 5.2, namely, Class I (suitable), Class II (basically suitable), and Class III (unsuitable); the weight of various indicators was considered as the highest score (Table 5.7), namely, Class I and so on, and the corresponding scores of various classes of indicators were given. The method was the same as that in Sect. 1.1.3.1.

Table 5.6 Class division of assignments of various varieties (gel type—theoretical analysis)

Classification	Standard	Number of samples	Sample name
Suitable	≥75	13	Bianhua 3, Shuangji 2, Shanhua 7, Luhua 11, Haihua 1, Shanhua 9, Yueyou 14, Fenghua 1, Fenghua 3, Ji 9814, Yuhua 15, Kainong 30, Honghua 1
Basically suitable	65–75	34	Zhonghua 8, Silihong, Fenghua 5, Guihua 771, Shanyou 250, Longhua 243, Pearl Red, Baisha 1016, Black Peanut, Zhongnong 108, Qinglan 8, Huayu 8, Luhua 14, Heyou 11, Yuanhua 8, Huayu 23, 034–256-1, Quanhua 551, Lufeng 2, Hongguan, Minhua 9, Zhanhua 82, White Peanut, Fenghua 4, Xuhua 13, Huayu 20, Colorful Peanut, Yuhua 9326, Yuhua 9327, Kainong 37, Huayu 16, Haiyu 6, Luhua 9, Xianghua 509–77
Unsuitable	≤65	14	Yueyou 45, Yueyou 86, Yuanza 9102, Xuhua 15, Zhonghua 4, Yuanza 9307, Fenghua 6, Xuhua 5, Huayu 28, Huayu 31, Zhengnong 7, Zhonghua 15, Luhua 15, Hua 17

Table 5.7 Class standards of various indicators (gel type—practical application)

Indicator		Class I	Class II	Class III
Fruit shape−	Classification value	≤3.10	3.10–6.69	≥6.69
	Score	17	11	5
Crude protein+	Classification value	≥27.42	24.27–27.42	≤24.42
	Score	21	15	9
Arginine+	Classification value	≥3.70	3.00–3.70	≤3.30
	Score	16	11	6
Conarachin I+	Classification value	≥29.37	23.82–29.37	≤23.82
	Score	23	16	9
23.5 kDa−	Classification value	≤20.80	20.80–23.72	≥23.72
	Score	23	16	9

1.2.2.2 Classification of Final Scores of Various Variety Resources

The sums of scores of various character indicators were considered as the final scores of various variety resources, and the final scores of various varieties were divided into three categories according to the Formulas 5.1 and 5.2, namely, Class I (suitable), Class II (basically suitable), and Class III (unsuitable) (Table 5.8). Through the comparative analysis between the result and that of Table 5.1, it was found that the suitable accuracy rate was 91.7%, the basically suitable accuracy rate was 90.3%, and the unsuitable accuracy rate was 63.6%. The results were excellent.

The gelation of 152 peanut varieties in China was evaluated by using the evaluation standard, and the peanut varieties suitable for gel-type protein processing were Quanhua 464, Luhua 11, Shuangji 2, Bianhua 3, Fenghua

Table 5.8 Class division of assignments of various varieties (gel type—practical application)

Classification	Classification standard	Number of samples	Sample name
Suitable	≥74.5	13	Luhua 11, Shuangji 2, Bianhua 3, Fenghua 1, Honghua 1, Luhua 14, Kainong 30, Zhanhua 82, Minhua 9, Yueyou 14, Yuhua 15, Ji 9814, Hua 7
Basically suitable	64.5–74.5	32	Haiyu 6, Shanhua 9, Shanyou 250, Pearl Red, Yuanhua 8, 034–256-1, Lufeng 2, Hongguan, Zhonghua 4, Fenghua 6, Fenghua 5, Haihua 1, Silihong, Quanhua 551, Huayu 16, White Peanut, Black Peanut, Zhongnong 108, Qinglan 8, Yuhua 9326, Yuhua 9327, Guihua 771, Xuhua 13, Yueyou 86, Heyou 11, Kainong 37, Luhua 9, Hua 17, Huayu 20, Zhonghua 8, Huayu 8, Longhua 243
Unsuitable	≤64.5	16	Yueyou 45, Fenghua 3, Huayu 23, Huayu 28, Baisha 1016, Xuhua 5, Xuhua 15, Yuanza 9102, Zhonghua 15, Luhua 15, Zhengnong 7, Huayu 31, Xianghua 509–77, Colorful Peanut, Yuanza 9307, Fenghua 4

1, Kainong 30, Feng Hua 3, Zhongkaihua 4, Huaguanwang, Shanhua 7, Minhua 9, Zhanhua 82, Yuhua 15, Haihua 1, Guihua 166, and other 31 varieties (Table 5.6, Appendix 3).

2 Quality Evaluation Standards of Peanut Suitable for High-Solubility Protein Processing

The solubility property of peanut protein is related to the variety. Different peanut varieties have different protein composition and structures and different protein solubility. In recent years, the quality evaluation technologies, methods, and standards of peanuts suitable for high-solubility protein processing have been also one of the research emphases in this field. This section focused on the introduction of the quality evaluation standards of peanuts suitable for soluble protein processing, including the theoretical analysis evaluation standards and practical application evaluation standards, so as to provide technical support for the selection of peanut varieties or raw materials for soluble peanut protein.

Table 5.9 Suitability analysis of solubility

Cluster no.	Standard	Number of samples	Sample name
Suitable	≥86	24	Yuhua 9326, Baisha 1016, Yuhua 15, Colorful Peanut, White Peanut, Luhua 15, Fenghua 6, Xuhua 15, Yuhua 9327, Luhua 9, Yuanhua 8, Xuhua 14, Haiyu 6, 034-256-1, Quanhua 551, Hua 17, Zhonghua 15, Zhonghua 4, Qinglan 8, Xuhua 13, Huaguanwang, Hongguan, Honghua 1, Huayu 23
Basically suitable	68–86	22	Kainong 37, Huayu 20, Zhengnong 7, Huayu 8, Ji 9814, Luhua 14, Lufeng 2, Kainong 30, Huayu 28, Yueyou 40, Yueyou 45, Huayu 31, Huayu 22, Yuanza 9307, Xianghua 509-77, Xuhua 5, Huayu 16, Huayu 19, Yueyou 86, Black Peanut, Shuangji 2, Shanhua 9
Unsuitable	≤68	8	Fenghua 1, Minhua 9, Fenghua 3, Heyou 11, Pearl Red, Zhanhua 82, Guihua 771, Fenghua 5

2.1 *Theoretical Analysis Evaluation Standards*

2.1.1 Classification Standards of Solubility

K-means cluster analysis was carried out for the solubility of 54 varieties of established model (34 varieties) and predicted model (20 varieties), and the classification number was three; according to the Formulas 5.1 and 5.2, these 54 peanut varieties were divided into three categories, suitable, basically suitable, and unsuitable (Table 5.9), and this classification was taken as the basis for the quality evaluation standards of soluble protein peanut.

2.1.2 Weight of Evaluation Indicator

The weight of various indicators was calculated by using the regression coefficients, and the coefficients of the relation between the solubility y_1 value in Chap. 4 which conformed to the normal distribution and various indicators (Formula 5.4) were analyzed. In the Formula 5.4, the indicators were standardized results, so the results could reflect the relation between the various indicators better. Formula 5.4 was obtained by the supervised principal component regression analysis, 11 quality indicators were involved, and these indicators were not only of theoretical significance but also of practical guiding significance.

$$\begin{aligned} Y_1 = {} & 0.770362 \times \text{crude fat} - 0.60393 \times \text{crude protein} \\ & - 0.91626 \times \text{total sugar} - 8.32449 \times \text{cystine} \\ & + 3.214817 \times \text{arginine} - 0.21846 \times \text{conarachin I} \\ & - 1.16885 \times 37.5\ \text{kDa} + 1.81934 \times 23.5\ \text{kDa} \\ & + 1.018139 \times 15.5\ \text{kDa} - 0.44476 \times \text{protein extraction rate} \\ & + 0.207081 \times \text{pure kernel rate} + 59.67507 \end{aligned} \tag{5.4}$$

The correlation analysis between peanut quality and protein solubility (Table 5.7, Appendix 3) showed that there was a negative correlation between crude fat and solubility ($r = -0.542$), indicating that the higher the crude fat content, the worse the protein solubility; there was a positive correlation between crude protein and solubility ($r = 0.662$), indicating that the higher the peanut protein content, the better the protein solubility. The results verified that crude protein and crude fat were negatively correlated ($r = -0.415$). There was a negative correlation between the solubility and cystine ($r = -0.513$), conarachin I ($r = -0.456$), and 37.5 kDa subunit contents ($r = -0.363$), and there was a positive correlation between the solubility and 23.5 kDa subunit content ($r = 0.384$), indicating that the higher the cystine, conarachin I, and 37.5 kDa subunit contents, the worse the solubility, whereas the higher the 23.5 kDa subunit content, the better the protein solubility, which indicated that the peanut protein composition and relative contents of various subunits were closely related to the processing quality of peanut protein. There was a positive correlation between pure kernel rate and protein extraction rate and solubility (Table 5.7, Appendix 3). Lin Jinhu (2006) found that there was a highly significant positive correlation between hundred fruit weight and hundred kernel weight, but the correlation coefficients between hundred fruit weight and hundred kernel weight and pure kernel rate were all negative and very nonsignificant. Through the path analysis, it was found that the correlation coefficients between the protein content and hundred kernel weight and pure kernel rate were negative, which indicated that the varieties with high protein content had thick pod shells, small kernels, and low pure kernel rate. The results were basically consistent with that in this book; therefore, it indicated that the pure kernel rate might be an important factor affecting the solubility. The protein extraction rate had a great effect on the quality of peanut protein, which was consistent with the research results of wheat and soybean. For example, Wei Yimin (1989) found that the flour yield was the main indicator of milling quality. Flour mill required the varieties with few milling times, high flour yield, and less energy consumption. The requirements for grain shape were that the well-developed, full, elliptic seeds with shallow ventral grooves had good milling characteristics. The research conducted by Zhang Guoquan et al. (2000) showed that there were highly significant negative correlations between the water absorption of noodles and the flour yield of wheat varieties and wet gluten content, and there were highly significant negative correlations between the grain hardness, water absorption of noodles, and ash. Guo Boli et al. (2001) analyzed the wheat quality and found that there was a significant positive correlation between flour yield and springiness of

Table 5.10 Weight of various indicators (soluble—theoretical analysis)

No.	Indicator	Weight	No.	Indicator	Weight
1	Arginine	13	7	Protein extraction rate	9
2	23.5 kDa	11	8	Crude protein	9
3	15.5 kDa	11	9	Total sugar	9
4	Crude fat	10	10	37.5 kDa	8
5	Pure kernel rate	10	11	Cystine	0.7
6	Conarachin I	9			

steamed bun. Generally, the flour yield of hard wheat was high, the protein content and quality were good, and the steamed bun made by hard wheat had good springiness. Du Wei et al. (2001) analyzed the quality of 25 varieties of wheat and noodles made and found that there was a highly significant negative correlation between the protein content and water absorption of cooking, and there was a highly significant negative correlation between the protein content and dry matter loss rate and protein loss rate, which indicated that the higher the protein content, the more the protein loss. In addition, there were close correlations between solubility and other indicators, such as total sugar, and it was to be further explained why these indicators would have an impact on the solubility.

It could be seen from the Formula 5.4 that the coefficients in the front of various indicators were positive or negative; in order to avoid the effect of negative on the calculation of weight, all the coefficients were increased by 9 for normalization processing so as to calculate the weight of indicators (Table 5.10).

2.1.3 Evaluation Indicator Standards

2.1.3.1 Class Standards of Evaluation Indicators

K-means cluster analysis was carried out for the 11 quality of 54 peanut varieties, and each indicator was divided into three categories according to the Formulas 5.1 and 5.2, namely, Class I (suitable), Class II (basically suitable), and Class III (unsuitable); the weight of various indicators determined in Sect. 2.1.2 was considered as the highest score (Table 5.11), namely, Class I and so on, and the corresponding scores of various classes of indicators were given.

2.1.3.2 Classification of Final Scores of Various Variety Resources

The sums of scores of various character indicators were considered as the final scores of various variety resources, and the final scores of various varieties were divided into three categories according to the Formulas 5.1 and 5.2, namely, Class I (suitable), Class II (basically suitable), and Class III (unsuitable) (Table 5.12).

Table 5.11 Class standards of evaluation indicators (soluble—theoretical analysis)

Indicator		Class I	Class II	Class III
Crude fat−	Classification value	≤46.95	46.95–52.83	≥52.83
	Score	10	7	4
Crude protein+	Classification value	≥27.58	24.49–27.58	≤24.49
	Score	9	6	3
Total sugar−	Classification value	≤5.14	5.14–8.99	≥8.99
	Score	9	6	3
Cystine−	Classification value	≤0.48	0.48–0.85	≥0.85
	Score	0.7	0.4	0.1
Arginine+	Classification value	≥4.40	3.04–4.40	≤3.04
	Score	13	9	5
Conarachin I−	Classification value	≤23.49	23.49–29.69	≥29.64
	Score	9	6	3
37.5 kDa−	Classification value	≤12.65	12.65–15.17	≥15.17
	Score	8	5	2
23.5 kDa+	Classification value	≥24.00	21.25–24.00	≤21.25
	Score	11	7	4
15.5 kDa−	Classification value	≤5.78	5.78–7.60	≥7.60
	Score	11	7	4
Protein extraction rate+	Classification value	≥85.38	73.22–85.38	≤73.22
	Score	9	6	3
Kernel yield+	Classification value	≥74.32	65.95–74.32	≤65.95
	Score	10	7	4

Table 5.12 Class division of assignments of various varieties (soluble—theoretical analysis)

Cluster no.	Classification standard	Number of samples	Sample name
Suitable	≥68	18	Colorful Peanut, Lufeng 2, Zhonghua 4, Luhua 9, Yuanhua 8, Xuhua 14, Xuhua 15, Xianghua 509-77, Fenghua 6, Luhua 15, White Peanut, Xuhua 13, Huaguanwang, Zhonghua 15, Haiyu 6, Hongguan, Hua 17, Huayu 23
Basically suitable	58–68	25	034-256-1, Yueyou 40, Yueyou 45, Huayu 20, Baisha 1016, Ji 9814, Honghua 1, Yuanza 9307, Huayu 19, Huayu 16, Zhengnong 7, Xuhua 5, Shanhua 9, Huayu 28, Kainong 37, Luhua 14, Huayu 22, Yueyou 86, Yuhua 9362, Zhanhua 82, Qinglan 8, Huayu 31, Yuhua 9327, Kainong 30, Black Peanut
Unsuitable	≤58	11	Minhua 9, Guihua 771, Huayu 8, Heyou 11, Fenghua 3, Yuhua 15, Pearl Red, Fenghua 1, Fenghua 5, Quanhua 551, Shuangji 2

The comparison analysis between the result and that in Table 5.9 was carried out, and their matching degrees were 89% for suitable, 72% for basically suitable, and 67% for unsuitable. The evaluation results were good and suitable as the quality evaluation standards of peanuts suitable for soluble protein processing.

2.2 *Practical Application Evaluation Standards*

2.2.1 Weight of Evaluation Indicator

The weight of evaluation indicators was calculated by using the regression coefficients, and the method was the same as that in Sect. 2.1.2.

2.2.2 Evaluation Indicator Standards

2.2.2.1 Class Standards of Evaluation Indicators

K-means cluster analysis was carried out for the four quality indicators of various peanut varieties, and each indicator was divided into three categories according to the Formulas 5.1 and 5.2, namely, Class I (suitable), Class II (basically suitable), and Class III (unsuitable); the weight of various indicators was considered as the highest score (Table 5.13), namely, Class I and so on, and the corresponding scores of various classes of indicators were given. The method was the same as that in Sect. 2.1.3.1.

2.2.2.2 Classification of Final Scores of Various Variety Resources

The sums of scores of various character indicators were considered as the final scores of various variety resources, and the final scores of various varieties were divided into three categories according to the Formulas 5.1 and 5.2, namely, Class I (suitable), Class II (basically suitable), and Class III (unsuitable) (Table 5.14). Through the comparative analysis between the result and that of Table 5.9, it was found that the suitable accuracy rate was 66.67%, the basically suitable accuracy rate was 72%, and the unsuitable accuracy rate was 80%.

The solubility of 152 peanut varieties in China was evaluated by using the evaluation standard, and the peanut varieties suitable for soluble protein processing were Shanhua 7, Yuhua 9326, Baisha 1016, Yuhua 15, Colorful Peanut, Hua 67, White Peanut, Luhua 15, Luhua 8, Guihua 35, Baisha 101, Yueyou 25, Lanhua 2, Yuhua 9327, Yuhua 11, and other 50 varieties (Table 5.8, Appendix 3).

Table 5.13 Class standards of various indicators (soluble—practical application)

Indicator		Class I	Class II	Class III
Crude protein+	Classification value	≥27.58	24.49–27.58	≤24.49
	Score	18	12	6
Cysteine−	Classification value	≤0.48	0.48–0.85	≥0.85
	Score	27	18	9
Conarachin I−	Classification value	≤23.49	23.49–29.69	≥29.64
	Score	31	20	9
15.5 kDa−	Classification value	≤5.78	5.78–7.60	≥7.60
	Score	25	17	9

Table 5.14 Class division of assignments of various varieties (soluble—practical application)

Cluster no.	Classification standard	Number of samples	Sample name
Suitable	≥74	18	Yueyou 86, Yuhua 9326, Yuanhua 8, Colorful Peanut, Yuhua 9327, Xuhua 14, Fenghua 3, Huayu 28, Ji 9814, White Peanut, Black Peanut, Shuangji 2, Xuhua 15, Honghua 1, Huayu 16, Luhua 14, Qinghua 8, Huayu 8
Basically suitable	57–74	21	Zhonghua 15, Zhengnong 7, Hua 17, Yuhua 15, Hongguan, Huayu 23, 034-256-1, Fenghua 1, Baisha 1016, Huaguanwang, Pearl Red, Zhonghua 4, Luhua 9, Shanhua 9, Kainong 37, Xuhua 13, Huayu 20, Minhua 9, Lufeng 2, Xianghua 509-77, Fenghua 6
Unsuitable	≤55	15	Haiyu 6, Luhua 15, Quanhua 551, Fenghua 5, Huayu 19, Guihua 771, Zhanhua 82, Heyou 11, Yuanza 9307, Xuhua 5, Huayu 22, Huayu 31, Kainong 30, Yueyou 40, Yueyou 45

3 Quality Evaluation Standards of Peanut Suitable for Peanut Oil Processing

Fifty-five percent of peanuts in China are used for oil expression, and peanut oil is one of the most important peanut processing products. The quality of peanut oil is closely related to the peanut varieties, and the quality of peanut oil produced by different peanut varieties is different. In recent years, the research on the quality evaluation technologies, methods, and standards of peanut suitable for peanut oil processing, proportion pattern of fatty acid of different peanut varieties, and stability of peanut oil has become the focus of attention in this field. This section focused on the introduction of quality evaluation standards of peanut suitable for high-quality peanut oil processing in order to provide the basis for the selection of peanut varieties or raw materials for high-quality peanut oil.

3.1 Class Standards of Comprehensive Evaluation Value of Peanut Oil

Equal-weighted addition was made for the original data of peanut oil of established model and validated model (Table 5.15), the suitability analysis was carried out for the results by using the K-means cluster analysis method, and the cluster number was three. There were 23 peanut varieties suitable for peanut oil processing, 7 peanut varieties basically suitable for peanut oil processing, and 15 peanut varieties unsuitable for peanut oil processing (Table 5.16).

3.2 Peanut Quality Evaluation Standards

3.2.1 Weight of Evaluation Indicator

The weight of various indicators was calculated by using the regression coefficients and the coefficients of the relation between the comprehensive value of peanut oil stability y_1 value in Chap. 4 which conformed to the normal distribution, and various indicators (Formula 5.5) were analyzed. Formula 5.5 was the normalized formula for the data in Formula 4.3, so the results could reflect the relationship between the indicators better.

$$\begin{aligned} Y_1 = {} & 5.9999982 - 1.3946738^{*}\text{standardized data of crude fat} \\ & + 0.6473054^{*}\text{standardized data of oleic acid/linoleic acid} \\ & - 1.9303204^{*}\text{ standardized data of unsaturated fatty acid} \end{aligned} \tag{5.5}$$

The coefficients in the front of various indicators in the quality evaluation model of peanut for oil were taken as the weight values.

In 1980, Sekhon et al. (1980) pointed out that the oleic acid/linoleic acid ratio was an important biochemical indicator to measure storability of peanut and its products and the greater the ratio, the better the storability of peanut and its products. The research results of Li Hongyan et al. (2010) showed that the oxidative stability of vegetable oil was affected by the content of unsaturated fatty acids and rapeseed oil was more unstable than sesame oil because the content of unsaturated fatty acids (88.53%) in rapeseed oil was much higher than that in sesame oil (81.92%). Li Linqiang et al. (2009) analyzed the fatty acid composition of *Andrias davidianus* oil. The results showed that the content of unsaturated fatty acids in *Andrias davidianus* oil was 71.3%, and the unsaturated fatty acids would lose easily when suffering oxidation. When the extraction temperature exceeded 50 °C, the double bonds in the unsaturated fatty acids were easily oxidized. After the literature review and actual situation analysis, it was confirmed that the greater the crude fat content, the better; the greater the oleic acid/linoleic acid content, the better; and the smaller the content of unsaturated fatty acids, the better (Table 5.17).

Table 5.15 Original data of comprehensive value of peanut oil

No.	Variety name	Original data	No.	Variety name	Original data	No.	Variety name	Original data
1	Zhonghua 8	−11.44	16	Xuhua 13	−10.35	31	Yuhua 9327	−8.71
2	Shanhua 7	17.58	17	Xuhua 14	8.65	32	Kainong 30	−10.22
3	Silihong	12.78	18	Huayu 19	10.46	33	Kainong 37	−9.27
4	Luhua 11	9.12	19	Huayu 20	11.29	34	Yuanza 9102	11.82
5	Bianhua 3	10.73	20	Huayu 22	10.32	35	Zhonghua 15	12.2193
6	Haihua 1	10.59	21	Huayu 23	−8.38	36	Fenghua 6	−14.6891
7	Shuangji 2	13.03	22	Huayu 28	−10.02	37	Huaguanwang	−10.4647
8	Shanhua 9	−8.41	23	Huayu 31	8.19	38	Zhongnong 108	8.9086
9	Fenghua 5	−14.45	24	Baisha 1016	8.55	39	Xuhua 15	8.10127
10	White Peanut	10.29	25	Colorful Peanut	8.77	40	Huayu 16	−11.3781
11	Fenghua 1	8.55	26	Black Peanut	11.55	41	Zhonghua 4	−9.74147
12	Fenghua 3	−10.39	27	034–256-1	−11	42	Luhua 9	13.87601
13	Fenghua 4	−14.16	28	Ji 9814	−12.1	43	Lufeng 2	15.13395
14	Xuhua 5	8.34	29	Yuhua 15	−13.73	44	Luhua 154	−1.4597
15	Yuanhua 8	−11.91	30	Yuhua 9326	−10.07	45	Luhua 14	−2.50608

Table 5.16 Peanut processing suitability

Classification	Standard	Number of samples	Sample name
Suitable	≥8.0	23	Shanhua 7, Lufeng 2, Luhua 9, Shuangji 2, Silihong, Zhonghua 15, Yuanza 9102, Black Peanut, Huayu 20, Bianhua 3, Haihua 1, Huayu 19, Huayu 22, White Peanut, Luhua 11, Zhongnong 108, Colorful Peanut, Xuhua 14, Fenghua 1, Baisha 1016, Xuhua 5, Huayu 31, Xuhua 15
Basically suitable	8.0~−10.0	7	Luhua 15, Luhua 14, Huayu 23, Shanhua 9, Yuhua 9327, Kainong 37, Zhonghua 4
Unsuitable	≤10.0	15	Huayu 28, Yuhua 9326, Kainong 30, Xuhua 13, Fenghua 3, Huaguanwang, 034-256-1, Huayu 16, Zhonghua 8, Yuanhua 8, Ji 9814, Yuhua 15, Fenghua 4, Fenghua 5, Fenghua 6

Table 5.17 Weight values of various indicators

No.	Indicator	Coefficient	Weight
1	Crude fat	1.3946738	35
2	Oleic acid/linoleic acid	0.6473054	16
3	Unsaturated fatty acid	1.9303204	49

Table 5.18 Evaluation indicator standards

Indicator		Class I	Class II	Class III
Crude fat+	Classification value	≥53.68	47.70–53.68	≤47.70
	Score	35	25	15
Oleic acid/linoleic acid+	Classification value	≥1.48	1.00–1.48	≤1.00
	Score	16	11	6
Unsaturated fatty acid	Classification value	≤37.63	37.63–44.66	≥44.66
	Score	49	34	19

3.2.2 Evaluation Indicator Standards

K-means cluster analysis was carried out for the 3 indicators of 45 varieties of established model and validated model, respectively, and each indicator was divided into three categories according to the Formulas 5.1 and 5.2, namely, Class I (suitable), Class II (basically suitable), and Class III (unsuitable); the weight of various indicators was considered as the highest score (Table 5.18), namely, Class I and so on, and the corresponding scores of various classes of indicators were given. The method was the same as that in Sect. 2.1.3.1.

The assignments of various varieties were added, and the cluster analysis was carried out for the adding results (Table 5.19). The comparison between the results and results in Table 5.16 showed that the accuracy rate of the first category was

Table 5.19 Processing suitability of peanut for oil

Classification	Standard	Number of samples	Sample name
Suitable	≥80	20	Yuanza 9102, Luhua 9, Xuhua 14, Black Peanut, Haihua 1, Lufeng 2, Yuhua 9326, Baisha 1016, Yuhua 15, Shuangji 2, Shanhua 7, Zhonghua 15, Luhua 11, Bianhua 3, Huayu 31, Zhongnong 108, Colorful Peanut, White Peanut, Xuhua 15, Huayu 19
Basically suitable	70–80	11	Huayu 28, Huayu 23, Shanhua 9, Luhua 14, Kainong 37, Luhua 15, Huayu 20, Yuhua 9327, Zhonghua 4, Kainong 30, Fenghua 1
Unsuitable	≤80	14	Silihong, 034-256-1, Huaguanwang, Xuhua 5, Huayu 22, Zhonghua 8, Ji 9814, Fenghua 6, Yuanhua 8, Fenghua 4, Huayu 16, Fenghua 3, Fenghua 5, Xuhua 13

90%, that of the second category was 63.6% and that of the third category was 78.6%, which could be used as the quality evaluation standard of peanut for oil.

4 Quality Evaluation Standards of Peanut Suitable for Peanut Butter Processing

4.1 Division of Peanut Butter Classification Standards

The cluster analysis (classification number is 3) was carried out for the comprehensive values of 26 varieties of peanut butter (established models and predicted models). The original data of comprehensive values of peanut butter is shown in Table 5.20. From the above analysis results, it was found that the greater the comprehensive value of peanut butter, the better; these 26 peanut varieties were divided into suitable, basically suitable, and unsuitable categories (Table 5.21), which were taken as the basis of characteristic evaluation standards of peanuts suitable for peanut butter processing.

4.2 Determination of Characteristic Evaluation Indicator Weight of Peanuts Suitable for Peanut Butter Processing

The characteristic evaluation model of peanuts suitable for peanut butter processing was obtained through correlation analysis, principal component analysis, and regression analysis, five peanut characteristic indicators were involved, and these indicators were of not only theoretical significance but also practical significance.

Table 5.20 Original data of comprehensive value of peanut butter

No.	Variety name	Original data	No.	Variety name	Original data	No.	Variety name	Original data
1	Huayu 22	0.3901	10	Luhua 17	0.05887	19	Yuhua 9414	−0.16839
2	New Huayu 22	0.63366	11	Fenghua 1	0.90228	20	Shangyan 9658	−0.2578
3	Weihua 8	0.52809	12	Guihua 17	−0.0532	21	Huayu 20	1.857132
4	13–2	−0.2511	13	High oleic acid	0.5352	22	Fuhua 16	−0.3874
5	13_3	−0.2267	14	Xiaobaisha	0.3262	23	Silihong	−0.59031
6	Guihua 22	−0.5574	15	Fuhua 18	−0.1857	24	Rose Red	2.680833
7	Guihua Red 95	−0.1466	16	Fuhua 12	−0.21408	25	Shanhua 9	−0.3979
8	P905	0.0873	17	Kai 17–15	−0.0203	26	Luhua 19	1.330101
9	Jihuatian 1	2.43755	18	Luhua 18	0.552			

Table 5.21 Suitability analysis of peanut butter

Classification	Standard	Number of samples	Sample name
Suitable	≥0.90	4	Jihuatian 1, Huayu 20, Rose Red, Luhua 19
Basically suitable	0.90~−0.39	18	Shandong Huayu 22, Weihua 8, 13-2, 13-3, Guihua Red 95, P905, Luhua 17, Fenghua 1, Guihua 17, high oleic acid, Xiaobaisha, Fuhua 18, Fuhua 12, Kai 17-15, Luhua 18, Yuhua 9414, Shangyan 9658, Xinjiang Huayu 22
Unsuitable	≤−0.38	4	Silihong, Shanhua 9, Fuhua 16, Guihua 22

The results of this experimental research showed that there was a significant negative correlation between fat and protein and there was a significant correlation between fat and protein and comprehensive value of peanut butter (Table 5.22). It indicated that the fat and protein had great effect on the quality of peanut butter. Navnitkumar et al. (2012) and Özcan (2010) found that fat and protein contents were important indicators of peanut butter and other related foods from the nutrition point of view and the higher the protein content, the better the quality. Ahmed and Ali (1986) found that the texture of peanut butter was affected by the fat content and peanut kernel content; when the fat content in peanut was about 50% and the peanut kernel content was above 95%, the texture of peanut butter was good, pure, and thick, and the peanut butter could be smeared smoothly. It indicated that the protein and fat contents were closely related to the quality of peanut butter.

The sensory quality of peanut was closely related to the processing quality. The research conducted by Misra et al. (2000, 2004) showed that the average size of peanut kernel was about 0.55 g and 40–50/oz peanut kernels was suitable for the

Table 5.22 Correlation analysis of 20 peanut butters

	Protein	Moisture	Ash	Fiber	Total sugar	Oleic acid	Linoleic acid	Hardness	Viscosity	L^*	C^*	H	POV	AV	IV	Eccentricity	Acceptability	Particle size
Fat	−0.280	−0.135	−0.008	−0.102	−0.094	0.205	−0.146	−0.497*	−0.512*	0.439	−0.031	0.343	−0.499*	0.145	−0.368	−0.175	0.078	−0.388
Protein	1	−0.069	−0.066	−0.352	−0.578**	0.186	0.036	−0.041	−0.305	−0.074	0.135	−0.110	0.240	0.203	−0.205	0.629**	−0.155	0.073
Moisture		1	−0.204	0.304	0.187	−0.073	−0.063	0.100	0.007	0.213	−0.413	0.336	−0.011	−0.171	0.232	−0.158	−0.145	−0.286
Ash			1	0.063	−0.146	0.275	−0.098	−0.092	0.099	−0.010	0.293	−0.082	0.330	−0.478*	−0.439	−0.072	0.200	−0.087
Fiber				1	0.279	−0.424	0.262	0.225	0.304	−0.072	−0.047	−0.045	−0.053	−0.094	0.363	−0.335	0.126	−0.059
Total sugar					1	−0.319	−0.262	0.554*	0.693**	−0.232	−0.206	−0.104	−0.212	−0.040	0.460*	−0.300	−0.238	0.185
Oleic acid						1	−0.664**	−0.421	−0.286	−0.089	0.415	−0.203	−0.245	−0.578**	−0.602**	0.030	0.226	−0.302
Linoleic acid							1	0.014	−0.057	0.302	−0.155	0.234	0.267	0.406	0.040	−0.200	0.074	0.168
Hardness								1	0.841**	−0.427	−0.045	−0.334	0.231	0.126	0.411	−0.011	−0.457*	0.427
Viscosity									1	−0.542*	0.118	−0.481*	0.154	−0.260	0.320	−0.223	−0.130	0.248
L^*										1	−0.634**	0.929**	−0.181	0.387	−0.404	−0.158	−0.014	−0.330
C^*											1	−0.835**	−0.136	−0.400	−0.146	0.056	0.270	0.193
H												1	−0.031	0.406	−0.197	−0.108	−0.160	−0.326
POV													1	−0.103	0.075	0.127	−0.037	0.077
AV														1	0.225	0.334	−0.425	0.333
IV															1	0.142	−0.324	0.466*
Eccentricity																1	−0.343	0.288
Acceptability																	1	−0.546*

Note: **Extremely significantly correlated ($P < 0.01$); *significantly correlated ($P < 0.05$)

production of peanut candies. The research conducted by Li Shaohua et al. (2004) and Yin Dongmei et al. (2010) showed that there was a significant positive correlation between hundred kernel weight and hundred fruit weight of peanut. At the same time, according to the research of peanut butter production company, it was found that the size of baked peanuts for preparation of peanut butter should be in the range of 28–34 mm to prevent that the baking degree was insufficient due to large size or the baking color was too deep due to small size, resulting in too light or dark color of peanut butter. It indicated that the hundred kernel weight had great effects on the baked peanuts for preparation of peanut butter.

Some research showed that the ratio of oleic acid and linoleic acid was closely related to the quality of peanut butter. This research showed that there was a significant correlation between oleic acid and comprehensive value of peanut butter quality. Riveros et al. (2010) found that the peanut butter prepared with high oleic acid peanut had better storage stability than that prepared by common oleic acid. Navnitkumar et al. (2012) also reported that the peanut product had better shelf life when the O/L ratio (namely, stability indicator) was greater than 2.0. At the same time, some research also showed that the increase of O/L value could improve the quality property and nutritional value of peanut. It indicated that the O/L value played an important role in the storage stability of peanut butter.

Vitamin E was closely related to the peanut product. It was reported that the peanuts and peanut products were the excellent sources of vitamin E. The data display of the US Department of Agriculture showed that the eighth largest source of vitamin E in the United States was silky peanut butter which could provide vitamin E2.3% individually (Chun et al. 2003). The research conducted by Chun et al. (2005) showed that vitamin E was a fat-soluble vitamin and potent antioxidant; in the storage process of peanut products, the vitamin E content significantly decreased with the increase of peroxide value. Due to the antioxidation of vitamin E, the variation of content could prevent the oxidation and rancidity of peanut butter to different degrees. At the same time, the V_E contents of peanut and peanut products were affected by variety, growth conditions, and maturity (Hashim et al. 1993).

Correlation analysis was carried out for the five quality determined in the quality evaluation model of peanut suitable for peanut butter processing (Table 5.23), and the weight of each indicator was the main basis of the importance of each indicator. Therefore, the correlation coefficient was used to further determine the weight of each indicator, which meant that the normalized processing was conducted for the correlation coefficients between the 26 peanut processing quality evaluation indicators to determine the weight.

Table 5.23 Calculated weight of correlation coefficient

	Hundred kernel weight	Crude protein	Crude fat	Total VE	O/L	Sum total
Hundred kernel weight	1	0.360	0.150	0.127	0.171	1.808
Crude protein	0.360	1	0.462	0.293	0.161	2.276
Crude fat	0.150	0.462*	1	0.362	0.081	2.055
Total VE	0.127	0.293	0.362	1	0.158	1.940
O/L	0.171	0.161	0.081	0.158	1	1.571
Sum total	1.808	2.276	2.055	1.940	1.571	9.650
Weight	18.74	23.58	21.30	20.10	16.28	100

4.2.1 Analysis of Characteristic Evaluation Standard of Peanut Suitable for Peanut Butter Processing

1. Division of evaluation indicator class standards

K-means cluster analysis was carried out for 5 characteristics of 26 peanut varieties, respectively, and each indicator was divided into three categories: Class I (suitable), Class II, (basically suitable), and Class III (unsuitable). The weight of each indicator determined in 4.3.2 was considered as the highest score (Table 5.24), namely, Class I and so on, and the corresponding scores of indicators at various classes were obtained.

2. Classification of final scores of various variety resources

The sum of scores of various characteristic indicators was considered as the final score, and the final scores of various varieties could be divided into three categories: Class I (suitable), Class II (basically suitable), and Class III (unsuitable) (Table 5.25).

The results in Table 5.25 were compared with that in Table 5.22, and the matching degrees were 80% for suitable category, 88% for basically suitable category, and 66% for unsuitable category. The evaluation result was good, and it suited as the quality evaluation standard of peanut suitable for peanut butter processing.

5 Quality Evaluation Standards of Peanut Suitable for Export Peanut Processing

Seven percent of peanuts in China are used for export, and the export volume is in the first place in the world's trade volume. Although there are differences in the quality indicators of peanuts exported to different countries or regions, the requirements of the main peanut exporting countries of China (such as the European Union, Japan, and other countries) for peanut kernel quality mainly include six

Table 5.24 Suitability of indicators at all classes

Indicator		Class I	Class II	Class III
Crude protein	Classification value	≥25.78	25.78–22.36	≤22.36
	Score	24	19	14
Crude fat	Classification value	≥53.27	53.27–42.64	≤42.64
	Score	21	16	11
Hundred kernel weight	Classification value	≥108	108–87.3	≤87.3
	Score	19	14	9
Total VE	Classification value	≥19.59	19.59–13.51	≤13.51
	Score	20	16	12
O/L	Classification value	≥13.15	13.15–1.00	≤1.00
	Score	16	12	8

Table 5.25 Division of assignment classes of various varieties

Classification	Standard	Number of samples	Sample name
Suitable	≥82	5	Jihuatian 1, P905, Rose Red, high oleic acid big peanut, Huayu 22
Basically suitable	82–73	15	Yuhua 9414, (Xinjiang) Huayu 22, 13-2, 13-3, Guihua Red 95, Luhua 17, Fenghua 1, Xiaobaisha, Kai 17-15, Luhua 18, Shangyan 9658, Guihua 17, Huayu 20, Luhua 19, Weihua 8
Unsuitable	≤73	6	Silihong, Guihua 22, Fuhua 18, Shanhua 9, Fuhua 12, Fuhua 16

indicators: seed coat color, water content, aflatoxin content, oleic acid/linoleic acid ratio, imperfect grain content, and grade; the quality of peeled export peanut kernel mainly includes water content, aflatoxin content, oleic acid/linoleic acid ratio, imperfect grain content, red skin residual rate, half grain content, and grade. This section focused on the introduction of evaluation standards of peanut suitable for export peanut kernel processing and that suitable for peeled export peanut kernel processing so as to provide the basis for the selection of peanut varieties or raw materials for export.

5.1 *Evaluation Standards of Peanut Suitable for Export Peanut Kernel Processing*

Sixty-five varieties were used for model establishment to obtain the model of three indicators (Formula 5.6), ten varieties were used to validate the model, and the correlation coefficient between the predicted value and true value was 0.87, which indicated that the predicted results of this equation was good. This model could be used to determine the quality evaluation indicators of export peanut kernel, predict the quality of unknown varieties of peanuts, and provide the basis for the selection of varieties suitable for export peanut kernel processing to the processing enterprises.

$$\begin{aligned}\text{Quality of export peanut kernel} = {} & 26.291 + 0.234 \text{ hundred kernel weight} \\ & + 14.609 \text{ oleic acid/linoleic acid ratio} \\ & - 2.964 \text{ aflatoxin content}\end{aligned} \tag{5.6}$$

5.1.1 Model Reliability

5.1.1.1 Significance

The regression analysis results of indicators in model and export peanut quality showed that (Table 5.20) there were significant correlations between these three indicators (hundred kernel weight, oleic acid/linoleic acid ratio, and aflatoxin content) and the quality of export peanut kernel at 0.01 level, which indicated that these indicators were important indicators affecting the quality of export peanut kernel (Table 5.26).

5.1.1.2 Correlation Between Indicators

The correlation analysis between three indicators showed that the correlation coefficients between the indicators were all less than or equal to 0.2 (Table 5.27), indicating that these three indicators were relatively independent and could reflect the relationship with export peanut quality individually.

5.1.2 Suitability

Sixty-five varieties were divided into three categories by using the K-means cluster analysis method (Table 5.28).

Table 5.26 Significance analysis of export peanut kernel quality and raw material indicators

	Nonstandard coefficient		Standard coefficient		
Model	Regression coefficient	Standard error	Trial	*t*	Sig.
Constant	26.291	4.861		5.409	0.000
Hundred kernel weight	0.234	0.038	0.566	6.097	0.000
Oleic acid/linoleic acid ratio	14.609	3.056	0.442	4.780	0.000
Aflatoxin content	−2.964	1.089	−0.253	−2.720	0.009

Table 5.27 Correlation analysis between hundred kernel weight, oleic acid/linoleic acid ratio, and aflatoxin content

Indicator	Hundred kernel weight	Oleic acid/linoleic acid ratio	Aflatoxin content
Hundred kernel weight	1.000		
Oleic acid/linoleic acid ratio	−0.037	1.000	
Aflatoxin content	0.120	−0.090	1.000

Table 5.28 Suitability class standards of export peanut kernel

Cluster no.	Number of varieties	Variety name
Suitable	15 (>66 分)	Huayu 28, Luhua 14, Lufeng 2, 9616, Baisha 1016, Zhongnong 108, Bianhua 3, Shuangji 2, Yuanhua 8, Xuhua 13, 034-256-1, 600-6, 365-1, L03-329-336, L03-601-604
Basically suitable	9 (60–66 分)	Huayu 19, Huayu 22, Luhua 9, Fenghua 3, Fenghua 5, Shanhua 7, Xuhua 14, 818, 780-15
Unsuitable	41 (<60 分)	Huayu 8, Huayu 16, Huayu 20, Huayu 23, Huayu 25, Huayu 31, Luhua 11, Luhua 15, Fenghua 1, Fenghua 4, Fenghua 6, Variety 060, Baisha 101, Zhonghua 4, Zhonghua 8, Zhonghua 15, Shanhua 9, Haihua 1, Haiyu 6, Qinglan 8, Honghua 1, Xuhua 5, Xuhua 15, Hongguan, Colorful Peanut, Kainongbai 2, White Peanut, Zhongyu 1, Black Peanut, Huaguanwang, Ji 9814, 060–18-w, Fuhua 1, Silihong, 8130, Hua 17, Xianghua 509-77, Zhengnong 7, Yuanza 9102, 606, L03-257-294

5.2 *Evaluation Standards of Peanut Suitable for Processing of Peeled Export Peanut Kernel*

Ten varieties were used to validate the Formula 5.7, and the correlation coefficient between the predicted value and true value was 0.741, which indicated that the predicted result of this equation was good. The model could be used to determine the quality evaluation indicator of peeled export peanut kernel, predict the quality of unknown varieties of peanuts, and provide the basis for the selection of peanut varieties suitable for processing of peeled export peanut kernel to processing enterprises.

$$\begin{aligned}&\text{Quality of peeled export peanut kernel}\\&\quad = 30.839 + 0.091 \times \text{hundred kernel weight} + 5.648\\&\qquad \times \text{oleic acid/linoleic acid ratio} - 1.909\ \text{aflatoxin content}\end{aligned} \tag{5.7}$$

Table 5.29 Correlation analysis between the hundred kernel weight, oleic acid/linoleic acid ratio, and aflatoxin content and the total score

		Hundred kernel weight	Oleic acid/ linoleic acid ratio	Aflatoxin content	Total score
Hundred kernel weight	Correlation coefficient	1.000	−0.037	0.120	0.396**
	Significance		0.781	0.354	0.001
Oleic acid/ linoleic acid ratio	Correlation coefficient	−0.037	1.000	−0.090	0.229
	Significance	0.781		0.501	0.084
Aflatoxin content	Correlation coefficient	0.120	−0.090	1.000	0.276*
	Significance	0.354	0.501		0.030
Total score	Correlation coefficient	0.396**	0.229	0.276*	1.000
	Significance	0.001	0.084	0.030	

5.2.1 Model Reliability

It could be seen from Table 5.29 that there was a highly significant correlation between the hundred kernel weight and aflatoxin content and the total score, while the correlation between oleic acid/linoleic acid ratio and total score was slightly low, but the significance was below 0.1, which indicated that these three indicators were important factors affecting the quality of peeled peanut kernel.

The significance of correlation coefficients between these three indicators (hundred kernel weight, oleic acid/linoleic acid ratio, and aflatoxin content) was greater than 0.3, indicating that these three indicators were relatively independent and could reflect the relationship with peeled peanut kernel quality individually.

5.2.2 Suitability

Sixty-five varieties were divided into three categories by using the K-means cluster analysis method (Table 5.30).

6 DNA Fingerprint of Special Variety for Peanut Processing

DNA molecular marker technology has played a significant role in genetic map construction, genetic diversity analysis, and target gene mapping. Since the creation of RFLP technology, more than ten kinds of new marker technologies have been developed, and SSR and AFLP have been widely used. SSR marker was a highly

Table 5.30 Suitability class standards of peanut suitable for peeled export peanut kernel

Cluster no.	Number of varieties	Variety name
Suitable	15 (>49 分)	Huayu 8, Huayu 16, Huayu 31, Baisha 101, Shanhua 7, Shanhua 9, Haihua 1, Honghua 1, Xuhua 14, Hongguan, Colorful Peanut, 034-256-1, Ji 9814, Zhengnong 7, Yuanza 9102
Basically suitable	30 (45–49 分)	Huayu 19, Huayu 22, Huayu 23, Huayu 25, Huayu 28, Luhua 9, Luhua 14, Fenghua 1, Fenghua 3, Fenghua 5, Lufeng 2, Variety 060, 9616, Baisha 1016, Zhongnong 108, Haiyu 6, Qinglan 8, Bianhua 3, Shuangji 2, Yuanhua 8, Xuhua 13, Zhongyu 1, Black Peanut, Huaguanwang, 818, 600-6, 780-15, 365-1, 03-329-336, L03-601-604
Unsuitable	20 (<45 分)	Huayu 20, Luhua 11, Luhua 15, Fenghua 4, Fenghua 6, Zhonghua 4, Zhonghua 8, Zhonghua 15, Xuhua 5, Xuhua 15, Kainongbai 2, White Peanut, 060-18-w, Fuhua 1, Silihong, 8130, Hua 17, Xianghua 509-77, 606, L03-257-294

conserved single sequence in the genome, the artificial synthetic primers and PCR amplification could be made through SSR marker, the changes in the length of amplified products would cause polymorphisms and could be used for resource polymorphism analysis and gene mapping of crops, and it had been used in rice (Zhao Hong et al. 2012), wheat (Gao Fengmei et al. 2012), and other grain crops maturely. Its advantage is the large quantity which exists in the entire genome with more allelic variation, and the heterozygotes and homozygous can be identified. At present, SSR marker had assisted in the preliminary research in the peanut breeding, and the research mainly focused on the peanut genetic map construction, peanut disease resistance, agronomic characters, quality characters, and other aspects (Moretzsohn et al. 2005; Wang et al. 2008; Hong Yanbin et al. 2009a, b; Zhao 2011). AFLP is used to detect the length polymorphism of cleavage fragment through selective amplification after it conducts restriction enzyme digestion for the genomic DNA. The number of bands of AFLP amplified fragments is related to the types of endonucleases and primers, and it has been used in germplasm polymorphism analysis, genetic map construction, and gene mapping. The advantages are that there are many types of restriction endonucleases analyzed, there are many bands in the reaction products each time, and multiple locia can be detected through analysis at one time with high polymorphism.

There has been preliminary research on the peanut disease resistance, agronomic characters and quality characters, and DNA genetic marker, and good effects have been achieved. It is preliminarily judged that gene is closely related to the peanut quality characteristic, but the research on the peanut processing characteristics does not involve the gene aspect, especially on the gelation and solubility of peanut protein, oxidation stability of peanut oil and export peanut, and other excellent processing characteristics. Therefore, on the basis of the research on the gelation and solubility of peanut protein, oxidative stability of peanut oil, and peanut varieties suitable for export peanut, the author's research team analyzed the genotypes of different peanut varieties by using DNA genetic marker technology so as to illuminate the existing intrinsic relationship between peanut processing characteristics and genes and provide the theoretical basis for the cultivation of special peanut varieties.

6.1 DNA Fingerprint of Special Variety Suitable for Peanut Protein Processing

Wang (Wang 2012; Wang et al. 2012) found that the protein gelation, solubility, and other functional properties of different peanut varieties were significantly different; the protein of Luhua 11, Huayu 22, Shuangji 2, and other varieties had good gelation; and the protein of Shanhua 7, Yuhua 9326, Baisha 1016, Yuhua 15, Colorful Peanut, Hua 67, White Peanut, Luhua 15, Luhua 8, Guihua 35, Baisha 101, Yueyou 25, Lanhua 2, Yuhua 9327, Yuhua 11, and other varieties had good solubility. However, the correlation between these good processing characteristics and peanut variety genes had been unclear. The in-depth research on the DNA fingerprints of special varieties and their correlation with gelation and solubility is of guiding significance to directionally select and cultivate the special varieties with good processing characters and guarantee the supply of peanut protein production raw materials.

6.1.1 Gelation and Solubility of Peanut Protein

Twenty-seven peanut varieties with good protein processing characteristics determined by the author's research team are shown in Table 5.31. One to 14 are the varieties with good gelation, and 16–27 are the varieties with good solubility.

6.1.2 Peanut DNA Extraction Method

Refer to the CTAB method researched by Haymes (1996) and make a little change. The specific steps are shown below:

1. Culture the peanut in the greenhouse for 2 weeks, take 0.2–0.3 g of peanut leaves after the seedling leaf grows and expands, and grind these leaves into powder quickly under the liquid nitrogen freezing conditions.
2. Transfer the powder to a 1.5 ml centrifuge tube in the frozen state, add 0.75 ml of 2×CTAB extraction buffer preheated at 65 °C, and mix them adequately.
3. Place the centrifuge tube in a water bath at 65 °C for 0.5 h and gently mix them once every 10 min in the water bath process.
4. Centrifuge the solution for 10 min at 10,000 r/min and transfer the supernatant to another 1.5 ml centrifuge tube.

Table 5.31 Peanut varieties suitable for processing of gel-type and soluble protein

No.	Variety	Gelation	Solubility	No.	Variety	Gelation	Solubility
1	Quanhua 464	1.27	82.82	15	Yuhua 9326	0.88	93.44
2	Luhua 11	1.27	88.65	16	Baisha 1016	0.81	93.02
3	Shuangji 2	1.26	87.72	17	Colorful Peanut	0.78	91.55
4	Bianhua 3	1.24	–	18	Hua 67	1.04	91.19
5	Fenghua 1	1.23	80.88	19	White Peanut	0.88	90.68
6	Kainong 30	1.23	77.71	20	Luhua 15	0.72	89.97
7	Fenghua 3	1.22	77.76	21	Luhua 8	1.10	89.96
8	Zhongkaihua 4	1.21	85.71	22	Guihua 35	0.84	89.68
9	Huaguanwang	1.20	83.71	23	Baisha 101	1.05	89.53
10	Shanhua 7	1.18	93.93	24	Yueyou 25	0.83	89.41
11	Minhua 9	1.17	80.80	25	Lanhua 2	0.96	89.18
12	Yuhua 15	1.16	91.72	26	Yuhua 9327	0.90	89.03
13	Haihua 1	1.15	86.79	27	Yuhua 11	0.93	88.91
14	Guihua 166	1.15	87.37				

5. Add an equal volume of Tris saturated phenol, mix them, and centrifuge the solution for 10 min at 10,000 r/min.
6. Transfer the supernatant to a 1.5 ml centrifuge tube, add an equal volume of phenol/chloroform (phenol/chloroform/isoamyl alcohol = 25:24:1) to each centrifuge tube, mix them gently, and centrifuge the solution for 10 min at 10,000 r/min.
7. Transfer the supernatant to a 1.5 ml centrifuge tube, add an equal volume of chloroform/isoamyl alcohol (24:1) to each centrifuge tube, mix them gently, and centrifuge the solution for 10 min at 10,000 r/min.
8. Transfer the supernatant to a 1.5 ml centrifuge tube, add 1/10 volume of 3 MNaAc, mix them gently, then add double volumes of absolute ethanol, mix them adequately, and place it at −20 °C for 20 min.
9. Centrifuge the solution for 10 min at 12,000 r/min, discard the supernatant, add 0.75 ml of 70% ethanol, and place it at 4 °C for 20 min.
10. Centrifuge the solution for 5 min at 12,000 r/min, discard the supernatant, and rinse it with 70% ethanol two to three times.
11. Dry it at 4 °C and add 0.75 ml 1×TE (pH = 8.0) to dissolve DNA.
12. Add 5 μl of Rnase (10 g/L) and make it react at 37 °C for 1 h to remove RNA.
13. Add an equal volume of chloroform/isoamyl alcohol (24:1), mix them gently, and centrifuge the solution for 10 min at 10,000 r/min.
14. Take the supernatant, add an equal volume (0.75 ml) of chloroform, mix them gently, and centrifuge the solution for 10 min at 10,000 r/min.
15. Take the supernatant, add 1/10 volume of 3 M NaAc, mix them adequately, add double volume of anhydrous ethanol, mix them adequately, and then place it at −20 °C for 20 min to make DNA form precipitation.
16. Centrifuge the solution for 10 min at 12,000 r/min, discard the supernatant, add 0.75 ml of 70% ethanol, and place it at 4 °C for 20 min.

17. Centrifuge the solution for 5 min at 12,000 r/min, discard the supernatant, and rinse it with 70% ethanol two to three times.
18. Dry it at 4 °C and make DNA redissolved in 0.1 ml of 1×TE. At 100 V voltage, detect the integrity of DNA by using the 1% agarose gel electrophoresis; determine the values of DNA extraction solution OD260, OD280, and OD230 and OD260/OD280 and OD260/OD230 values by using the nucleic acid protein tester; and then store them in refrigerator at −20 °C for future use.

6.1.3 Amplification Polymorphism of SSR Primers

Thirty-five pairs of SSR primers were used to conduct amplification analysis for DNAs of 27 peanut varieties, and the results showed that 15 pairs of primers did not have amplification products, 13 pairs of primers had amplification products, but the products did not show polymorphism. The polymorphism was displayed by seven pairs of primers. The polymorphism amplification results of these seven pairs of primers are shown in Table 5.32.

It could be seen from Table 5.32 that 17 bands were amplified by 7 pairs of primers; among them, 15 bands were polymorphic, 2.1 bands were produced by each pair of primers averagely, and the polymorphic frequency was 88.24%; 4 special bands were amplified by the No. 13 primer with the best polymorphism, and 1 band was amplified by No. 8, No. 25, or No. 35 with the worst polymorphism, which indicated that there was abundant variability among the 27 peanut varieties. Figures 5.1 and 5.2 showed the electrophoretograms of different primers. From Fig. 5.1 , it could be seen that No. 4 primer could divide these nine peanut varieties and other varieties into two categories and these nine varieties were Quanhua 464, Minhua 9, Haihua 1, Yuhua 9326, Baisha 1016, White Peanut 67, Luhua 8, Guihua 35, and Guihua 35. Through the comparison of Fig. 5.1, it was found that the 11 varieties had larger molecular weight fragments in the SSR marker electrophoretograms of No. 1–14 varieties. It was also found that the peanut varieties with small fragments were mainly concentrated in No. 16–27 varieties. Therefore, it could be inferred that these large-fragment SSR markers might be related to the gelation of peanut protein, while the small fragments might be related to the solubility of peanut protein. It could be seen from Fig. 5.2 that there were significant differences in the bands between various varieties, for example, there were three different types of genotype bands among No. 1–14 peanuts, which indicated that, among the varieties with good gelation, the gelation difference showed the

Table 5.32 Polymorphism of SSR marker

Primer no.	Amplified band	Polymorphism band	Primer no.	Amplified band	Polymorphism band
4	3	2	25	1	1
5	3	3	33	3	3
8	1	1	35	1	1
13	5	4	Total	17	15

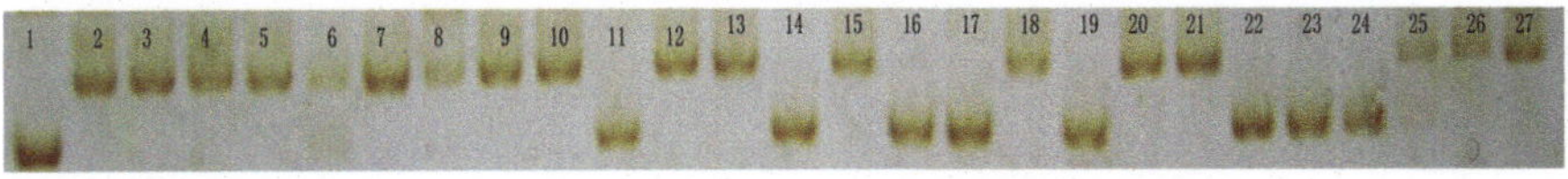

Fig. 5.1 Electrophoretogram of SSR amplification products of No. 4 primer

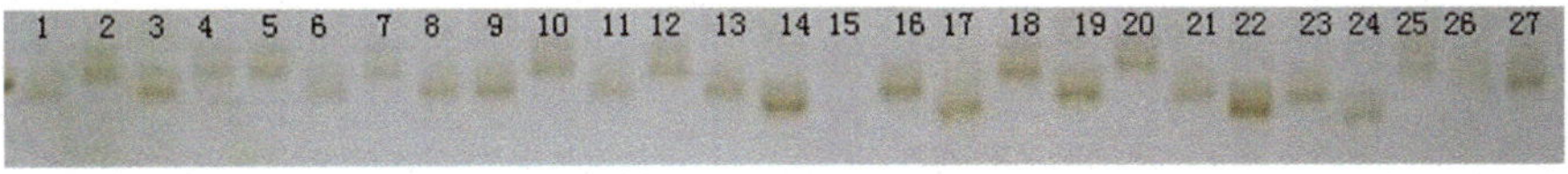

Fig. 5.2 Electrophoretogram of SSR amplification products of No. 33 primer

differences in genotype bands. The differences in SSR marker amplification products of peanut varieties might be related to the processing characteristics of peanut protein, and it could be concluded that the differences of these bands might lead to the differences of processing characteristics.

6.1.4 Cluster Analysis of SSR Amplification Polymorphism

According to the electrophoregrams of amplification products of seven pairs of primers, at the same movement position, the varieties with clear bands were recorded as 1, and those without clear bands were recorded as 0. 0–1 EXCEL matrix was established (Zhao 2011), cluster analysis was carried out for the data by using the NTSYS-2.1 software, and the tree diagram was drawn (Fig. 5.3). At the similarity coefficient of 0.66, 27 peanut varieties could be divided into four groups. Kainong 30 was divided as the first group individually; Lanhua 2 was divided as the second group individually; the third group included Chunhua 464, Minhua 9, Baisha 1016, Baisha 101, White Peanut, Guihua 166, Colorful Peanut, Guihua 35, and Yueyou 25; and the rest were divided as the fourth group. From the cluster results, we could generally see the variety characteristics and genetic relationship; it could be seen from Fig. 5.3 that the protein solubility of all the varieties in the third group was good; in the fourth group, when the similarity coefficient was 0.9, it could be seen that No. 2, 5, and 12 varieties with good gelation were gathered into one category. However, it was also found that these seven pairs of primers selected in this research could distinguish the most peanut varieties and cluster the varieties with the similar processing characteristics together, but a few peanut varieties could not be distinguished, and the varieties with the similar processing characteristics could not be clustered, such as Quanhua 464, Minhua 9, Baisha 1016, and Baisha 101; the similarity coefficient of these four peanut varieties was 1.00, which indicated that these seven pairs of primers could not distinguish the differences between the four peanut varieties. The molecular polymorphism in peanut cultivars was relatively deficient, there were less SSR markers of peanuts developed at present, the number of primers selected was not enough, the amount of genetic map information of peanut that had been established was small (Hong Yanbin et al.

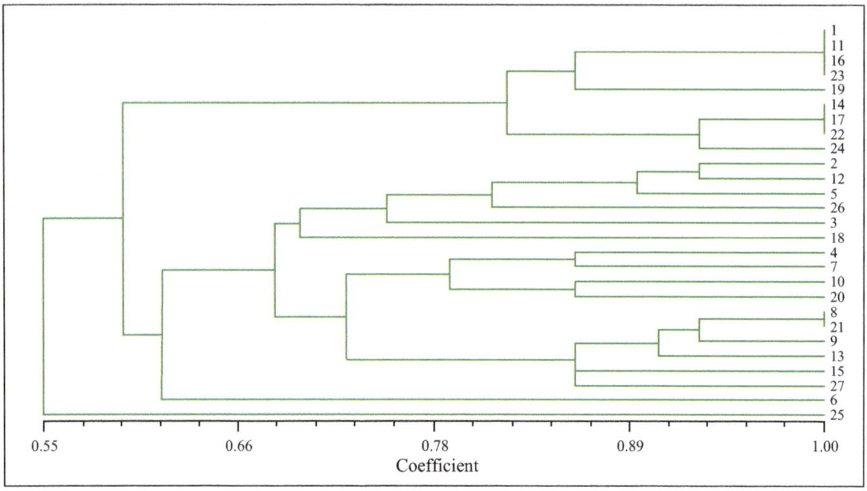

Fig. 5.3 SSR cluster diagram of 27 peanut varieties (Note: 1–27 refers to the first to the 27th peanut varieties and the variety names are shown in Table 5.26)

2009a, b), and the functions of many genes had not yet been determined; therefore, it could be guaranteed that the whole genomic information of peanut was covered only through mass selection during the polymorphic analysis. In the further research, the number of primers would be increased, the primers with rich polymorphism would be found by mass selection, multiple pairs of primers would be fit to differentiate the peanut varieties, the fingerprints of peanut varieties with good processing characteristics would be established, and the primers obtained through selection would be used to identify the new peanut varieties and predict their processing characteristics.

6.1.5 Correlation Analysis Between SSR Amplification Polymorphism and Processing Characteristics

The electrophoretogram bands of each primer were assigned, the bands with the largest movement position were recorded as 1, the band assignment was increased by 1 successively with the reduction of movement position, and the sum of band assignment of each peanut variety was considered as the genotype data of the variety (Hong Yanbin et al. 2009a, b); the correlation analysis was carried out for SSR genotype and functional characteristic data of peanut by using the SAS 9.1 software, and the results were shown in Table 5.33. The correlation coefficient of amplification result of No. 33 primer and gelatin was 0.477, which showed a significant correlation. It could be inferred that there might be a certain correlation between the amplified fragment of No. 33 primer and gelation of peanut protein. In addition, the amplification results of No. 4, 5, and 33 primers were positively correlated with the gelation and negatively correlated with the solubility, and

Table 5.33 Correlation between SSR results and functional characteristics

Primer no.	Correlation coefficient		Primer no.	Correlation coefficient	
	With gelation	With solubility		With gelation	With solubility
4	0.350	−0.095	25	0.267	0.054
5	0.344	−0.174	33	0.477	−0.222
8	0.081	0.111	35	−0.384	0.258
13	0.138	0.228			

No. 35 primer was positively correlated with the gelation and negatively correlated with the solubility, so it could be inferred that there was a negative correlation between gelation and solubility, which was consistent with the correlation between gelation and solubility shown in Table 5.26. The correlation between the primer amplification results and peanut processing characteristics did not reach the highly significance level, because the number of primers selected was too small, the correlation was greatly affected by the composition of peanut group structure, and the germplasm group with high genetic diversity was the premise of correlation analysis between phenotype and SSR marker (Hong Yanbin et al. 2009a, b). The number of peanut varieties selected in this book was only 27, and these 27 varieties had good performance in the aspect of processing characteristics (protein solubility and gelation). Therefore, these 27 peanut varieties might not contain all the genetic variations of this species. In the further research, the number of peanut varieties would be increased, and the varieties with poor performance in processing characteristics would be added to increase the diversity of peanut group structure.

In this book, the preliminary analysis was carried out for the peanut varieties suitable for processing of gel-type and soluble protein by using the SSR molecular marker method, and it was found that there was a certain correlation between processing characteristics and SSR molecular markers. Cluster analysis was carried out for 27 peanut varieties, and most of the varieties with the similar processing characteristics were divided as one category. Finally, the correlation analysis between SSR markers and processing characteristics of peanut was carried out, and the gene mapping was conducted preliminarily. At present, the research on peanut gene mapping mainly involves agronomic characters, insect and disease resistance, oil and protein contents, and quantitative characters, and there are few research reports on the gene mapping of peanut processing characteristics, for example, it has not been verified whether the solubility and gelation of peanut protein belong to the quantitative characters. There are many factors that affect the peanut processing characteristics, and the genotype, growing environment, and even processing methods will affect the processing characteristics. Therefore, in the further research, it will be verified whether the processing characteristics belong to the quantitative characters and meet the law of inheritance through the analysis of these complex factors, and then the processing characteristics will be positioned in a gene fragment by using the SSR marker method, so as to guide the breeding to directively cultivate the desired peanut processing characteristics after production and solve the problems encountered in the later modification of peanut protein.

Table 5.34 Oxidation stability of different varieties of peanut oil

No.	Variety	Oxidation stability of peanut oil	No.	Variety	Oxidation stability of peanut oil	No.	Variety	Oxidation stability of peanut oil
1	Luhua 11	9.12	6	Fenghua 3	−10.39	11	Baisha 1016	8.55
2	Shuangji 2	13.03	7	Shanhua 7	17.58	12	Colorful Peanut	8.77
3	Bianhua 3	10.73	8	Yuhua 15	−13.73	13	White Peanut	10.29
4	Fenghua 1	8.55	9	Haihua 1	10.59	14	Yuhua 9327	−10.22
5	Kainong 30	−10.22	10	Yuhua 9326	−10.07			

6.2 *DNA Fingerprint of Special Variety Suitable for Peanut Oil Processing*

6.2.1 Processing Characteristics of Peanut Oil

There were significant differences in the oil content and oxidation stability of different peanut varieties (Liu Dachuan et al. 2008; Bockisch 1998; Moore and Knauft 1989; Bovi 1983), and there was a lack of special peanut varieties suitable for peanut oil processing, so it was urgently needed to directively select and cultivate the high-quality new varieties suitable for peanut oil processing through searching the correlation between peanut oil quality characteristics and genes so as to meet the growing demand for high-quality peanut oil of China. Table 5.34 showed the varieties suitable for peanut oil processing and oil oxidation stability data which had been determined by the author's research team.

6.2.2 Amplification Polymorphism of SSR Primer

The analysis of SSR amplification polymorphism was the same as that in Sect. 6.1.3.

6.2.3 Cluster Analysis of SSR Amplification Polymorphism

According to the electrophoregrams of amplification products of seven pairs of primers, at the same movement position, the varieties with clear bands were recorded as 1, and those without clear bands were recorded as 0. 0–1 EXCEL matrix was established, cluster analysis was carried out for the data by using the NTSYS-2.1 software, and the tree diagram was drawn (Fig. 5.4). When the similarity coefficient was 0.67, 14 peanut varieties could be divided into three groups.

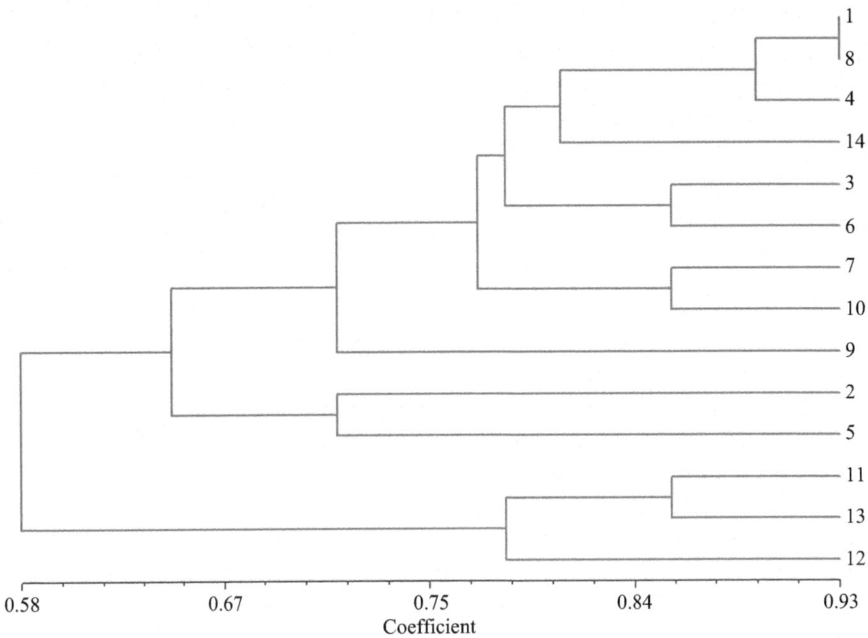

Fig. 5.4 SSR cluster diagram of 14 peanut varieties (Note: 1–14 refers to the first to the 14th peanut varieties and the variety names are shown in Table 5.23)

Baisha 1016, Colorful Peanut, and White Peanut were the first group, Shuangji 2 and Kainong 30 were the second group, and the rest were the third group. It could be seen from Fig. 5.4 that the peanut oil processing suitability of the varieties in the first group was good, but the rest of the peanut varieties were difficult to classify, which indicated that these seven pairs of primers could not distinguish the differences between the four peanut varieties. The molecular polymorphism in peanut cultivars was relatively deficient, there were less SSR markers of peanuts developed at present, the number of primers selected was not enough, the amount of genetic map information of peanut that had been established was small (Hong Yanbin et al. 2009a, b), and the functions of many genes had not yet been determined; therefore, it could be guaranteed that the whole genomic information of peanut was covered only through mass selection during the polymorphic analysis.

6.2.4 Correlation Analysis Between SSR Amplification Polymorphism and Processing Characteristics

The electrophoretogram bands of each primer were assigned, the bands with the largest movement position were recorded as 1, the band assignment was increased by 1 successively with the reduction of movement position, and the sum of band assignment of each peanut variety was considered as the genotype data of the

Table 5.35 Correlation between SSR results and oxidation stability of peanut oil

Primer no.	Correlation coefficient	Primer no.	Correlation coefficient	Primer no.	Correlation coefficient
4	−0.23	13	−0.18	33	0.07
5	−0.3	25	0.41	35	0.31
8	−0.25				

variety; the correlation analysis was carried out for SSR genotype and peanut oil processing suitability of peanut by using the SAS 9.1 software. The results are shown in Table 5.35. The correlation coefficient of amplification result of No. 25 primer and peanut oil processing suitability was 0.41, which showed a significant correlation. It could be inferred that there might be a certain correlation between the amplified fragment of No. 25 primer and peanut oil processing suitability. The correlation between the amplification results of other primers and oxidation stability of peanut oil did not reach the highly significance level, because the number of primers selected was too small, the correlation was greatly affected by the composition of peanut group structure, and the germplasm group with high genetic diversity was the premise of correlation analysis between phenotype and SSR marker (Hong Yanbin et al. et al. 2009a, b).

6.3 *DNA Fingerprint of Special Variety Suitable for Export Peanut Processing*

Ten kinds of peanut varieties with good export performance were identified by using AFLP molecular marker technology in order to establish the DNA fingerprints of export peanut varieties. The number of these ten kinds of peanut varieties was as follows: 1, Huayu 20; 2, Huayu 22; 3, Luhua 9; 4, Fenghua 6; 5, White Peanut; 6, 8130; 7, Black Peanut; 8, Baisha 1016; 9, Fenghua 1; and 10, Luhua 15.

6.3.1 Extraction Method of Young Leave Genomic DNA of Peanut

The peanut genomic DNA was extracted by CTAB method. As shown in Fig. 5.5, the bands were clear, and the molecular weight was normal (more than 5000 bp) without obvious degradation phenomenon, so they could be used for subsequent experiments.

6.3.2 Enzyme-Digested Product

The peanut genomic DNA was digested by using the restriction endonucleases EcoR I and Mse I successively, and then the agarose gel electrophoresis detection was carried out. Figure 5.6 showed the EcoR 1 enzyme-digested products of ten

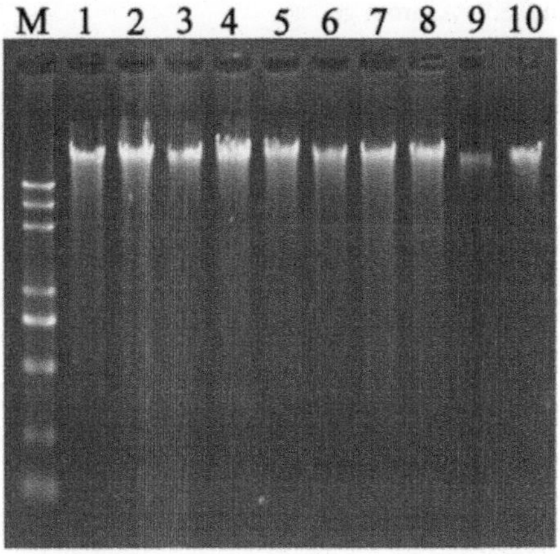

Fig. 5.5 Agarose gel electrophoretogram of peanut genomic DNA

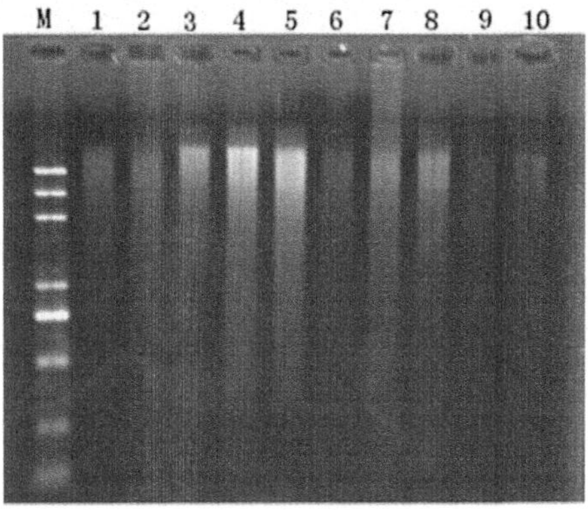

Fig. 5.6 Agarose gel electrophoretogram of EcoR 1 enzyme-digested products

groups of peanuts, and continuous bands occurred in ten lanes with large molecular weight. After the further Mse I digestion, the fragment size was significantly reduced, the concentration was lowered, and continuous bands occurred only in Lane 4 at about 100–500 bp. See Fig. 5.7.

6.3.3 Primer Selection

The double enzyme-digested products of genomic DNA were recovered, and PCR amplification was carried out after the linkers were connected. The designed and selected primers and linkers are shown in Table 5.36.

The T4 DNA ligation products of ten groups of peanuts were detected by the agarose gel electrophoresis, and vague bands occurred in ten lanes below 100 bp, as shown in Fig. 5.8.

The pre-amplification products of ten groups of peanuts were detected by the agarose gel electrophoresis, and except that the bands in Lane 6 were vague, clear

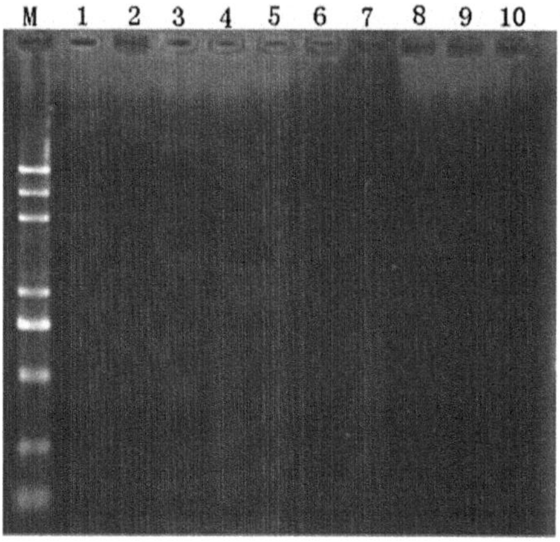

Fig. 5.7 Agarose gel electrophoretogram of double enzyme-digested products

Table 5.36 Linker and sequence of selective amplification primer

Linker	Sequence
1	5-CTCGTAGACTGCGTACC-3
2	5-AATTGGTACGCAGTCTAC-3
3	5-GACGATGAGTCCTGAG-3
4	5-TACTCAGGACTCAT-3
Selective amplification primer	Sequence
E-1	5-GACTGCGTACCAATTCAAC-3
E-2	5-GACTGCGTACCAATTCAAG-3
E-3	5-GACTGCGTACCAATTCACA-3
E-4	5-GACTGCGTACCAATTCACT-3
E-5	5-GACTGCGTACCAATTCACC-3
M-1	5-GATGAGTCCTGAGTAACAA-3
M-2	5-GATGAGTCCTGAGTAACAC-3
M-3	5-GATGAGTCCTGAGTAACTT-3

and bright bands occurred in the other Lanes 1–10 below 100 bp, as shown in Fig. 5.9.

The selective amplification products of ten groups of peanuts were detected by the agarose gel electrophoresis, and clear and bright bands occurred in ten lanes below 100 bp, as shown in Fig. 5.10.

In the AFLP analysis of peanut varieties, 169 bands were recorded in the amplification products of nine pairs of primer combinations; among them, 55 bands were polymorphic, and the polymorphic band ratio was 32.54%. The results showed that the polymorphism of amplified bands of primer combination of E-ACT/M-CAA was the highest (71.43%), followed by that of E-AA G/M-CAC (64.71%), and that of E-AA G/M –CAA was the lowest (6.67%). E-ACC/M-CAA and E-ACT/M-CAA had high variety identification efficiency (80%). The polymorphism of amplified bands of different primer combinations was different, and the variety identification efficiency was also different.

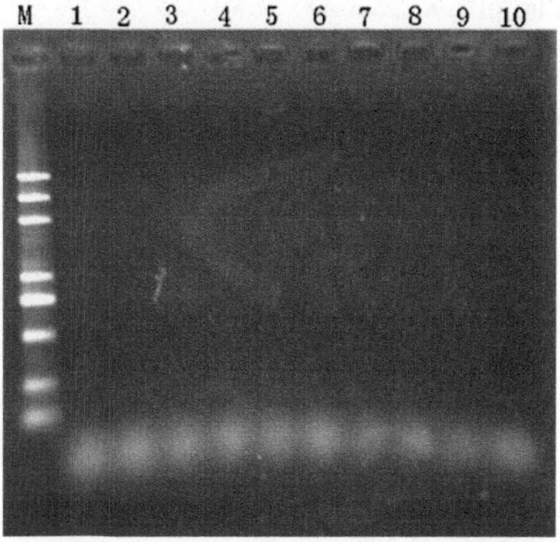

Fig. 5.8 Agarose gel electrophoretogram of T4 DNA ligation products of peanut

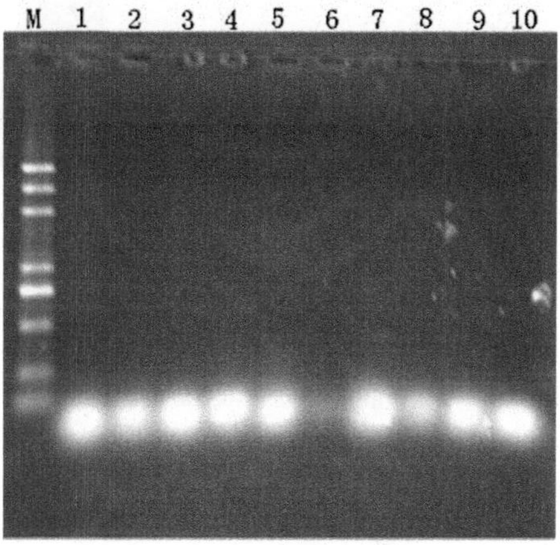

Fig. 5.9 Agarose gel electrophoretogram of pre-amplification products of peanut

6.3.4 Polyacrylamide Electrophoresis

The corresponding band of each variety was observed through the polyacrylamide gel electrophoresis, and the peanut DNA fingerprint library was established. The coordination of ACT/M-CAA primer combination could completely distinguish the different varieties with processing suitability which included E-ACT/ M-CAA/ / E-AA G/ M-CAC, E-ACT/M-CAA/ / E-ACC/ M-CAA, and E-ACT/ M-CAA/ / E-AAC/ M-CTT; in addition, the other two groups of primer combinations (E-ACC/ M-CAA/ /E-A G/ M-CAC and E-ACC/ M-CAA/ / E-A GC/ M-CAA) could also completely distinguish the different varieties, as shown in Fig. 5.11 and Table 5.37.

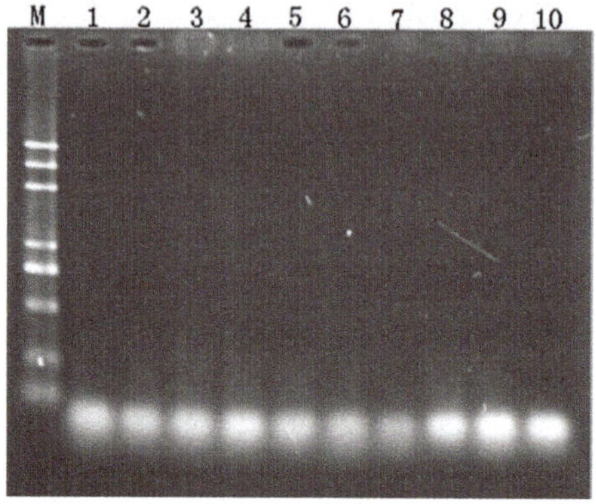

Fig. 5.10 Agarose gel electrophoretogram of selective amplification products of peanut

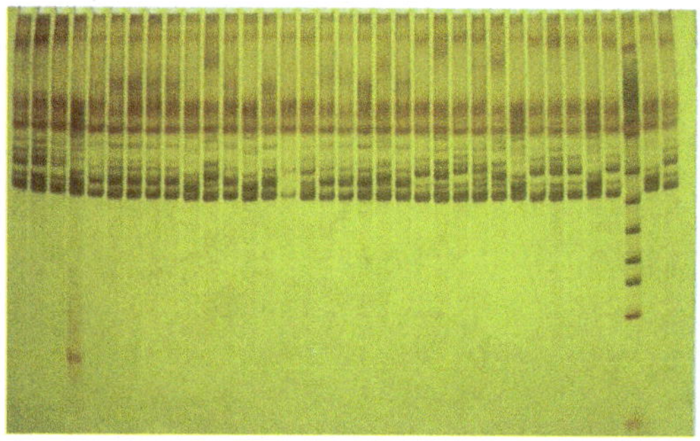

Fig. 5.11 DNA agarose gel electrophoretogram

Table 5.37 Amplification effects of different primer combinations

Primer combination	Number of bands	Number of polymorphic bands	Polymorphic band rate (%)	Identification efficiency (%)
E-AAG/M-CAA	15	1	6.67	10
E-AAG/M-CAC	17	11	64.71	60
E-ACA/M-CAA	14	2	14.29	0
E-ACA/M-CAC	11	4	36.36	30
E-ACT/M-CAA	21	15	71.43	80
E-ACC/M-CAA	24	6	25.00	80
E-ACG/M-CAA	16	4	25.00	20
E-AGC/M-CAA	21	5	23.81	40
E-AAC/M-CTT	30	7	23.33	50
Total	169	55	32.54	–

References

Ahmed EM, Ali T. Textural quality of peanut butter as influenced by peanut seed and oil contents. Peanut Sci. 1986;13(1):18–20.

Bockisch M. Vegetable fats and oils. Fats and oils handbook. Champaign: AOCSPress; 1998. p. 174–326.

Bovi M.L.A. Genotypic and environment effects on fatty acid composition, iodine value, and oil content of peanut (*Arachis hypogaea* L.). NT:University of Florida, 1983,44:406.

Chun J, Ye L, Lee J, Eitenmiller RR. Effect of peanut butter manufacture on vitamin E. J Food Sci. 2003;68(7):2211–4.

Chun J, Lee J, Eitenmiller RR. Vitamin E and oxidative stability during storage of raw and dry roasted peanuts packaged under air and vacuum. J Food Sci. 2005;70(4):C292–7.

Du W, Wei YM, Zhang GG, et al. Study on the relation of wheat property and noodle quality. J Northwest A&F Univ. 2001;29(3):24–8.

Gao FM, Shao LG, Wang Y, et al. SSR molecular characterization and applied research of wheat scab major resistance genes in spring wheat. Heilongjiang Agric Sci. 2012;11:15–20.

Guo BL, Wei YM, Zhang GQ, et al. Study on the relationship between wheat quality and their food quality. J Northwest A&F Univ (Natur Sci Ed). 2001;29(5):61–5.

Guohao H, Ronghua M, Melanie N, et al. Microsatellites as DNA markers in cultivated peanut (*Arachis hypogaea* L.). BMC Plant Biol. 2003;3:3.

Hashim IB, Koehler PE, Eitenmiller RR. Tocopherols in runner and Virginia peanut cultivars at various maturity stages. J Am Oil Chemists' Soc. 1993;70(6):633–5.

Haymes K. Mini-prep method suitable for a plant breeding program. Plant Mol Biol Report. 1996;14:280–4.

Hong YB, Li SX, Liu HY, et al. Correlation analysis of SSR markers and host resistance to aspergillus flavus infection in peanut(Arachis Hypogaea L.). Mol Plant Breed. 2009a;7 (2):360–4.

Hong YB, Liang XQ, Chen XP, et al. Construction of genetic linkage map in peanut (Arachis Hypogaea L.) cultivars. Acta Agron Sin. 2009b;35(3):395–402.

Jiyeon Chun, Junsoo Lee, Ronald R. Eitenmiller, Vitamin E and Oxidative Stability During Storage of Raw and Dry Roasted Peanuts Packaged under Air and Vacuum. Journal of Food Science 70 (4):C292-C297

Kanungo T, Netanyahu NS, Wu AY. An efficient k-means clustering algorithm: analysis and implementation. IEEE Trans Pattern Anal Mechine Intell. 2002;24(7):881–92.

Li SH, Dong SP, Guo YJ, et al. The relationship between the main characters and yield of peanut. Hubei Agric Sci. 2004;(1):49–50.

Li JJ. Study on Chinese soybean varieties for special-use and its suitability on processing. Shi Jiazhuang: Hebei Agricultural University, 2009a.

Li WQ. Selection of SSR Markers specific to Fenghua cultivars and genetic analysis of physiological trait in peanut. Shandong: Shandong Agricultural University, 2009b.

Li ZX, Wei YM, Zhang JG. Study on the effects of wheat protein Components on dough properties and baking quality. J Chin Cereals Oils Assoc. 1998;13(3):1–5.

Li LQ, Zan LS, Ren ZD. Fatty acid composition, Antioxidation and rheological behavior of Chinese Giant salamander (Andrias Davidianus) oil. J Chin Cereals Oils Assoc. 2009;24 (9):83–5.

Li HY, Zheng ZY, Li J, et al. Study on oxidative stability of plant oils with different fatty acid composition. Sci Technol Food Ind. 2010;31(1):173–82.

Liang ZF. Study on the relationship of oil and protein contents and some qualitative characters of soybeans. Sci Agric Sin. 1982;5:48–56.

Lin JH. Correlation and path analysis of protein content and major quantitative characters in peanut seeds. Acta Agric Jiangxi. 2006;18(4):38–9.

Liu ZQ, Zeng YL, et al. Effect of enzymatic limited hydrolysis on functional properties of peanut protein. Food Sci. 2004;25(1):69–72.

Liu DC, Sun W, Yu BQ, et al. Simultaneous preparation of peanut oil and defatted protein powder by low-temperature prepressing, extraction and low-temperature desolventizing. China Oils Fats. 2008;12:13–5.

Misra JB. A mathematical approach to comprehensive evaluation of quality in groundnut. J Food Compos Anal. 2004;17:69–79.

Moore KM, Knauft DA. The inheritance of high oleic acid in peanut [J]. Heredity. 1989;80(3):8–10.

Moretzsohn MC, Leoi L, Proite K, et al. A microsatellite-based ,gene-rich linkage map for the AA genome of A rachis (Fabaceae). Theor Appl Genet. 2005;111:1060–71.

Murphy PA, Chen HP, Hauck CC, et al. Soybean protein composition and tofu quality. Food Technol. 1997;51(1):86–66.

Navnitkumar K, Dhamsaniya, Patel NC, Dabhi MN. Selection of groundnut variety for making a good quality peanut butter. J Food Sci Technol [J]. 2012;49(1):115–8.

Özcan MM. Some nutritional characteristics of kernel and oil of peanut (Arachis hypogaea L.) [J]. J Oleo Sci. 2010;59(1):1–5.

Pu B, Li JF, Zhou F. Study on characteristics and adaptability of material for frozen- dried instant Rice. J Chin Cereals Oils Assoc. 2007;22(1):9–15.

Ramesh V, Ramar K, Babu S. Paraiiel k-means algorithm on agricultural databases. Int J Comput Scie Issues. 2013;1(1):710–3.

Relkin P. Reversibility of heat-induced conformational changes and surface exposed hydrophobic clusters of β-lactoglobulin: their role in heat-induced sol-gel state transition. Int J Biol Macromol. 1998;22:59–66.

Riveros CG, Mestrallet MG, Gayol MF. Effect of storage on chemical and sensory profiles of peanut pastes prepared with higholeic and normal peanuts. J Sci Food Agric. 2010;90:2694–9.

Sekhon KS, Gupta SK, Ahuja KL, et al. Variability in fatty acid composition in semi-spreading peanut types. Oleagineux. 1980;35(8):409–12.

Shi JL, Wei YM, Zhang GQ. Influence of protein and starch content on doodles quality. J Henan Univ Technol. 2001a;22(1):32–5.

Shi JL, Wei YM, Zhang GQ. Influence of wheat protein and starch on Chinese-style white salted noodle and fried instant noodle quality and microstructure. J Northwest A&F Univ (Natur Sci Ed). 2001b;29(1):42–50.

Shi YG, Cheng CL, Zhu XQ. The effect of soybean variety on gelling function of SPI. J Chin Cereals Oils Assoc. 2005;3:58–60, 69.

Su QQ. Different peanut several qualities characters and main quantity character correlation analysis. Chin Agric Sci Bull. 2006;22(6):192–4.

Tang RH. Genetic diversity and molecular classification of germplasm in genus arachis. Fujian: Fujian Agriculture and Forestry University, 2004.

Wang L. Research on processing characteristics and quality evaluation of protein suitability of peanuts. Beijing: Chinese Academy of Agricultural Sciences, 2012.

Wang L, Zhang YH. Research progress on gel forming property and application of soybean protein isolate. J Chin Cereals Oils Assoc. 2010;25(4):96–9.

Wang XS, Yang XQ, Gao WR. Gel texture property of soybean protein isolates from cultivars with different protein subunit contents. J Chin Cereals Oils Assoc. 2006;3:116–21.

Wang H, Shi MY, Ren Y, et al. Development of SSR markers for root-knot nematode resistance in peanut. J Peanut Sci. 2008;37(2):14–7.

Wang L, Wang Q, Liu HZ, et al. Evaluation methods of gel property of peanut protein isolate with different varieties. China Oils Fats. 2012;7:20–3.

Wei YM. Quality evaluation and quality improvement of wheat. Shaanxi J Agric Sci. 1989;2:46–7.

Yang XQ, et al. Separation, purification and characterization of 2S protein from peanut (Arachis hypogaea L.) seeds. J S China Univ Technol. 1998;4:1–5.

Yin DM, Li SZ, Cui DQ. Agronomic character and cluster analysis of peanut cultivars. Chin J Oil Crop Sci. 2010;32(2):212–6.

Zhang GQ, Wei YM, Ou Yang SH, et al. Noodle quality characters of wheat varieties. J Chin Cereals Oils Assoc. 2000;15(3):5–8.
Zhao XY. Association analysis of SSR markers with oil content in wild arachis accessions. Beijing: Chinese Academy of Agricultural Sciences, 2011.
Zhao H, Zhang HF, Lin M, et al. Study on detection of seed purity of two-line hybrid Rice by simple sequence repeat(SSR) V.The veracity of purity detection on the same variety with different Parimers. Hubei Agric Sci. 2012;51(20):4–7.

Chapter 6
Functional Improvement of Peanut Protein Concentrate

The natural peanut protein has solubility, emulsibility, foamability, gelation, and other functional properties. But the modification is still needed to improve the functional properties of protein so as to meet its application requirements in different processed products. For example, the solubility is improved to meet the processing needs of plant protein beverage, and the gelation is improved to meet the application of peanut protein in meat products. Therefore, Wu Haiwen et al. (2009) prepared the peanut protein concentrate (PPC) by using the alcohol precipitation process and modified the protein by using physical, chemical, and biological enzymatic methods; Ma (2009) modified the PPC by using physical method and protease, prepared the peanut protein concentrate products with strong gelatinous and high solubility, and researched the factors that affect the modification.

1 Optimization of Peanut Protein Concentrate Preparation Process

The common protein extraction methods include isoelectric point precipitation method, ethanol precipitation extraction method, isoelectric point precipitation and ethanol precipitation combination method, and alkali extraction and acid precipitation method. The ethanol precipitation extraction method is to use the insolubility of peanut protein in ethanol to remove the alcohol-soluble impurities in raw materials through ethanol extraction and then obtain the peanut protein concentrate through precipitation and drying. Compared with the alkali extraction and acid precipitation method, the ethanol precipitation extraction method could reduce the environmental pollution caused by acid and alkali. And the peanut protein concentrate prepared by using this method also had relatively good functional characteristics, such as solubility, water absorption, oil absorption, emulsibility,

Q. Wang, *Peanut Processing Characteristics and Quality Evaluation*,
DOI 10.1007/978-981-10-6175-2_6

foamability, and rheological properties (Wu et al. 2009). In this research, the process of extracting peanut protein concentrate from defatted peanut protein powder was optimized, and the impact of each process parameter on the product protein content was investigated.

1.1 Single-Factor Experiment of Ethanol During the First Extraction

1.1.1 Solid-Liquid Ratio

The impact of solid-liquid ratio on the protein content during the first extraction is shown in Fig. 6.1. When the solid-liquid ratio increased from 1: 4 to 1: 10, the protein content increased gradually. The protein content decreased slightly when the solid-liquid ratio increased to 1: 16. Because the excess solvent would increase the pressure of subsequent centrifugal separation and result in protein loss, thus the protein content would decrease. Therefore, 1: 10 was determined as the optimal single-factor solid-liquid ratio during the first extraction.

1.1.2 Ethanol Concentration

The impact of ethanol concentration on protein content during the first concentration is shown in Fig. 6.2. The experiment results showed that the protein content increased when the ethanol concentration increased from 60% to 75%. However, when the ethanol concentration continued to increase to 80%, the protein content decreased, because too high ethanol concentration would make the dissolution rate of water-soluble sugar and other impurities in peanut protein powder decrease and a number of residual impurities reduced the protein content in peanut protein concentrate. Therefore, 75% was determined as the optimal single-factor ethanol concentration during the first extraction.

1.1.3 Extraction Time and Temperature

The impact of extraction temperature on protein content during the first extraction is shown in Fig. 6.3. The experiment results showed that the protein content gradually increased when the extraction temperature increased from 30 to 50 °C, because the relatively high temperature would increase the solubility of impurities in the peanut protein powder in ethanol and thus promote the improvement of protein content of product made. The high temperature of more than 50 °C would lead to protein denaturation, and the gel structure formed due to denaturation might bind the dissolution of some small-molecule weight impurities and thus result in a

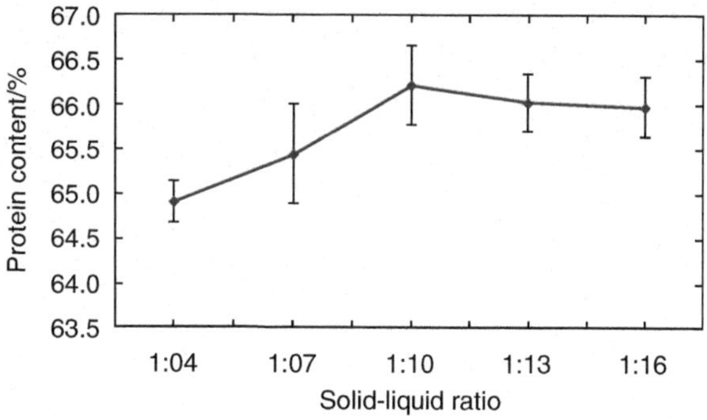

Fig. 6.1 Impact of solid-liquid ratio of ethanol on the protein content during the first extraction

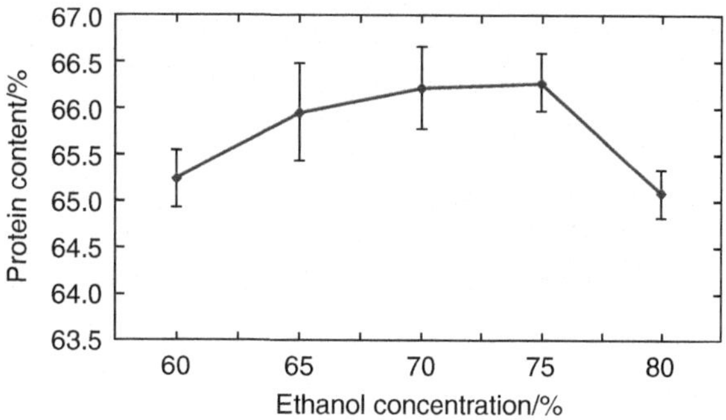

Fig. 6.2 Impact of ethanol concentration on protein content during the first concentration

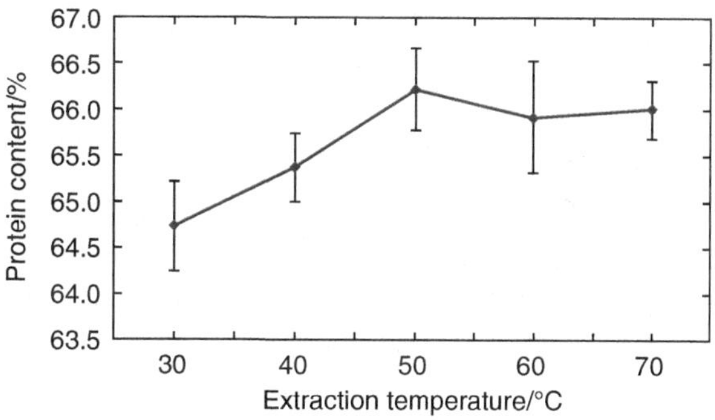

Fig. 6.3 Impact of extraction temperature on protein content during the first extraction

slight decrease in protein content. Therefore, 50 °C was determined as the optimal single-factor extraction temperature during the first extraction.

The impact of extraction time on protein content during the first extraction is shown in Fig. 6.4. The experiment results showed that the protein content increased significantly when the extraction time increased from 20 to 60 min, and the protein content did not change significantly when the time continued to increase. This showed that the complete infiltration of solvents and dissolution of impurities could not be completed instantaneously and a certain time was needed; 60 min was

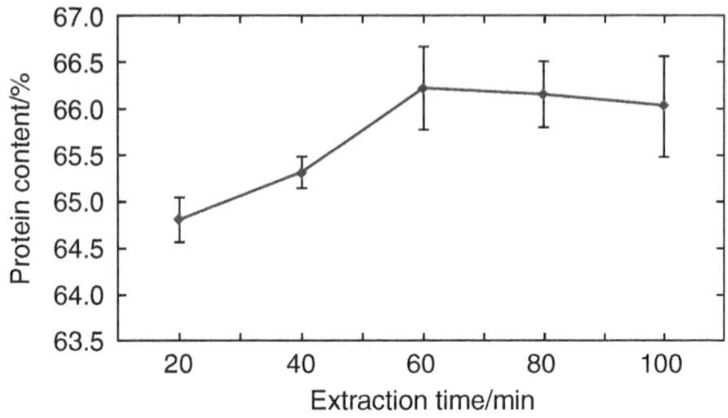

Fig. 6.4 Impact of extraction time on protein content during the first extraction

needed for the first extraction of peanut protein powder, and the dissolution system could not be changed by adding the extraction time because a balance between the solution and solute was formed. Therefore, 60 min was determined as the optimal single-factor extraction time during the first extraction.

1.2 Orthogonal Rotational Combination Experiment of Ethanol During the First Extraction

Through the single-factor experiment, the solid-liquid ratio, ethanol concentration, and extraction temperature were selected as the three factors of response surface optimization process during the first extraction. And the extraction time was directly determined as 60 min. By combining the results of simple orthogonal preexperiment and response surface preexperiment, p_1 was selected as 10 and q_1 was selected as 3; p_2 was selected as 75 and q_2 was selected as 10; p_3 was selected as 50 and q_3 was selected as 10.

The regression equation was obtained by analyzing the experimental determination value during the first extraction. The equation was an empirical correlation between the independent variable coding value and protein content value, and it could be used to predict the change of response value within the given variation range of independent variable. **The regression equation of the protein content $\mathbf{Y_1}$** during the first extraction is shown below:

$$
\begin{aligned}
Y_1 = {} & 66.51554 + 0.30080x_1 - 0.24214x_2 + 0.19607x_3 \\
& - 0.22916x_1^2 - 0.39175x_2^2 + 0.06274x_3^2 \\
& + 0.26449x_1x_2 - 0.19226x_1x_3 + 0.01381x_2x_3
\end{aligned}
$$

The design parameters of central synthetic experiment of response surface during the first extraction are shown in Table 6.1. The experimental determination values and predicted values of model of each experiment were compared in the table.

The significance of each parameter was measured by the Student *t* test and *p* value, as shown in Table 6.2. The greater the absolute value of *t* or the smaller

Table 6.1 Response surface experiment design and experimental determination value and predicted value of model during the first extraction of protein content optimization

No.	Coded value of independent variable level (actual value)			Experimental value			Predicted value
	x_1(w/v)	x_2 (%)	x_3 (°C)	I	II	Mean	
1	1 (1:13)	1 (85)	1 (60)	65.583	66.114	65.849	66.298
2	1 (1:13)	1 (85)	−1 (40)	66.304	66.304	66.304	66.263
3	1 (1:13)	−1 (65)	1 (60)	65.383	67.167	66.275	66.226
4	1 (1:13)	−1 (65)	−1 (40)	66.366	65.642	66.004	66.246
5	−1 (1:7)	1 (85)	1 (60)	66.176	65.236	65.706	65.552
6	−1 (1:7)	1 (85)	−1 (40)	64.559	64.662	64.611	64.748
7	−1 (1:7)	−1 (65)	1 (60)	65.986	66.832	66.409	66.538
8	−1 (1:7)	−1 (65)	−1 (40)	65.802	66.498	66.150	65.789
9	−1.682 (1:4.95)	0 (75)	0 (50)	64.738	65.604	65.171	65.361
10	1.682 (1:15.05)	0 (75)	0 (50)	66.314	67.061	66.688	66.373
11	0 (1:10)	−1.682 (58.18)	0 (50)	65.748	65.748	65.748	65.815
12	0 (1:10)	1.682 (91.82)	0 (50)	65.101	65.280	65.190	65.000
13	0 (1:10)	0 (75)	−1.682 (33.18)	66.033	66.581	66.307	66.363
14	0 (1:10)	0 (75)	1.682 (66.82)	67.148	67.259	67.203	67.023
15	0 (1:10)	0 (75)	0 (50)	65.665	66.786	66.225	66.516
16	0 (1:10)	0 (75)	0 (50)	66.378	66.447	66.412	66.516
17	0 (1:10)	0 (75)	0 (50)	65.962	66.609	66.286	66.516
18	0 (1:10)	0 (75)	0 (50)	66.871	65.922	66.396	66.516
19	0 (1:10)	0 (75)	0 (50)	66.312	67.022	66.667	66.516
20	0 (1:10)	0 (75)	0 (50)	66.696	66.853	66.774	66.516
21	0 (1:10)	0 (75)	0 (50)	66.398	66.869	66.634	66.516
22	0 (1:10)	0 (75)	0 (50)	66.771	66.673	66.722	66.516
23	0 (10:1)	0 (75)	0 (50)	66.743	66.263	66.503	66.516

Table 6.2 Significance analysis of regression equation coefficient during the first extraction of protein content optimization

Variable	Regression coefficient	Standard deviation	Calculated value of t	p value
Constant term	66.51554	0.091195	729.3733	0.000000
x_1	0.30080	0.074076	4.0607	0.001349**
x_2	−0.24214	0.074076	−3.2687	0.006104**
x_3	0.19607	0.074076	2.6469	0.020130*
x_1x_1	−0.22916	0.068665	−3.3373	0.005350*
x_2x_2	−0.39175	0.068665	−5.7052	0.000072**
x_3x_3	0.06274	0.068665	0.9137	0.377529
x_1x_2	0.26449	0.096790	2.7326	0.017097*
x_1x_3	−0.19226	0.096790	−1.9864	0.068488
x_2x_3	0.01381	0.096790	0.1427	0.888711
R	0.9348			
R^2	0.8738			

the p value, the higher the significance of corresponding variable (Amin and Anggoro 2004). It could be seen that the quadratic term (x_2x_2) of ethanol concentration was the most significant, followed by the first term (x_1) and quadratic term (x_1x_1) of solid-liquid ratio. According to the t test and p value, there was no statistically significance in the quadratic term (x_3x_3) of extraction temperature, interaction term (x_1x_3) between solid-liquid ratio, and extraction temperature and interaction term (x_2x_3) between the ethanol concentration and extraction temperature (p value was greater than 0.05). The analysis results showed that the factors influencing the protein content during the first extraction were in the order of priority: ethanol concentration>solid-liquid ratio>extraction temperature.

The correlation coefficient of the regression model was $R = 0.9348$ and the total determination coefficient was $R^2 = 0.8738$, so it was reliable to use the model to predict the protein content during the first extraction.

The response surface of interaction between the solid-liquid ratio and ethanol concentration on the protein content during the first extraction is shown in Fig. 6.5. The experiment results showed that both solid-liquid ratio and ethanol concentration had quadratic effects on the response value in their interactions on the protein content. And the effect of ethanol concentration was greater than that of solid-liquid ratio. The interaction between the solid-liquid ratio and ethanol concentration was significant.

The response surface of interaction between the solid-liquid ratio and extraction temperature on the protein content during the first extraction is shown in Fig. 6.6. The experiment results showed that, in the interaction between the solid-liquid ratio

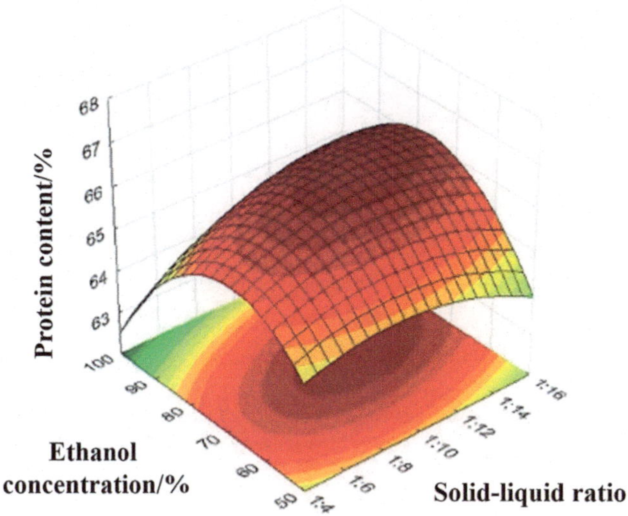

Fig. 6.5 Response surface of impact of solid-liquid ratio and ethanol concentration on protein content during the first concentration

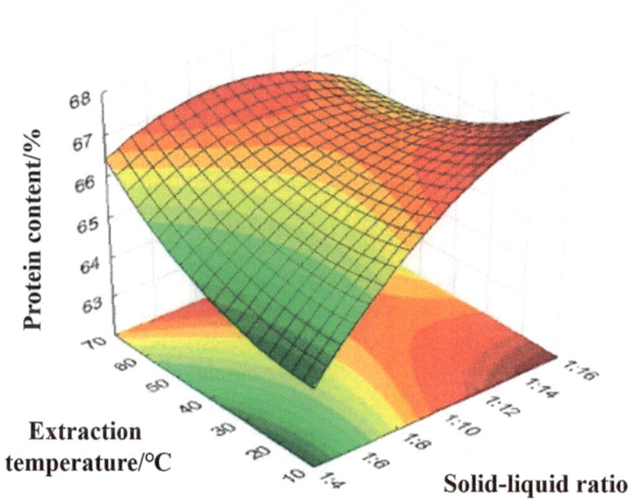

Fig. 6.6 Response surface of impact of solid-liquid ratio and extraction temperature on protein content during the first concentration

and extraction temperature on the protein content, the protein content increased with the increase of extraction temperature under the low solid-liquid ratio, while it was the opposite under the high solid-liquid ratio. In the range of low extraction temperature, the solid-liquid ratio had a linear effect on the protein content; in the

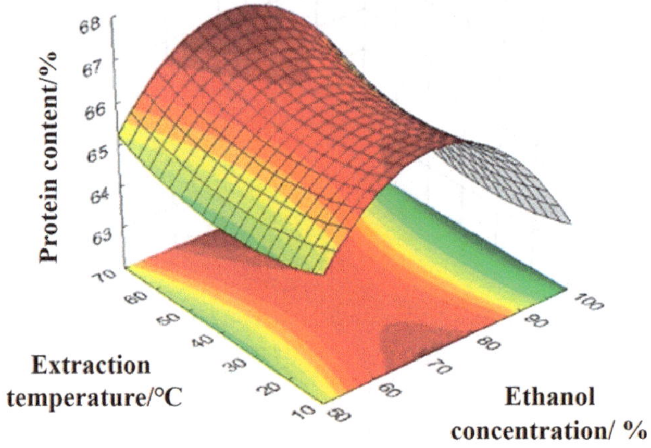

Fig. 6.7 Response surface of impact of ethanol concentration and extraction temperature on protein content during the first extraction

range of high extraction temperature, the solid-liquid ratio had a quadratic effect on the protein content. The interaction between the solid-liquid ratio and ethanol concentration was not significant.

The response surface of interaction between the solid-liquid ratio and extraction temperature on the protein content during the first extraction is shown in Fig. 6.7. The experiment results showed that the ethanol concentration had a quadratic effect on the response value and the extraction temperature had little effect in the interaction between the ethanol concentration and extraction temperature on the protein content. There were no interactions between the ethanol concentration and extraction temperature basically.

In order to obtain the exact optimum process parameters, by finding the partial derivatives of these three independent variables of the regression equation of the protein content Y_1 during the first extraction, respectively, and making them equal to zero, the following three equations could be obtained:

$$0.30080 - 0.45832x_1 + 0.26449x_2 - 0.19226x_3 = 0$$
$$-0.24214 + 0.26449x_1 - 0.78350x_2 + 0.01381x_3 = 0$$
$$0.19607 - 0.19226x_1 + 0.01381x_2 + 0.12548x_3 = 0$$

Through the simultaneous equations, the optimal process parameters during the first extraction could be obtained:

$$x_1 = 0.77856 \quad x_2 = -0.05264 \quad x_3 = -0.36385$$
$$X_1 = 1:12.34 \quad X_2 = 74.47\% \quad X_3 = 46.36^{\circ}\mathrm{C}$$

The optimal processes for protein content improvement during the first extraction have been researched above. Through the model calculation, the protein content of peanut protein concentrate obtained under the optimal process conditions during the first extraction was 66.60%, and the experimental determination value was 66.49 ± 0.43% which was consistent with the predicted value of model.

1.3 Process Optimization During the Second Extraction of Ethanol

1.3.1 Single-Factor Experiment

The impact of solid-liquid ratio on the protein content during the second extraction is shown in Fig. 6.8. The experiment results showed that the protein content increased gradually when the solid-liquid ratio increased from 1:4 to 1:10. And the protein content did not change significantly when the solid-liquid ratio continued to increase to 1:12. Therefore, by combining with the economy, 1: 10 was determined as the optimal single-factor solid-liquid ratio during the second extraction.

The impact of ethanol concentration on the protein content during the second extraction is shown in Fig. 6.9. The experiment results showed that there was no change basically in the protein content of the product obtained when the ethanol concentration increased from 80% to 95%. And the protein content was significantly reduced when the ethanol concentration increased from 95% to 100%. Because the absolute ethanol could not dissolve the residual impurities in the protein, a small amount of water is needed to be added. From the point of view of cost reduction, 80% was selected as the optimal single-factor ethanol concentration during the second extraction.

The impact of extraction temperature on the protein content during the second extraction is shown in Fig. 6.10. The experiment results showed that the extraction temperature had no impact on the protein content between 15 and 55 °C. From the point of view of the actual process operation, 25 °C which was close to normal water temperature was selected as the optimal single-factor extraction temperature during the second extraction.

The impact of extraction time on the protein content during the second extraction is shown in Fig. 6.11. The experiment results showed that the protein content increased slightly when the extraction time increased from 10 to 30 min. And the protein content did not change when the extraction time continued to increase. Therefore, 30 min was determined as the optimal single-factor extraction time during the second extraction.

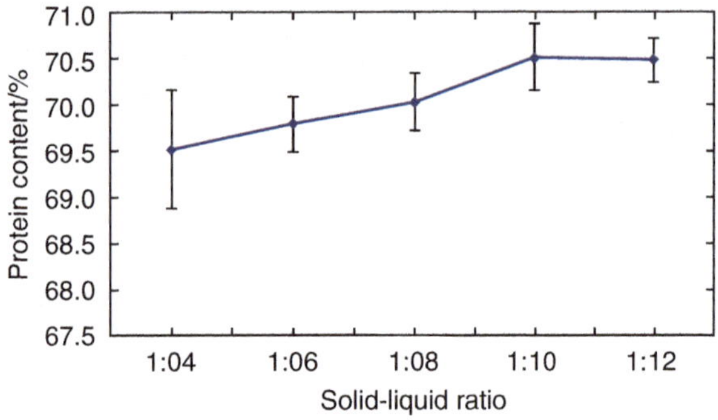

Fig. 6.8 Impact of solid-liquid ratio on the protein content during the second extraction

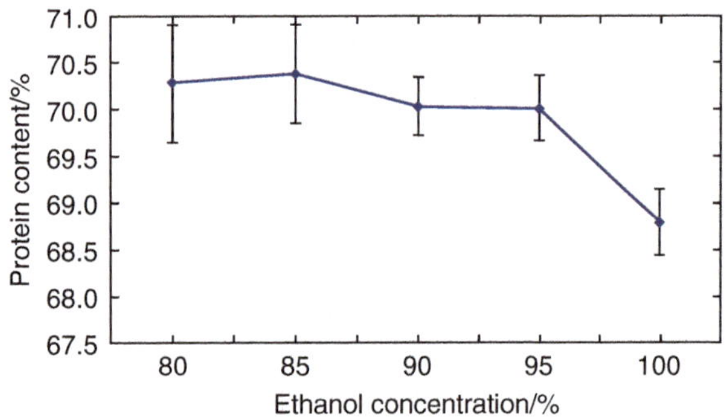

Fig. 6.9 Impact of ethanol concentration on the protein content during the second extraction

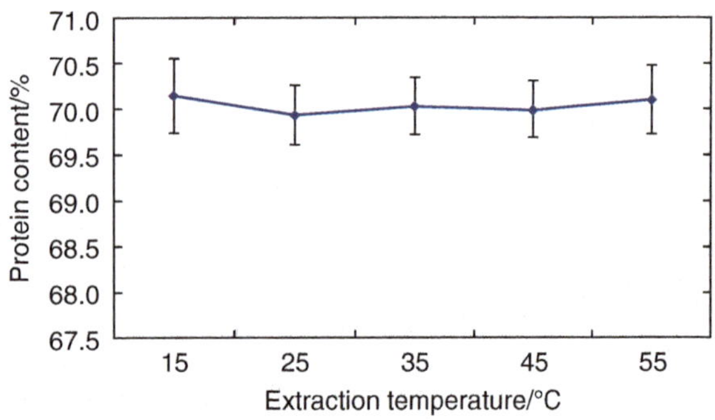

Fig. 6.10 Impact of extraction temperature on the protein content during the second extraction

1.3.2 Process Establishment

As with the first extraction, the extraction time during the second extraction was directly determined as 30 min. The ethanol concentration (in the range of 80–95%) and extraction temperature (in the range of 15–45 °C) had little effect on the protein content. So they could be directly selected in combination with the operational economy. Among the four factors, only the solid-liquid ratio had a significant impact on the protein content, and there was no need to carry out orthogonal experiments. Therefore, the experiment was carried out when all the single factors

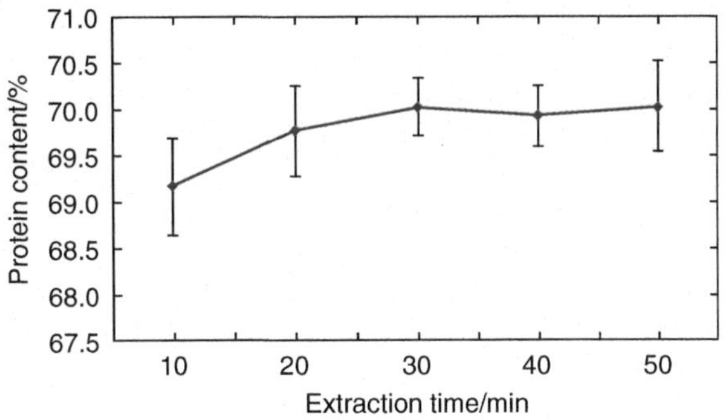

Fig. 6.11 Impact of extraction time on the protein content during the second extraction

were determined (the solid-liquid ratio was 1:10, ethanol concentration was 80%, extraction temperature was 25 °C, and extraction time was 30 min), and the protein content of the peanut protein concentrate prepared was 70.41 ± 0.55%, which fully met the content requirements of protein concentrate of high-quality product.

The protein content process optimization experiment results showed that, among the three factors (solid-liquid ratio (x_1), ethanol concentration (x_2), and extraction temperature (x_3) during the first extraction), the principal factor affecting the protein content was the ethanol concentration, followed by the solid-liquid ratio. The single-factor experiment during the second extraction showed that the solid-liquid ratio had the largest effect among these three factors as same as those during the first extraction. In the end, the preparation process of peanut protein concentrate with high protein content was determined as follows: during the first extraction, the solid-liquid ratio was 1:12.34, ethanol concentration was 74.47%, extraction temperature was 46.36 °C, and extraction time was 60 min. During the second extraction, the solid-liquid ratio was 1:10, ethanol concentration was 80%, extraction temperature was 25 °C, and extraction time was 30 min. The mean of protein content of product was 70.41% which met the requirements of protein concentrate of product.

2 Gelation Improvement of Peanut Protein Concentrate

Gelation of peanut protein concentrate is a complex process which usually includes the expansion, fracture, binding, and aggregation processes of protein molecule chain. And the protein forms gel through the cross-linking between the fully extended peptide chains. The quick-setting characteristics of peanut protein concentrate were mainly manifested as the expansion and fracture processes of protein molecule chain. After the protein was modified, the globular protein molecules began to stretch out, and the functional groups embedded inside the crimped molecule chain, such as disulfide and hydrophobic groups, were exposed (Batista et al. 2005), so that the disulfide bonds, hydrogen bonds, hydrophobic effect, electrostatic attraction, and van der Waals force among and inside the protein molecules changed in different degrees, and thus the protein had the potential property to form gel. Protein modification is to modify the protein structure artificially so as to improve the functionality of product. The current protein modification methods used commonly included physical, chemical, enzymatic, and genetic engineering methods (Petruccelli and Anon 1994).

2.1 *Modification Technology*

Wu et al. (2009) mainly used the physical, chemical, and biological enzyme methods to modify the peanut protein concentrate. The physical modification mainly involves heating, mechanical action, ultrasonic wave, microwave, and other methods. The chemical modification is to use the reductant to conduct structure modification for the protein. The enzymatic modification is to use the TG enzyme to improve the gelation of peanut protein concentrate. By determining the free sulfydryl and disulfide bond contents in protein and hydrophobicity index changes of protein surface, the change of protein gelation was determined in order to obtain the protein with high gelation.

2.1.1 Heat Treatment

It could be seen from Fig. 6.12 that after the peanut protein concentrate was treated in 40–100 °C, the free sulfydryl content in protein reached 10.66 μmol/g at 50 °C, the -SH content decreased slightly with the increase of temperature, the free sulfydryl content was 9.98 μmol/g at 80 °C, and then the -SH content in protein decreased rapidly; the change trend of disulfide bond content in protein was opposite to that of sulfydryl, the disulfide bond content was only 12.84 μmol/g at 50 °C, the disulfide bond content increased slowly with the increase of temperature, and it increased to 15.25 μmol/g at 80 °C and then increased sharply. The variance analysis showed that there were no significant differences in the changes of free sulfydryl and disulfide bond contents when the temperature was 50–80 °C ($P > 0.05$). It could be seen from Table 6.3 that the surface hydrophobicity index (So) of protein was increased gradually with the increase of heat treatment temperature, it reached the maximum (9.0845) at 90 °C, and then the So value decreased rapidly with the increase of temperature; the variance analysis showed that there were no significant differences between the surface hydrophobicity index value (9.005) at 80 °C and that at 90 °C. Therefore, 80 °C was selected as the optimum condition of heat treatment intensity according to the changes of surface hydrophobicity index of free sulfydryl and disulfide bond of protein in different heat treatment intensities.

It could be seen from Fig. 6.13 and Table 6.3 that the free sulfydryl content in protein increased gradually. The disulfide bond content decreased gradually and So value increased when the treatment time was 0–6 min. The ideal values (9.97 μmol/

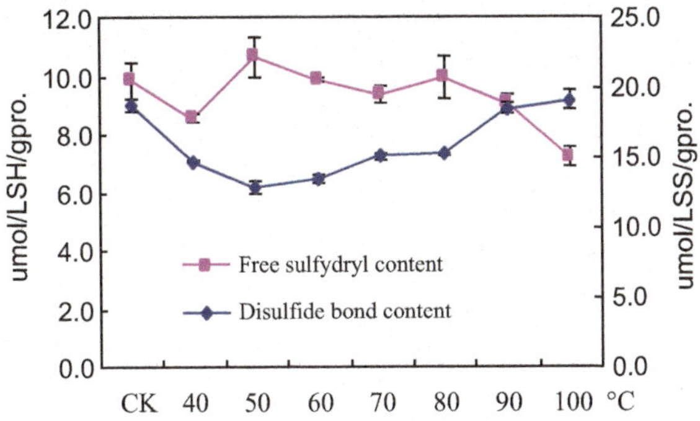

Fig. 6.12 Impact of heat treatment intensity on -SH/-S-S- content in protein

Table 6.3 Impact of heat treatment on surface hydrophobicity index So of protein

Heat treatment intensity (°C)	Surface hydrophobicity index So	Heat treatment time (min)	Surface hydrophobicity index So
40	7.77	2	8.13
50	7.55	4	8.82
60	8.06	6	9.52
70	8.53	8	9.08
80	9.01	10	8.53
90	9.08	12	8.23
100	7.41	14	8.19
CK	7.46	CK	7.46

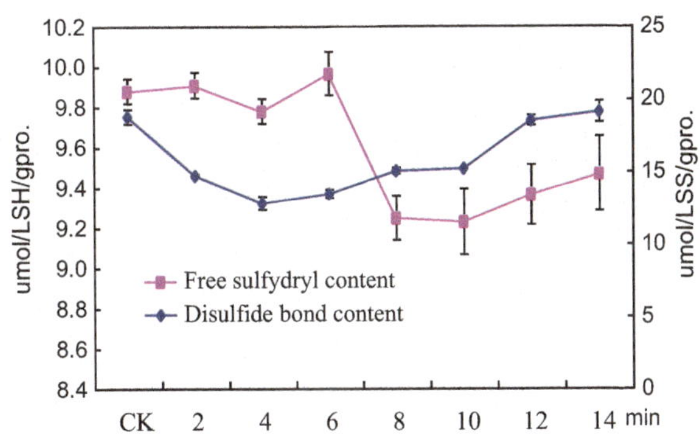

Fig. 6.13 Impact of heat treatment time on -SH/-S-S- content in protein

g, 13.48 μmol/g, and 9.52, respectively) were reached when the treatment time was 6 min. The free sulfydryl content in protein decreased rapidly, and the disulfide bond content increased with the time extension, which indicated that the exchange reaction occurred between the free sulfydryl and disulfide bond in protein. The intermolecular and intramolecular disulfide bonding was formed with the extension of action time, and thus the surface hydrophobic group of protein was wrapped again, resulting in the reduction of surface hydrophobicity index. This also verified the conclusion of Friedman et al. (1982) that the thermal denaturation of plant protein had two aspects of impacts on its structure: firstly, the globular structure of globulin developed into a linear macromolecule, and the hydrophobic group originally embedded inside the coil structure was exposed; secondly, with the increase of action intensity or time, the interaction of hydrophobic groups or -SS- generated by SH oxidation led to the formation of dense network structure, which made the originally exposed groups be re-wrapped.

2.1.2 High-Speed Stirring

From Fig. 6.14 and Table 6.4, it could be seen that the intensity change of high-speed stirring had a great impact on the quick-setting property of protein. The free sulfydryl content and surface hydrophobicity index So of protein increased first and then decreased with the increase of stirring speed, while the change of disulfide bond content was opposite to that of free sulfydryl content. The specific manifestations are shown below: when the rotate speed was 3000–12,000 r/min, the free sulfydryl content in protein increased gradually; when the rotate speed was 12,000 r/min, the free sulfydryl content reached the maximum (11.57 μmol/g), the disulfide content was the lowest (11.83 μmol/g), and the So value of protein was

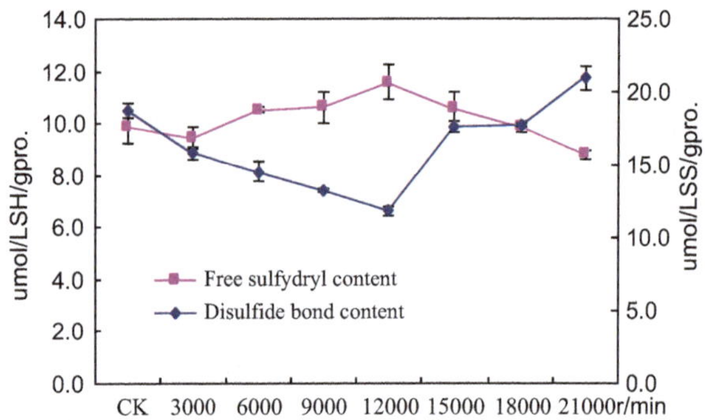

Fig. 6.14 Impact of high-speed stirring intensity on -SH/-S-S- content in protein

Table 6.4 Impact of high-speed stirring on surface hydrophobicity index So of protein

High-speed stirring (r/min)	Surface hydrophobicity index So	High-speed stirring treatment (s)	Surface hydrophobicity index So
3000	7.94	12	7.28
6000	7.88	24	7.68
9000	8.17	36	7.89
12,000	8.43	48	8.27
15,000	8.65	60	8.89
18,000	7.86	72	8.87
21,000	7.24	CK	7.46

8.425; when the rotate speed was 12,000–21,000 r/min, the free sulfydryl content in protein decreased rapidly, the disulfide bond content increased, and the So value increased slightly and then decreased rapidly at 15,000 r/min. The variance analysis showed that there were no significant differences in the So value at 12,000 r/min and that at 15,000 r/min ($P > 0.05$). Therefore, according to the results of comprehensive investigation, 12,000 r/min was selected as the optimal condition of high-speed stirring intensity.

From Fig. 6.15 and Table 6.4, it could be seen that, when the protein was treated at 12,000 r/min for 12–72 s, the free sulfydryl content in protein increased slowly with the increase of stirring time; when it was stirred for 60 s, the free sulfydryl content in protein reached the maximum (9.99 μmol/g), the disulfide bond content was the lowest (11.83 μmol/g), and the So value of protein was 8.89; the extension of stirring time reduced the gelation of protein. This was mainly because the high-speed shear action made the protein macromolecule aggregation be broken into protein molecule bonds in the high-speed stirring process of protein; the weak acting force, such as hydrogen bonds and van der Waals force between the protein molecules, was interrupted; and the opportunity to expose the hydrophobic groups originally embedded inside the molecule increased simultaneously.

2.1.3 Ultrasonic Treatment

From Fig. 6.16 and Table 6.5, it could be seen that the ultrasonic treatment had a significant impact on the gelation of peanut protein concentrate. The ultrasonic power increased from 600 to 1000 W, the free sulfydryl content increased significantly to 11.91 μmol/g ($P < 0.01$), while the disulfide bond content decreased to 19.47 μmol/g and the So value of protein was 12.89; then the increase of ultrasonic power made the quick-setting property of protein decrease rapidly. Therefore, 1000 W was determined as the optimal action power of ultrasonic power. The power was fixed simultaneously, and it was obtained that the ultrasonic time could significantly affect the gelation of protein by investigating the impacts of different ultrasonic treatment time (1–8 min) on the quick-setting property of protein

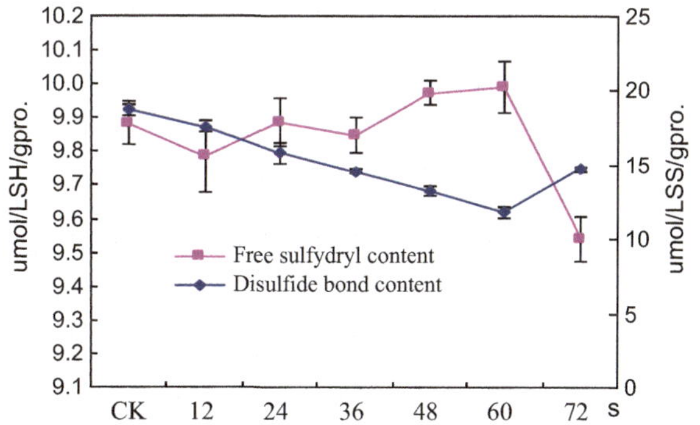

Fig. 6.15 Impact of high-speed stirring time on -SH/-S-S- content in protein

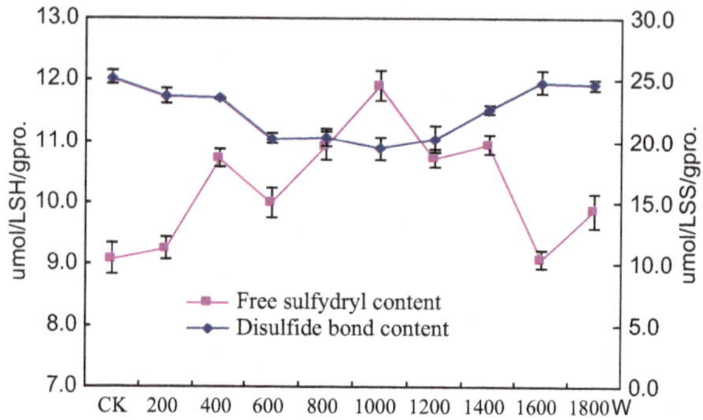

Fig. 6.16 Impact of ultrasonic treatment intensity on -SH/-S-S- content in protein

Table 6.5 Impact of ultrasonic treatment on surface hydrophobicity index So of protein

Ultrasonic treatment (H_Z)	Surface hydrophobicity index S_0	Ultrasonic treatment (min)	Surface hydrophobicity index S_0
200	10.13	1	10.12
400	10.74	2	10.54
600	11.26	4	10.38
800	11.59	6	11.72
1000	12.89	8	11.08
1200	11.68	CK	7.46
1400	10.09		
1600	10.60		
1800	10.49		

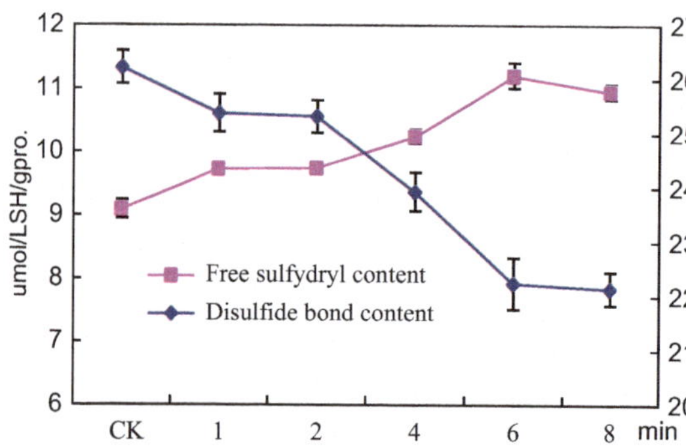

Fig. 6.17 Impact of ultrasonic treatment time on -SH/-S-S- content in protein

(Fig. 6.17). When the ultrasonic time was 1–6 min, the free sulfydryl content in protein showed an increasing trend, and the disulfide bond content always showed a decreasing trend. When the ultrasonic time was 6 min, the free sulfydryl content and disulfide bond content were 11.21 μmol/g and 22.25 μmol/g, respectively, and the action time was extended to 8 min; the free sulfydryl content and disulfide bond content in protein did not change significantly, compared with that at 6 min, so it was found that the time extension would not facilitate the improvement of protein gelation.

Ultrasonic wave is a kind of elastic mechanical wave in the material medium and its frequency range is 2×10^4–2×10^9 Hz. Ultrasonic wave would produce heat effects, mechanical effects, or cavitation effects when propagating in the medium and thus cause changes of some characteristics of medium (Mason and Paniwnyk 1996). The research showed that, for the ultrasonic wave with a certain frequency and a certain emission surface, if the power increased, the sound intensity also increased; if the sound intensity increased, the sound pressure amplitude and the pressure in the liquid also increased, and the time required for cavitation bubble collapse would become shorter, that is, the amount of cavitation events generated by ultrasound in the unit time increased (Huang 2004), which were conducive to the dissolution of solute in the liquid system. But the sound intensity could not be increased without limit, because the sound intensity increased, the sound pressure amplitude also increased, and the cavitation bubbles were too late to collapse in the sonic expansion phase; in addition, too high sound intensity would produce a large number of cavitation bubbles which reduced the energy transfer through the reflected sound waves. This effectively explained the changing situations of surface hydrophobicity index of protein during the ultrasonic treatment from one side (Singh et al. 1990; Singh and MacRitchie 2001; Ng et al. 1992; Mecham 1968; Tanaka and Bushuk 1973; MacRitchie 1975).

2.1.4 Microwave Treatment

From Fig. 6.18 and Table 6.6, it could be seen that the change of gelation property of peanut protein concentrate was obvious with the increase of microwave power. Under the treatment condition of 160–800 W, the free sulfydryl content in protein always showed an increasing trend, and the disulfide bond content showed a decreasing trend; when the power was 800 W, the free sulfydryl content in protein was 10.04 μmol/g, the disulfide bond content was 20.49 μmol/g, and the So value of protein was 8.15, so the microwave power of 800 W was selected as the optimum

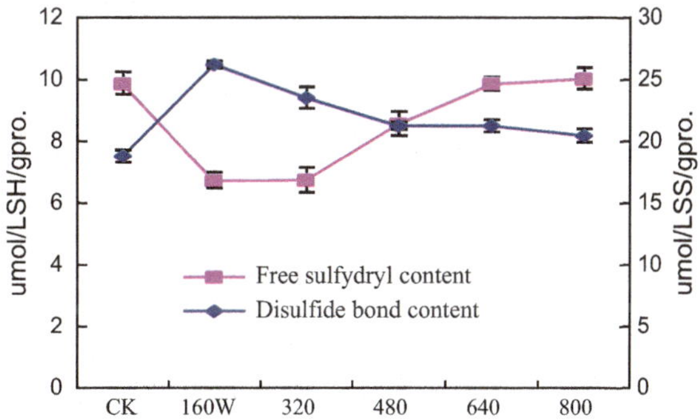

Fig. 6.18 Impact of microwave treatment intensity on -SH/-S-S- content in protein

Table 6.6 Impact of microwave treatment on surface hydrophobicity index So of protein

Microwave treatment (W)	Surface hydrophobicity index S_0	Microwave treatment (s)	Surface hydrophobicity index So
160	7.41	20	7.58
320	7.27	30	7.98
480	7.93	40	8.09
640	7.93	50	7.46
800	8.15	60	7.61
CK	7.46	CK	7.46

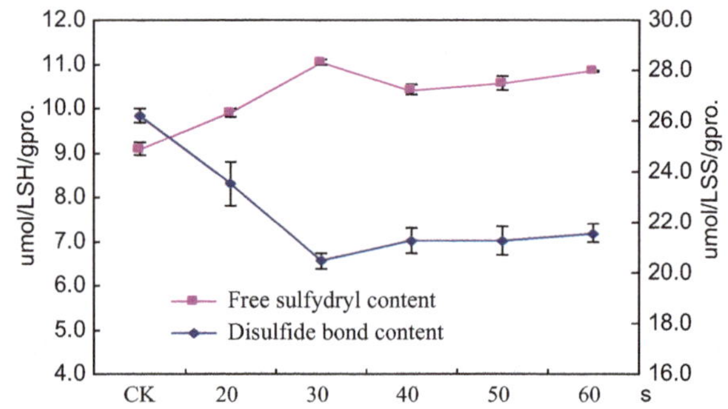

Fig. 6.19 Impact of microwave treatment time on -SH/-S-S- content in protein

condition of microwave treatment intensity. Under this power, the impact of treatment time of 20–60 min on the protein gelation was investigated. From Fig. 6.19 and Table 6.6, it could be seen that the free sulfydryl content in protein was the highest, the disulfide bond content was the lowest, and the So value of protein was 7.98 when the treatment time was 30 s; this So value was not the highest (So value was the maximum (8.09) when the treatment was 40 s), but the variance analysis showed that there were no significant differences in the So values of protein when the treatment time was 30 s and 40 s; if the treatment time was extended, the protein solution boiled over, which was not good for the improvement of quick-setting property of protein, and took much time and energy. In general, the microwave power selected should be able to most effectively improve the dissolution rate and would not cause the material denaturation, the microwave power should not be too large, and the microwave radiation time should not be too long so as not to cause the system overheating, resulting in inaccurate temperature control and even dangers (Liompart and Lorenzo 2007). Therefore, the microwave treatment conditions selected in this research were power 800 W and time 30 s.

Microwave energy was a nonionizing radiation energy that caused molecular motion through ion transport and dipole rotation. When it acted on the molecule, the conversion of molecule could be promoted, or the molecule had a certain polarity, that is, the transient polarization could be generated under the action of microwave field and the polarization inversion movement was conducted at a rate of 2.45 billion times (Zhou and Liu 2006). And thus the vibration and tearing of bond and friction and collision among the particles would be caused, and a lot of heat would be generated rapidly to make the intermolecular and intramolecular noncovalent bonds and -SS- break, so that some insoluble protein molecules embedded inside the molecules were released and diffused into the solvent, which was why the microwave treatment could promote the increase of free sulfydryl content and surface hydrophobicity index of protein.

2.1.5 Reductant-1 Treatment

Aminlari et al. (1977) found that the addition of a certain amount of reductant, such as $NaSO_3$ and Cys (cysteine) in the protein, was an effective method to improve the protein PDI (protein dispersibility index, PDI). Fukushima and Van Buren (1970) also achieved the similar conclusion in the research. Both $NaSO_3$ and Cys were reductants which were specific for disulfide bonds, so some scholars interpreted this result as that the utilization of reductants broke the disulfide bond in protein, that is, the disulfide cross-linking caused the decrease in protein solubility. Therefore, this book had investigated the impact of the addition of reductants on the gelation of protein.

From Fig. 6.20 and Table 6.7, it could be seen that the free sulfydryl content in protein increased rapidly when the addition amount of reductant was low (0–0.625 mmol/L); when the addition amount was 0.625 mmol/L, the free sulfydryl content in protein was 15.89 μmol/g which was 74.82% higher than that of the original PPC (peanut protein concentrate, PPC), and the So value of protein was 13.48 which was 80.65% higher than that of the original PPC. There was no significant change in the free sulfydryl, disulfide bond contents, and So value when the dosage of reductant increased ($P > 0.05$). Therefore, the addition amount of reductant was determined as 0.625 mmol/L. Under this condition, the impact of treatment time on the gelation of protein was investigated. From Fig. 6.21 and Table 6.7, it could be seen that the free sulfydryl content and So value of protein increased slowly when the treatment time was 15–90 min. When the treatment time was 90 min, the free sulfydryl content in protein reached 13.01 μmol/g, and the So value was 12.81 which were 43.16% and 71.72% higher than those of the original PPC, respectively; when the treatment time was extended, there was no significant change in the free sulfydryl and disulfide contents and So value of protein ($P > 0.05$). Therefore, the treatment conditions of reductant-1 were determined as addition amount, 0.625 mmol/L and treatment time, 90 s.

2.1.6 Reductant-2 Treatment

The impact of addition amount of reductant-2 on the gelation of protein is shown in Fig. 6.22 and Table 6.8. The free sulfydryl content in protein increased as the addition amount of reductant increased gradually; when the addition amount was 0.75 mmol/L, the free sulfydryl content in protein reached the maximum

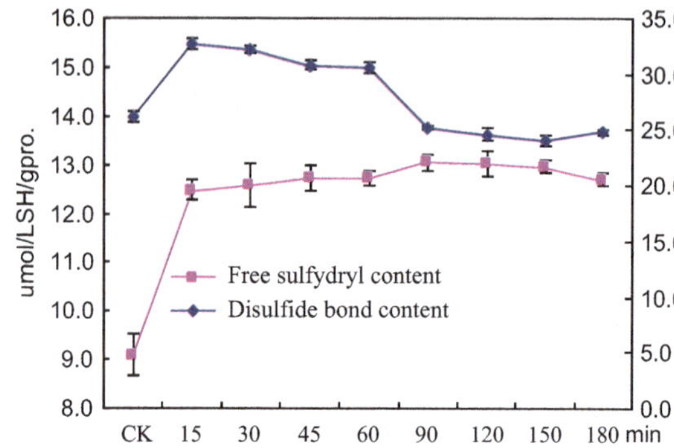

Fig. 6.20 Impact of action time of reductant-1 on -SH/-S-S- content in protein

Table 6.7 Impact of reductant-1 treatment on surface hydrophobicity index So of protein

Reductant-1 (mmol/L)	Surface hydrophobicity index So	Reductant-1 treatment (min)	Surface hydrophobicity index So
0.25	8.42	15	10.86
0.375	9.43	30	11.97
0.5	118	45	12.06
0.625	13.48	60	12.49
0.75	12.90	90	12.81
0.875	12.92	120	11.61
1.0	13.00	150	11.83
CK	7.46	180	11.86

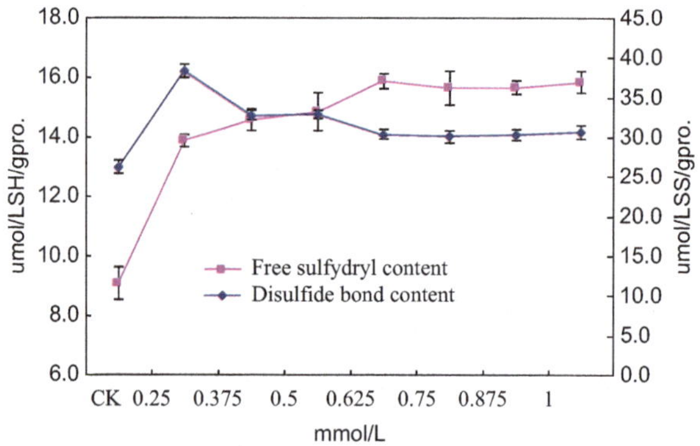

Fig. 6.21 Impact of addition amount of reductant-1 on -SH/-S-S- content in protein

(15.91 μmol/g) which was 75.06% higher than that of the original PPC, and it was not significantly different from the increase of free sulfydryl content when the addition amount of reductant −1 was 0.625 mmol/L ($P > 0.05$), but the addition amount increased by 20%; later, the increase of addition amount of reductant had no significant impact on the quick-setting property of protein; therefore, the addition amount of reductant-2 was determined as 0.75 mmol/L. Under this condition, the impacts of different treatment time (0–180 min) on the quick-setting property of

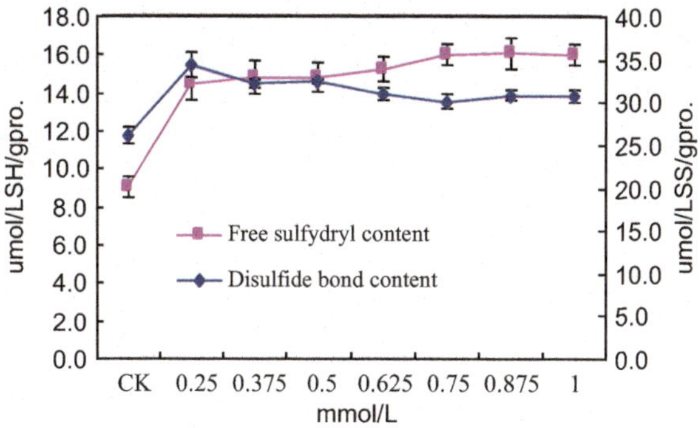

Fig. 6.22 Impact of addition amount of reductant-2 on -SH/-S-S- content in protein

Table 6.8 Impact of reductant-2 treatment on surface hydrophobicity index So of protein

Reductant-2 (mmol/L)	Surface hydrophobicity index S_o	Reductant-2 treatment (min)	Surface hydrophobicity index S0
0.25	8.55	15	8.03
0.375	8.69	30	8.40
0.5	8.65	45	8.85
0.625	9.39	60	8.93
0.75	12.33	90	10.24
0.875	11.02	120	11.09
1	10.20	150	10.22
CK	7.46	180	10.22

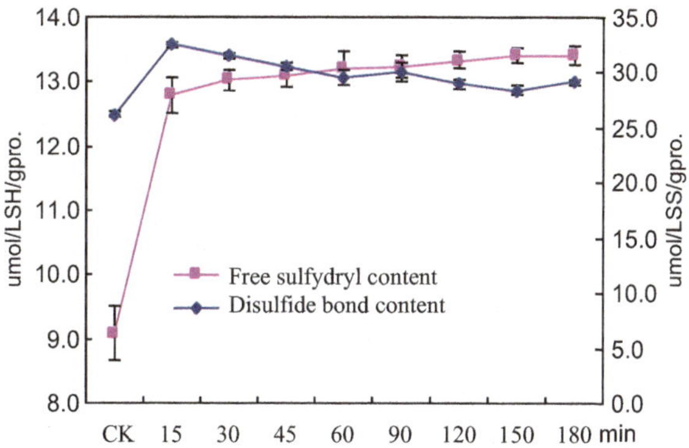

Fig. 6.23 Impact of action time of reductant-2 on -SH/-S-S- content in protein

protein were investigated. From Fig. 6.23 and Table 6.8, it was found that, when the treatment time was 0 min and 15 min, there was a significant difference in the quick-setting property of protein ($P < 0.01$), the free sulfydryl content in protein increased rapidly with an increase of 40.72% compared with that of the original PPC, but the disulfide bond content also increased simultaneously with an increase of 24.27%. This may be because the reductant was used to treat the protein in the state of oscillation or stirring. And oscillation or stirring increased the opportunities of mutual collision and friction between the protein molecules in the protein solution, and the sulfydryl was easy to form disulfide cross-linking after drifting away from the protein, which caused the increase of disulfide bond content, but if the increasing speed of free sulfydryl content was much greater than the disulfide cross-linked speed, the free sulfydryl content determined finally was still increased, while the disulfide bond content could be relatively reduced. This hypothesis was verified in the following tests: the free sulfydryl content in protein was slowly increased with the extension of treatment time, while the disulfide bond content decreased slowly, and a dynamic equilibrium was achieved when the treatment time was 120 min, that is, the free sulfydryl content was the maximum, the disulfide bond content was the lowest, the time was extended continuously, and there was no significant difference in the changes of free sulfydryl content, disulfide bond content, and So value ($P > 0.05$).

On one hand, the addition of reductant to the protein solution could not only increase the -SH content in the protein, improve the surface hydrophobicity of protein, and increase the solubility of protein, but also save the -SH content in protein and allow it to act in the quick-setting stage; on the other hand, in the protein treatment process, the free sulfydryl was easy to form disulfide cross-linking due to oscillation or stirring. Therefore, it was important to control the oscillation or stirring speed when the reductant was used to treat the protein. In summary, the appropriate amount of reductant treatment was conducive to the improvement of protein gelation.

2.1.7 TG-B Enzyme (Transglutaminase-B) Treatment

1. TG-B enzyme (Transglutaminase-B) modified single-factor experiment

(a) Substrate concentration

In the range of substrate concentration of 5–25%, the peanut protein—TG-B enzyme system showed obvious substrate inhibition, that is, with the increase of substrate concentration, the gel strength increased to a certain value and then decreased (Fig. 6.24). When the substrate concentration was 20%, the gel strength was 9.87 g which was 0.41 g lower than that when the substrate concentration was 15%, 0.51 g higher than that when the substrate concentration was 25%, which showed that the gel strength decreased significantly when the substrate concentration increased from 20% to 25%. In theory, if the substrate concentration of system was too large, the too low concentration of effective water in the system would be caused; therefore the substrate and enzyme diffusion and movement would be reduced so as to inhibit the cross-linking; when the substrate concentration was too low, the probability of collision of enzyme and active substrate would be reduced so that the cross-linking reaction was inhibited. From the test results, it could be seen that the substrate concentration of 5% had reduced the probability of

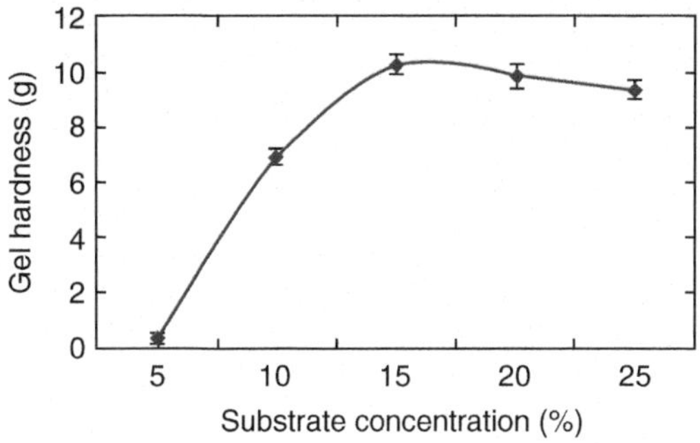

Fig. 6.24 Impact of substrate concentration on gel strength

collision between enzyme and substrate, while the substrate concentration of 20–25% had reduced the diffusion and movement of substrate and enzyme system. In the actual production, the too low or too high protein concentration would increase the energy consumption and reduce equipment utilization, so the too low or too high substrate concentration did not have practical significances and the substrate concentration of the final test was selected as 15%.

(b) Enzyme dosage

Wu et al. (2009) researched the impact of enzyme dosage on protein gel strength based on the enzyme addition amount of 1 U/g. The results are shown in Fig. 6.25. It could be seen that the gel strength increased rapidly with the increase of enzyme dosage when the enzyme dosage was in the range of 1–3 U/g; when the enzyme dosage reached 3 U/g, the gel strength reached 10.35 g which was increased by 7.79 g and 1.94 g, respectively, compared with that when the enzyme dosage was 1 U/g and 2 U/g, and the gel strength did not change any longer when the enzyme dosage was in the range of 3–5 U/g, so the enzyme dosage was determined as 3 U/g.

(c) Reaction temperature

The enzyme action temperature is closely related to the stability of enzyme molecule, because the enzyme molecule has a specific spatial structure; if the reaction temperature exceeds a certain limit, the dissociation of the secondary bond will be caused easily, resulting in the enzyme loss or partial loss of catalytic activity; if the reaction temperature is too low, the intense degree of molecular movement in the system will be reduced greatly, and thus the probability of collision between enzymes and substrates will be reduced. The author's research team researched the changing situations of protein gel strength in the temperature range of 20–60 °C. The results are shown in Fig. 6.26. When the temperature increased from 20 to 40 °C, the gel strength of the system increased gradually; when the temperature was 40 °C, the gel strength of system reached the maximum (10.56 g); when the temperature exceeded 40 °C, the gel strength of system began to decrease. Therefore, it was determined that the TG-B enzyme action temperature was 40 °C.

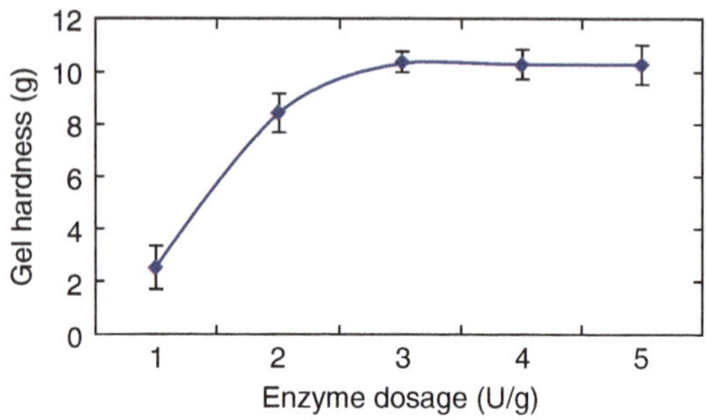

Fig. 6.25 Impact of enzyme dosage on gel strength

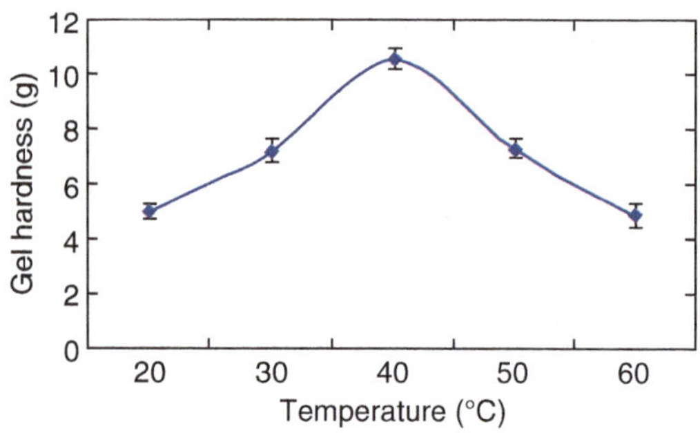

Fig. 6.26 Impact of reaction temperature on Fig. 6.26 gel strength

(d) Reaction time

The changing trend of the gel strength of the system for different enzyme action time within the range of 30–150 min was analyzed (Fig. 6.27). In the first 120 min of enzymatic action, the gel strength increased, and the value reached 9.89 g when the hydrolysis was conducted for 120 min. When the cross-linking was conducted for 150 min, the gel strength continued to rise, but the increase amplitude was only 3.24%. Therefore, when the action time was 120 min, the protein gel strength of the system had been close to the maximum, and the increase amplitude was not significant with the extension of action time, so it was determined that the primary action time is 120 min.

(e) pH value

The impact of pH on the enzymatic reaction was mainly reflected in the impact on the enzyme activity. Wu et al. (2009) analyzed the changes of protein gel strength of system in the pH range of 4–8. The results are shown in Fig. 6.28. With the increase of pH value, the protein gel strength increased rapidly; when the pH of system reached 7.0, the gel strength value reached 9.26 g; when the pH of system was 8.0, the gel strength value decreased compared with that when the pH was 7.0, so pH 7.0 was selected as the primary pH of TG-B enzyme.

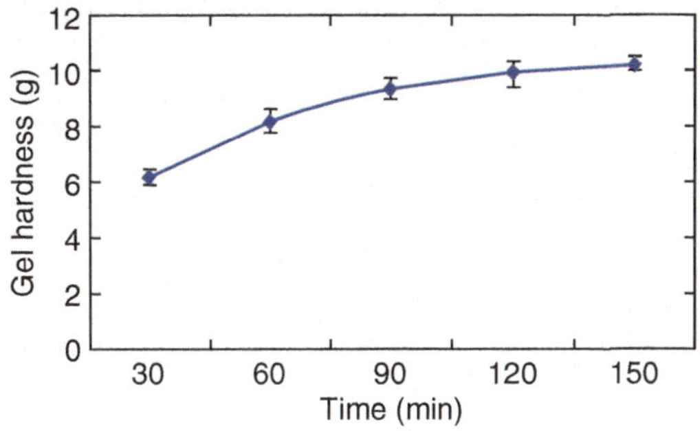

Fig. 6.27 Impact of reaction time on gel strength

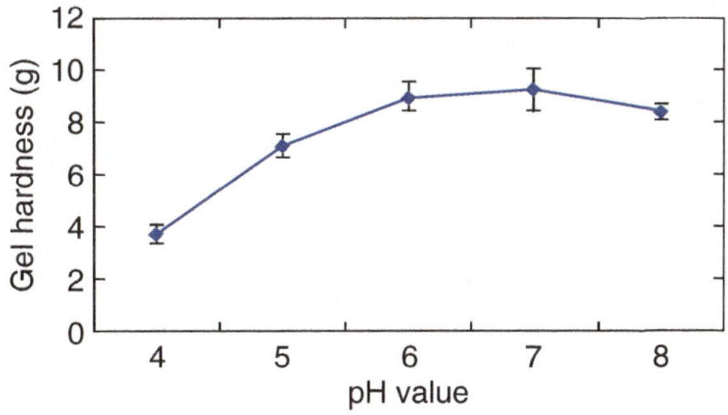

Fig. 6.28 Impact of pH value on gel strength

2. Orthogonal experiment of TG-B enzyme-modified peanut protein concentrate

From the range analysis results of gel strength of orthogonal experiment in Table 6.9, it could be seen that the impact of each factor on gel strength was in the order of B>A>C>E>D. In the factor A, the mean of level 3 was the highest; in the factor B, the mean of level 3 was the highest; in the factor C, the mean of level 2 was the highest; in the factor D, the mean of level 4 was the highest; in the factor E, the mean of level 3 was the highest. Therefore, the optimal process combination of enzyme-modified peanut protein was A3B3C2D4E3, that is, the enzyme dosage was 3.0 U/g substrate, the substrate concentration was 15%, the reaction temperature was 40 °C, the reaction time was 140 min, and the pH value was 7, and the gel hardness of product at this time was 11.85 g. The impact of reaction time in the whole process was the smallest, and it could be seen that there was no significant difference in the change of gel strength after the reaction time exceeded 60 min according to the variance analysis of single-factor results. Therefore, 60 min was determined as the reaction time by combining with the actual production.

The functional determination was carried out for the high-gel peanut protein concentrate product prepared under the optimal conditions. The results are shown in Table 6.10. In the comparison between the gel hardness of high-gel peanut protein concentrate and that of gel-type soybean protein concentration, the former

Table 6.9 Analysis of orthogonal experiment results

Treatment No.	Enzyme dosage (U/g substrate) A	Substrate concentration (%) B	Reaction temperature (°C) C	Reaction time (min) D	pH value E	Gel strength (g)
1	1	1	1	1	1	0.563
2	1	2	2	2	2	8.989
3	1	3	3	3	3	11.658
4	1	4	4	4	4	10.638
5	2	1	2	3	4	0.684
6	2	2	1	4	3	9.564
7	2	3	4	1	2	7.365
8	2	4	3	2	1	10.564
9	3	1	3	4	2	1.321
10	3	2	4	3	1	6.541
11	3	3	1	2	4	11.412
12	3	4	2	1	3	11.674
13	4	1	4	2	3	1.562
14	4	2	3	1	4	10.514
15	4	3	2	4	1	11.847
16	4	4	1	3	2	11.548
Maximum	8.5143	11.106	8.8677	8.3425	8.615	
Minimum	7.0443	1.0325	6.5265	7.529	7.306	
Range R	1.9878	10.0735	1.8235	0.8135	1.309	
Adjustment R′	1.6412	9.0662	1.789	0.7322	1.178	
Quadratic sum	6.7969	262.3061	10.2431	1.8901	5.22	
Degree of freedom	3	3	3	3	3	
Mean square	2.2656	87.4354	3.4144	0.6301	1.74	
Sum	286.4565					

(11.85 g $\pm$ 0.56 g) was higher than the latter; their water absorption and oil absorption were similar, and they were 176.43% $\pm$ 1.33% and 205.12% $\pm$ 1.43%, respectively; while the water absorption and oil absorption of high-gel peanut protein concentrate were slightly worse than that of gel-type soybean protein isolation. There were no standards on the functionality of gel-type protein product at home and abroad currently, so the enterprise standards (i.e., the gel hardness of gel-type soybean protein concentrate is $\geq$10 g, the water binding capacity is $\geq$150%, and the oil holding capacity is $\geq$200%) were taken as the basis to measure the gelation of the product. It could be seen that the quality of this product had exceeded the gelation index of the similar soybean products sold on the market.

Table 6.10 Functional evaluation results

Sample	Nitrogen solubility index (%)	Water absorption (%)	Oil absorption (%)	Emulsibility (%)	Emulsion stability (%)	Foamability (%)	Foam stability (%)	凝胶硬度 (g)
Gel-type peanut protein concentrate	39.13 ± 1.13	176.43 ± 1.33	205.12 ± 1.43	56.73 ± 1.74	90.54 ± 1.22	40.00 ± 2.05	11.80 ± 0.82	11.85 ± 0.56
Gel-type*soybean protein concentrate	25.64 ± 1.02	159.23 ± 2.01	193.62 ± 1.95	61.89 ± 1.63	80.45 ± 1.12	42.13 ± 1.18	10.56 ± 1.34	10.56 ± 1.23
Gel-type*soybean protein isolate	90.23 ± 2.51	203.45 ± 1.28	216.12 ± 1.32	62.16 ± 1.03	90.13 ± 1.54	41.55 ± 1.76	10.84 ± 0.59	19.55 ± 0.85

Note: The relevant parameters of gel-type soybean protein concentrate and gel-type soybean protein isolate are provided by the company, and the experiment data is measured by the author's research team

2.2 *Physicochemical Characteristics*

2.2.1 Molecular Weight Distribution of Protein Treated by Physicochemical Methods

Wu et al. (2009) treated the peanut protein concentrated by using the different modification methods. The molecular weight distribution of protein after SDS-PAGE electrophoresis detection is shown in Fig. 6.29.

It could be seen from Fig. 6.29 that there was no obvious difference in the molecular weight distribution of the modified and original PPCs, and most of the protein molecule weight was distributed between 25 and 80 kDa, which indicated that it was insufficient to make the protein molecule degrade to small-molecule short peptides by any treatment method and PPC protein molecules were still composed of macromolecules.

The molecular weight distribution and occupied peak area of protein product were calculated, respectively, according to the differences of the appearance time of the standard molecular weight curve $\log \text{mol wt} = 6.66e^{0} – 2.11e^{-001}\ T^{1}$, as shown in Figs. 6.30, 6.31, 6.32, and 6.33 and Table 6.11. It could be seen from the figures that the protein molecule weight of original PPC was >1800 Da and the occupied peak area proportion was 78.78% which was reduced to 72.26% after ultrasonic treatment and further reduced to 68.89% after reductant-1 treatment. At the same time, after the ultrasonic treatment and reductant-1 modification treatment, the peak areas occupied by the protein molecules which were smaller than 10,000 Da were increased by 6.52% and 9.89%, respectively, compared with the original PPC. It could be seen that the modified treatment could increase the content of small-molecule protein, but it could not make most of the protein degrade to small-molecule peptide. The results were consistent with the observation results of SDS-PAGE atlas.

2.2.2 Gelation of Modified PPC and Original PPC in Different pHs and Ionic Strength

The food system is a complex system with a certain pH (3–7) and ionic strength ($I = 0.02$–0.2), and these substances can also improve the functionality of the food protein so as to give the desired texture to the food (Lakemond 2003). Therefore, this research analyzed the impact of pH and ionic strength on protein gelation. These two pH values selected (3.0 and 7.0) were not only in the range of food system but also on both sides of PPC isoelectric point, respectively, so they were representative.

1. Linear viscoelastic region

The measurement principle of rheometer is to apply a sinusoidal vibration shear force to the sample between the parallel plates or cone plates to determine the

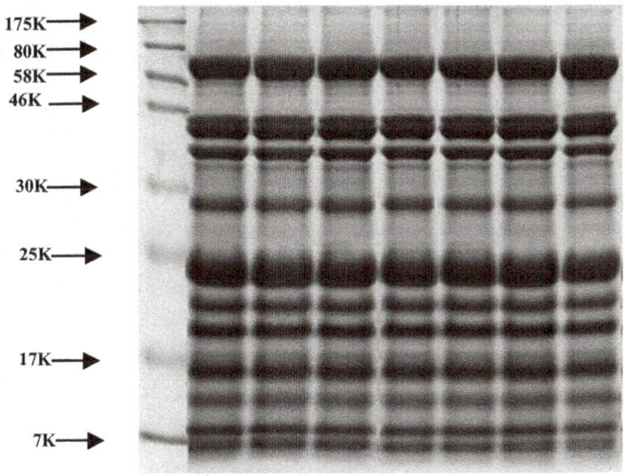

Fig. 6.29 SDS-PAGE atlas of peanut protein concentrate before and after treatment. Note: *1* MW markers, *2* PPC, *3* heat treatment, *4* high-speed stirring treatment, *5* ultrasonic radiation treatment, *6* microwave radiation treatment, *7* reductant-1 treatment, *8* reductant-2 treatment

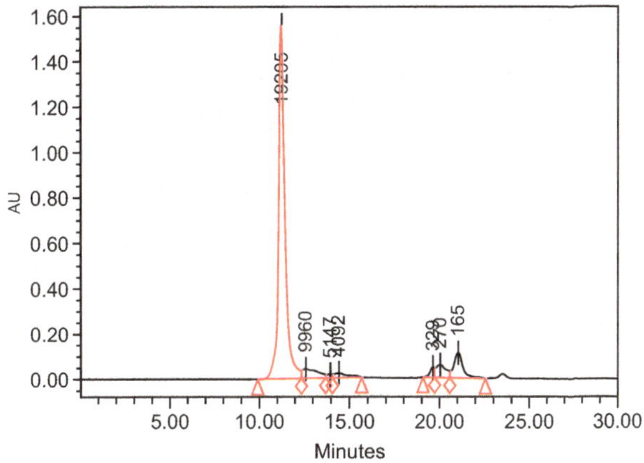

Fig. 6.30 Molecular weight distribution of PPC protein

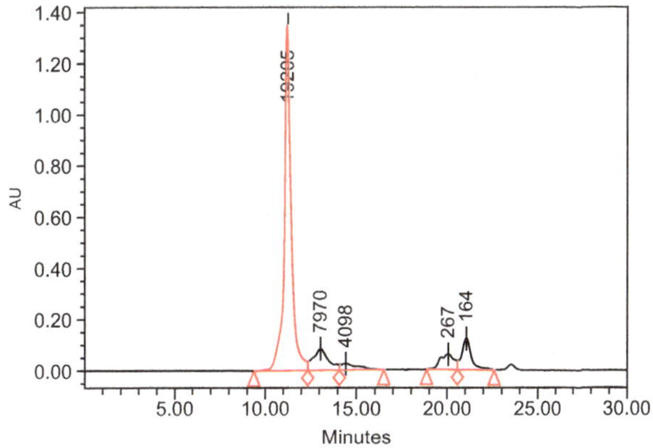

Fig. 6.31 Molecular weight distribution of PPC protein treated with ultrasound

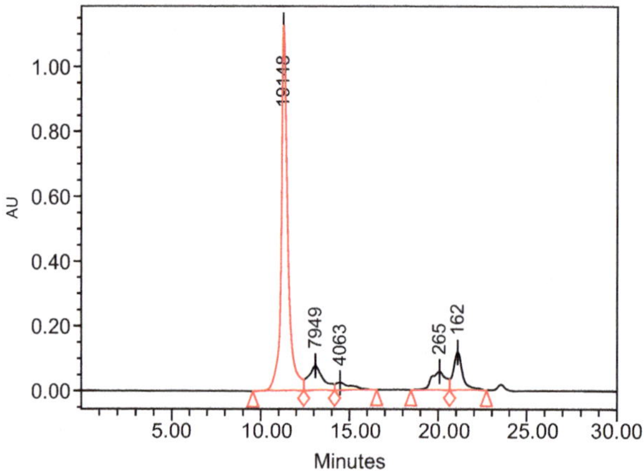

Fig. 6.32 Molecular weight distribution of PPC protein treated with reductant-1

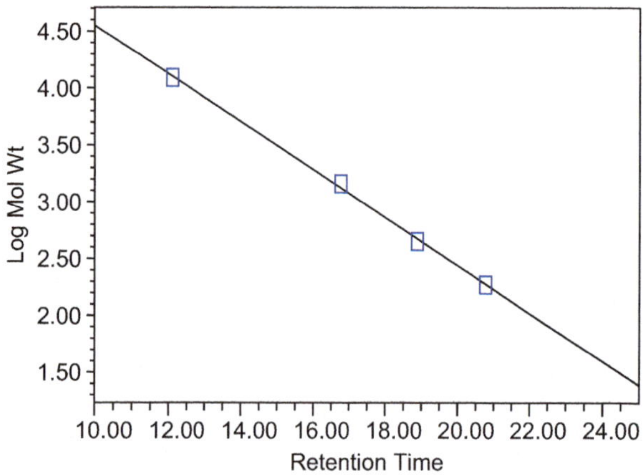

Fig. 6.33 Standard molecular weight determination curve

Table 6.11 Molecular weight distribution of peanut protein concentrate sample

No.	PPC		PPC treated with ultrasound		PPC treated with reductant-1	
1	% Area	Mw(Da)	% area	Mw(Da)	% area	Mw(Da)
2	78.78	18,601	72.26	18,900	68.89	18,449
3	4.82	8520	9.53	7882	10.31	7808
4	0.63	5248	3.02	3521	3.34	3504
5	2.08	3688	6.52	279	7.53	279
6	1.66	337	8.67	160	9.92	158
7	4.33	258				
8	7.70	161				

$R = 0.999451$
$R^2 = 0.998902$
Standard error: 3.191800e^{-002}
Equation = Log Mol Wt = 6.66e^{0}–2.11e^{-001} T^{1}

dynamic storage modulus (G′) and loss modulus (G″), namely, the energy storage and dissipation condition of object during the deformation. G′ refers to the springiness of sample, G″ refers to the viscosity of sample, and the tangent tan (δ) = G″/G′ of the loss angle refers to the viscosity and springiness ratio of the sample measured. But these results were unrelated to the stress or strain applied during the determination, and also they reflected the real nature of the system only within the linear viscoelastic region of sample (Daubert and Foegeding 1998). Under ideal conditions, the variation range of shear deformation should be proportional to that of the applied shear stress in the linear viscoelastic region, and the viscoelastic modulus (G′ and G″) of substances was unrelated to the applied shear stress, as well as the shear deformation. The change of viscoelastic modulus could be expressed by the change of composite modulus G* (G* = G′ + i G″).

The stress sweep test was conducted between 0.12 and 6.0 Pa. The results are shown in Fig. 6.34. It could be seen that the variation of composite modulus was very small in the range of 1.0–6.0 Pa, and the shear deformation was linear with the shear stress. Therefore, it could be considered that the viscoelasticity performed by PPC was unrelated to the applied stress in the range of 1.0–6.0 Pa. So the stress was fixed to 1.0 Pa to analyze the viscoelasticity of modified/original PPC in the following experiments.

3. Impact of pH and NaCl concentration on PPC rheological property

Frequency scanning could reflect the relationship between the viscosity and springiness of sample and the applied stress or change rate of strain and the impact of different conditions or processing methods on the viscoelasticity of product, and it was the most commonly used method in oscillatory testing (Nishinari et al., 1991). Figure 6.35 showed the changes in the storage modulus (G′) and loss modulus (G″) of 12% PPC solution with the oscillating frequency under the temperature of 90 °C and fixed stress of 1.0 Pa. From Figs. a, b, c, and d, it could be seen that the G′ of protein was always greater than G″ and increased with the increase of frequency when pH was 7.0. When NaCl was not added and the protein was at low frequency (<0.1 Hz), G′ and G″ were intersectant, the orderly protein chains had begun to entangle with each other, G′ was significantly greater than G″ with the increase of frequency, and the protein existed in the shape of gel at this time; when NaCl was added, from the ordinate, it could be seen that the G′ and G″ values decreased compared with those when NaCl was not added and the order was C<D<B<A; the increase of G′ was not stable at low frequency and G″ increased slowly with the increase of frequency. This indicated that the addition of salt ions was not good for the neutral PPC protein to form elastic gel. From Figs. e, f, g, and h, it could be seen that the change of G′ and G″ of protein with the addition of salt ions was very complex when the pH was 3.0. When the NaCl was not added, it could be seen from the ordinate that when the pH was 3.0, the G′ value of protein was increased by 110.6%, compared with the G′ value of protein (8.5) when the pH was 7.0 and oscillation frequency was 0.1 Hz, and also it was multiplied with the increase of oscillation frequency, but G″ did not change greatly; when the salt concentration (0.1 mol/L) was low and oscillation frequency was 0.1 Hz, the G′ and

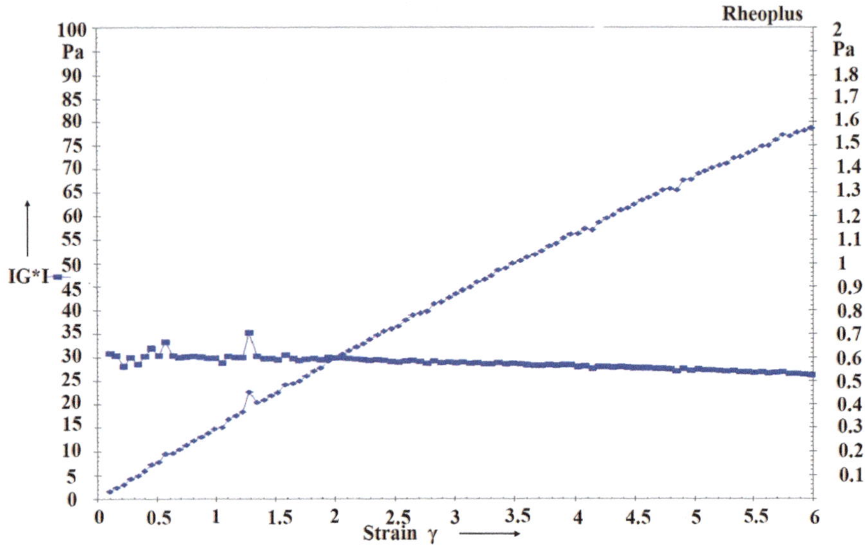

Fig. 6.34 Changes of composite modulus and strain with stress. Note: pH 7.0, none Na Cl, oscillation frequency 1 Hz, temperature 25 °C

G″ values of protein were 192 Pa and 32.7 Pa which were 10.73 and 6.62 times, respectively, higher than those when NaCl was not added. With the increase of oscillation frequency, G′ and G″ values increased slowly; when the concentration was 0.5 mol/L, G′ and G″ values of protein were the lowest (G″ > G′), which indicated that the viscosity of protein solution was greater than the springiness and the samples existed in the state of dilute solution at this time; when the concentration was up to 1.0 mol/L, the G′ and G″ values of protein returned to the level when the concentration was 0.1 mol/L.

The above results indicated that the addition of pH and salt ions had a great impact on the protein rheology. When the pH of protein solution was 3.0, the pH of protein was generally between 3.8 and 5.5 due to its buffering effect, just near the isoelectric point of protein, so as to induce the protein gel. The denatured aggregation of protein was a prerequisite for gel formation, so the samples with great aggregate contents could form gels with high strength rapidly. When the temperature increased from 30 to 90 °C at the rate of 20 °C/min rate rapidly, the heating temperature instantly rose above the denaturation temperature of the natural protein, so the denaturation of natural protein almost simultaneously occurred in the denaturation process. In the acidic environment, the repulsion among the protein molecules was small, the aggregation process was very active, and the natural protein could act with the surrounding denatured proteins or aggregates quickly simultaneously during the denaturation, so that the system could form a network structure in a very short period, and the undenatured protein would participate in the network structure after denaturation to avoid the further aggregation (Renkema

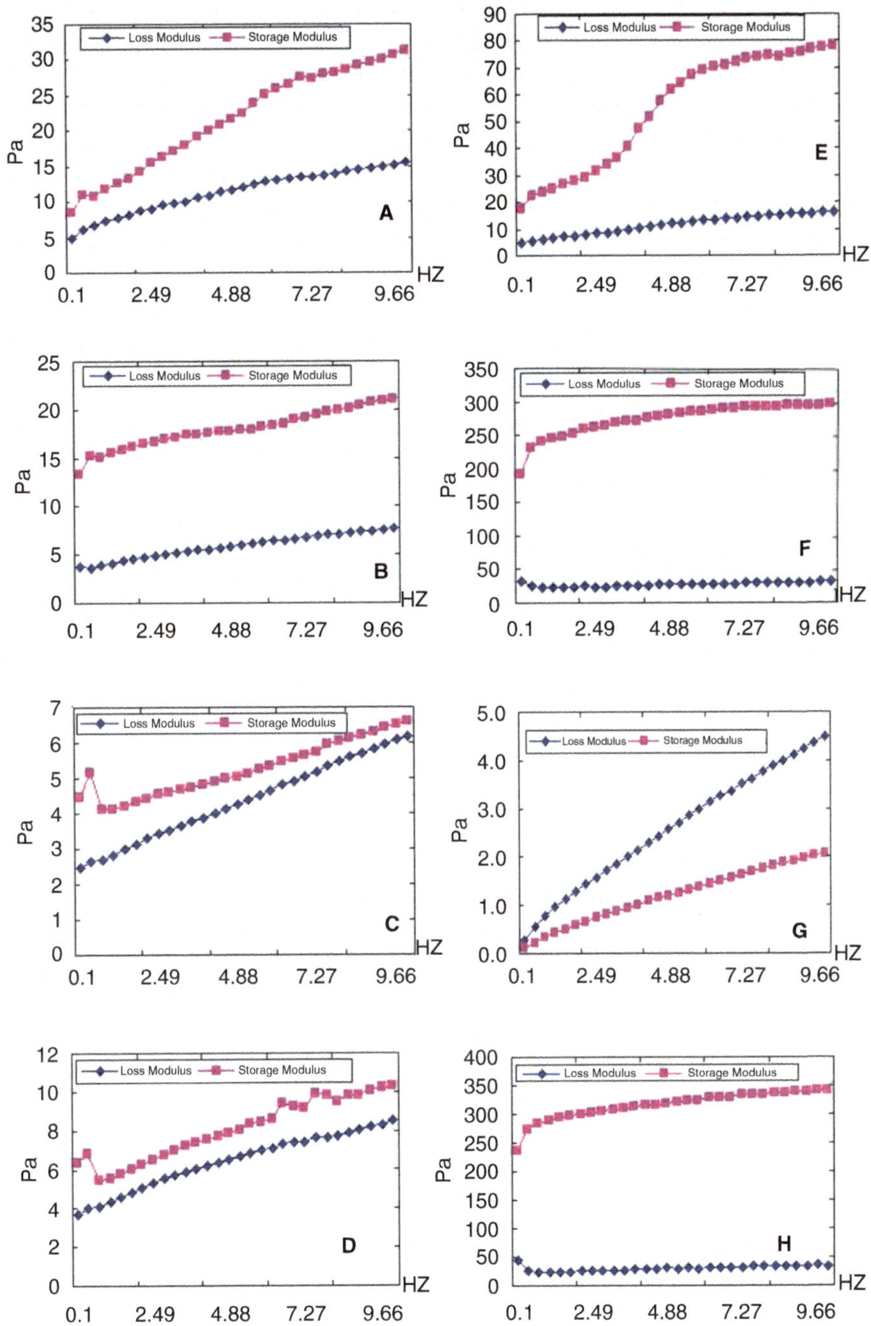

Fig. 6.35 Relationship between storage modulus and loss modulus of PPC solution and oscillation frequency. Note: pH 7.0, NaCl 0, 0.1, 0.5, 1.0 mol/L (**a–d**); pH 3.0, NaCl 0, 0.1, 0.5, 1.0 mol/L (**e**, **f**, **g**, and **h**)

2004). In general, the pH value affected the electrostatic interaction between protein molecules by changing the charged state of protein, while the inorganic salt affected the protein-protein interaction through the electrostatic shielding and disturbance of hydrophobic interaction and the electrostatic shielding was especially obvious when the salt concentration was low (less than 0.1–0.2 mol/L). It was generally believed that the hydrophobic interaction would be disturbed when the salt concentration was greater than 0.2 mol/L (Babaji et al. 1983).

4. Impact of pH and NaCl concentration on rheological property of PPC treated with ultrasound

It could be seen from Fig. 6.36 that the addition of salt ions had a significant impact ($P < 0.05$) on the rheological property of protein when the solution pH was 7.0 (A, B, C, and D). When NaCl was not added, G' and G'' values of protein were the lowest ($G'' > G'$), which indicated that the viscosity of protein solution was greater than the springiness and G'' increased linearly and G' increased slowly with the increase of oscillation frequency. The sample existed in the state of dilute solution at this time; the addition of salt ions made the G' of protein be greater than G'' all the time, and the increase of salt concentration had no effect on the G' and G'' of protein, which indicated that the protein solution existed in the state of gel. When the solution pH was 3.0 (E, F, G, and H), the rheological property of protein was basically consistent with that when pH was 7.0. When the salt concentration was 0.5 mol/L, the protein existed in the state of dilute solution, $G'' > G'$ at this time, the viscosity was greater than the springiness, the ion concentration continued to increase, and the protein returned to the level of protein elastic gel when the concentration was 0.1 mol/L; by observing the ordinate only, it could be seen that the springiness of protein gel (G') when pH was 3.0 was five times as much as that when pH was 7.0. In simple terms, acidic conditions were easy for protein to form elastic gels. Therefore, the author inferred that the electrostatic shielding effect of salt ions played a dominant role in the gel formation process of neutral protein treated with ultrasound, but the hydrophobic region was partially exposed, and the exposed hydrophilic groups and charged groups contacted with each other in the solution with the performance of electrostatic repulsion effect because the ultrasonic radiation could only partially destroy the chemical bonds in the protein. When the salt ions of different concentrations were added, the charged groups were constantly neutralized, they attracted and aggregated with each other with the non-charged or electrostatic attraction as the main force, and then they formed elastic gel finally with the rise of temperature and continuous increase of protein denaturation degree. When the pH was 3.0, the protein was near the isoelectric point. At this time, the protein molecules were positively charged, the electrostatic shielding effect generated after the addition of salt ions made the electrostatic interaction between the protein molecules weaken, the protein-protein interaction was mainly the hydrophobic interaction, and the stronger the hydrophobic interaction, the lower the gelation temperature, the shorter the gelation time. The cavitation, mechanical shear, and thermal effects caused by ultrasonic wave had damage effects on the secondary bond, hydrogen bonds, and van der Waals force in the

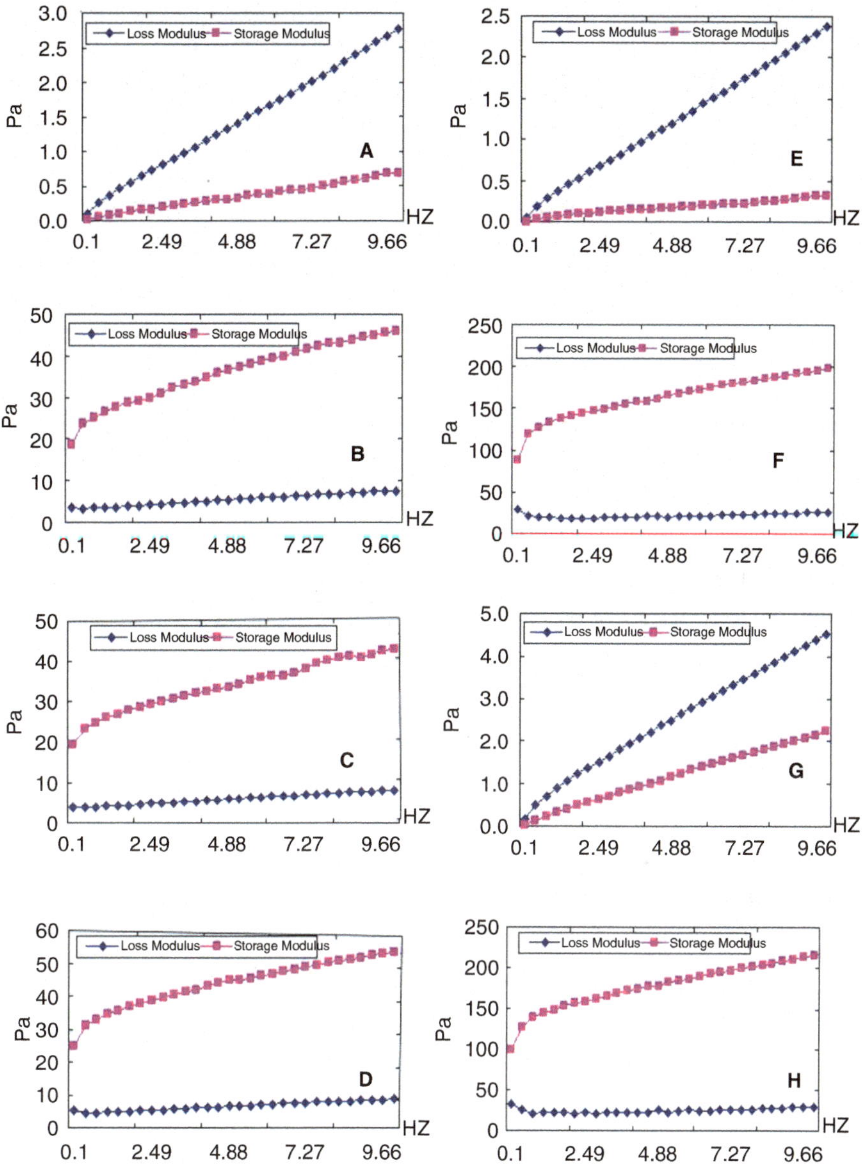

Fig. 6.36 Relationship between storage modulus and loss modulus of PPC solution treated with ultrasound and oscillation frequency. Note: pH 7.0, NaCl 0, 0.1, 0.5, 1.0 mol/L (**a–d**); pH 3.0, NaCl 0, 0.1, 0.5, 1.0 mol/L (**e–h**)

protein used to maintain the high-level structure of protein, so the protein after ultrasonic treatment was in the solution, the protein-protein hydrophobic interaction and hydrogen bonding between protein and water were enhanced (Sanchez and

Burgos 1997), the stable network structure could be formed quickly, the protein gelation time was reduced, and the gel springiness was increased.

5. Impact of pH and NaCl concentration on rheological property of PPC treated with reductant-1

It could be seen from Fig. 6.37 that the increase of salt ion concentration caused the gradual decrease of G′ and G″ of protein in the PPC solution treated with the reductant-1 when pH was 7.0, but G′ was always greater than G″, which indicated that the protein existed in the state of gel; when NaCl was not added and pH was 3.0, the G″ of protein was >G′ and the protein existed in the state of dilute solution at this time; the addition of salt ions enhanced the springiness of protein significantly, and G′ increased gradually with the increase of ion concentration. The results were consistent with the research results of Zhang et al. (2006). The above results indicated that the increase of ionic strength was not conducive to the gelation of PPC treated with reductant-1 when pH was 7.0; the increase of ionic strength could enhance the gelation of protein when pH was 3.0. According to the statement of Wang et al. (1991), the α-helix changed into a β-sheet or random structure in the process of heat gelation, the β-sheet conformation occupied a certain proportion, and it was a necessary condition for gel formation. The quantitative analysis results of secondary structure of protein molecule treated with reductant-1 were as follows: 0.23% was α-helix, 98.68% was β-sheet, and 1.09% was turn structure. It could be seen that the PPC treated with reductant-1 had strong gelation. Zhang (2006) found that the characteristic band of the amide I′ of the protein after gelation moved in the increasing direction of β-sheet with the increase of ionic strength at pH was 3.0 when he researched the gelation of chickpea protein isolate, which indicated that the gelation was enhanced. The results after gelation when pH was 7.0 were just the opposite. The characteristic band of amide I′ moved from 1662 cm^{-1} when there was no NaCl to 1660 cm^{-1} when there was 0.1 mol/L NaCl, so it was inferred that this was the main reason for the weakening of gel strength with the increase of ionic strength under this pH condition.

6. Impact of pH and CaCl2 concentration on rheological property of modified/original PPC

It could be seen from Fig. 6.38 that the addition of $CaCl_2$ had a great impact on the rheological property of protein when pH was 7.0 (A, B, C, D, E and F). In the PPC solution, the protein gradually formed an ordered cross-linked structure (Fig. B, 0.1–2.0 Hz low-frequency oscillation) from the state of dilute solution (Fig. A, G″ > G′) and then formed the elastic gel (Fig. B, G′ > G″) with the increase of salt concentration; in the PPC solution treated with ultrasound, the addition of salt ions was not conducive to the formation of elastic gel, and the protein always existed in the state of dilute solution (Figs. C and D, G″ > G′); in the PPC solution treated with reductant-1, when the concentration of CaCl2 was 0.1 mol/L (Fig. E), the protein formed an ordered cross-linked structure at low frequency, which was not affected by the increase of frequency and could form stable elastic gel; when the concentration increased to 0.3 mol/L (Fig. F), the ordered protein chains entangled

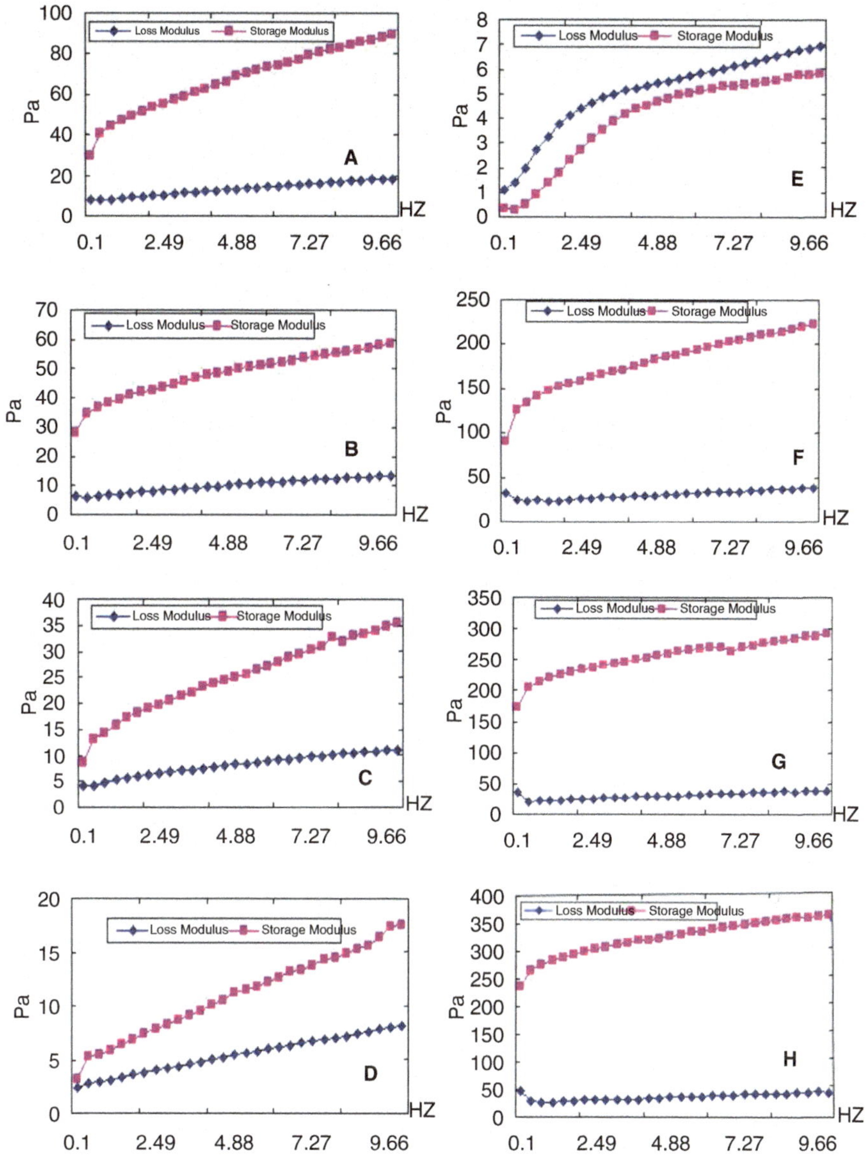

Fig. 6.37 Relationship between storage modulus and loss modulus of PPC solution treated with reductant −1 and oscillation frequency. Note: pH 7.0, NaCl 0, 0.1, 0.5, 1.0 mol/L (**a**–**d**); pH 3.0, NaCl 0, 0.1, 0.5, 1.0 mol/L (**e**–**h**)

with each other, but stable elastic gel could not be formed, which indicated that the increase of salt concentration was not good for the PPC treated with reductant-1 to form elastic gel. When pH was 3.0 (a, b, c, d, e, and f), the addition of $CaCl_2$ had little impact on the rheological property of protein and G′ was always greater than G″.

It could be seen from ordinate that the increase of ionic concentration weakened the elastic strength of PPC gel greatly and enhanced the elastic strength of the PPC gel treated with ultrasound, but it did not have a significant impact on the gel formation of PPC treated with reductant-2.

According to the above experiment results, it could be inferred that the increase of divalent salt ion concentration made the gel structure of protein change. When the solution pH was 7.0, the PPC protein was negatively charged, and the electrostatic repulsion between the molecules was dominant, so it was difficult to approach each other or combine with each other. After the addition of salt ions, the positive ions of salt shielded the partial negative charges of protein, so that the electrostatic repulsion between the proteins was weakened; when the salt ions in the solution reached the solidification concentration, the repulsion (electrostatic effect) and gravitation (hydrophobic effect and hydrogen bond) between the protein molecules achieved a balance (Matsumura and Mori 1996), and the protein in the solution

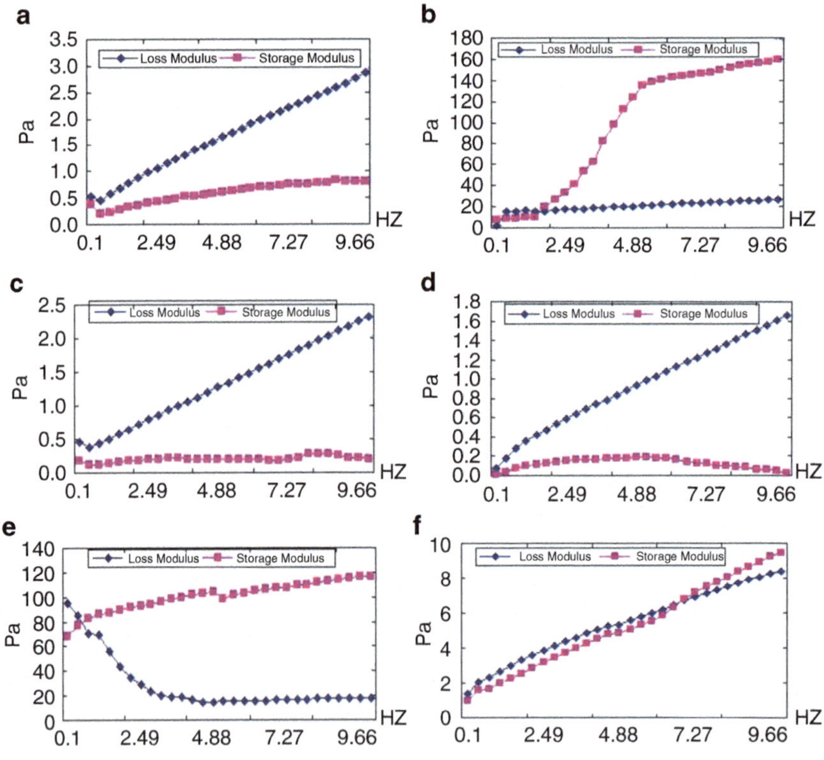

Fig. 6.38 Relationship between storage modulus and loss modulus of protein in solutions with different pH and CaCl2 concentration and oscillation frequency. Note: pH 7.0, CaCl2 0.1, 0.3 mol/L, PPC (**A** and **B**), PPC treated with ultrasound (**C** and **D**), PPC treated with reductant-1 (**E** and **F**); pH 3.0, CaCl2 0.1, 0.3 mol/L, PPC (**a** and **b**), PPC treated with ultrasound (**c** and **d**), PPC treated with reductant-1 (**e** and **f**)

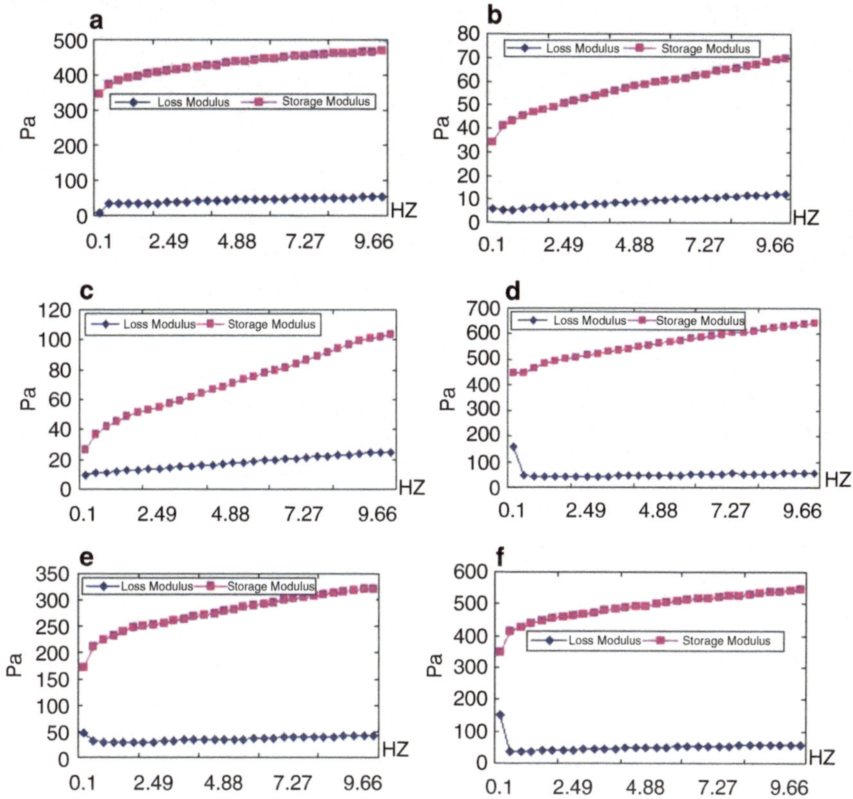

Fig. 6.38 (continued)

bound into an ordered network structure; when pH was 3.0 and the protein was in the low salt concentration, the acting force between molecules quickly reached a balance and the increase of salt concentration destroyed this balance. The protein could form gel, but the strength was significantly reduced.

7. Relationship between tanδ (G″/G′) and pH and salt concentration

The change of tanδ value could visually indicate that the addition of different salt ions caused the change of viscoelasticity of sample, and it had the opposite result depending on the difference of pH. It could be seen from Fig. 6.39, when pH was 7.0 (a, c, and e), the tanδ value of protein was greater than 1.0 in the solution with $CaCl_2$. The tanδ values of PPC and PPC treated with reductant-1 were greater than 1 when the $CaCl_2$ concentration was 0.3 mol/L, and the tanδ value of PPC treated with ultrasound was greater than 1 when the $CaCl_2$ concentration was 0.1 mol/L, which indicated that the sample changed from the gel shape when there was no salt to the macromolecule solution shape; when pH was 3.0 (b, d, and f), the tanδ value of protein was greater than 1.0 in the salt-free or NaCl

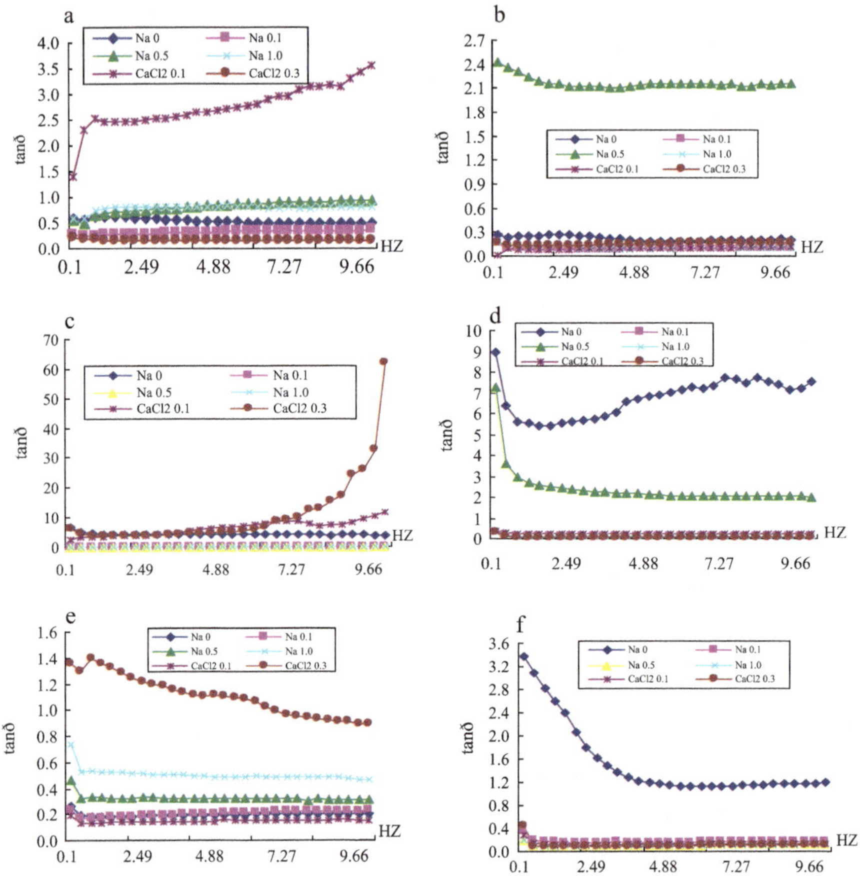

Fig. 6.39 Relationship between tanδs of different protein solutions and pH and salt concentration. Note: PPC solution (**a** pH 7.0, **b** pH 3.0); PPC solution treated with ultrasound (**c** pH 7.0, **d** pH 3.0); PPC treated with reductant-1 (**e** pH 7.0, **f** pH 3.0)

0.5 mol/L solution, it was smaller than 0.2 in the solution with other salt concentration, and the sample existed in the state of gel at this time.

8. Impact of pH and salt concentration and original protein texture property

Figure 6.40 showed that the hardness value of the modified/original protein in the 0.1 mol/L NaCl solution was the largest when pH was 7.0 (Ia, IIa, and IIIa), and the hardness value of gel decreased with the increase of salt concentration; the gel springiness and cohesiveness of the original PPC in the salt solution were higher than that of the modified PPC; the results showed that the presence of ions was not conducive to the formation of strong gel when pH was 7.0. When pH was 3.0 (Ib, IIb, and IIIb), the gel hardness value of protein was significantly higher than that when pH was 7.0, especially that of the PPC treated with ultrasound; the addition of

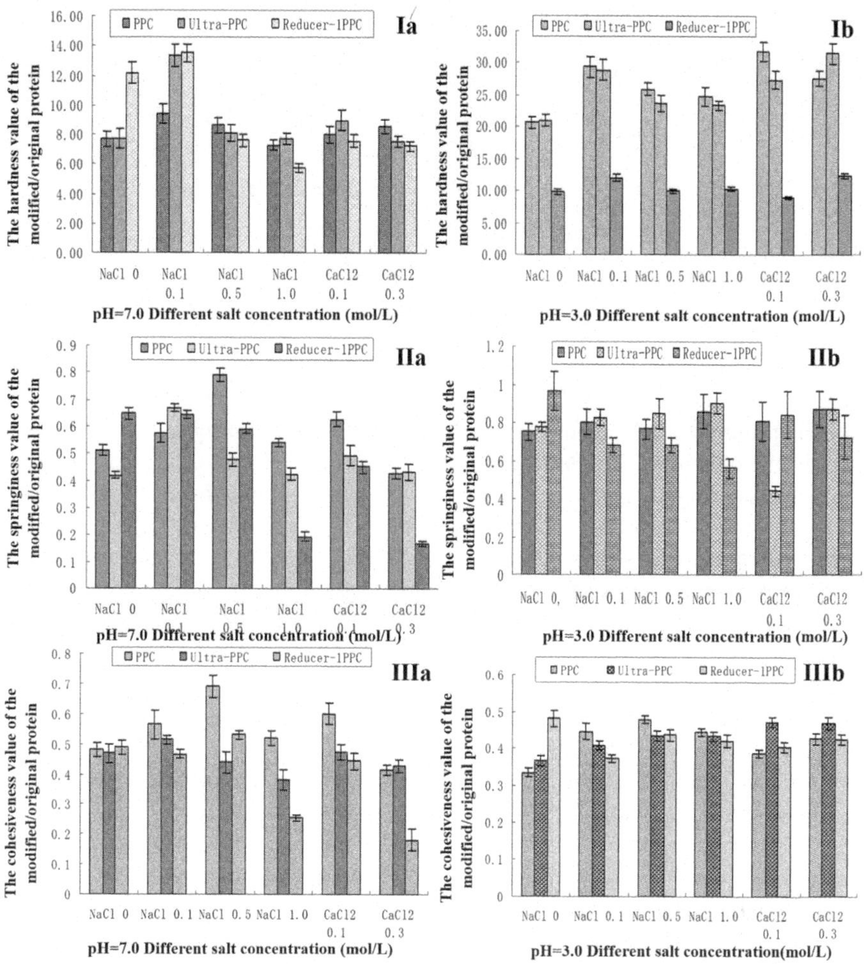

Fig. 6.40 Impact of pH and salt concentration on protein texture property. Note: strength value of gel (*Ia*, pH 7.0; *Ib*, pH 3.0); springiness value of gel (*IIa*, pH 7.0; *IIb*, pH 3.0); cohesiveness of gel (*IIIa*, pH 7.0; *IIIb*, pH 3.0)

salt ions had no significant impacts on the springiness and cohesiveness of protein, and the protein solutions with and without salt could form gel with high springiness.

9. Correlation between rheological property and texture property of modified and original protein

The G* (X_1), G″ (X_2), and G′ (X_3) values of protein solution when the oscillation frequency was 1.0 Hz were taken to analyze the correlation between these values and the gel strength (X_4), springiness (X_5), and cohesiveness (X_6) values of protein by using the DPS7.55 statistical analysis software. The results are shown in Table 6.12.

Table 6.12 Correlation between rheological property and texture property of protein

Correlation coefficient	X_1	X_2	X_3	X_4	X_5	X_6
X_1	1.00					
X_2	0.92**	1.00				
X_3	1.00**	0.89**	1.00			
X_4	0.79**	0.57*	0.81**	1.00		
X_5	0.56*	0.47*	0.57*	0.61**	1.00	
X_6	0.40	0.40	0.40	0.32	0.92**	1.00

Note: * $p < 0.05$; ** $p < 0.01$

It could be seen from Table 6.12 that the G*, G″, and G′ of protein solution were highly significantly correlated with the gel strength and significantly correlated with the gel springiness value, but they were not significantly correlated with the gel cohesiveness; there were highly significant correlations between G″, G*, and G′ and between gel springiness value, gel hardness, and cohesiveness; the gel cohesiveness was only highly significantly correlated with the gel springiness and not significantly correlated with other parameters. This simple correlation analysis considered the three protein products as a whole and investigated the correlation between the rheological property and texture property. For the specific correlation between the factors in the rheological property and texture property, a detailed factor analysis was required to draw a conclusion.

Considering that the difference in the variability of the variables (evaluation indicators) in this research was too large, the normalized processing was conducted for the data (Table 6.13), and then the factor analysis was performed. According to the factor analysis principle, the factor loading matrix of the rheological property and texture property of the sample was established, and the eigenvalues, eigenvalue contribution rates, and accumulative contribution rates of eigenvalues of the correlation coefficient matrices were calculated. The results are shown in Table 6.14. According to the accumulative contribution rates of eigenvalues, it could be seen that the accumulative contribution rate of variance had been 90.90% when the first two factors were taken. Therefore, the first two common factors were used to establish the factor loading matrix, and the simplified factor loading matrix was obtained by using the varimax rotation method (Table 6.15).

It could be seen from Table 6.15 that the common factor f_1 played a dominant role in the G* and G′ in the sample, and these indicators mainly reflected the rheological characteristics of sample, so f_1 could be called the comprehensive factor; the common factor f_2 played a dominant role in the cohesiveness in the sample and mainly reflected the texture characteristics of product, so the common factor f_2 could be called the texture factor. In other words, for the protein solutions with different salt concentrations, it could be seen that the G* and G′ values in all the indicators of rheological property of sample were representative through the detailed factor analysis, while the hardness value in the texture property was representative. Therefore, only these three indicators needed to be selected from

Table 6.13 Standardized data processing results

Salt concentration (mol/L)	Sample	G*	G″	G′	Hardness	Springiness	Cohesiveness
Na0	PPC	0.24	0.19	0.24	0.25	0.55	0.59
Na0.1	PPC	0.48	0.43	0.48	0.47	0.65	0.75
Na0.5	PPC	0.27	0.19	0.27	0.37	1.00	1.00
Na1.0	PPC	0.19	0.14	0.19	0.20	0.60	0.67
Ca0.1	PPC	0.75	1.00	0.69	0.29	0.74	0.82
Ca0.3	PPC	0.23	0.18	0.23	0.36	0.42	0.45
Na0	Ultra-PPC	0.31	0.21	0.31	0.25	0.40	0.57
Na0.1	Ultra-PPC	1.00	0.75	1.00	0.99	0.81	0.65
Na0.5	Ultra-PPC	0.52	0.32	0.52	0.30	0.50	0.51
Na1.0	Ultra-PPC	0.11	0.10	0.10	0.25	0.40	0.39
Ca0.1	Ultra-PPC	0.89	0.94	0.86	0.41	0.52	0.57
Ca0.3	Ultra-PPC	0.34	0.22	0.35	0.23	0.42	0.49
Na0	Reducer-1PPC	0.66	0.50	0.66	0.83	0.77	0.61
Na0.1	Reducer-1PPC	0.82	0.61	0.83	1.00	0.76	0.57
Na0.5	Reducer-1PPC	0.26	0.16	0.26	0.24	0.68	0.69
Na1.0	Reducer-1PPC	0.00	0.00	0.00	0.00	0.03	0.14
Ca0.1	Reducer-1PPC	0.17	0.29	0.15	0.23	0.45	0.51
Ca0.3	Reducer-1PPC	0.16	0.13	0.16	0.19	0.00	0.00

many indicators to analyze the gelation of sample to obtain the results with the accuracy rate of 90.90%.

10. Impact of salt concentration on thermal denaturation temperature of modified and original PPC protein

Wu et al. (2009) used DSC to detect the varieties and concentrations of different salt ions, and the thermal denaturation temperature of modified/original PPC changed. The results are shown in Table 6.16. The thermal denaturation temperature of PPC occurred at 67.09 °C and 91.48 °C, respectively, when there was no salt. With the gradual increase of NaCl concentration, the thermal denaturation temperature (T_d) and ΔH significantly increased ($P < 0.05$). With the increase of $CaCl_2$ concentration, the thermal denaturation temperature had a nonsignificant decrease ($P > 0.05$), and ΔH did not change significantly ($P > 0.05$); the PPC treated with ultrasound and PPC treated with reductant −1 showed the same trend. The research on the impact of salt concentration on the thermal denaturation temperature and enthalpy change of peanut protein had not been

Table 6.14 Eigenvalue and accumulative contribution rate of matrix related to target parameter

No.	Eigenvalue	Percentage %	Accumulative percentage %	No.	Eigenvalue	Percentage %	Accumulative percentage %
1	4.2643	71.0711	71.0711	4	0.0535	0.8915	99.7304
2	1.1899	19.8312	90.9023	5	0.0161	0.2682	99.9987
3	0.4762	7.9366	98.8389	6	0.0001	0.0013	100.0000

Table 6.15 Varimax orthogonal rotation factor loading matrix

Index	Factor f_1	Factor f_2	Index	Factor f_1	Factor f_2
G^*	0.9478	−0.2965	Springiness	0.7925	0.5910
G''	0.8655	−0.2787	Cohesiveness	0.6432	0.7456
G'	0.9468	−0.2976	Eigenvalue	4.2643	1.1899
Hardness	0.8238	−0.1747	Accumulative contribution	0.7107	0.9090

reported, but Zheng et al. (1993) found that the addition of NaCl significantly improved the thermal denaturation temperature of broad bean protein; Meng and Ma (2001) and Arnfield et al. (1986) obtained that the addition of thermal stability of protein was because the presence of NaCl increased the intermolecular hydrophobic association (Arnfield et al. 1990a, b) and electrostatic interaction or changed the hydration state around the protein, and all these changes could increase the hydration of protein molecules when researching the impact of NaCl on the thermal denaturation temperature of red bean globulin and broad bean globulin (Meng and Ma 2001).

In conclusion, with the increase of Na salt ion concentration, the thermal denaturation temperature and absorption enthalpy value of protein increased; in Ca salt solution, the thermal denaturation temperature of protein decreased, and the absorption enthalpy increased with the increase of concentration; this phenomenon was basically consistent with the change trend of tanδ with pH and salt concentration. In Na salt solution, with the increase of salt concentration, the protein tended to form macromolecule solution which needed to absorb more heat, and the viscosity increased, so the absorption enthalpy increased and tanδ increased simultaneously; in the Ca salt solution, the protein was easy to absorb high-valence salt ions, the tan δ was always greater than 1, the protein existed in the state of dilute fluid, the viscosity was always greater than the springiness, and the protein formed the macromolecular viscous liquid during the continuous endothermic process.

2.2.3 Solubility of Modified PPC and Original PPC in Different Solvents

Through the analysis of solubility changed of protein modified by physicochemical method in different solvents (see Fig. 6.41), it could be seen that the solubility of the modified protein was significantly higher than that of the original protein in the phosphate buffer solution (PBS), and the protein solubility increased from 44.33% to 60.71% and 63.72% with the increase of 36.95% and 43.74%, respectively. In the 0.025MSDS solvent (hydrogen bond and hydrophobic effect breaker), if only the changes of protein solubility before and after treatment were compared, the solubility of the protein after treatment was higher than that before treatment and the order was solubility of PPC treated with reductant −1>solubility of PPC treated with ultrasound>solubility of PPC; but the solubility of protein before and after treatment was significantly increased by 74.90%, 32.01%, and 42.64%,

Table 6.16 Thermal denaturation temperature (*Td*) and ΔH of protein solution in different salt concentrations

Sample	PPC		PPC treated with ultrasound		PPC treated with reductant −1	
Salt concentration (mol/L)	*Td* (°C)	ΔH (J/g)	*Td* (°C)	ΔH (J/g)	*Td* (°C)	ΔH (J/g)
Peak value 1	**	**	**	**	**	**
0 NaCl	67.09 ± 1.03^{c}	0.06 ± 0.001^{d}	65.74 ± 0.96^{c}	0.01 ± 0.001^{d}	67.85 ± 1.02^{b}	0.01 ± 0.001^{c}
0.1 NaCl	71.69 ± 0.56^{b}	0.08 ± 0.002^{cd}	73.18 ± 1.42^{b}	0.02 ± 0.001^{c}	73.43 ± 1.47^{a}	0.02 ± 0.002^{c}
0.5 NaCl	74.02 ± 2.01^{a}	0.104 ± 0.007^{c}	75.95 ± 1.33^{ab}	0.03 ± 0.003^{b}	73.88 ± 2.11^{a}	0.06 ± 0.011^{b}
1.0 NaCl	77.78 ± 1.58^{a}	0.362 ± 0.011^{a}	75.22 ± 1.55^{a}	0.05 ± 0.004^{a}	78.47 ± 2.58^{a}	0.11 ± 0.015^{a}
0.1 $CaCl_2$	75.93 ± 1.47^{a}	0.302 ± 0.023^{b}	76.00 ± 1.17^{a}	0.01 ± 0.001^{d}	72.54 ± 1.14^{a}	0.07 ± 0.012^{b}
0.3 $CaCl_2$	74.11 ± 1.31^{a}	0.061 ± 0.003^{d}	74.77 ± 1.02^{ab}	0.03 ± 0.002^{d}	63.83 ± 1.05^{b}	0.16 ± 0.024^{a}
Peak value 2	**	**	**	**	**	*
0 NaCl	91.48 ± 0.55^{c}	0.03 ± 0.001^{b}	92.70 ± 0.36^{c}	0.01 ± 0.001^{d}	95.65 ± 0.56^{a}	0.01 ± 0.002^{b}
0.1 NaCl	91.70 ± 0.84^{c}	0.01 ± 0.001^{d}	93.92 ± 0.44^{b}	0.01 ± 0.002^{d}	87.69 ± 0.48^{c}	0.02 ± 0.007^{ab}
0.5 NaCl	92.24 ± 0.34^{c}	0.02 ± 0.001^{c}	93.94 ± 0.50^{b}	0.02 ± 0.001^{c}	87.46 ± 0.41^{c}	0.03 ± 0.005^{ab}
1.0 NaCl	94.13 ± 0.93^{b}	0.04 ± 0.002^{a}	96.06 ± 0.21^{a}	0.03 ± 0.003^{a}	92.12 ± 0.55^{b}	0.04 ± 0.018^{ab}
0.1 $CaCl_2$	95.65 ± 0.36^{a}	0.02 ± 0.001^{c}	91.01 ± 0.48^{c}	0.02 ± 0.001^{c}	95.52 ± 0.28^{a}	0.02 ± 0.006^{ab}
0.3 $CaCl_2$	94.63 ± 1.01^{ab}	0.01 ± 0.002^{e}	87.80 ± 0.11^{d}	0.03 ± 0.001^{b}	92.84 ± 0.37^{b}	0.05 ± 0.019^{n}

Note: The same letter in the same column means nonsignificant difference ($P > 0.05$); during treatment, $*P<0.05$, $**P<0.01$

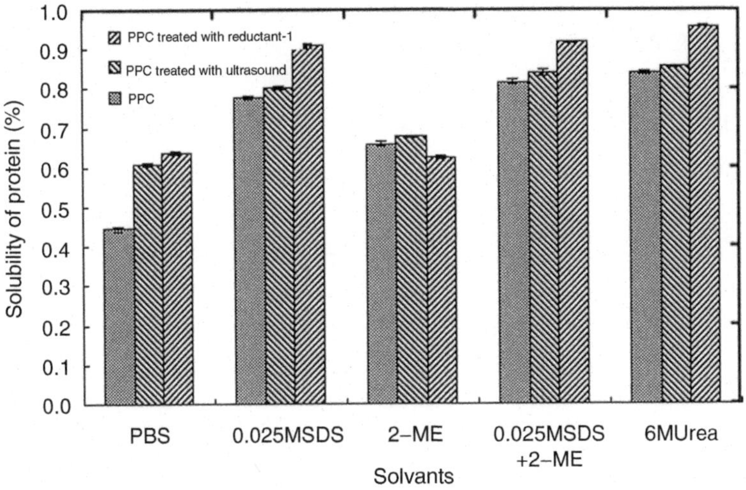

Fig. 6.41 Impact of different treatment on solubility of peanut protein concentrate

respectively, compared with the solubility in PBS buffer solution. It indicated that the hydrogen bonds and chemical bonds of hydrophobic interaction between the protein molecules in the 0.025MSDS buffer solution were damaged in different degrees, and the hydrophobic interaction and hydrogen bond contents in the protein molecules before and after treatment were relatively large. In the 2-ME (disulfide breaker) buffer solution, the protein solubility of PPC increased significantly compared with that in PBS, with the increase of 48.09%, while the increase in the solubility of PPC treated with ultrasound and PPC treated with reductant-1 was 11.34% and −2.06%, which indicated that the content of free disulfide bonds in PPC was high, ultrasonic treatment could break some disulfide bonds in the protein, and the free disulfide bonds did not exist in PPC basically after reduction agent-1 treatment. The increase in negative value might be caused by errors. In the 0.025MSDS+2-ME buffer solution, the protein solubility was higher than that in the 0.025 MSDS buffer solution, but there was no significant difference ($P > 0.05$) through the variance analysis. In the 6 M urea buffer solution, the protein solubility was the highest (83.82%, 85.38%, and 95.61%, respectively). Urea was a strong denaturant, and it could destroy all secondary bonds (noncovalent bonds) other than the disulfide bonds, which indicated that the hydrophobic interaction and hydrogen bond contents in the protein treated with reductant-1 were higher than the contents of free disulfide bonds.

The above results showed that the protein in PPC mainly contained hydrophobic interaction hydrogen bonds and disulfide bonds; after the ultrasonic treatment, the hydrophobic interaction and hydrogen bond in protein were reduced relatively, and the content of free disulfide bonds was reduced greatly; after the reductant-1 treatment, the content of disulfide bonds in protein was small, and the main acting force was the noncovalent bond (hydrophobic interaction and hydrogen bond). The

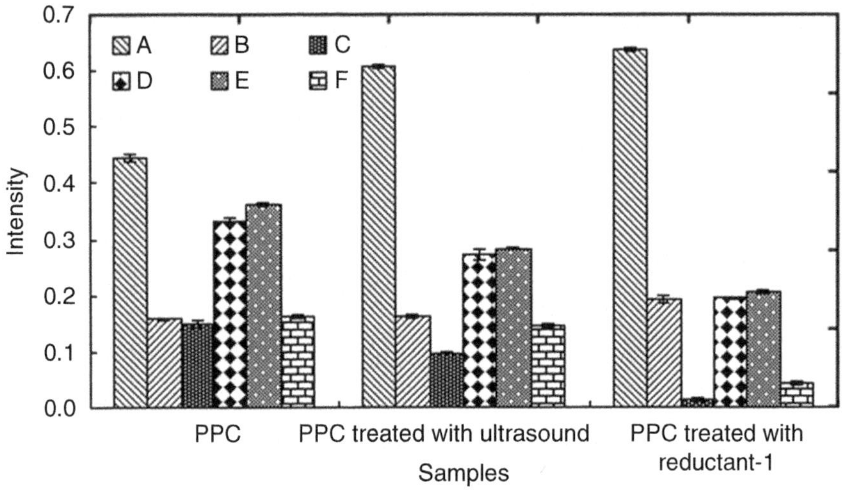

Fig. 6.42 Mode of intermolecular forces of protein in various solvents. Note: (**a**) ph 7.0 phosphate buffer solution, soluble protein (*); (**b**) protein which is insoluble in buffer solution and is combined by noncovalent bonds=[soluble in (SDS + 2-ME)-soluble in 2-ME]; (**c**) protein which is insoluble in buffer solution and is combined by disulfide bonds= [soluble in (SDS + 2-ME)-soluble in SDS]; (**d**) protein which is insoluble in buffer solution and is combined by disulfide bonds and noncovalent bonds= [soluble in SDS-[*]; (**e**) protein which is insoluble in buffer solution and is combined by intermolecular secondary bonds (including hydrophobic bonds and electrostatic and hydrogen bond interactions in the hydrophobic region)=[soluble in 6 M urea-soluble in SDS]; (**f**) protein which is insoluble in buffer solution and is combined by covalent bonds-[completely solubility-soluble in 6 M urea]

impact of different treatment on protein solubility was significant, but there were no significant differences in the solubility between the protein treated with ultrasound and that treated with reductant-1 ($P > 0.05$).

2.2.4 Intermolecular Force of Modified PPC and Original PPC in Different Solvents

It could be seen from Fig. 6.42 that about more than 50% of the peanut protein concentrate was soluble protein, and the soluble protein content was significantly increased after treatment ($P < 0.05$). In the protein before and after treatment, the contents of proteins which were insoluble in buffer solution and were combined by noncovalent bonds account for 15.77%, 16.92%, and 19.30% of the total protein, respectively; the contents of proteins which were insoluble in buffer solution and were combined by disulfide bonds account for 14.86%, 9.74%, and 1.35% of the total protein, respectively; the contents of proteins which were insoluble in buffer solution and were combined by disulfide bonds and noncovalent bonds account for 33.28%, 27.17%, and 19.44% of the total protein, respectively; the contents of proteins which were insoluble in buffer solution and were combined by intermolecular secondary bonds (including hydrophobic bonds and electrostatic and hydrogen bond interactions in the hydrophobic region) account for 36.11%, 28.23%, and 20.72% of the total protein, respectively; the contents of proteins which were insoluble in buffer solution and were combined by covalent bonds account for 16.18%, 14.62%, and 4.39% of the total protein, respectively.

The above results showed that the disulfide bond, noncovalent bond, intermolecular secondary bond, and covalent bond were taken as the main intermolecular interaction bonds in the original PPC protein; after the ultrasonic treatment, the chemical bonds in protein were damaged in different degrees; especially the disulfide bond content was significantly reduced and the hydrophobic region of protein was partially exposed, so the hydrophobic bond in protein and electrostatic and hydrogen-bond interaction in the hydrophobic region became the main intermolecular interaction bonds; after the reductant-1 treatment, there was almost no disulfide bond because the treatment with the solution containing 2-ME could release the peptides involved in the disulfide bonds, and the content of protein combined by disulfide binds which was measured in the test was very low, but there were still many intermolecular secondary bonds and noncovalent bonds. In conclusion, after treatment, the hydrophobic region of protein was partially exposed and the disulfide bond content was significantly reduced. Therefore, it could be inferred that the free -SH content in protein increased greatly, while partial hydrogen bonds and noncovalent bonds were destroyed, and these results could effectively improve the quick-setting properties of protein to achieve the expected purpose of test.

2.2.5 Secondary Structures of Modified PPC and Original PPC

Fourier transform infrared spectroscopy (FTIR) is a method that can be used to carry out quantitative research for the secondary structure of protein in recent decades. The infrared band mostly used for the secondary structure analysis is the amide I band which is located between 1600 and 1700 cm^{-1} and mainly involves the stretching vibration of C=O. According to the research of predecessors (Byler and Susi 1986; Susi and Byler 1986; Krimm and Bandekar 1980; Jackson and Mantsch 1995), it was thought that, when the hydrogen bonds (C=O...H-) formed between the amino acid residues of protein were strong, the electronic cloud density of C=O was low, only less energy was needed to make it vibrate, and the absorption peak of C=O moved in the direction of low wave number and was close to 1600 cm^{-1}. Based on this principle, the correspondences between various peak positions and components of secondary structure were shown below: (1694 ± 2) cm^{-1}, (1689 ± 2) cm^{-1}, (1683 ± 2) cm^{-1}, (1671 ± 3) cm^{-1}, and (1663 ± 4) cm^{-1} were turn structures, (1653 ± 4) cm^{-1} was a α-helix, (1645 ± 4) cm^{-1} was a free coil structure (disorder structure), and (1675 ± 5) cm^{-1} and (1630 ± 10) cm^{-1} were stretched peptide chains (β-sheets). This research further identified the 1600–1625 cm^{-1} as β_1 (viz., the strong interaction hydrogen bonds between the protein molecules) by combining with the research results of Karren and

Carrasquillo (2001); 1600–1700 cm^{-1} was identified as β_2, namely, the non-plane weak hydrogen bond effect (caused by the dipole transformation) formed between the protein molecules; $\beta_1 + \beta_2$ was identified as the overall interaction hydrogen bond between the protein molecules; 1626–1639 cm^{-1} was identified as the intra-molecular β-sheet hydrogen bond; $\beta + \alpha$ was identified as the overall interaction hydrogen bond between the protein molecules and represented the closeness degree of protein molecules.

The FTIR spectrograms of different protein samples were processed as shown in Fig. 6.43. To get the second derivative, the spectrograms were fit as shown in Figs. 6.44, 6.45, and 6.46. The peak area was fit for statistics according to the second derivative, the identifications and areas of the relevant peak positions were listed in Table 6.17, and the percentage content of peak area was calculated as the percentage content of the corresponding structure. The results are shown in Table 6.18.

From Figs. 6.44, 6.45, and 6.46 and Table 6.17, it could be seen that the strong peak of the original PPC was located near 1635 cm^{-1}, there was also strong absorption near 1685 cm^{-1} and 1690 cm^{-1}, and the peak position near 1610 cm^{-1} had little relationship with the secondary structure and belonged to the vibration of amino acid side chain. The quantitative analysis results of the secondary structure were shown below: 5.13% was α-helix, 56.02% was β-sheet, and 38.86% was turn structure. The strong peak of PPC after ultrasonic treatment was located near 1638 cm^{-1} and 1690 cm^{-1}, and there was strong absorption near 1680 cm^{-1} and 1655 cm^{-1}. The quantitative analysis results of the secondary structure were shown below: 15.01% was α-helix, 81.66% was β-sheet, and 3.34% was turn structure. The PPC after reductant-1 treatment was located near 1635 cm^{-1}, and there was strong absorption near 1680 cm^{-1} and 1690 cm^{-1}. The quantitative analysis results of the secondary structure were shown below: 0.23% was α-helix, 98.68% was β-sheet, and 1.09% was turn structure.

The above results indicated that the content of β-sheet in the protein treated with ultrasound increased significantly ($P < 0.01$), the turn structure decreased rapidly, and the α-helix structure increased significantly at the same time, which might be because the ultrasonic radiation effect made the partial disulfide bonds in protein dissociate rapidly to form the free -SH. The thermal effect, mechanical effect, or cavitation effect might be caused when the ultrasonic wave was transmitted in the medium, which increased the probability of collision of molecules, so the free -SH would be cross-linked in the media to form more disulfide bonds; the ultrasonic effect could interrupt some intermolecular hydrogen bonds and destroyed the hydrophobic region of protein, and the hydrophobic bonds were exposed so that the tight protein structure was loose, so the content of β-sheet increased significantly. After the protein was treated with the reductant-1, almost all the structures in protein were β-sheet structures, which indicated that the reductant-1 almost completely reduced all disulfide bonds in protein. The reaction conditions were mild, so the free -SH did not have the chance to form disulfide bond again and the protein structure was loose. These characteristics were all easy for quick-setting and instant solution of protein.

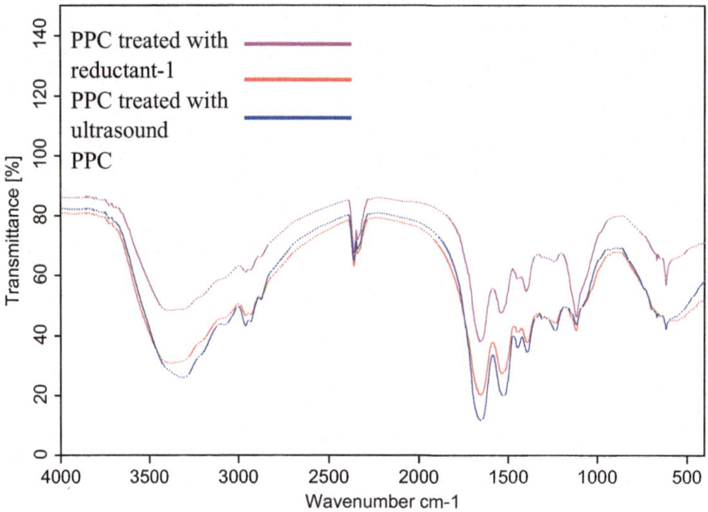

Fig. 6.43 FTIR spectrogram of protein sample (400–4000 cm-1)

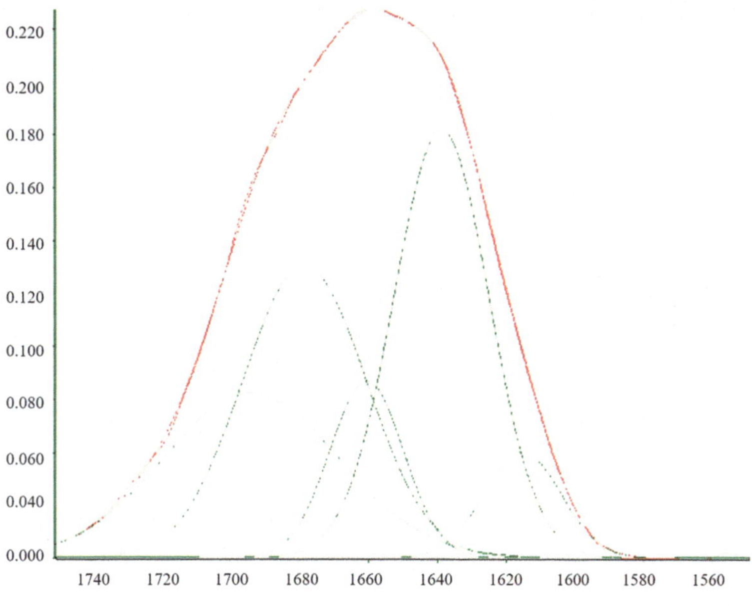

Fig. 6.44 1600–1700 cm-1 fitting spectrogram of PPC

Based on the above analysis, this research further analyzed the molecular acting force between the secondary structures of protein. It could be seen from Table 6.18 that the strong interaction hydrogen bond (β1) between the protein molecules in the PPC treated with ultrasound and reductant-1 decreased to 0, and the overall

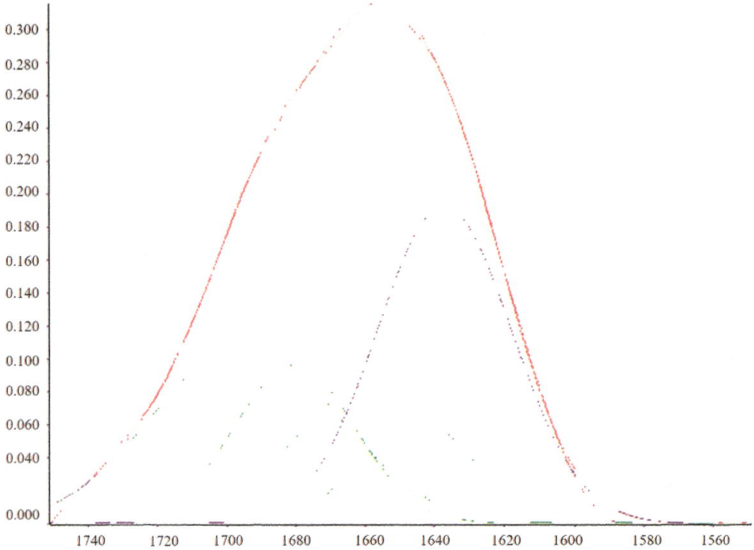

Fig. 6.45 1600–1700 cm-1 fitting spectrogram of PPC treated with ultrasound

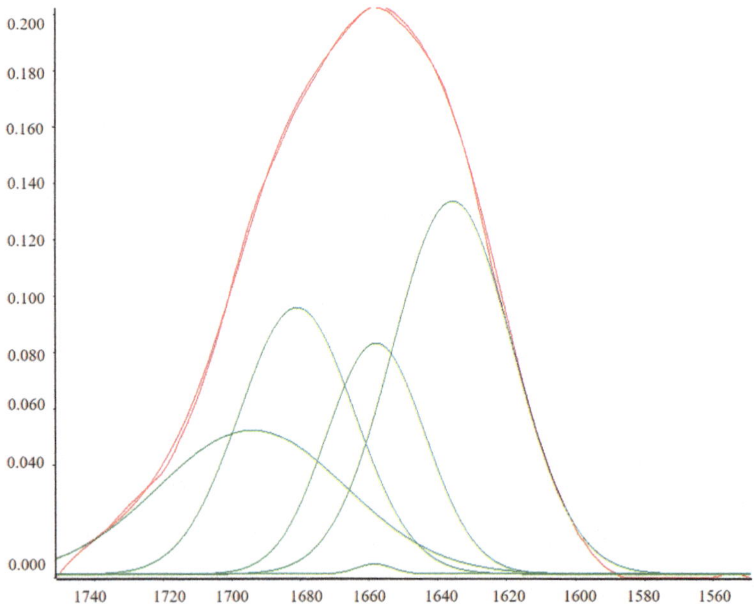

Fig. 6.46 1600–1700 cm-1 fitting spectrogram of PPC treated with reductant-1

interaction hydrogen bonds (β1 + β2) between the main chains decreased to 27.07% and 14.06%, respectively, with the decrease of 28.33% and 62.77%; the different treatment methods made the content of β-sheet hydrogen bond in the protein

Table 6.17 FTIR fitting spectrogram of protein sample and wave number and area related to secondary structure

Fig. 6.42	cm^{-1}	1637.08	1679.53	1659.08	1694.65	1657.39
	%	27.97	28.05	27.01	11.85	5.13
Fig. 6.43	cm^{-1}	1637.08	1680.42		1694.65	1657.53
	%	35.17	46.49		3.34	15.01
Fig. 6.44	cm^{-1}	1635.52	1680.64		1693.98	1657.75
	%	42.54	56.14		1.09	0.23
Assignment		E	E	T	T	H

Note: *H* total helix, *E* extended chain, *T* turns

molecules reduce, and the overall interaction hydrogen bond (α + β) in the protein molecules also decreased rapidly, which indicated that the original tight protein molecule structure became loose, but the random structure was always zero, so it could be inferred that the protein molecules became loose, but they were still ordered structures, rather than the randomly stretched peptide chains.

2.2.6 Fluorescence Intensity and Maximum Fluorescence Emission Wavelength (λem) of Modified PPC and Original PPC

The changes in fluorescence intensity and maximum emission wavelength of protein samples treated in different methods in different solvents (PBS buffer solution, 0.025 MSDS buffer solution, and 6 M urea buffer solution) are shown in Table 6.19.

It could be seen from Table 6.19 that in the PBS buffer solution, when the excitation wavelength was 290 nm, the maximum emission wavelengths of protein samples were 331.5, 332.0, and 330.5 nm, respectively; in the 0.025 MSDS buffer solution, the maximum emission wavelengths of protein samples were 338.5, 337.5, and 337.5, respectively; in the 6 M urea buffer solution, the maximum emission wavelengths of protein samples were 337.5, 336.5, and 335.5, respectively. In either buffer solution, the fluorescence intensity of the treated protein sample was always higher than that of the original protein. The order of their fluorescence intensity was PPC treated with reductant-1>PPC treated with ultrasound>original PPC.

Under normal circumstances, only Phe, Tyr, and Trp groups in protein could emit fluorescence, but the fluorescence emission intensity of Phe was very low, and the fluorescence of Tyr was easy to be extinguished by the carboxyl and amino in protein or it was easy to transfer the energy to Trp through non-radiation approaches, so the fluorescence of protein itself was essentially from the Trp group. The maximum emission wavelength λem was closely related to the position of the chromophore in the protein molecule. If the excitation wavelength was 280 nm and the emission wavelength λem was 330–333 nm, it indicated that the chromophore was located in the hydrophobic region of protein; if λem was

Table 6.18 Impact of different treatment on secondary structure of protein

	β_1(%)	β_2(%)	$\beta_1 + \beta_2$	β(%)	α(%)	$\alpha + \beta$	T(%)	R (%)
Sample	1600–1625	1690–1700	(%)	1626–1639	1651–1660	(%)	1661–1689	1640–1650
PPC	8.27	29.5	37.77	41.87	18.14	60.01	49.8	0
PPC treated with ultrasound	0	27.07	27.07	25.08	29.03	54.11	15.7	0
PPC treated with reductant-1	0	14.06	14.06	18.95	0.062	19.01	18.95	0

Table 6.19 Fluorescence intensity and maximum emission wavelength of protein sample in different solvents

Buffer solution	Sample	Start (nm)	λem (nm)	Fluorescence intensity
PBS	PPC	290.0	331.5	617.5
	PPC treated with ultrasound	291.0	332.0	717.0
	PPC treated with reductant-1	290.5	330.5	798.7
0.025MSDS	PPC	290.5	338.5	842.6
	PPC treated with ultrasound	290.5	337.5	851.4
	PPC treated with reductant-1	290.5	337.5	899.3
6 M urea	PPC	292.0	337.5	796.4
	PPC treated with ultrasound	292.0	336.5	855.6
	PPC treated with reductant-1	292.0	335.5	948.5

340–342 nm, it indicated that the chromophore was located in the polar region of protein and the group might be surrounded by water; if the emission wavelength was 350–353 nm, it indicated that the group was similar to the free Trp in the microenvironment.

It could be seen that the maximum emission wavelengths of protein samples in PBS buffer solution were all in the range of 330–333 nm, indicating that the chromophore (at least part) of protein was located in the hydrophobic region of protein; in the 0.025 MSDS buffer solution and 6 M urea buffer solution, the different degrees of red shifts occurred in the maximum emission wavelengths of protein samples, indicating that the chromophore in the hydrophobic region was exposed partially.

However, the fluorescence intensity was related to different modification methods and different solvents. In contrast to the fluorescence emission wavelength, the factors influencing the fluorescence emission intensity were very complex, and they depended on temperature, viscosity, pH, solvent polarity, quencher, and other conditions. According to the general rules, if the solvent viscosity increased and the polarity decreased, the fluorescence intensity would increase. In the same buffer solution, the different modification methods led to different improvement of protein solubility, namely, continuous increase, so the solution viscosity increased, the polarity decreased, and the fluorescence intensity of protein increased; different buffer solutions had different viscosity and their viscosity in order was 6 M urea >0.025MSDS>PBS, which could explain the reasons for the gradual increase in fluorescence intensity well. In addition to this, the other factor for the increase of fluorescence intensity was the increase of Tyr→Trp energy transfer in solutions. The energy transfer efficiency was

$$E = \frac{A}{A + n^4 r^6} \text{且} A = \frac{161.9K^2\phi_D}{\pi^5 \cdot N} \int_0^\infty \frac{F_D(\overline{\upsilon})\varepsilon_a}{\overline{\upsilon}^4} d\overline{\upsilon}$$

where K denotes the directional coefficient, φD is the (Tyr) quantum yield provided, π is the pi, N is the Avogadro constant, n is the refractive index of solvent, r is the average distance between the donor (Tyr) and receptor (Trp), and the integral term denotes the spectral overlap degree of donor and receptor which is a constant for Tyr and Trp.

It could be seen that when the protein concentration increased gradually, the quantum yield (φD) of Tyr increased and thus the E value increased.

2.2.7 Surface Property of Protein Molecule of Modified PPC and Original PPC

The light absorption of uncharged acrylamide at 295 nm would cause the changes in the fluorescence intensity of protein, so the quencher in the solution must react with the Trp group during fluorescence quenching, so that the fluorescence intensity of the fluorescent substance Trp was reduced. When the fluorescence of Trp in the protein molecules could not be quenched by acrylamide and other fluorescent quenchers, the Trp group must be located inside the protein molecule where the quencher molecule could not enter. The principle of quenching fluorescence by acrylamide was deduced based on the Stern-Volmer equation.

The concentrations of added acrylamide 1–12 were 0, 0.04, 0.08, 0.12, 0.16, 0.20, 0.24, 0.28, 0.36, 0.40, 0.44, and 0.48 mol/L, respectively. From Figs. 6.47, 6.48, and 6.49, it could be seen that there were changes in the λem and fluorescence intensity (F_0) of protein before and after modification when the quencher was not added; the λem and F_0 values of the original PPC were 363 nm and 52.54 nm, respectively. After the ultrasonic treatment and reductant-1 treatment, λem shifted 4 nm and 7 nm toward the blue end of spectrum, respectively, and the fluorescence intensity values increased by 57.19% and 90.52%. [Q] was mapped with F_0/F. See Fig. 6.50, where F_0 and F were the fluorescence intensity with and without quencher and [Q] was the concentration of quencher. It could be seen from the figure that the F_0/F of acrylamide had a linear relationship with the [Q] mapping, so the quenching effect of acrylamide was the diffusion control mechanism which conformed to the Stern-Volmer equation:

$$\mathrm{F0/F} = 1 + \mathrm{KQ}[Q],$$

where K_Q was the quenching constant; the quenching constants of acrylamide could be calculated as follows according to the figure: PPC, 0.2874 L/mol; PPC treated with ultrasound, 0.3326 L/mol; and PPC treated with reductant-1, 0.3542 L/mol. The formula below could be derived from the above formula:

$$\mathrm{F0}/\Delta\mathrm{F} = 1/\mathrm{fa} + 1/(\mathrm{faK}) \times 1/[\mathrm{Q}]$$

K is the constant of the Stern-Volmer of the chromophore which can be approached by the quencher Q. 1/[Q] was mapped with $F_0/\Delta F$, 1/[Q] was

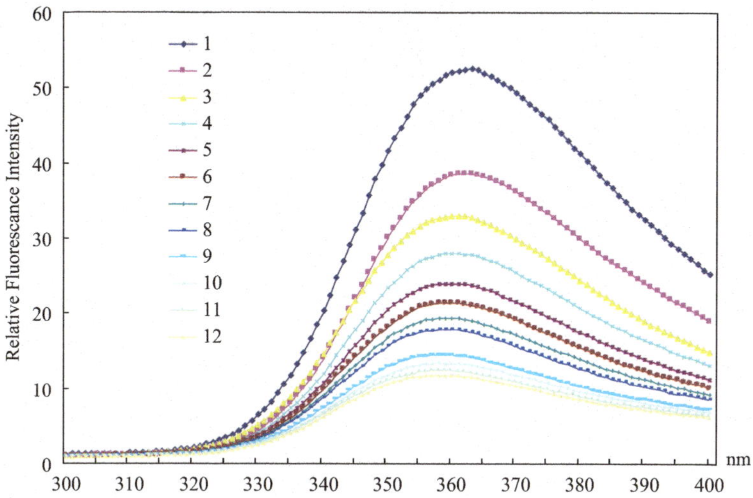

Fig. 6.47 Quenching of acrylamide (Acr) on fluorescence of PPC

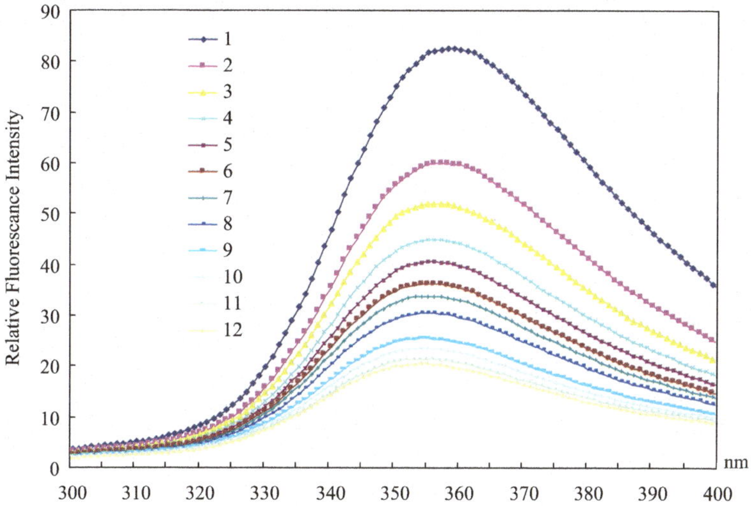

Fig. 6.48 Quenching of acrylamide on fluorescence of PPC treated with ultrasound

extrapolated to 0 with the intercept of $1/f_a$, and the reciprocal f_a was the percentage of the chromophore which could be approached by the fluorescence quencher accounting for the total chromophore. Through calculation, the f_a obtained was shown below: PPC, 0.6809; PPC treated with ultrasound. 0.7319; and PPC treated with reductant-1, 0.7944. It could be seen that acrylamide could not completely quench the Trp fluorescence in the protein.

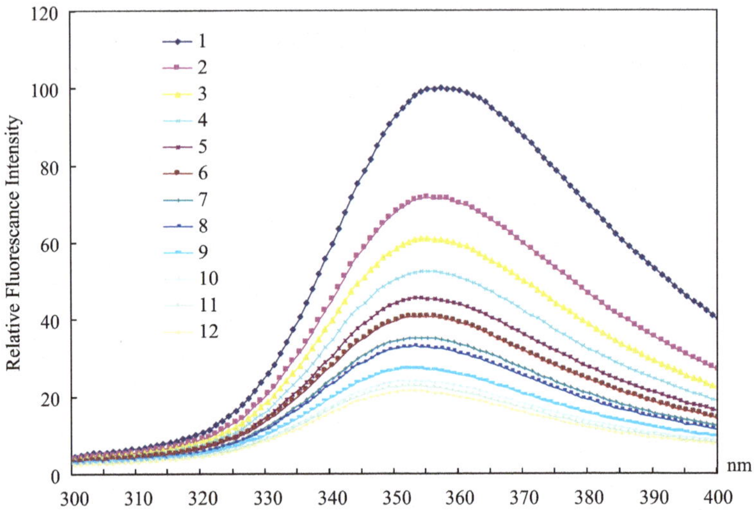

Fig. 6.49 Quenching of acrylamide on fluorescence of PPC treated with reductant-1

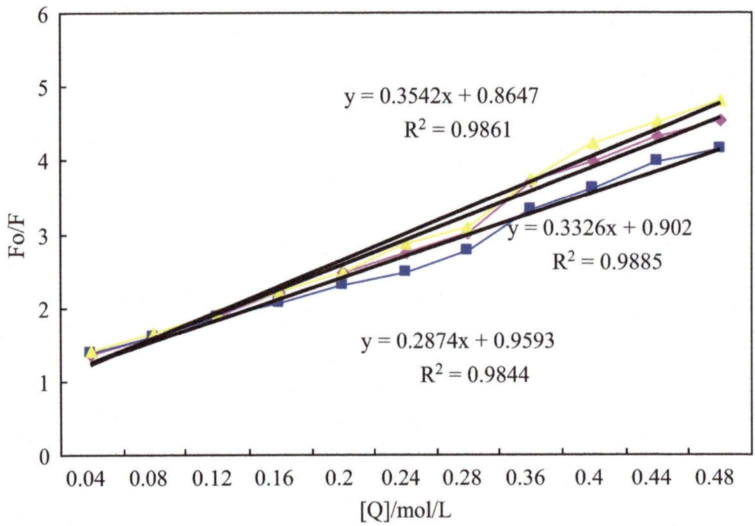

Fig. 6.50 Quenching of acrylamide on endogenous fluorescence of protein

The above results showed that, although the R groups of Trp had large hydrophobicity, half of the Trp in the original PPC molecule was located on the surface of protein molecule, and more Trp was exposed on the surface of protein molecule after treatment, which was mainly because it occupied a large space, and thus it could not enter the hydrophobic core completely. Nakai and Li (1988) divided the protein hydrophobicity into "aromatic hydrophobicity" produced by Trp, Tyr, Phe, and other aromatic nucleus groups and "aliphatic hydrophobicity" produced by Val, Leu, Ilu, and other fatty chain groups, and they thought that the protein solubility was mainly related to the aromatic hydrophobicity and had nothing to do with the aliphatic hydrophobicity and the main reason was that the former was located on the molecular surface, while the latter was located inside the molecule.

2.2.8 Ultraviolet Absorption Spectrum of Modified PPC and Original PPC

The samples were scanned in the wavelength range of 190–400 nm to capture the ultraviolet absorption spectrum values (250–300 nm) of the protein samples for analysis. From Figs. 6.51, 6.52, and 6.53 and Table 6.20, it could be seen that the peak values of the same protein in different buffer solutions were different, and the rules were as follows: the peak values of all the protein samples in the urea were the maximum; in the SDS buffer solution, the peak value of the PPC treated with ultrasound was the minimum; in the PBS buffer solution, the peak value of the original PPC was the minimum, and the maximum absorption peak position shifted toward the blue end; in the same buffer solution, the maximum absorption peak positions of different proteins were the same, but the peak values were different, and the variation rule was: PPC treated with reducing agent-1>PPC treated with ultrasound>PPC, which was consistent with the fluorescence measurement results. The characteristic absorption peaks of the protein at 276–282 nm were mainly caused by the transition of indole ring of Trp residue and benzene ring π-π' of Tyr residue on the peptide chain. The peak position shifted toward the blue end, and the microenvironment of Trp residue was enhanced from the view of solvent effect, which might be caused by the hydrogen bond effect (i.e., the ionization of phenol group of Tyr residue was blocked in different buffer solutions, and the excited-state charge transfer effect between the lone-pair electrons of phenol oxygen atoms and aromatic rings was partially inhibited) (Sharma et al. 1996).

2.2.9 Amino Acid Composition of Modified PPC and Original PPC Protein Samples

Table 6.21 showed the changing situations of composition and contents of amino acid and free amino acid in the PPC modified by physicochemical method/original PPC. The composition and contents of amino acid in the modified/original PPCs were almost the same, and there was no significant difference in their contents of sulfur-containing amino acids (cysteine (Cys-s) and methionine (Met)); however, the content of some free amino acids in the modified protein was different from that in the original protein, for example, the content of cysteine (Cys-s) in the original PPC was only 4.34×10^{-6} mg/g, the content of cysteine increased to 6.26×10^{-4} mg/g after ultrasound treatment, and the content increased to 9.36×10^{-3} mg/g after the reductant-1 treatment; it was obvious that the content of sulfur-containing

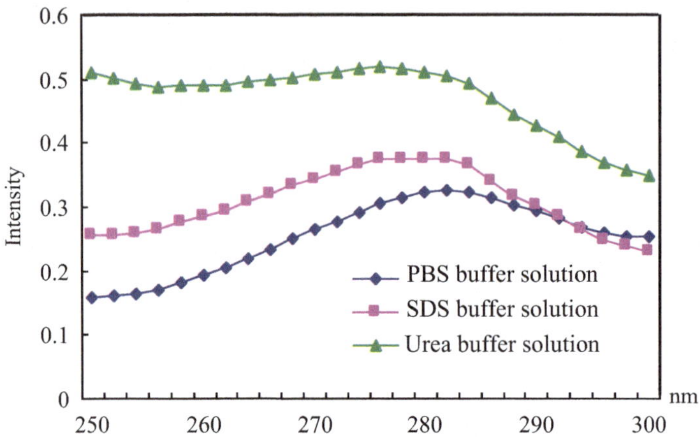

Fig. 6.51 Ultraviolet absorption spectrum of PPC in different buffer solutions

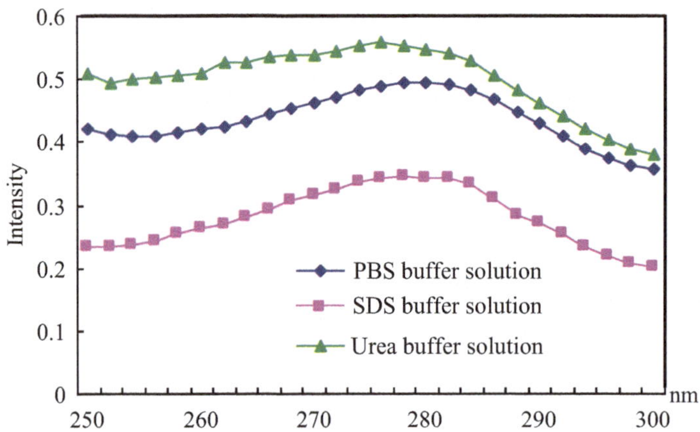

Fig. 6.52 Ultraviolet absorption spectrum of PPC treated with ultrasound in different buffer solutions

amino acids increased after the original PPC was modified, which also confirmed that the free sulfydryl content of the modified protein increased. However, although the modified/original PPC contained free amino acids and short peptides, it was mainly composed of macromolecular proteins; therefore, the content of free amino acids was very low.

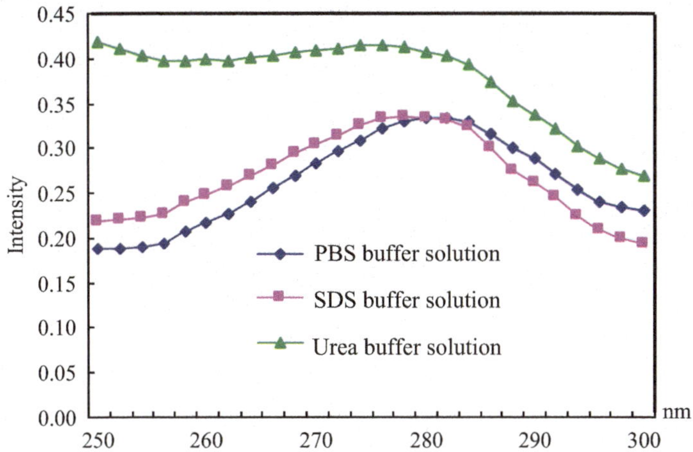

Fig. 6.53 Ultraviolet absorption spectrum of PPC treated with reductant-1 in different buffer solutions. Note: *1* PPC, *2* PPC treated with ultrasound, *3* PPC treated with reductant-1

Table 6.20 λapex and peak values of protein samples in different buffer solutions

Buffer solution	Sample	Apex(nm)	Height
PBS	1	282	0.3258
	2	282	0.4921
	3	282	0.3341
SDS	1	278	0.3758
	2	278	0.3443
	3	278	0.3348
Urea	1	276	0.5191
	2	276	0.5590
	3	276	0.4156

2.2.10 Microstructures of Modified PPC and Original PPC Protein Samples

The microstructure changes of the modified/original PPC were observed by using a scanning electron microscope, as shown in Fig. 6.54. Figures (A–C) showed the structure of the original PPC, the protein structure was presented as tight and disorderly aggregates, the protein molecules were cross-linked, and the sheet structure was thick. Figures (D–F) showed the structure of the PPC treated with ultrasound; compared with the structure of the original protein, the ultrasonic treatment could really open the tight structure of protein to make it become sheets to scatter; after the sheet proteins were amplified, it could be seen that there were fractures between the protein sheets, but the whole was still a complete structure, indicating that the ultrasonic treatment was not enough to destroy the protein molecular structure so as to make it become small-molecule short peptides. Figures (G–I) showed the structure of the PPC treated with reductant-1. The protein

Table 6.21 Analysis of composition of amino acid and free amino acid in protein sample

Sample	PPC		Ultra-PPC		Reducer-1PPC	
Amino acid	Total amino acid	Free amino acid	Total amino acid	Free amino acid	Total amino acid	Free amino acid
Asp	7.79	0.26	7.69	0.27	7.62	0.35
Glu	13.94	1.51	13.49	1.44	13.41	1.33
Ser	3.42	8.26×10^{-4}	3.40	6.44×10^{-4}	3.45	2.11×10^{-3}
His	1.66	4.81^{-2}	1.65	4.44×10^{-2}	1.55	5.61×10^{-2}
Gly	3.68	1.64^{-1}	3.68	0.16	3.69	0.24
Thr	1.77	4.95^{-2}	1.78	4.64×10^{-2}	1.78	7.67×10^{-2}
Arg	7.79	0.61	7.77	0.61	7.69	0.55
Ala	2.52	0.12	2.53	0.10	2.56	0.12
Tyr	2.21	0.34	2.20	0.32	2.14	0.28
Cys-s	**0.44**	$\mathbf{4.34 \times 10^{-6}}$	**0.43**	$\mathbf{6.26 \times 10^{-4}}$	**0.46**	$\mathbf{9.36 \times 10^{-3}}$
Val	2.52	1.31×10^{-3}	2.61	5.59×10^{-4}	2.53	4.41×10^{-3}
Met	**0.65**	$\mathbf{3.01 \times 10^{-3}}$	**0.72**	$\mathbf{2.62 \times 10^{-3}}$	**0.65**	$\mathbf{2.21 \times 10^{-3}}$
Phe	3.49	0.36	3.56	0.34	3.56	0.28
Ile	1.89	2.09×10^{-2}	1.99	1.26×10^{-2}	1.90	3.47×10^{-2}
Leu	4.20	3.06×10^{-2}	4.28	1.34×10^{-2}	4.30	1.31×10^{-2}
Lys	2.33	8.45×10^{-2}	2.33	8.16×10^{-2}	2.32	8.20×10^{-2}
Pro	1.91	4.35×10^{-2}	1.82	7.46×10^{-2}	2.75	8.15×10^{-2}
Totals	62.21	3.64	61.94	3.54	62.35	3.51

Note: Unit of total amino acid content, g/100 g; unit of free amino acid content, mg/g

was still in the form of clusters, but the structure was obviously looser than that of the original PPC; the sheet structure was in the fragile fibrous shape which was brittler than the sheet molecular structure of the PPC treated with ultrasound. The results obtained by microstructure analysis were consistent with those obtained by the determination by using infrared spectroscopy, fluorescence spectroscopy, and ultraviolet spectroscopy methods, that is, the protein molecules changed from a tight globular structure to a loose and ordered structure.

2.2.11 Acting Force Involved in Protein Gelation in Modified PPC and Original PPC

The forces involved in the formation of protein gels included hydrogen bonds, electrostatic interactions, hydrophobic interactions, disulfide bonds, etc. (translated by Wang et al. 1991). The formation of hydrogen bond was mainly because the groups formed by electronegative atoms and hydrogen (such as N-H and O-H) had large dipole moments, the bonding electron cloud distribution biased the heavy nuclei with strong electronegativity, the positive charged hydrogen nucleus (proton) was exposed outside when the electrons distributed around the hydrogen nuclei was few, and the electrostatic attraction would occur when this positively charged

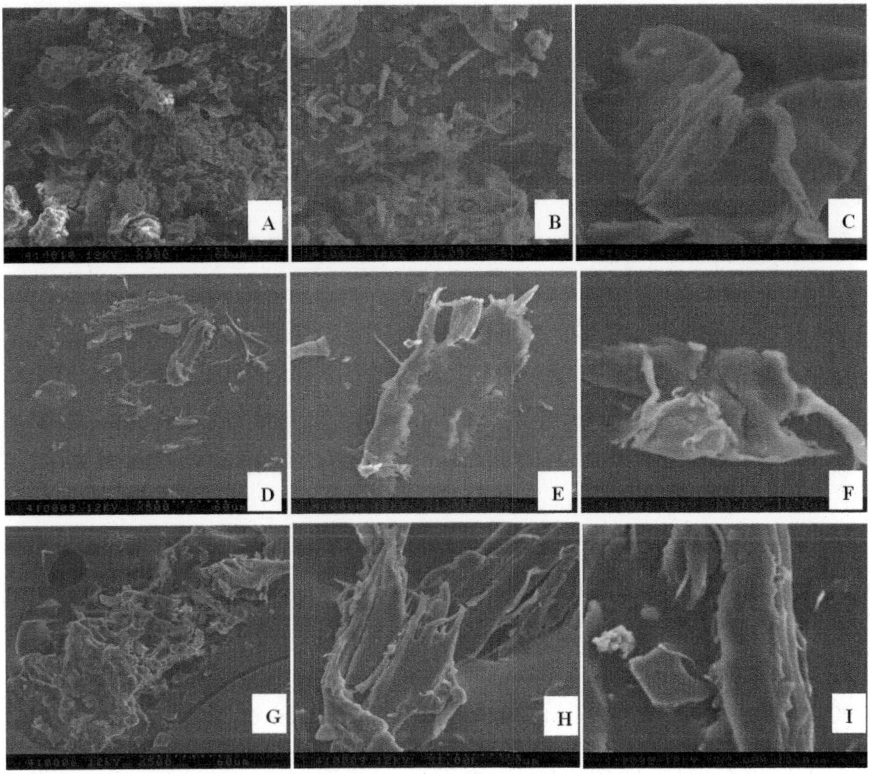

Fig. 6.54 Microstructure of modified/original protein sample. Note: original PPC, 500 × (**a**), 1000 × (**b**), 3000 × (**c**); PPC treated with ultrasound, 500 × (**d**), 1000 × (**e**), 3000 × (**f**); PPC treated with reductant-1, 500 × (**g**), 1000 × (**h**), 3000 × (**i**)

hydrogen nucleus encountered another atom with strong electronegativity. The hydrogen bonds could be formed between the carbonyl oxygen and amide hydrogen on the main chain of polypeptide, between the side chains, between the side chain and medium water, between the main-chain peptide group and side chain, or between the main-chain peptide group and water. Hydrophobic interaction referred to the hydrophobic bond. The nonpolar groups (hydrophobic groups) or hydrophobic side chains were forced to aggregate together in order to avoid the need for aqueous phase, and when the hydrophobic compounds or groups entered the water, the surrounding water molecules would be arranged in a rigid ordered structure. The salt bond was an electrostatic interaction between positive and negative charges. The hydrogen bond, hydrophobic interaction, salt bond, and van der Waals force were all noncovalent bonds; the disulfide bond was a covalent bond with great bond energy, and it had reversibility under normal circumstances once it was formed (Tao et al. 1995).

In the protein gels in different forms, the dominant acting forces were not the same. From Figs. 6.55, 6.56, and 6.57, it could be seen that the changes occurred in

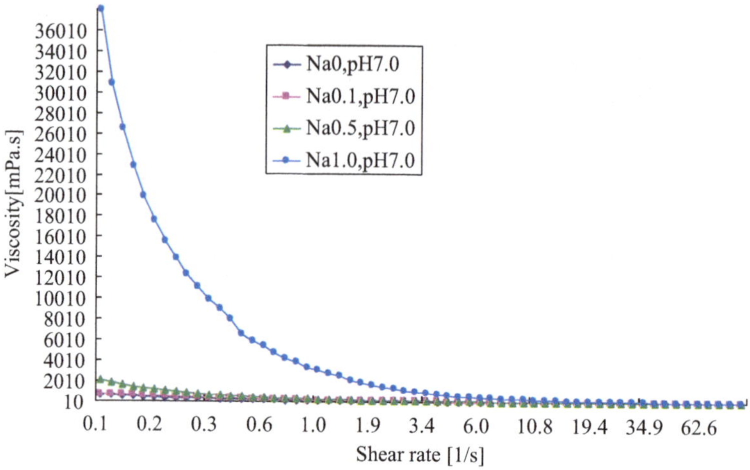

Fig. 6.55 Impact of different NaCl concentrations on apparent viscosity of PPC

the apparent viscosity of protein with different rules in the different NaCl concentrations of solutions. The apparent viscosity of the original PPC increased with the increase of NaCl concentration, and the protein shear became thin and presented pseudoplasticity with the increase of shear rate; the changes of apparent viscosity of PPC treated with ultrasound with the increase of NaCl concentration were complex, the order was 0.1 mol/L NaCl solution >0 mol/L NaCl solution >1.0 mol/L NaCl solution >0.5 mol/L NaCl solution, and the protein viscosity decreased with the increase of shear rate; the changes of apparent viscosity of PPC treated with reductant with the increase of NaCl concentration were also complex, the order was 0.5 mol/L NaCl solution >0.1 mol/L NaCl solution >0 mol/L NaCl solution >1.0 mol/L NaCl solution, and the protein viscosity decreased rapidly with the increase of shear rate. In the different $CaCl_2$ concentrations of solutions (Figs. 6.58, 6.59, and 6.60), the changes of apparent viscosity of different protein samples were different, the change trends of the apparent viscosity of the original PPC and PPC treated with reductant with the increase of $CaCl_2$ concentration were consistent, and the trend was 0.1 mol/L $CaCl_2$ solution >0.3 mol/L $CaCl_2$ solution >0 mol/L $CaCl_2$ solution; the apparent viscosity of PPC treated with ultrasound increased regularly with the increase of $CaCl_2$ concentration. In simple terms, the addition of $CaCl_2$ could increase the viscosity of protein. Low concentration of NaCl could help to increase the apparent viscosity of protein, but when the concentration exceeded 0.5 mol/L, it was not helpful to improve the apparent viscosity of protein.

The above results showed that the low shear viscosity of the modified protein was higher than the apparent viscosity of the original protein and had low dependence on the shear rate. The fluid viscosity was derived from the intermolecular friction, so the intermolecular acting force was small, the molecular chain was highly compliant, and the substances with relatively small molecular mass had good fluidity (Jian and Weiyuan 2003). The cavitation, mechanical shear, and thermal

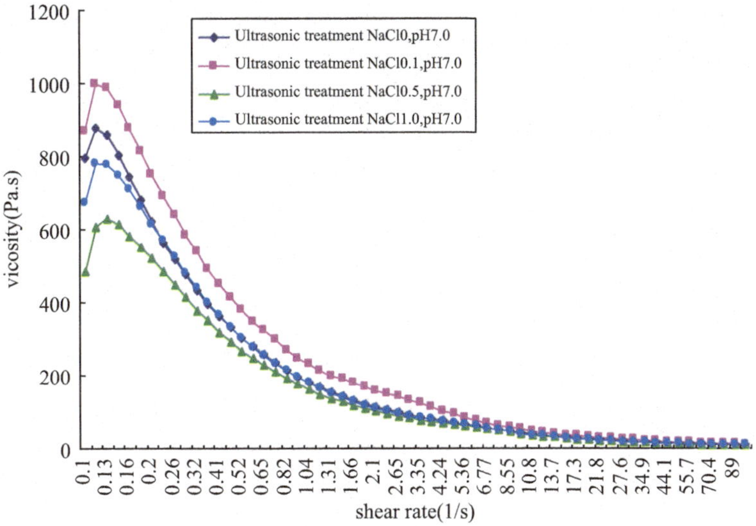

Fig. 6.56 Impact of different NaCl concentrations on apparent viscosity of PPC treated with ultrasound (25 °C)

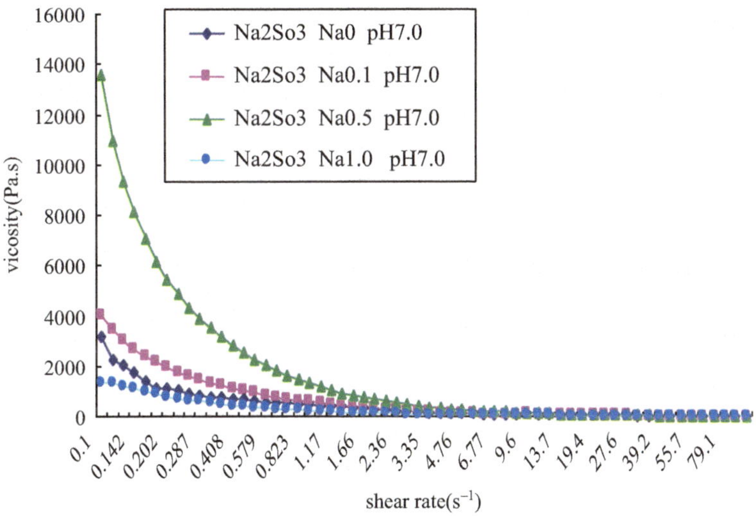

Fig. 6.57 Impact of different NaCl concentrations on apparent viscosity of PPC treated with reductant-1 (25 °C)

effects caused by ultrasonic wave would produce damage effects on the secondary bonds in protein used to maintain the high-level structure of protein (including hydrogen bond, hydrophobic bond, van der Waals force, etc.) and changed the tight structure of PPC; the ultrasonic wave would produce stripping, etching, and comminution effects on the solid particles on the surface to create a new active surface

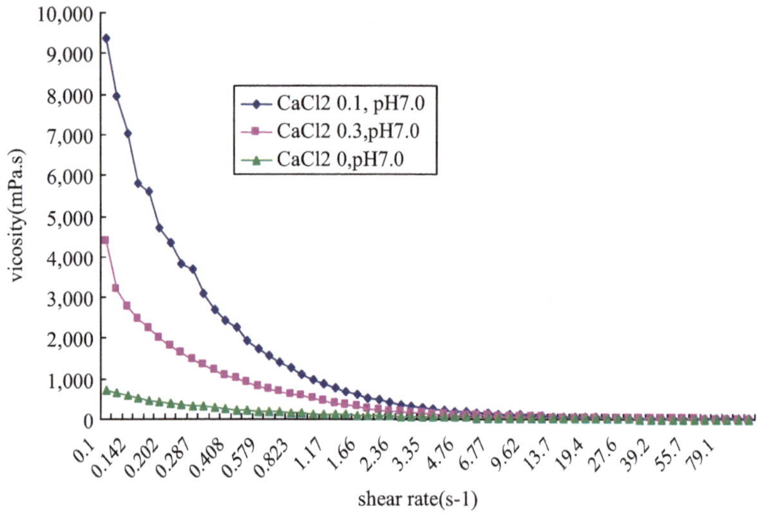

Fig. 6.58 Impact of different CaCl2 concentrations on apparent viscosity of PPC

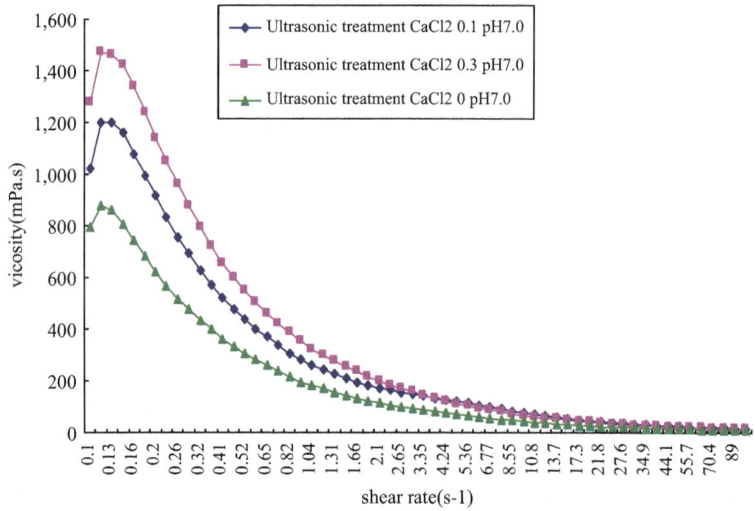

Fig. 6.59 Impact of different CaCl2 concentrations on apparent viscosity of PPC treated with ultrasound (25 °C)

so as to enhance the hydrophilicity of protein surface (Suslick 1990); ultrasonic treatment made the protein molecules become disorderly, the Brownian motion of molecules increase, and the viscosity of protein decreases slowly with the increase of shear rate. It could be inferred that, in the gel formation process of certain concentration of PPC solution treated with ultrasound, the protein molecules

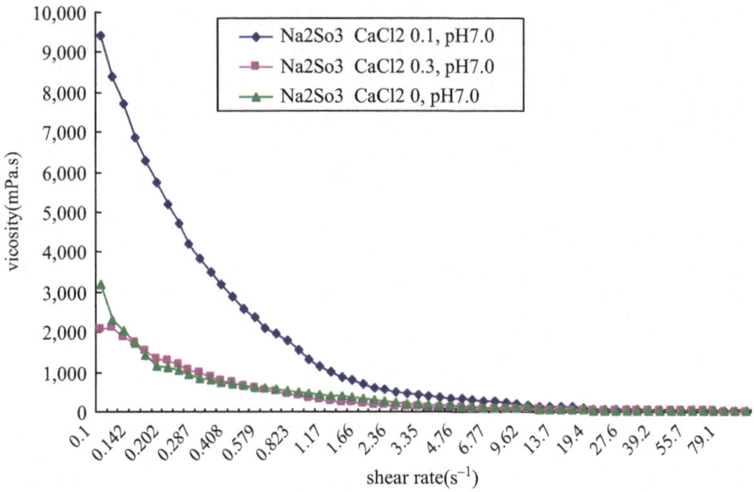

Fig. 6.60 Impact of different CaCl2 concentrations on apparent viscosity of PPC treated with reductant-1 (25 °C)

contacted with each other, and the intermolecular forces involved in the gelation might be the electrostatic interactions, hydrophobic interactions, and van der Waals forces between protein-water, protein-protein, and protein-ion phase. The reductant-1 acted on PPC and reduced the disulfide bonds to allow the free sulfydryl in protein to be fully exposed. The reductant also attacked the hydrophobic core of protein to make the hydrophobic group to be partially exposed, the α-helix in protein be dissociated, and many macromolecule peptide fragments spread freely. Therefore, when a certain concentration of reductant-1 PPC dissolved in water to form gels, the probability of contact between the protein molecules increased due to the stirring action, and the free sulfydryl in the peptide fragment was re-cross-linked to form disulfide bonds, which significantly increased the apparent viscosity of protein at low shear rate. However, with the increase in the shear rate, the other molecular bonds involved in protein gelation, such as hydrogen bond and van der Waals force, were damaged again, so the apparent viscosity of protein decreased rapidly. Nevertheless, the apparent viscosity of PPC solution treated with reductant-1 was significantly greater than that of the other two proteins.

The research showed that salt concentration could break the electrostatic interaction between the proteins or peptides (translated by Wang et al. 1991) and increase the salt ion concentration, and especially calcium ions could increase the gel rate and gelation strength. The above results showed that the apparent viscosity of PPC increased with the increase of NaCl concentration, so it could be deduced that the interactions between the original PPC protein molecules were the acting forces of other covalent bonds and secondary bonds other than electrostatic interaction. The apparent viscosity of PPC protein treated with ultrasound and PPC protein treated with reductant-1 was large at low NaCl concentration (0.1–0.5 mol/L), and the apparent viscosity of protein decreased with the increase of salt concentration (0.5–1.0 mol/L). This phenomenon showed that the interaction force between the modified protein molecules was led by the electrostatic interaction. When the salt concentration in the solution system was low, the electrostatic attraction between protein-protein and between protein-ion and protein-water interaction made the acting force between the protein bonds increase; thus the solution viscosity increased and the liquidity decreased; the salt concentration increased and the salt ions continued to neutralize the charge on the surface of protein, resulting in the charge shielding effect; the salt concentration continued to increase, the electrostatic repulsion in the protein solution became dominant, and the fluorescence scanning results showed that the chromophore of protein was located in the hydrophobic region, so it could be inferred that the protein contained a high proportion of nonpolar region, and then the charge shielding effect and electrostatic repulsion generated by the effect of salt ions would make the protein solubility reduce and viscosity decrease. The addition of divalent salt ions (Ca^{2+}) made the apparent viscosity of protein increase, and the reason was that the salt ions acted as "salt bridges" to make the protein-salt-protein form macromolecular aggregates and the protein molecules were cross-linked to promote the gelation.

2.2.12 Scanning Electron Microscope of TG-B Enzyme-Modified PPC/Original PPC (Peanut Protein Concentrate)

From the scanning electron microscope picture (Fig. 6.61), it was clear that the natural peanut protein concentrate was mostly in the shape of olive with smooth edge and compact structure. The peanut protein concentrate after enzyme action was connected with each other into a large whole in the shape of honeycomb, the external was in the irregular shape, and there were many holes inside the whole. It was confirmed that the TG-B-modified proteins were cross-linked with good effect. By comparing the scanning electron microscope pictures of modified protein product textures with different enzyme dosages in the same modification condition, it could be seen that when the enzyme dosage was 2–3 U/g protein (i.e., Figs. d and e), the section of gel protein product formed a clear network structure with compact structure, and the cross-linking of section when the enzyme dosage was 3 U/g protein was the best. When the enzyme dosage was 1 U/g protein, the product section was in the shape of honeycomb, but the clear network structure was not formed; when the enzyme dosage was 4 U/g protein, the network structure of product section was clear, but the cross-linked surface is very crisp. This phenomenon also proved the speculation that the increase of enzyme dosage would cause the inhibiting effect of substrate-enzyme system in the single-factor experiment from the side.

Fig. 6.61 Scanning electron microscope picture of peanut protein concentrate. Note: (**a**) Peanut protein concentrate molecule; (**b**) Modified peanut protein concentrate under the optimal conditions; (**c**) Peanut protein concentrate product (substrate concentration is 15% and enzyme dosage is 1U/g.pro); (**d**) Peanut protein concentrate product (substrate concentration is 15% and enzyme dosage is 1U/g.pro); (**e**) Peanut protein concentrate product (substrate concentration is 15% and enzyme dosage is 2U/g.pro); (**f**) Peanut protein concentrate product (substrate concentration is 15% and enzyme dosage is 3 U/g.pro)

3 Solubility Improvement of Peanut Protein Concentrate

3.1 Modification Process

3.1.1 Physical Solubilization Modification Process

The protein content of peanut protein concentrate obtained by ethanol extraction was significantly higher than that of peanut protein powder, but the denaturation of the organic solvent (ethanol) could affect the electrostatic force, hydrogen bond, and hydrophobic effect of protein and thus led to the conformational change of protein. The partial denaturation of protein caused the solubility reduction and affected its application in the food industry, so the solubilization modification was imperative.

The physical modification methods related to solubility mainly included high-pressure homogenization, moderate heating, high-speed stirring, ultrasonic treatment, microwave treatment, electrolytic reduction, etc. (Yufei 1993). Among them, the method most commonly used for protein solubilization was high-pressure homogenization, it was usually used as a posttreatment method of enzymatic solubilization, and the research on the pretreatment method of enzymatic solubilization was relatively few currently.

The principles of high-speed stirring and high-pressure homogenization were similar, and the principles were to cut down and break up the material particles or droplets so that the molecular structure was dissolved to improve the solubility (Yang et al. 2002). Heat treatment is to break the covalent bonds between protein molecules, so that the original tight molecular structure becomes loose, the internal structure spreads gradually, and the solubility increases. Microwave treatment changes the traditional heat conduction method from the surface to inside, the extremely strong rapid transformation of microwave field changes the electrical properties of protein molecules, the protein molecules are denatured partially, and the solubility changes accordingly. Ultrasonic wave could produce cavitation in the liquid media to cause deformation and rupture of protein molecules, so as to increase the hydration and increase the solubility (Rooney 1988; Jambrak et al. 2008).

1. Process of solubilization modification through high-speed stirring treatment

(a) Single-factor experiment

The impact of stirring speed on the nitrogen solubility index of peanut protein concentrate is shown in Fig. 6.62. The results showed that the nitrogen solubility index of protein increased with the increase of stirring speed. This was because the structure of aggregated proteins caused by ethanol extraction due to stirring was redispersed and the solubility also increased. The increase of stirring speed would make the friction and collision between the materials strong and the dispersion effect of protein particles good. The experimental equipment could achieve the maximum design speed of 28,000 r/min, but the most suitable working speed should be no more than 22,000 r/min; if the speed was exceeded, the excessive equipment load would be caused, the motor heat increased dramatically, and it was difficult to maintain the longtime running of more than 60 s. Therefore, according to the actual situations, 22,000 r/min was selected as the optimal single-factor stirring speed for high-speed stirring solubilization modification.

The impact of stirring time on the nitrogen solution index of peanut protein concentrate solubilized and modified through high-speed stirring is shown in Fig. 6.63. The results showed that the nitrogen solubility index of protein increased with the increase of stirring time from 12 to 66 s, which was because the appropriate extension of stirring time could make the protein particles disperse fully and thus the protein solubility increased. When the stirring time continued to rise to 84 s, the nitrogen solubility index did not rise, but fell, which was because excessive stirring made the protein denature excessively; the changes could be seen from the color of protein solution, the original milky white protein solution had a tendency to become black, and this phenomenon indicated that the denaturation degree had exceeded the need. Therefore, 66 s was determined as the optimal single-factor stirring time for high-speed stirring solubilization modification.

Fig. 6.62 Impact of stirring speed on nitrogen solubility index of solubilized and modified PPC

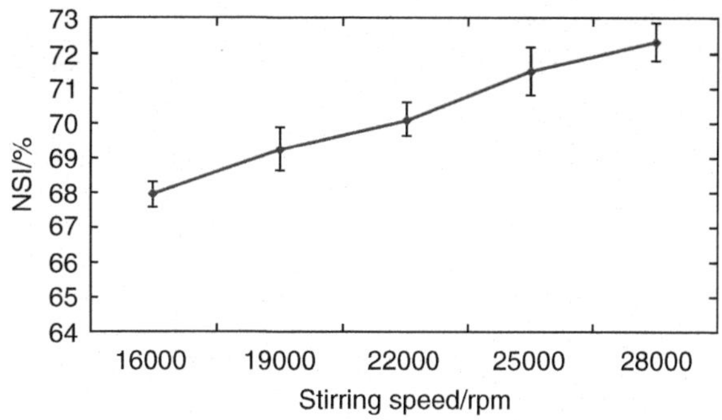

Fig. 6.63 Impact of stirring time on nitrogen solubility index of solubilized and modified PPC

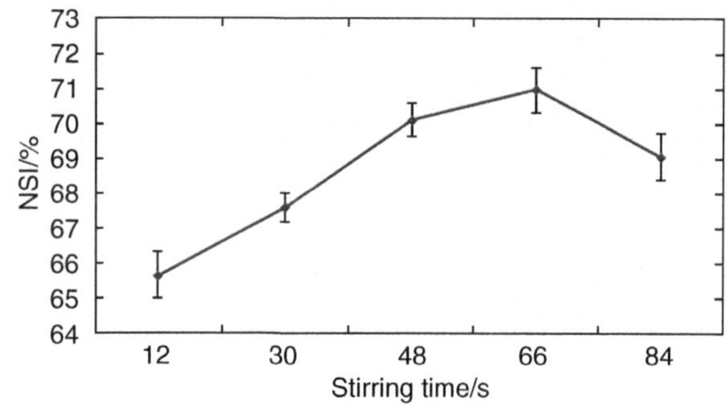

The impact of solid-liquid ratio on the nitrogen solubility index of peanut protein concentrate solubilized and modified through high-speed stirring is shown in Fig. 6.64. The results showed that the nitrogen solubility index of protein increased when the solid-liquid ratio increased from 1: 6 to 1: 7. The solid-liquid ratio continued to increase from 1: 7 and the nitrogen solution index decreased rapidly. When the solid-liquid ratio increased to 1: 10, the nitrogen solubility index of protein had little difference with that of the original ethanol extraction protein. The stirring solubilization was caused by the collision and friction between the materials, too thick liquid would increase the movement resistance of solid particles, and the collision and friction were not enough to dissolve the particle structures; the probability of collision and friction between the particles in the too thin liquid was greatly reduced, and the stirring could only make them flow around, but could not make them be broken and dispersed. For the samples used in this experiment, the appropriate solid-liquid ratio of high-speed stirring solubilization modification was 1: 7.

(b) Process conditions

The confirmatory experiment was carried out according to the stirring speed of 22,000 r/min determined directly and the optimal stirring time of 66 s and optimal solid-liquid ratio of 1:7 obtained in the single-factor experiment at this speed. The nitrogen solubility index of peanut protein concentrate solubilized and modified

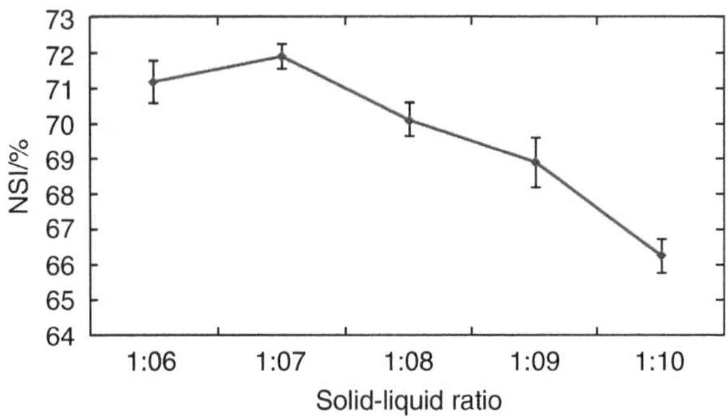

Fig. 6.64 Impact of solid-liquid ratio on the nitrogen solubility index of PPC solubilized and modified through high-speed stirring

through high-speed stirring reached 72.39% ± 0.42% which was significantly higher than that of the ethanol extracting peanut protein concentrate (65.30%) and close to that of the raw material peanut protein powder (72.89%). This indicated that the aggregation formed by the ethanol extraction denaturation of peanut protein concentrate was broken up after the high-speed stirring treatment, so the side effect of ethanol extraction on the protein solubility was basically eliminated.

2. Process of solubilization modification through microwave treatment

(a) Single-factor experiment

The impact of microwave power on the nitrogen solubility index of modified peanut protein concentrate is shown in Fig. 6.65. With the increase of microwave power, the nitrogen solubility index of protein was on the rise, but it rose slowly after the power was more than 640 W. The main effect of microwave treatment on protein modification was the thermal denaturation, and the high microwave power was favorable to the increase of solution temperature, so that the protein structures aggregated through the ethanol extraction denaturation were dissociated. The high microwave power would have better solubilization effect, but during the actual operation, the protein solution under the high power of 800 W was easy to boil and overflow the container, which would interrupt the operation several times. And when the microwave modification was used in the actual industrial production, it was impossible to keep the liquid boiled in long time. Therefore, according to the actual situations, 640 W was selected as the optimal single-factor power for microwave treatment solubilization modification.

The impact of microwave time on the nitrogen solubility index of peanut protein concentrate is shown in Fig. 6.66. The results showed that the nitrogen solubility index of protein increased significantly with the increase of microwave time from 10 to 40 s, especially in the range of 20–30 s, and the nitrogen solubility index did not change after the microwave time exceeded 40 s. During the actual operation,

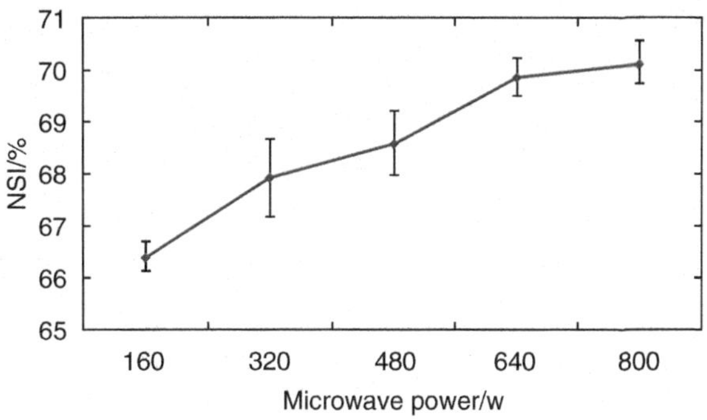

Fig. 6.65 Impact of microwave power on nitrogen solubility index of solubilized and modified PPC

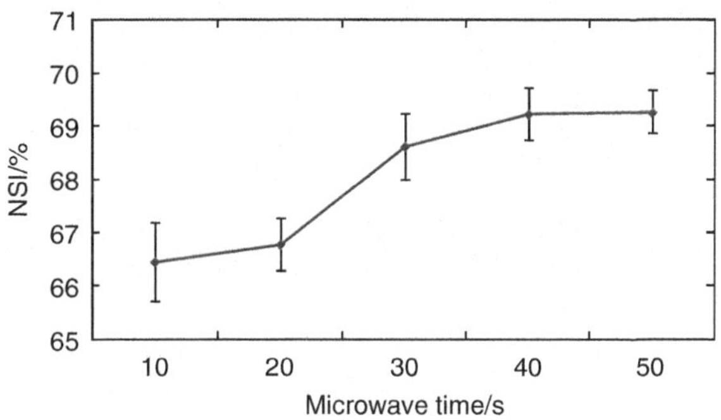

Fig. 6.66 Impact of microwave time on nitrogen solubility index of solubilized and modified PPC

when the microwave time exceeded 30 s, the protein solution was easily boiled and overflowed the container, which would cause the interruption of experiment. According to the actual situations, 30 s was selected as the optimal single-factor operation time for microwave treatment solubilization modification.

The impact of microwave treatment solution height on the nitrogen solubility index of peanut protein concentrate is shown in Fig. 6.67. The results showed that the nitrogen solubility index of protein did not change when the solution height increased from 0.5 to 1.0 cm, the nitrogen solubility index began to decrease when the solution height continued to increase, and the nitrogen solubility index did not change after the solution height exceeded 2.0 cm. This was mainly because the penetration of microwave was limited, too thick liquid layer could not play a good heating effect, and a large number of proteins which were not in the surface of solution would not be denatured by the microwave effect. However, during the actual operation, the solution height of less than 1.5 cm would also make the solution be boiled rapidly during the microwave heating, and thus the experiment was interrupted. Therefore, 1.5 cm was selected as the optimal single-factor solution height for microwave treatment solubilization modification after comprehensive consideration.

The impact of solid-liquid ratio on the nitrogen solubility index of peanut protein concentrate solubilized and modified through microwave treatment is shown in

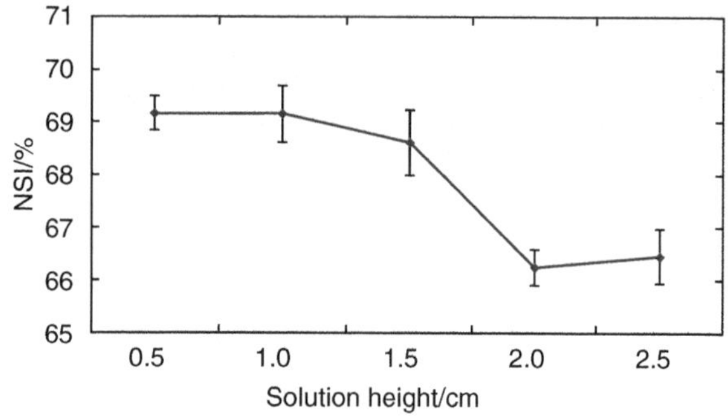

Fig. 6.67 Impact of solution height on nitrogen solubility index of solubilized and modified PPC

Fig. 6.68. The results showed that the impact on the nitrogen solubility index of peanut protein concentrate was not significant when the solid-liquid ratio changed from 1: 7 to 1: 11. From the point of view of the actual process operation, the low solid-liquid ratio was beneficial to the improvement of treatment efficiency. Therefore, 1: 7 was selected as the optimal single-factor solid-liquid ratio for microwave treatment solubilization and modification.

(b) Process conditions

According to the above experiment results and analysis, the microwave power of 640 W, microwave time of 30s, solution height of 1.5 cm, and solid-liquid ratio of 1:7 selected in accordance with the solubilization effect and actual experiment situations, the confirmatory experiment was carried out. The nitrogen solubility index of the peanut protein concentrate solubilized and modified through microwave treatment reached 69.83% ± 0.80% which was higher than that of the ethanol extracting peanut protein concentrate (65.30%), but lower than that of the raw material peanut protein powder (72.89%). This showed that the solubility of peanut protein concentrate lost due to the ethanol extraction denaturation was recovered to a certain degree after microwave treatment.

3. Process of solubilization modification through heat treatment

(a) Single-factor experiment

The impact of heating temperature on the nitrogen solubility index of the peanut protein concentrate solubilized and modified through water-bath heating treatment is shown in Fig. 6.69. When the water-bath temperature increased from 70 to 75 °C, the nitrogen solubility index of protein increased slightly. This indicated that the covalent bond between the protein molecules was broken, the original tight structure became loose, and the solubility increased. If the water-bath temperature continued to increase after it exceeded 75 °C, the nitrogen solubility index decreased gradually, and it decreased rapidly when the temperature exceeded 85 °C. This indicated that the proteins in water solution were over-denatured and formed aggregations under the high temperature of more than 75 °C, and the

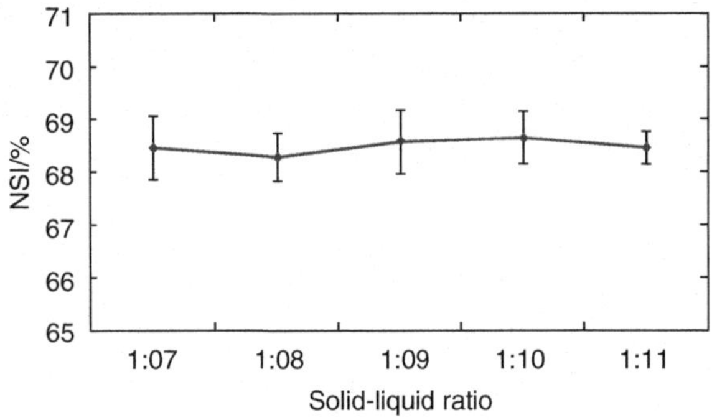

Fig. 6.68 Impact of solid-liquid ratio on nitrogen solubility index of solubilized and modified PPC

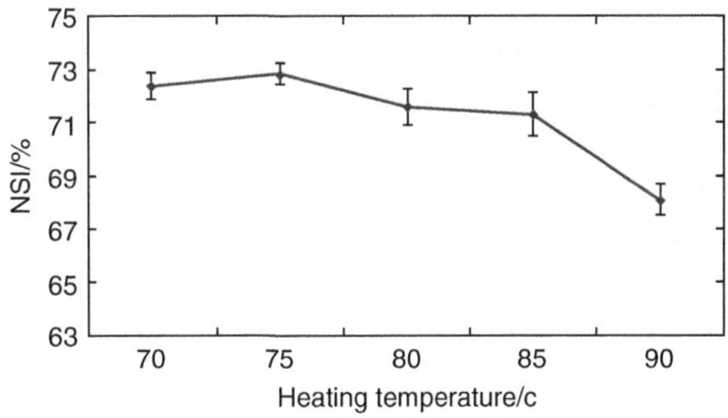

Fig. 6.69 Impact of heating temperature on the nitrogen solubility index of solubilized and modified PPC

solubility was lost gradually. Therefore, 75 °C was determined as the optimal single-factor heating temperature for the heat treatment solubilization modification.

The impact of heating time on the nitrogen solution index of the peanut protein concentrate solubilized and modified through water-bath heating treatment is shown in Fig. 6.70. When the heating time was 1 min, the nitrogen solubility index of protein was consistent with that of the original ethanol extracting protein concentrate, indicating that the water-bath heat had not yet entered the solution. When the heating time increased from 1 to 3 min, the nitrogen solubility index jumped and it reached the maximum when the heating time was 7 min. At this time, the dissolution of internal structure of protein reached the state which was the most suitable to solubility. Thereafter, when the heating time continued to increase, the nitrogen solubility index began to decrease, and there had been protein aggregations unfavorable to solubility at this time. Therefore, 7 min was determined as the optimal single-factor heating time for heat treatment solubilization modification.

The impact of solid-liquid ratio on the nitrogen solution index of the peanut protein concentrate solubilized and modified through water-bath heating treatment is shown in Fig. 6.71. When the solid-liquid ratio increased from 1: 7 to 1: 9, the nitrogen solubility index of protein increased, the index began to decrease when the solid-liquid ratio continued to increase, and it decreased rapidly when the index exceeded 1:10. It indicated that too low or too high solid-liquid ratio was not conducive to the heat treatment modification of protein. Too few solvent might

Fig. 6.70 Impact of heating time on the nitrogen solubility index of solubilized and modified PPC

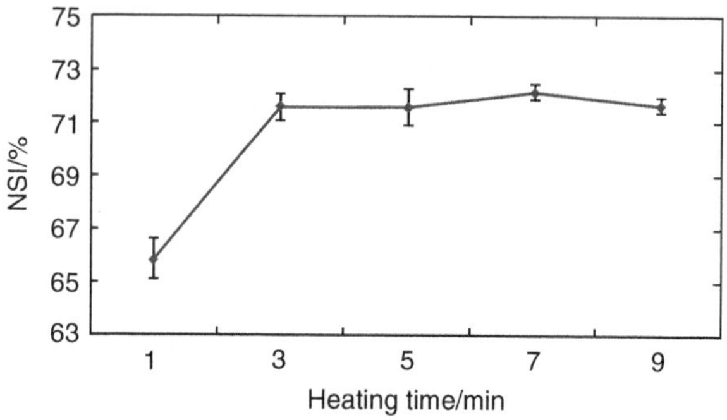

Fig. 6.71 Impact of solid-liquid ratio on nitrogen solution index of PPC solubilized and modified through water-bath heating treatment

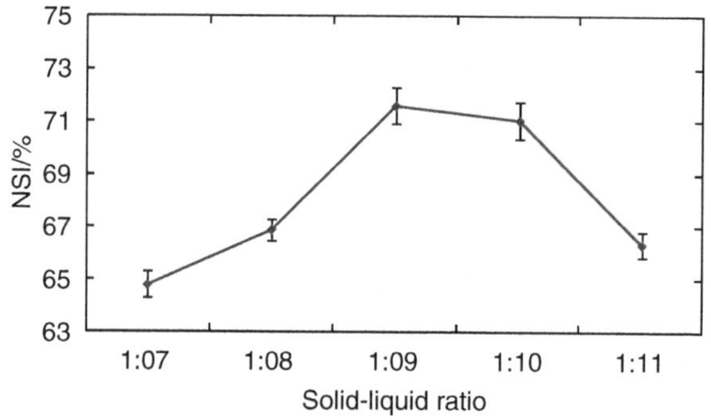

increase the spatial steric hindrance of extension of protein molecules, which was not conducive to the dissolution of structure. Excessive solvent might accelerate the heat transfer and thus result in the aggregations of proteins that were not conducive to solubility. Therefore, 1:9 was determined as the optimal single-factor solid-liquid ratio for heat treatment solubilization and modification.

(b) Simple orthogonal experiment

According to the results of single-factor experiment, the heating temperature (A), heating time (B), and solid-liquid ratio (C) were selected as the three factors of simple orthogonal experiment of water-bath heating treatment. The experiment was performed according to Table 6.22. The results are shown in Table 6.23 and Fig. 6.72. The results of orthogonal experiment were analyzed by using the Duncan's new multiple range method, as shown in Table 6.24.

The larger the F value or the smaller the p value, the higher the significance of the corresponding variable. It could be seen that the heating temperature had the greatest impact in the solubilization process, followed by the solid-liquid ratio, and both of them had significant impacts on the nitrogen solubility index. There was no statistically significance of heating time. The analysis results showed that the combination of the optimal factor level was A2B2C2, and when the heating temperature was 75 °C, the heating time was 7 min, and the solid-liquid ratio was 1: 9, the solubilization and modification effect was the best. When the confirmatory

Table 6.22 Orthogonal experiment condition factor level of peanut protein concentrate solubilized and modified through heat treatment

Level	Factor		
	A Heating temperature/°C	B Heating time/min	C Solid-liquid ratio
1	70	5	1:8
2	75	7	1:9
3	80	9	1:10

Table 6.23 Design and determination result of orthogonal experiment of peanut protein concentrate solubilized and modified through water-bath heating treatment

No.	Factor			NSI determined value (%)		
	Heating temperature (A)/°C	Heating time (B)/min	Solid-liquid ratio (C)	I	II	Mean
1	70	5	1:08	69.9281	69.6265	69.7773
2	70	7	1:09	71.2638	70.4920	70.8779
3	70	9	1:10	70.4334	71.0575	70.7454
4	75	5	1:09	72.9866	72.3540	72.6703
5	75	7	1:10	72.4985	72.6248	72.5616
6	75	9	1:08	71.3201	70.4395	70.8798
7	80	5	1:10	70.3897	71.4909	70.9403
8	80	7	1:08	70.3320	70.0544	70.1932
9	80	9	1:09	70.6010	71.2724	70.9367
K1	282.891	283.5462	283.5252			
K2	287.3136	286.9986	287.0154			
K3	285.1338	284.7936	284.7978			
R	1.4742	1.1508	1.1634			

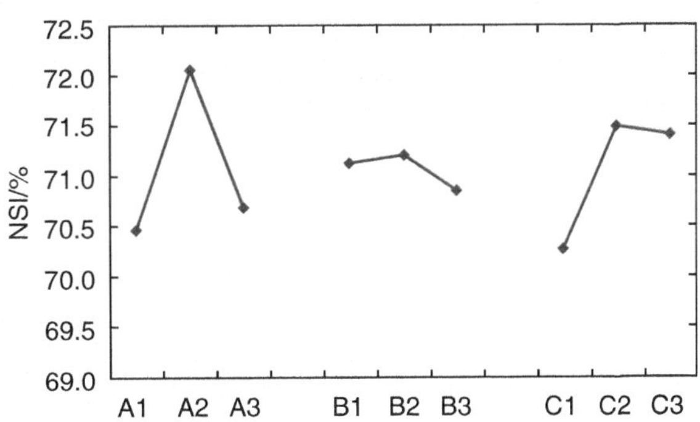

Fig. 6.72 Impact of various factors on nitrogen solution index of PPC solubilized and modified through water-bath heating treatment

experiment was carried out under these process parameters, the nitrogen solubility index of the peanut protein concentrate solubilized and modified through heat treatment was 72.94 ± 0.61% which was significantly higher than that of the ethanol extracting peanut protein concentrate (65.30%) and reached the level of

Table 6.24 Variance analysis of orthogonal experiment results of peanut protein concentrate solubilized and modified through water-bath heating treatment

Source of variance	Quadratic sum	Degree of freedom	Mean square	F value	*p* value
Heating temperature	4.3307	2	2.1654	30.5803	* 0.0317
Heating time	0.2098	2	0.1049	1.4818	0.4029
Solid-liquid ratio	2.7563	2	1.3782	19.4630	* 0.0489
Error	0.1416	2	0.0708		
Sum	7.4385				

that of the raw peanut protein powder (72.89%). This indicated that the solubility of peanut protein concentrate lost due to the ethanol extraction denaturation was completely recovered after the heat treatment, and the adverse effect of the ethanol extraction denaturation was offset. In the actual production, it can be considered to appropriately shorten the heating time to improve the production efficiency and reduce the production cost.

4. Process of solubilization modification through ultrasonic treatment

(a) Single-factor experiment

The impact of ultrasonic power on the nitrogen solubility index of peanut protein concentrate solubilized and modified through ultrasonic treatment is shown in Fig. 6.73. The experiment results showed that the nitrogen solubility index of protein increased when the ultrasonic power increased from 180 to 210 W. This indicated that the proper increase of ultrasonic power could increase the energy density of solution to generate stronger cavitation effect so that the protein solubility increased accordingly. When the ultrasonic power was more than 210 W, the nitrogen solubility index would significantly decrease with the increase of ultrasonic power, which might be because too high energy density produced too high local heat so that the protein conformation was not conducive to the solubility improvement. Therefore, 210 W was determined as the optimal single-factor power for ultrasonic treatment solubilization modification.

The impact of water-bath temperature on the nitrogen solubility index of peanut protein concentrate solubilized and modified through ultrasonic treatment is shown in Fig. 6.74. The experiment results showed that the water-bath temperature increased from 60 to 80 °C during the ultrasonic treatment, the nitrogen solubility index of protein also increased accordingly, and it increased slowly when the temperature was more than 70 °C. This might be because the relatively high temperature was conducive to the dissolution of protein structure, so that the ultrasonic wave could act on more protein molecules to promote that the protein

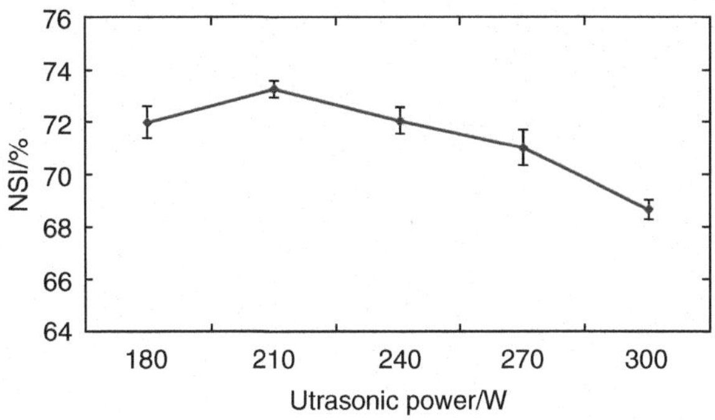

Fig. 6.73 Impact of ultrasonic power on the nitrogen solubility index of solubilized and modified PPC

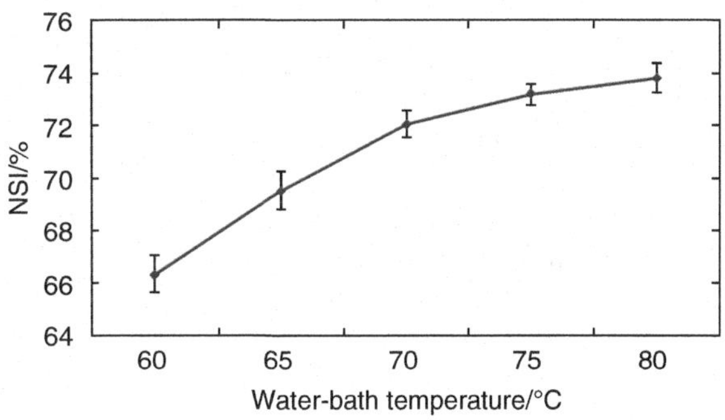

Fig. 6.74 Impact of water-bath temperature on the nitrogen solubility index of solubilized and modified PPC

denaturation conducive to solubility could appear. At this time, the protein was over-denatured when the temperature was higher than 80 °C which was higher than that of the heat treatment (75 °C), which might be related to the ultrasonic effect. The highest setable water-bath temperature of the ultrasonic cleaner used in this experiment was 80 °C. When the water-bath temperature reached 81 °C, the machine would automatically shut down for protection, and the water-bath temperature would continue to rise with the extension of operation time when the machine operated. The temperature would increase by 1–2 °C after the machine operated for 5 min, and the machine would shut down for protection quickly if the initial temperature was too high. Therefore, according to the actual situations, 75 °C was selected as the optimal single-factor water-bath temperature for ultrasonic treatment solubilization modification, and this factor would not be involved in the simple orthogonal experiment design any longer.

The impact of ultrasonic time on the nitrogen solubility index of the peanut protein concentrate solubilized and modified through ultrasonic treatment is shown in Fig. 6.75. The experiment results showed that the nitrogen solubility index of protein increased with the increase of ultrasonic time from 1 to 7 min, it increased rapidly before 3 min, and then it increased slowly. This was because a certain time was needed to complete the improvement of protein hydration by ultrasonic wave. When the ultrasonic time exceeded 7 min, the nitrogen solubility index decreased

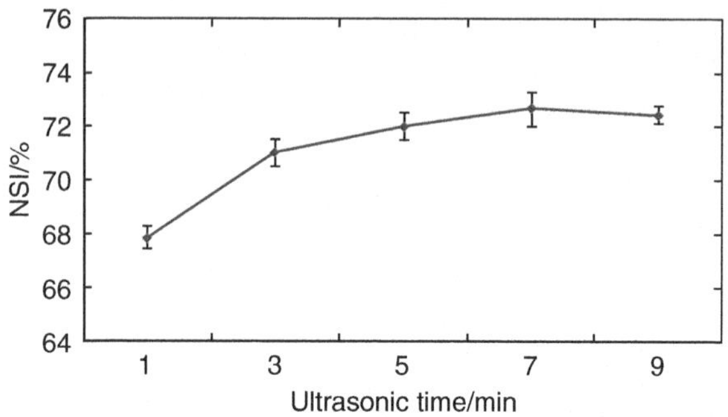

Fig. 6.75 Impact of ultrasonic time on nitrogen solubility index of solubilized and modified PPC

slightly, which might be because the ultrasonic treatment of too long time caused the excessive denaturation of protein or the water-bath heating of too long time made the aggregations unfavorable to the solubility occur. Therefore, 7 min was determined as the optimal single-time time for ultrasonic treatment solubilization modification.

The impact of soli-liquid ratio on the nitrogen solubility index of peanut protein concentrate solubilized and modified through ultrasonic treatment is shown in Fig. 6.76. The experiment results showed that the nitrogen solubility index of protein increased with the increase of solid-liquid ratio from 1: 7 to 1: 10, and the nitrogen solubility index would be reduced if the solid-liquid ratio exceeded 1: 10 and continued to increase. This indicated that too low or too high solid-liquid ratio was not conducive to the ultrasonic treatment modification of protein. If the concentration of solution was too high, the increase of steric hindrance of protein molecular extension would be caused, which was not conducive to the dissolution of structure; when the concentration was too low, the action effect of cavitation effect might be reduced, which was not conducive to the improvement of solubility. Therefore, 1: 10 was determined as the optimal single-factor solid-liquid ratio for the ultrasonic treatment solubilization modification.

(b) Orthogonal experiment

According to the results of single-factor experiment, the ultrasonic power (A), ultrasonic time (B), and solid-liquid ratio (C) were selected as the three factors of simple orthogonal experiment of ultrasonic treatment. The experiment was performed according to Table 6.25. The results are shown in Table 6.26 and Fig. 6.77. The results of orthogonal experiment were analyzed by using the Duncan's new multiple range method, as shown in Table 6.27.

The larger the F value or the smaller the p value, the higher the significance of the corresponding variable. It could be seen that the solid-liquid ratio had the greatest impact in the solubilization process, followed by the ultrasonic power, and both of them had significant impacts on the nitrogen solubility index. There was no statistically significance of ultrasonic time. The analysis results showed that the combination of the optimal factor level was A2B2C2, and when the ultrasonic power was 210 W, the ultrasonic time was 7 min, and the solid-liquid ratio was 1:

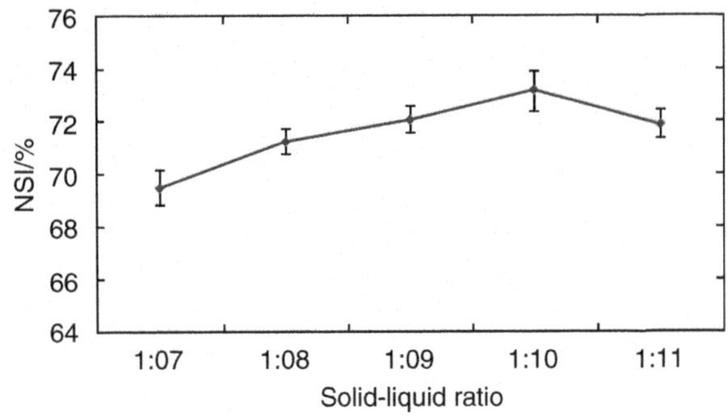

Fig. 6.76 Impact of solid-liquid ratio on nitrogen solubility index of solubilized and modified PPC

Table 6.25 Orthogonal experiment condition factor level of peanut protein concentrate solubilized and modified through ultrasonic treatment

Level	Factor		
	A Ultrasonic power/W	B Ultrasonic time/min	C Solid-liquid ratio
1	180	5	1:9
2	210	7	1:10
3	240	9	1:11

Table 6.26 Design and determination result of orthogonal experiment of peanut protein concentrate solubilized and modified through ultrasonic treatment

No.	Factor			NSI determined value (%)		
	Ultrasonic power (A)	Ultrasonic time (B)	Solid-liquid ratio (C)	I	II	Mean
1	180	5	1:09	71.1449	71.6309	71.3879
2	180	7	1:10	72.8553	73.6717	73.2635
3	180	9	1:11	71.4611	71.8607	71.6609
4	210	5	1:10	73.5904	73.0230	73.3067
5	210	7	1:11	72.7881	73.5837	73.1859
6	210	9	1:09	71.9963	71.7425	71.8694
7	240	5	1:11	71.8821	71.6349	71.7585
8	240	7	1:09	71.7096	71.0024	71.356
9	240	9	1:10	71.7835	71.7111	71.7473
K1	216.3123	216.4531	214.6133			
K2	218.362	217.8054	218.3175			
K3	214.8618	215.2776	216.6053			
R	1.1667	0.8426	1.2347			

10, the solubilization and modification effect was the best. When the confirmatory experiment was carried out under these process parameters, the nitrogen solubility index of the peanut protein concentrate solubilized and modified through ultrasonic treatment was 73.72% ± 0.57% which was significantly higher than that of the ethanol extracting peanut protein concentrate (65.30%) and exceeded that of the

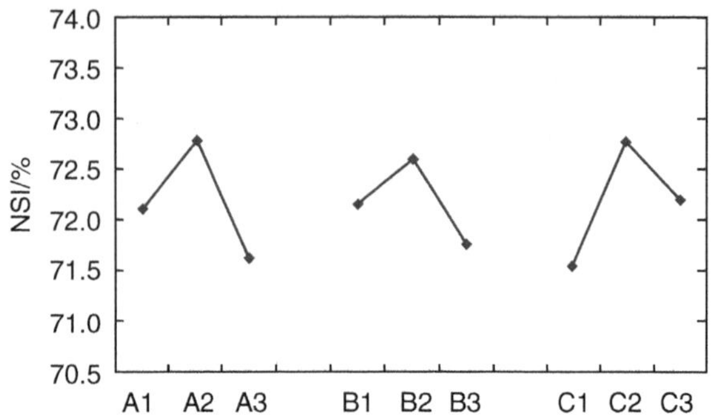

Fig. 6.77 Impact of various factors on nitrogen solubility index of PPC solubilized and modified through ultrasonic treatment

Table 6.27 Variance analysis of orthogonal experiment results of peanut protein concentrate solubilized and modified through ultrasonic treatment

Source of variance	Quadratic sum	Degree of freedom	Mean square	F value	*p* value
Ultrasonic power	2.0618	2	1.0309	28.6524	*0.0337
Ultrasonic time	1.0667	2	0.5333	14.8234	0.0632
Solid-liquid ratio	2.2912	2	1.1456	31.8396	*0.0305
Error	0.0720	2	0.0360		
Sum	5.4917				

raw peanut protein powder (72.89%). This indicated that the solubility of peanut protein concentrate lost due to the ethanol extraction denaturation was completely offset and improved on the basis of that of the raw material protein powder after the ultrasonic treatment. In the actual production, it can be considered to shorten the heating time to improve the production efficiency and reduce the production cost.

5. Process of combination of physical solubilization and modification methods

After the ethanol extracting peanut protein concentrate was modified by using four physical methods, respectively, the nitrogen solubility index of protein was improved at a certain degree. Under the selected optimal process parameters, the solubilization effect in descending order was ultrasonic treatment>heat treatment>high-speed stirring>microwave treatment. Due to the different mechanisms of different solubilization modification methods, if the protein treated by using one modification method was further modified by using other methods, it was possible to further improve the solubility. The experiment was to investigate the changes of solubility by combining the four methods to select the optimal combination of physical modification methods. All the process parameters used in the compound experiment were the optimal parameters obtained through optimization, as shown in Table 6.28.

Microwave treatment had no impact on the solubilization modification effect in the range of solid-liquid ratio between 1: 7 and 1: 11, but the other modification methods had the optimum solid-liquid ratios which were different from each other. If the compound modification process contained the microwave treatment, the

Table 6.28 Optimal process parameters of peanut protein concentrate solubilized and modified by using different physical methods

Modification method	NSI after modification	Optimum solid-liquid ratio	Other parameters of optimal process
High-speed stirring	72.39 ± 0.42%	1:7	Stirring speed, 22,000 r/min; stirring time, 66 s
Microwave treatment	69.83 ± 0.80%	1:7 ~ 11	Microwave power, 640 W; microwave time, 30 s; solution height, 1.5 cm
Heat treatment	72.94 ± 0.61%	1:9	Heating temperature, 75 °C; heating time, 7 min
Ultrasonic treatment	73.72 ± 0.57%	1:10	Ultrasonic power, 210 W; ultrasonic time, 7 min; water-bath temperature, 75 °C

solid-liquid ratio should be in accordance with the optimum ratio of the other treatment. If the compound modification process did not contain the microwave treatment and the solid-liquid ratio of the process operation conducted first was smaller than that conducted later, the solvent needed to be supplemented between the two operations so as to meet the solid-liquid ratio requirements of the latter. If the solid-liquid ratio of the process operation conducted later was larger than that conducted first, the solid-liquid ratio of the latter should not be used as the optimum parameter. Therefore, the solid-liquid ratio of 1: 10 would be used by the high-speed stirring treatment and microwave treatment of ultrasonic treatment, and the solid-liquid ratio of 1: 9 would be used by the high-speed stirring treatment of heat treatment. The experiment determination results of compound solubilization and modification by using four different physical methods are shown in Table 6.29. The results in bold indicated that the nitrogen solubility index of protein after compound treatment was higher than that of protein treated by using any method, which meant that this compound process had a partial superposition effect on the solubility improvement.

The experiment results showed that high-speed stirring was the most suitable method to combine with other physical modification treatment methods both before and after microwave treatment, heat treatment, and ultrasonic treatment, and their combination could further improve the solubility of protein on the original basis and achieve good effects that a single treatment could not reach. This was mainly because the solubilization mechanism of high-speed stirring treatment was different from that of other methods, the high-speed stirring treatment was to promote the mutual collision and friction between materials through mechanical forces to make the protein particles broken so as to promote the solubility, while the microwave treatment and heat treatment were to destroy the intramolecular and intermolecular covalent bonds of protein through heat energy, and the water-bath ultrasonic treatment also realized the solubilization with the help of thermal denaturation. Although the long-term high-speed stirring would make the protein solution heat up, the optimal stirring time selected by this research institute was only 66 s which

Table 6.29 Experiment determination results of peanut protein concentrate solubilized and modified through combining different physical methods

	Process operation conducted later			
Process operation conducted first	High-speed stirring	Microwave treatment	Heat treatment	Ultrasonic treatment
High-speed stirring	–	73.34 ± 0.74%	76.20 ± 0.67%	78.18 ± 0.54%
Microwave treatment	74.82 ± 0.77%	–	72.81 ± 1.03%	72.73 ± 0.83%
Heat treatment	73.83 ± 0.50%	71.35 ± 0.90%	–	72.44 ± 0.65%
Ultrasonic treatment	75.06 ± 0.36%	74.25 ± 0.61%	74.78 ± 0.46%	–

almost could not play a heating effect. Therefore, the mechanical force effects of high-speed stirring treatment and the heat effects of other treatments were well combined, and the excessive thermal denaturation or excessive mechanical force denaturation would not occur.

The operation order also had impacts on the effects of compound modification involved by the high-speed stirring treatment, and the effect generated when the high-speed stirring was carried out before the other treatments was better than that after them. For the reasons, first, the solid-liquid ratio was not the optimal condition of 1:7 when the high-speed stirring treatment was carried out after other treatments, and thus the solubilization effect was affected; second, when the high-speed stirring was carried out before them, the protein structures aggregated due to ethanol extraction were dissolved, which could make all molecules be heated uniformly. When the high-speed stirring and microwave treatment were combined, there was an abnormal situation that the effect was better if the microwave treatment was carried out first. Because the microwave treatment was not affected by the solid-liquid ratio, the high-speed stirring was carried out under the optimal condition of 1:7 even it was carried out later; in addition, different from the traditional heating methods, microwave heating had strong penetrating power and quick and even heating ability so that the protein could be heated without stirring and dissolution. Therefore, the high-speed stirring did not have the advantages owned when it was carried out first. It was inferred that the changes of microwave treatment to the electrical property of protein might be able to make it be broken effectively in the centrifugal force field generated by stirring, so that there would be a better superposition effect when the microwave treatment was carried out first.

The combination of ultrasonic treatment (carried out first) and the other two treatment processes which took thermal denaturation as the main solubilization mechanisms also played a certain superposition effect, but the effect was not significant, and only the nitrogen solubility index of protein was increased by at most 1% on the basis of original ultrasonic treatment. Because the water-bath condition of 75 °C was used during the ultrasonic treatment, the effect had been equivalent to that generated by the combination of ultrasonic modification and thermal modification, and it was limited to further improve the solubility if other thermal modification treatment methods were complemented. The reason for the superposition effect generated when the ultrasonic treatment was carried out first may be that the cavitation effect generated by the ultrasonic wave could deform and break the protein particles so as to make the particles be heated uniformly and efficiently. When the ultrasonic treatment was carried out later, the part with the high denaturation degree in the protein due to uneven thermal denaturation would be denatured excessively, which reduced the solubility.

The combination of microwave treatment and heat treatment did not have superposition effects. Because the heat treatment realized solubilization completely through thermal denaturation, the microwave treatment also realized solubilization mainly through this method, and the difference was that the microwave heating was more quick and even. The processes used in the compound experiments have been optimized, the degrees of thermal denaturation of protein modified through microwave treatment and heat treatment were appropriate for solubilization, and the reduction of solubility would be caused inevitably if continuing heating and modification.

In summary, through the combination of high-speed stirring and heat treatment and the mechanical energy, heat energy, and ultrasonic energy acted on the peanut protein concentrate successively, the structures aggregated during the ethanol extracting denaturation were dissolved, the adverse effects of ethanol denaturation on solubility were offset, the intramolecular and intermolecular covalent bonds of protein which were not conducive to the solubility were broken due to the compound denaturation effect, the solubility was further improved, and the excessive denaturation which reduced the solubility did not occur. The operation order (the high-speed stirring was carried out first and then the ultrasonic treatment was carried out) was also more conducive to the collaboration of a variety of denaturation effects, so that the physical solubilization modification could achieve the best effects.

3.1.2 Solubilization Modification Through Restriction Enzymolysis

1. Enzymolysis pretreatment method

The impact of different enzymolysis pretreatment methods on the nitrogen solubility index of peanut protein concentrate solubilized and modified through restriction enzymolysis is shown in Fig. 6.78. The experiment results showed that the nitrogen solubility index of peanut protein concentrate after the compound high-speed stirring and ultrasonic pretreatments was the highest after restriction enzymolysis, while the restriction enzymolysis effect of high-speed stirring pretreatment was the worst. The restriction enzymolysis effect of peanut protein concentrate after microwave and heating pretreatments was significantly better than that of the high-speed stirring pretreatment and slightly lower than that of the ultrasonic pretreatment and compound pretreatment. It could be seen that the sufficient action of thermal denaturation on protease was necessary, the destruction of covalent bonds made the protein structure dissolve, and the protease could find more enzyme cutting sites to play a better solubilization modification effect (Achouri et al. 1998). In the

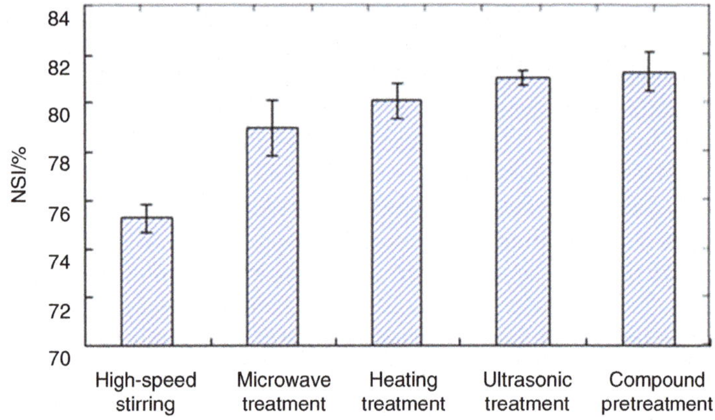

Fig. 6.78 Impact of different enzymolysis pretreatment methods on nitrogen solubility index of PPC solubilized and modified through enzymolysis

experimental operation, because the effects of ultrasonic pretreatment and compound pretreatment were the same basically, from the perspective of simplifying the experimental operation, the ultrasonic pretreatment could be selected as the restriction enzymolysis solubilization modification pretreatment method. However, in the industrial production, after a large amount of powdered solid peanut protein concentrate was added into the water, it was difficult to disperse it quickly and evenly if the stirring, shear, and other methods were not used, so it was necessary to use the high-speed stirring or the treatment process operation with the same effect before ultrasonic treatment. In order to make this experiment achieve the result similar to that of the actual industrial production, the combination of high-speed stirring and ultrasonic treatments was used for pretreatment in the following restriction enzymolysis solubilization experiments.

2. Process of solubilization modification through restriction enzymolysis

(a) Protease

The impact of different proteases on the nitrogen solubility index of peanut protein concentrate solubilized and modified through restriction enzymolysis is shown in Fig. 6.79. The experiment results showed that the solubilization modification effect of papain was significantly better than that of the other four neutral proteases. The action effects of N120P, Neutrase, and Protamex on protein solubility were similar, and the action effect of Flavourzyme was the worst; the nitrogen solubility indexes of restriction enzymolysis products of these four enzymes were lower than that after the compound physical solubilization treatment, and the nitrogen solubility index of Flavourzyme after enzymolysis was even lower than that of the raw material ethanol extracting peanut protein concentrate. For the reasons, first, the high temperature of 95 °C during enzyme deactivation might cause the excessive denaturation of protein, and the slight improvement of protein solubility by restriction enzymolysis could not offset the solubility reduction caused

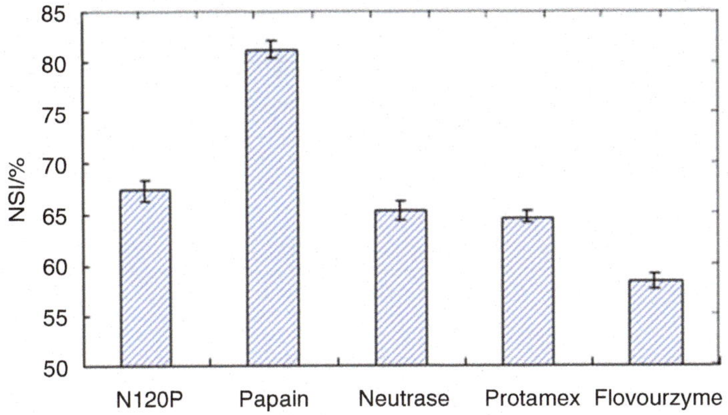

Fig. 6.79 Impact of different proteases on the nitrogen solubility index of PPC solubilized and modified through restriction enzymolysis

by thermal denaturation; second, the process parameters selected by reference to the information provided by the manufacturers were not suitable for solubilization modification of peanut protein concentrate. During the solubilization through restriction enzymolysis, the aggregations between the proteins would occur under the low degree of hydrolysis if the enzymolysis was relatively insufficient, and thus the solubility would be reduced; the relatively excessive enzymolysis would cause the exposure of more hydrophobic amino acid residues, the surface hydrophobicity of protein would increase, and the solubility would also be reduced.

A simple sensory evaluation was carried out for the enzymolysis products of the above five proteases. The results showed that only the enzymolysis products of papain did not have bitterness, and the enzymolysis products of Neutrase, Protamex, and Flavourzyme all had weak bitterness, while that of N120P had more obvious bitterness. The main purpose of high-solubility peanut protein concentrate products is the functional food additive, especially in the protein drinks, and bitterness will certainly damage the edible quality of the finished products. Therefore, according to the solubilization modification effects and sensory evaluation results, papain was selected as the optimum protease for solubilization modification of peanut protein concentrate, and its enzymolysis process parameters were further optimized.

(b) Single-factor experiment

The impact of solid-liquid ratio on the nitrogen solubility index of peanut protein concentrate solubilized and modified through restriction enzymolysis is shown in Fig. 6.80. The results showed that the nitrogen solubility index of restriction enzymolysis products did not change basically when the solid-liquid ratio was increased from 1: 10 to 1: 12, and it began to decrease when the solid-liquid ratio continued to increase. The optimum solid-liquid ratio of the ultrasonic treatment as the restriction enzymolysis pretreatment method was 1: 10, so the enzymolysis

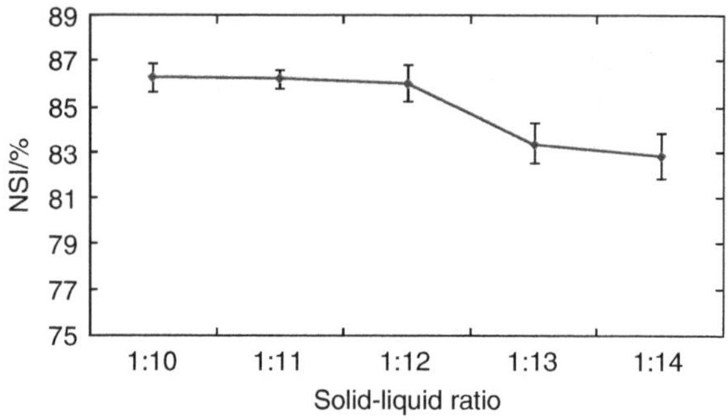

Fig. 6.80 Impact of solid-liquid ratio on nitrogen solubility index of PPC solubilized and modified through restriction enzymolysis

solid-liquid ratio could not be reduced. From the point of view of the actual process operation, the lower solid-liquid ratio was beneficial to the improvement of treatment efficiency. Therefore, 1: 10 was selected as the optimal single-factor solid-liquid ratio for restriction enzymolysis solubilization modification.

The impact of enzyme dosage on the nitrogen solubility index of peanut protein concentrate solubilized and modified through restriction enzymolysis is shown in Fig. 6.81. The experiment results showed that the nitrogen solubility index increased with the increase of enzyme dosage from (E: S, the same below) 0.1–0.3%, which indicated that when the enzyme dosage was less than 0.3%, the substrate was not saturated by enzyme, so the substrate could be sufficiently involved in the reaction and the protein aggregation at the low degree of hydrolysis caused by insufficient enzymolysis could be avoided by increasing the enzyme dosage. When the enzyme dosage exceeded 0.3%, the nitrogen solubility index decreased, which may be because the relatively excessive restriction enzymolysis resulted in the exposure of more hydrophobic amino acid residues, the surface hydrophobicity of protein increased, and thus the solubility was reduced. Therefore, 0.3 was determined as the optimal single-factor enzyme dosage for restriction enzymolysis solubilization modification.

The impact of enzymolysis temperature on the nitrogen solubility index of peanut protein concentrate solubilized and modified through restriction enzymolysis is shown in Fig. 6.82. The experiment results showed that the nitrogen solubility index of enzymolysis products slightly increased when the enzymolysis temperature increased from 40 to 50 °C, which indicated that the relatively high temperature was beneficial to the full and effective action of protease. When the enzymolysis temperature exceeded 50 °C, the nitrogen solubility index of enzymolysis products would decrease rapidly if the enzymolysis temperature was increased continuously, because the temperature exceeded the tolerance range of protease and the protease lost the activity. Therefore, 50 °C was determined as the optimal single-factor action temperature for restriction enzymolysis solubilization modification.

The impact of enzymolysis time on the nitrogen solubility index of peanut protein concentrate solubilized and modified through restriction enzymolysis is shown in Fig. 6.83. The experiment results showed that the nitrogen solubility

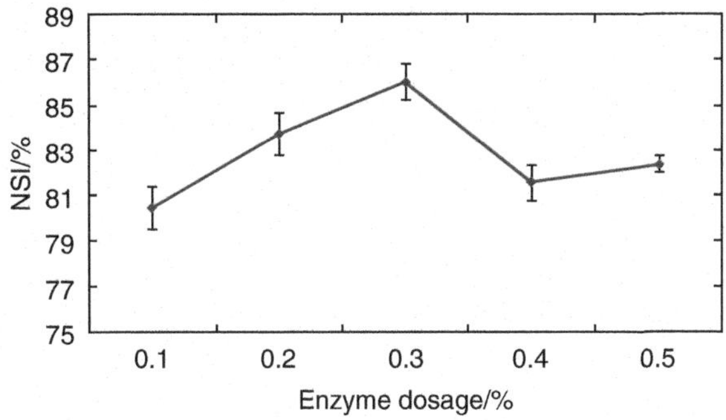

Fig. 6.81 Impact of enzyme dosage on nitrogen solubility index of PPC solubilized and modified through restriction enzymolysis

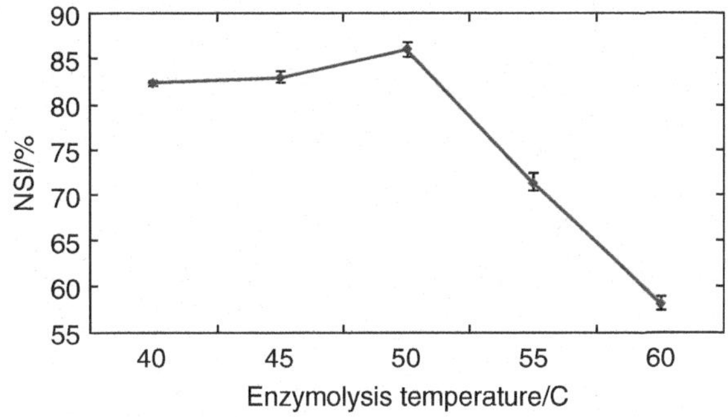

Fig. 6.82 Impact of temperature on the nitrogen solubility index of PPC solubilized and modified through restriction enzymolysis

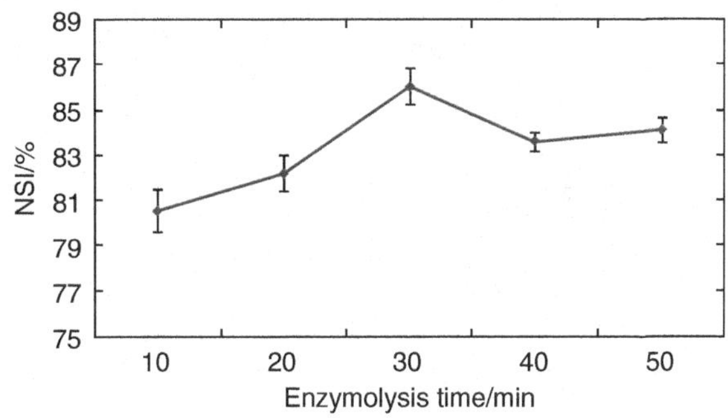

Fig. 6.83 Impact of enzymolysis time on the nitrogen solubility index of PPC solubilized and modified through restriction enzymolysis

index of enzymolysis products increased significantly when the enzymolysis time increased from 10 to 30 min, which indicated that certain time was needed by protease to make full action to play a better solubilization effect. After the time exceeded 30 min, the nitrogen solubility index began to decrease, this situation was similar to that of the excessive enzyme dosage, the relatively excessive restriction enzymolysis would be caused, and the solubility of products would be decreased. Therefore, 30 min was determined as the optimal single-factor enzymolysis time for restriction enzymolysis solubilization modification.

(c) Response surface optimization test

Through the single-factor experiment, the enzyme dosage, enzymolysis time, and enzymolysis temperature were selected as the three factors of restriction enzymolysis solubilization modification, and 1: 10 was determined as the solid-liquid ratio directly. By combining with the results of simple orthogonal preexperiment and response surface preexperiment, p_1 was selected as 0.2, and q_1 was selected as 0.1; p_2 was selected as 20 and q_2 was selected as 10; p_3 was selected as 45 and q_3 was selected as 5.

Through the extraction experiment and the analysis of experimental determination values, the regression equation was obtained. The equation was the empirical correlation between the independent variable coding value and nitrogen solubility index of protein. It could be used to predict the change of the response value within the given variation range of independent variable. The regression equation of the nitrogen solubility index Y of restriction enzymolysis solubilization modification product is shown below:

$$\begin{aligned} Y &= 86.11474 + 1.10366x_1 + 0.26755x_2 - 0.12853x_3 \\ &- 0.77358x_1^2 - 1.61479x_2^2 - 0.55189x_3^2 \\ &- 0.90887x_1x_2 + 0.62703x_1x_3 + 0.65869x_2x_3 \end{aligned}$$

The design parameters of central synthetic experiment of restriction enzymolysis solubilization modification response surface are shown in Table 6.30. The experimental determination values and predicted values of model of each experiment were compared in the table.

The significance of each parameter was measured by the Student t test and p value, as shown in Table 6.2. The greater the absolute value of t or the smaller the p value, the higher the significance of corresponding variable. It could be seen that the quadratic term (x_2x_2) of enzymolysis time was the most significant, followed by the first term (x_1) and quadratic term (x_1x_1) of enzyme dosage. According to the t test and p value, there was no statistically significance in the quadratic term (x_2) of enzymolysis time and quadratic term (x_3) of enzymolysis temperature (p value was greater than 0.05). The analysis results showed the factors influencing the nitrogen solubility index in the order of priority: enzyme dosage>enzymolysis time>enzymolysis temperature.

The correlation coefficient of the regression model was $R = 0.9560$, and the total determination coefficient was $R^2 = 0.9140$, so it was reliable to use the model to predict the nitrogen solubility index of restriction enzymolysis products (Table 6.31).

The impact of enzyme dosage and enzymolysis time of restriction enzymolysis solubilization modification on the nitrogen solubility index of protein of product is shown in Fig. 6.84. The experiment results showed that in the interactions of enzyme dosage and enzymolysis time on the nitrogen solubility index, the enzyme dosage had not only a quadratic effect but also a linear effect on the response value,

Table 6.30 Restriction enzymolysis solubilization modification response surface experiment design and experimental determination values and predicted values of model

No.	Coded value of independent variable level (actual value)			Experimental value			Predicted value
	x_1 (%)	x_2 (min)	x_3 (°C)	I	II	Mean	
1	1 (0.3)	1 (30)	1 (50)	85.018	85.364	85.191	84.794
2	1 (0.3)	1 (30)	–1 (40)	81.629	81.085	81.357	82.480
3	1 (0.3)	–1 (10)	1 (50)	83.979	83.211	83.595	84.759
4	1 (0.3)	–1 (10)	–1 (40)	85.964	86.073	86.019	85.080
5	–1 (0.1)	1 (30)	1 (50)	82.239	81.396	81.817	83.150
6	–1 (0.1)	1 (30)	–1 (40)	83.548	84.681	84.114	83.344
7	–1 (0.1)	–1 (10)	1 (50)	80.087	80.330	80.209	79.480
8	–1 (0.1)	–1 (10)	–1 (40)	81.773	81.262	81.518	82.309
9	−1.682 (0.032)	0 (20)	0 (45)	82.563	81.942	82.252	82.071
10	1.682 (0.368)	0 (20)	0 (45)	85.952	86.364	86.158	85.783
11	0 (0.2)	−1.682 (3.18)	0 (45)	81.596	80.561	81.079	81.097
12	0 (0.2)	1.682 (36.82)	0 (45)	83.036	82.111	82.573	81.997
13	0 (0.2)	0 (20)	−1.682 (36.59)	84.317	85.086	84.701	84.770
14	0 (0.2)	0 (20)	1.682 (53.41)	84.732	85.194	84.963	84.338
15	0 (0.2)	0 (20)	0 (45)	86.387	85.777	86.082	86.115
16	0 (0.2)	0 (20)	0 (45)	86.394	85.829	86.112	86.115
17	0 (0.2)	0 (20)	0 (45)	85.926	86.407	86.167	86.115
18	0 (0.2)	0 (20)	0 (45)	86.421	85.954	86.188	86.115
19	0 (0.2)	0 (20)	0 (45)	86.039	86.150	86.095	86.115
20	0 (0.2)	0 (20)	0 (45)	86.385	85.762	86.074	86.115
21	0 (0.2)	0 (20)	0 (45)	86.041	86.317	86.179	86.115
22	0 (0.2)	0 (20)	0 (45)	86.369	85.643	86.006	86.115
23	0 (0.2)	0 (20)	0 (45)	86.367	85.704	86.036	86.115

Table 6.31 Significance of regression equation coefficient of restriction enzymolysis solubilization modification

Variable	Regression coefficient	Standard deviation	Calculated value of t	p value
Constant term	86.11474	0.263210	327.1653	0.000000
x_1	1.10366	0.213820	5.1617	**0.000183
x_2	0.26755	0.213816	1.2513	0.232873
x_3	−0.12853	0.213816	−0.6011	0.558082
x_1x_1	−0.77358	0.198225	−3.9025	**0.001818
x_2x_2	−1.61479	0.198225	−8.1462	**0.000002
x_3x_3	−0.55189	0.198230	−2.7842	*0.015492
x_1x_2	−0.90887	0.279360	−3.2534	**0.006287
x_1x_3	0.62703	0.279360	2.2445	*0.042842
x_2x_3	0.65869	0.279360	2.3578	*0.034715
R	0.9560			
R^2	0.9140			

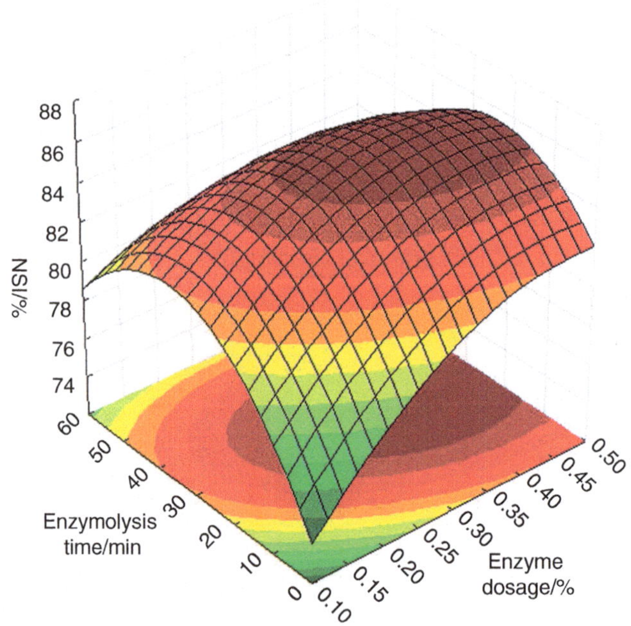

Fig. 6.84 Response surface of impact of enzyme dosage and enzymolysis time on nitrogen solubility index of solubilized and modified PPC

while the enzymolysis time had a quadratic effect on the response value. The interaction between the enzyme dosage and enzymolysis time was significant.

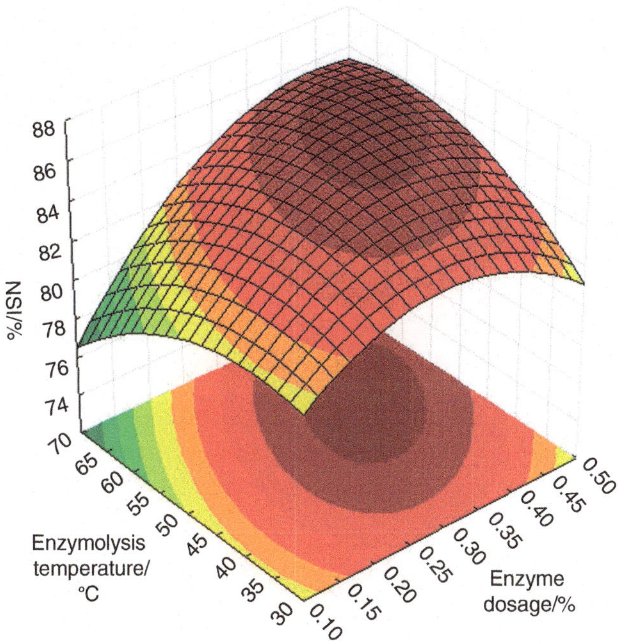

Fig. 6.85 Response surface of impact of enzyme dosage and enzymolysis temperature on nitrogen solubility index of solubilized and modified PPC

The impact of enzyme dosage and enzymolysis temperature of restriction enzymolysis solubilization modification on the nitrogen solubility index of protein of product is shown in Fig. 6.85. The experiment results showed that in the interactions of enzyme dosage and enzymolysis temperature on the nitrogen solubility index, the enzyme dosage mainly had a quadratic effect on the response value when the enzymolysis temperature was low and had a linear effect on the response value when the enzymolysis temperature was high, and the enzymolysis temperature had a quadratic effect on the response value. The interaction between the enzyme dosage and enzymolysis temperature was significant.

The impact of enzymolysis time and enzymolysis temperature of restriction enzymolysis solubilization modification on the nitrogen solubility index of protein of product is shown in Fig. 6.86. The experiment results showed that in the interactions of enzymolysis time and enzymolysis temperature on the nitrogen solubility index, both of them had quadratic effects on the response value, and the effect of enzymolysis time was larger than that of enzymolysis temperature. The interaction between the enzymolysis time and enzymolysis temperature was significant.

In order to obtain the exact optimum process parameters, by finding the partial derivatives of these three independent variables of the regression equation of the nitrogen solubility index Y of restriction enzymolysis solubilization modification

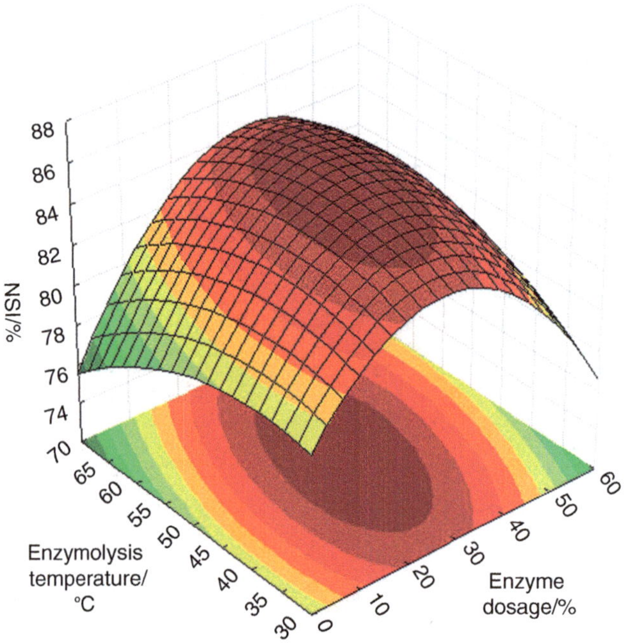

Fig. 6.86 Response surface of impact of enzymolysis time and enzymolysis temperature on nitrogen solubility index of PPC solubilized and modified through restriction enzymolysis

products and making them equal to zero, the following three equations could be obtained:

$$\begin{aligned} 1.10366 - 1.54716x_1 - 0.90887x_2 + 0.62703x_3 &= 0 \\ 0.26755 - 0.90887x_1 - 3.22958x_2 + 0.65869x_3 &= 0 \\ -0.12853 + 0.62703x_1 + 0.65869x_2 - 1.10378x_3 &= 0 \end{aligned}$$

Through the simultaneous equations, the optimal process parameters of restriction enzymolysis solubilization modification could be obtained:

$$\begin{aligned} x_1 &= 0.91340\ x_2 = -0.10471\ \ x_3 = 0.33938 \\ X_1 &= 0.2913\% X_2 = 18.95\text{min}\ X_3 = 46.70^{\circ}\text{C} \end{aligned}$$

Through the model calculation, the nitrogen solubility index of peanut protein concentrate was 86.58%, and the verification value was 86.34% ± 0.69% under the optimal process conditions, which were consistent with the predicted value of model and significantly higher than the nitrogen solubility index (78.18%) of peanut protein concentrate solubilized and modified through compound physical method. The following enzymolysis posttreatment solubilization modification experiments were established under the optimal process parameter conditions of restriction enzymolysis solubilization modification obtained here.

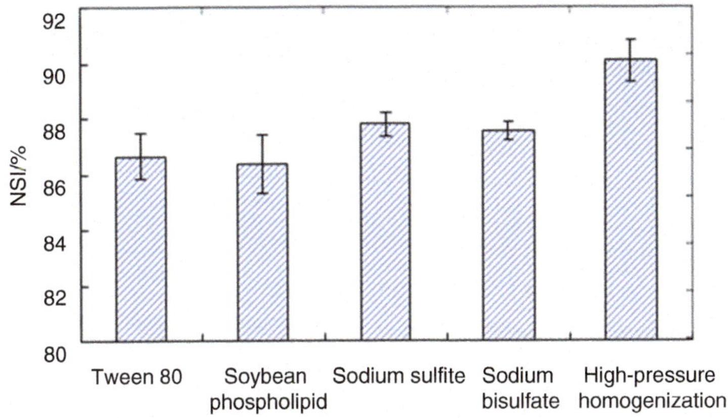

Fig. 6.87 Impact of different enzymolysis posttreatment methods on nitrogen solubility index of PPC solubilized and modified through enzymolysis

3. Process of solubilization modification through enzymolysis posttreatment

(a) Posttreatment method

The impact of different restriction enzymolysis posttreatment methods on the nitrogen solubility index of peanut protein concentrate solubilized and modified through enzymolysis is shown in Fig. 6.87. The experiment results showed that the solubilization effect of high-pressure homogeneous treatment was superior to that of addition of surfactants or reductants. In the actual production, the high-pressure homogenization was also the most widely used process in the food industry. The addition of Tween 80 and soybean phospholipid did not increase the solubility of protein basically, which was because the hydrophilicity of surfactant was poor and the peanut protein concentrate could not be solubilized effectively. The addition of sodium sulfite and sodium bisulfate made the protein solubility have a certain increase, which was because the reductant could effectively inhibit the formation of intramolecular and intermolecular disulfide bonds of protein; the addition of reductant and other chemical modification methods faced a lot of safety questions in the aspect of food processing, so only a preliminary attempt of reductant solubilization was carried out in this experiment, and the additive amount and other specific process parameters were not optimized. Therefore, by combining the solubilization effects and safety, the high-pressure homogeneous treatment was selected as the posttreatment method of restriction enzymolysis solubilization modification of peanut protein concentrate, and the process parameters were further optimized.

(b) Process of solubilization modification through high-pressure homogenization

(i) Single-factor experiment

The impact of homogenization pressure on the nitrogen solubility index of peanut protein concentrate is shown in Fig. 6.88. The nitrogen solubility index of

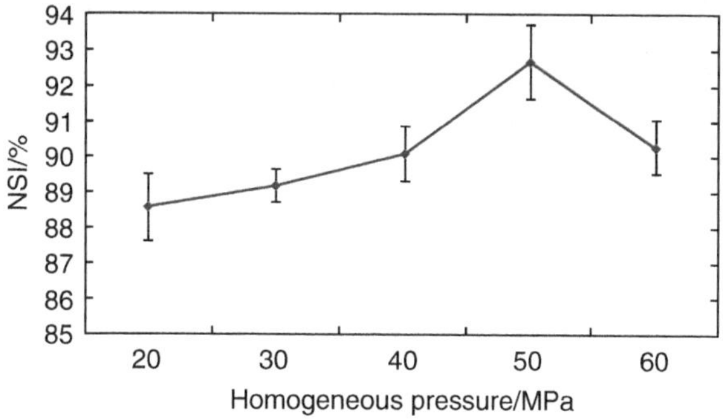

Fig. 6.88 Impact of homogenization pressure on solubilized and modified PPC

protein increased when the homogenization pressure increased from 20 to 50 MPa, indicating that the relatively high homogenization pressure was favorable to the generation of stronger shear and impact forces so as to break the protein more effectively and increase its solubility. When the homogenization pressure exceeded 50 MPa, the nitrogen solubility index began to decrease if the pressure continued to increase, which might be because the thermal energy generated by too high pressure made the solution temperature increase rapidly and the thermal denaturation of protein which was not conductive to solubility occurred. Therefore, 50 MPa was determined as the optimal single-factor homogenization pressure for high-pressure homogeneous treatment solubilization modification.

The impact of solid-liquid ratio on the nitrogen solubility index of peanut protein concentrate solubilized and modified through high-pressure homogenization is shown in Fig. 6.89. The experiment results showed that the nitrogen solubility index of protein increased significantly when the solid-liquid ratio increased from 1: 10 to 1: 13, and the nitrogen solubility index began to decrease when the solid-liquid ratio continued to increase to 1: 14. This was because too high concentration of solution would increase the movement resistance of protein and thus weaken the effects of shear force, collision force, and cavitation explosion force caused by high-pressure homogenization; if the concentration of solution was too low, the interaction between the proteins was affected, the effect of shear force was dispersed at certain extent, and the effect of solubilization modification was weakened. Therefore, 1: 13 was determined as the optimal single-factor solid-liquid ratio for high-pressure homogeneous treatment solubilization modification.

The impact of homogenization time on the nitrogen solubility index of peanut protein concentrate solubilized and modified through high-pressure homogenization is shown in Fig. 6.90. The experiment results showed that the nitrogen concentration index increased significantly when the homogenization time increased from 1 to 2 min, and the nitrogen solubility index decreased rapidly after the homogenization time exceeded 2 min. This indicated that the proper extension of homogenization time was conducive to the full effect of shear force and impact force on materials and was beneficial to the improvement of protein solubility. If the homogenization time was too long, the solution temperature increased significantly and the undesirable excessive thermal denaturation of

Fig. 6.89 Impact of solid-liquid ratio on solubilized and modified PPC

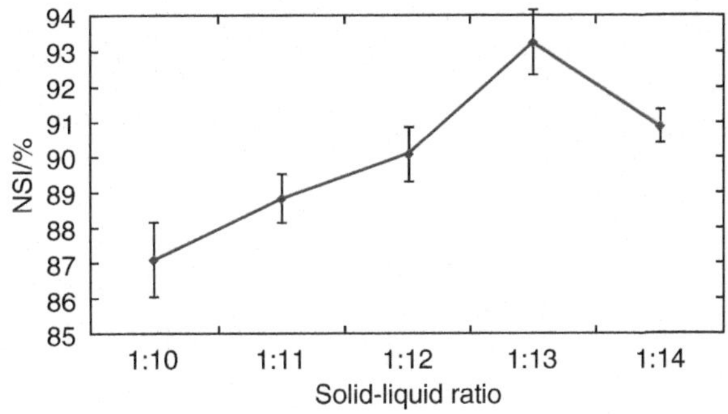

Fig. 6.90 Impact of homogenization time on nitrogen solubility index of PPC solubilized and modified through high-pressure homogenization

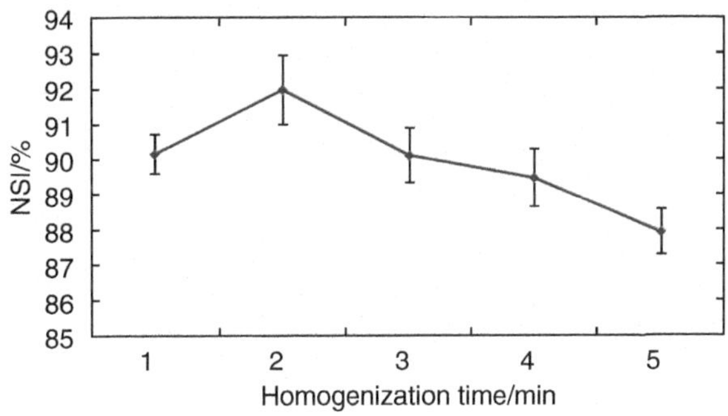

protein occurred, which damaged its solubility. Therefore, 2 min was determined as the optimal single-factor homogenization time for high-pressure homogenous treatment solubilization modification.

(ii) Orthogonal experiment

According to the results of single-factor experiment, the homogenization pressure(A), solid-liquid ratio (B), and homogenization time (C) were selected as the three factors of simple orthogonal experiment of high-pressure homogenization solubilization treatment. The experiments were performed according to Table 6.32. The results are shown in Table 6.33 and Fig. 6.91. The results of orthogonal experiment were analyzed by using the Duncan's new multiple range method, as shown in Table 6.34.

The larger the F value or the smaller the *p* value, the higher the significance of the corresponding variable. It could be seen that the homogenization pressure had the greatest impact in the solubilization process, followed by the homogenization time, and then the solid-liquid ratio, and these three factors all had significant impacts on the nitrogen solubility index. The analysis results showed that the combination of the optimal factor level was A2B2C2, and when the homogenization pressure was 50 MPa, the solid-liquid ratio was 1: 3, and the homogenization time was 2 min, the solubilization and modification effect was the best. When the confirmatory experiment was carried out under these process parameters, the

Table 6.32 Orthogonal experiment condition factor level of peanut protein concentrate solubilized and modified through high-pressure homogenization

Level	Factor A Homogenization pressure/MPa	B Solid-liquid ratio	C Homogenization time/min
1	40	1:12	1
2	50	1:13	2
3	60	1:14	3

Table 6.33 Design and determination result of orthogonal experiment of peanut protein concentrate solubilized and modified through high-pressure homogenization

No.	Factor Homogenization pressure (A)/MPa	Solid-liquid ratio (B)/min	Homogenization time (C)	NSI determined value (%) I	II	Mean
1	40	1:12	1	93.0540	93.7200	93.3870
2	40	1:13	2	95.3441	95.6299	95.4870
3	40	1:14	3	94.3760	93.6580	94.0170
4	50	1:12	2	96.0255	95.6037	95.8146
5	50	1:13	3	96.1688	96.7036	96.4362
6	50	1:14	1	94.7592	95.3664	95.0628
7	60	1:12	3	94.5277	94.1615	94.3446
8	60	1:13	1	95.2017	94.9491	95.0754
9	60	1:14	2	95.3912	96.0364	95.7138
K1	211.4006	213.3879	210.8503			
K2	216.1117	213.6327	214.4849			
K3	212.0702	212.5620	214.2474			
R	1.5704	0.3569	1.2115			

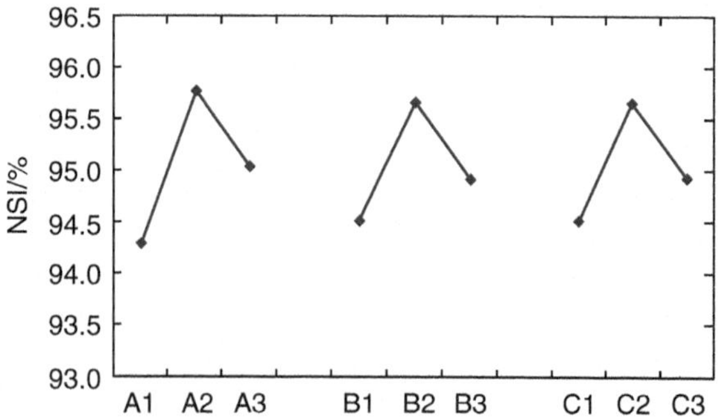

Fig. 6.91 Impact of various factors on nitrogen solubility index of PPC solubilized and modified through high-pressure homogenization

nitrogen solubility index of the peanut protein concentrate solubilized and modified through high-pressure homogeneous treatment was 96.57% ± 0.40% which was significantly higher than that (86.34%) of the peanut protein concentrate solubilized and modified through restriction enzymolysis.

Table 6.34 Variance analysis of orthogonal experiment results of peanut protein concentrate solubilized and modified through high-pressure homogenization

Source of variance	Quadratic sum	Degree of freedom	Mean square	F value	*p* value
Homogenization pressure	3.2601	2	1.6301	36.2238	*0.0269
Solid-liquid ratio	2.0375	2	1.0187	22.6386	*0.0423
Homogenization time	2.0799	2	1.0399	23.1098	*0.0415
Error	0.0900	2	0.0450		
Sum	7.4674				

The impacts of different physical pretreatment methods on the enzymolysis effect were investigated by taking the peanut protein concentrate obtained by ethanol extraction as the raw materials. The protease suitable for the restriction enzymolysis solubilization of peanut protein concentrate was selected, and then the response surface method was used to optimize the enzymolysis process. Finally, different enzymolysis posttreatment methods were investigated, and the high-pressure homogenization posttreatment process was optimized.

The comparison of restriction enzymolysis pretreatment methods showed that the most suitable physical treatment conducted before restriction enzymolysis solubilization modification was ultrasonic treatment. In this research, in order to simulate the actual industrial production, the enzymolysis pretreatment method combining high-speed stirring and ultrasonic treatment was used. The solubilization modification effect and sensory evaluation results were considered comprehensively during the selection of enzyme for solubilization, the papain was selected, and the principle factor that affected the nitrogen solubility index of product was enzyme dosage, followed by the enzymolysis time in these three factors (enzyme dosage (x_1), enzymolysis time (x_2), and enzymolysis temperature (x_3)). The optimal process of restriction enzymolysis solubilization was determined as follows: the enzyme dosage (E: S) was 0.2913%, the enzymolysis time was 18.95 min, and the enzymolysis temperature was 46.70 °C. Through the comprehensive consideration of solubilization effect and safety, the enzymolysis posttreatment method was determined as high-pressure homogeneous treatment, and the factors influencing the high-pressure homogenization solubilization effects in the order of priority were as follows: homogenization pressure>homogenization time>solid-liquid ratio. The optimal process parameters of high-pressure homogenization were determined as follows: the homogenization pressure was 50 MPa, the solid-liquid ratio was 1: 13, and the homogenization time was 2 min.

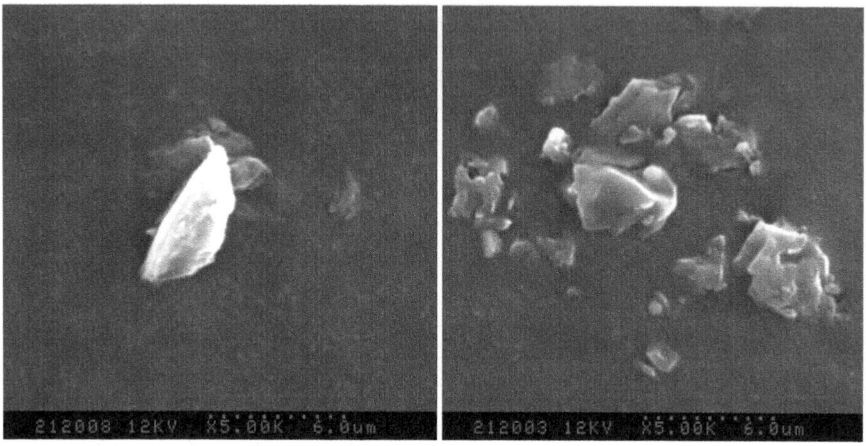

Fig. 6.92 Scanning electron microscope microstructure of peanut protein powder

3.2 Physicochemical Characteristics

3.2.1 Evaluation of Structural Characteristics of Protein

1. Scanning electron microscope of protein

The microstructures of raw material peanut protein powder and ethanol extracting peanut protein concentrate were observed through the scanning electron microscopy, as shown in Figs. 6.92 and 6.93, respectively. The experiment results showed that the peanut protein powder particles which were not through the ethanol solution denaturation were dispersed, there was no aggregation between the particles, and the surfaces of particles were smooth. The size of the single particle of the peanut protein concentrate obtained by ethanol extraction was equivalent to the size of about three to four peanut protein powder particles. It was obvious that the aggregations occurred between the protein particles. It also could be seen that the surfaces of these particles were very rough and the structures were tight, which was also caused by ethanol denaturation. In addition, the ethanol extracting peanut protein concentrate particles were independently different from the peanut protein powder particles, and there were aggregations between the larger particles. The increase of average volume of protein particles and tightness of structures made the opportunity of water molecules to contact the protein molecules decrease. At the same time, the ethanol extraction denaturation on the surfaces of protein particles improved the surface hydrophobicity of protein molecules, and the reasons for solubility reduction after ethanol extraction could also be explained from this.

The microstructure of the peanut protein concentrate modified through compound physical method is shown in Fig. 6.94. The experiment results showed that the size of peanut protein particle did not change basically after high-speed stirring and ultrasonic treatment, but the structure was looser than that of ethanol extracting

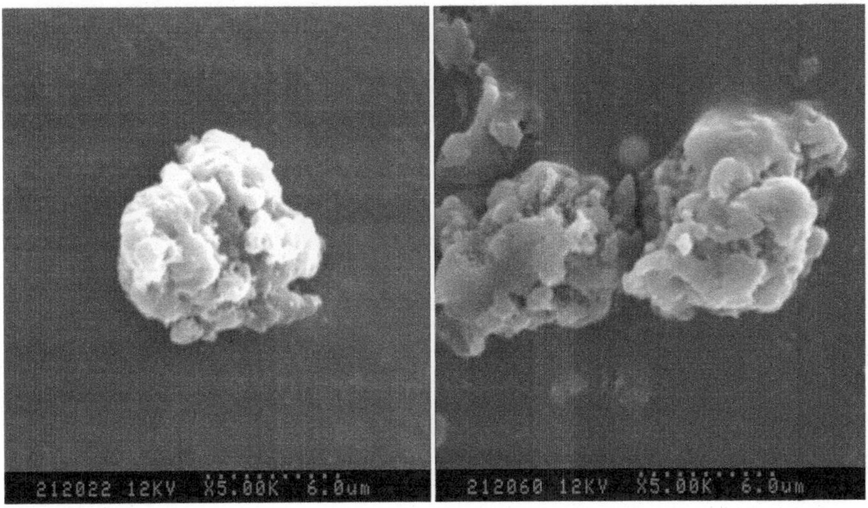

Fig. 6.93 Scanning electron microscope microstructure of ethanol extracting peanut protein concentrate

Fig. 6.94 Scanning electron microscope microstructure of peanut protein concentrate modified through compound physical method

peanut protein concentrate; some lamellar structures appeared, and there were still aggregation and cross-linking between the protein particles, but the tightness degree was far less than that before physical modification. These changes also provided good conditions for more adequate contact between protein molecules and water molecules, and the reasons for the solubility improvement of peanut protein after compound physical modification also could be explained.

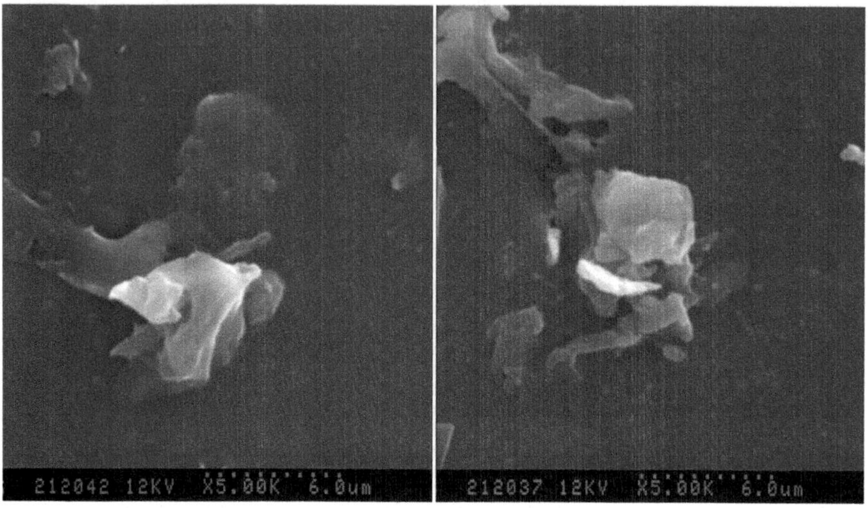

Fig. 6.95 Scanning electron microscope microstructure of peanut protein concentrate modified through restriction enzymolysis

The microstructure of the peanut protein concentrate modified through restriction enzymolysis is shown in Fig. 6.95. The experiment results showed that the size of peanut protein molecular significantly decreased through the restriction enzymolysis, the cross-linking and aggregation between the particles disappeared basically, the protein structure was looser than that after the compound physical modification, the lamellar structure was more obvious, and the lamellas were dissociated significantly. These changes provided an explanation for the solubility improvement of peanut protein after restriction enzymolysis modification, the molecular weight decreased significantly, the structure was looser, and more complete interactions could be made between protein molecules and water molecules.

The microstructure of peanut protein concentrate modified through high-pressure homogenization is shown in Fig. 6.96, and the sample is the final product. In order to show its structural characteristics clearly, four different forms of pictures were selected and analyzed. The experiment results showed that most of the lamellar structures of proteins disappeared after the high-pressure homogenization modification treatment, the structures of peanut protein were further dissociated into strips, the length of these strip proteins was different, and there were a small number of lamellar structures which had not been completely dissolved on some stripes. There was also some small protein molecular debris between the strip molecules of protein, which showed that the size of protein molecules treated by high-pressure homogenization had been further reduced. As with the restriction enzymolysis, the high-pressure homogeneous treatment also caused the dissolution of protein structures and further reduction of molecular weight. Therefore, the solubility of protein was further improved.

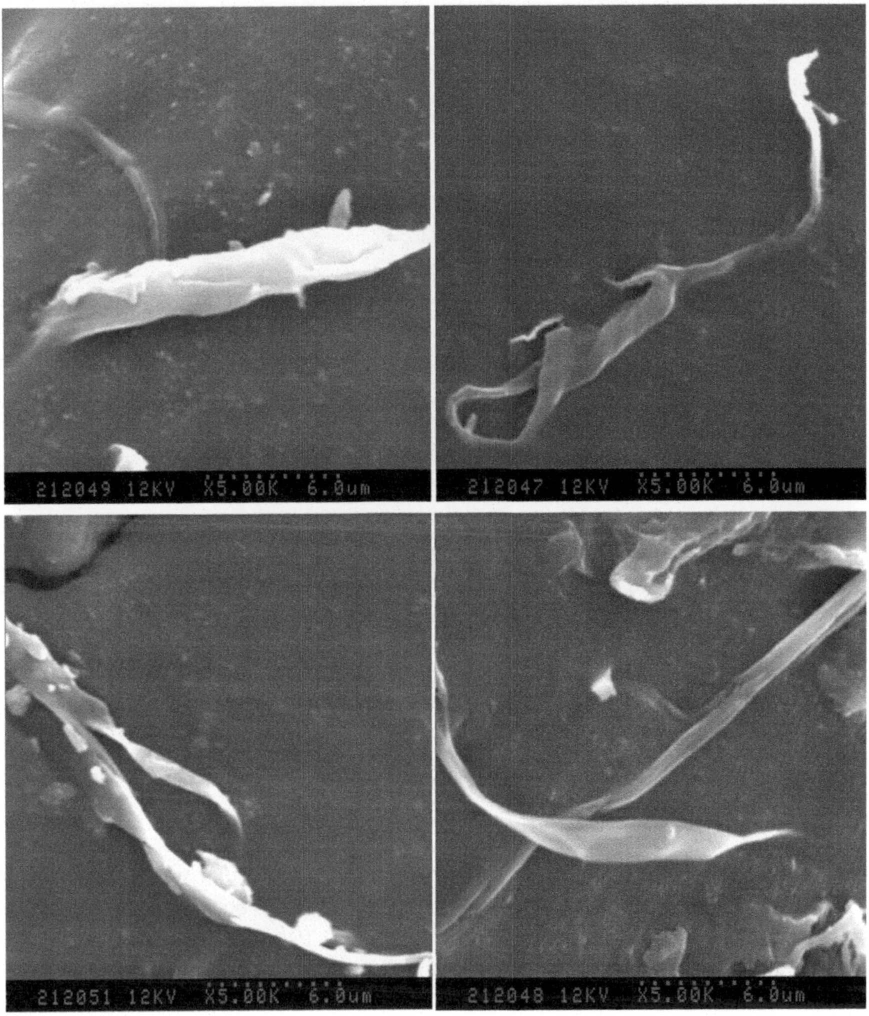

Fig. 6.96 Scanning electron microscope microstructure of peanut protein concentrate modified through high-pressure homogenization

2. Analysis of protein electrophoretogram

The electrophoretograms of peanut protein powder, ethanol extracting peanut protein concentrate, and peanut protein concentrate modified through compound physical method, restriction enzymolysis, and high-pressure homogenization are shown in Fig. 6.97. The experiment results showed that there were no significant differences in the electrophoretic bands of raw protein peanut protein powder, ethanol extracting peanut protein concentrate, and peanut protein concentrate modified through compound physical method, which indicated that the ethanol denaturation and compound physical modification only changed the conformation

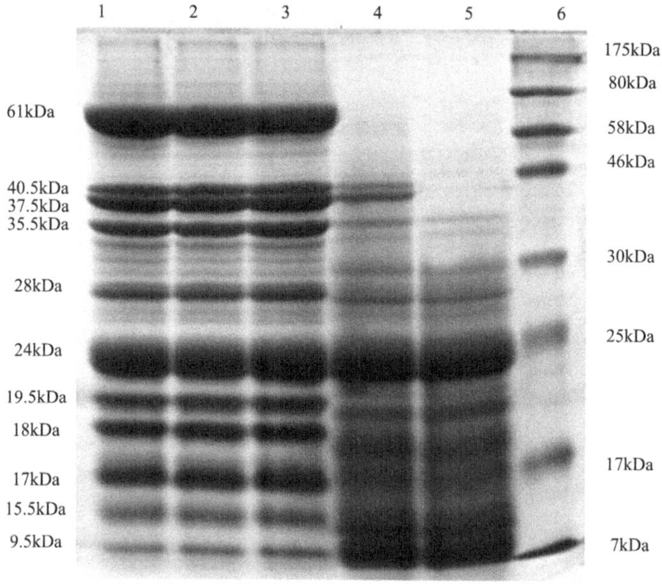

Fig. 6.97 Electrophoretograms of peanut protein powder, ethanol extracting peanut protein concentrate, and peanut protein concentrate through various modification treatments. Note: The above six electrophoretic bands from left to right were: *1*, raw material peanut protein powder; *2*, ethanol extracting peanut protein concentrate; *3*, peanut protein concentrate modified through compound physical method; *4*, peanut protein concentrate solubilized and modified through restriction enzymolysis; *5*, peanut protein concentrate modified through high-pressure homogenization; and *6*, low molecular weight standard protein

of protein in different degrees, and the size of protein molecular was not changed. The bands of these three protein samples from top to bottom were 61 kDa, 40.5 kDa, 37.5 kDa, 35.5 kDa, 28 kDa, 24 kDa, 19.5 kDa, 18 kDa, 17 kDa, 15.5 kDa, and 9.5 kDa successively. According to the existing literature information, 61 kDa was the component of conarachin; 40.5 kDa, 37.5 kDa, 35.5 kDa, and 19.5 kDa were the components of arachin; 18 kDa, 17 kDa, and 15.5 kDa were the components of 2S protein of peanut; 28 kDa, 24 kDa, and 9.5 kDa might be the components of subunits or peptide chains constituting the peanut protein composition (Yamada 1979; Wei et al. 1998; Yang et al. 2001; Zhang 2007).

The electrophoretic bands of the peanut protein concentrate were significantly changed after restriction enzymolysis solubilization modification; all the conarachin II was digested by papain and disappeared; the arachin, conarachin I, and other subunits and peptide chain components were also digested into smaller molecular weight protein components in different degrees; and thus the small-molecule components below 25 kDa were significantly increased. The decrease in the average molecular weight enabled the protein to be easy to dissolve, and thus the nitrogen solubility index was enhanced.

After the high-pressure homogenization modification, there were some changes in the electrophoretic bands. After restriction enzymolysis, the contents of 40.5 kDa and 37.5 kDa components of arachin were greatly reduced, and they disappeared basically after high-pressure homogenization modification. The content of components above 30 kDa in the final product had been very little, and the decrease of average molecular weight of protein made the solubility increase again.

3. Analysis of infrared spectrum in protein

Fourier transform infrared spectroscopy (FTIR) is a method that can be used to carry out quantitative research for the secondary structure of protein in recent decades. The infrared band mostly used for the secondary structure analysis is the amide I band which is located between 1600 and 1700 cm^{-1} and mainly involves the stretching vibration of C=O. According to the existing literature information, the correspondences between various peak positions of intermediate infrared spectrum detection fitting curve and components of secondary structure were shown below: $(1694 \pm 2)cm^{-1}$, $(1689 \pm 2)cm^{-1}$, $(1683 \pm 2)cm^{-1}$, $(1671 \pm 3)cm^{-1}$, and $(1663 \pm 4)cm^{-1}$ were β-turn structures; $(1653 \pm 4)cm^{-1}$ was a α-helix structure; $(1645 \pm 4)cm^{-1}$ was a freely coil structure; $(1675 \pm 5)cm^{-1}$ and $(1630 \pm 10)cm^{-1}$ were β-sheet structures (Krimm and Bandekar 1980; Byler and Susi 1986; Susi and Byler 1986).

The intermediate infrared spectrum detection fitting curves of raw material peanut protein powder, ethanol extracting peanut protein concentrate, and peanut protein concentrate modified through compound physical method, restriction enzymolysis, and high-pressure homogenization are shown in Figs. 6.98, 6.99, 6.100, 6.101, and 6.102. The peak areas were fit for further statistics according to the second reciprocal and the corresponding peak positions were confirmed. The results are shown in Table 6.35. The peaks without correspondence with secondary structure were not included in the table.

The sheet experiment results showed that the α-helix structure was not detected in the peanut protein powder and four kinds of peanut protein concentrate. After the ethanol extraction was carried out for the raw material peanut protein powder, the ratio of β-sheet structure decreased and the ratio of β-turn structure increased. The ratio of β-sheet in the raw material peanut protein powder was more than 90%, indicating that the structure was very regular; the ethanol extraction made its natural structure change, and a certain proportion of free coil structures occur, which also indicated that the ethanol extraction caused the partial denaturation of protein. After the compound physical modification, the ratio of free coil structure almost did not change, the ratio of β-sheet structure was further reduced, and the ratio of β-turn structure was further improved, which indicated that high-speed stirring and ultrasonic treatment could effectively loosen the tight protein structure. After the restriction enzymolysis modification, the ratio of β-sheet structure increased significantly, the ratio of β-turn decreased significantly, and the ratio of free coil structure also increased, which may be because the small-molecule proteins obtained due to enzymolysis were easier to recover and form orderly sheet structures, and some peptide fragments with small molecular weight tend to form unordered free coil structures. After high-pressure homogenization modification,

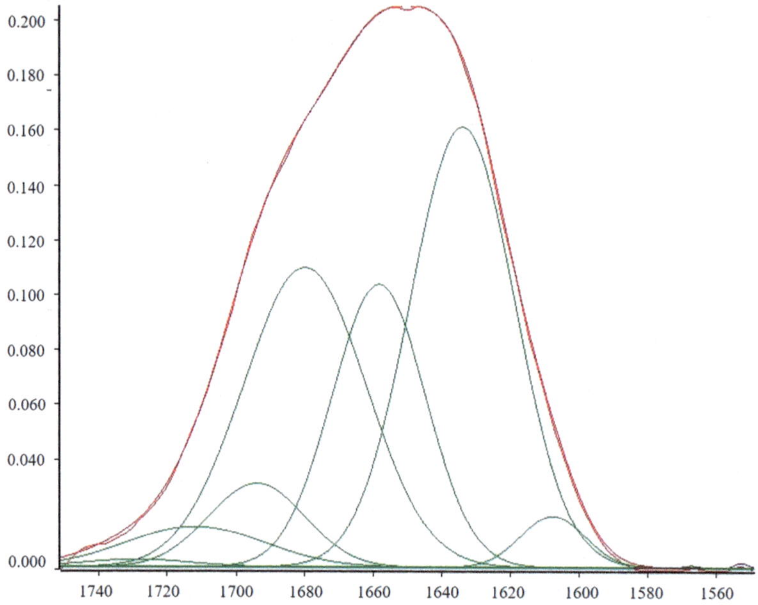

Fig. 6.98 Intermediate infrared spectrum detection fitting curve of peanut protein powder

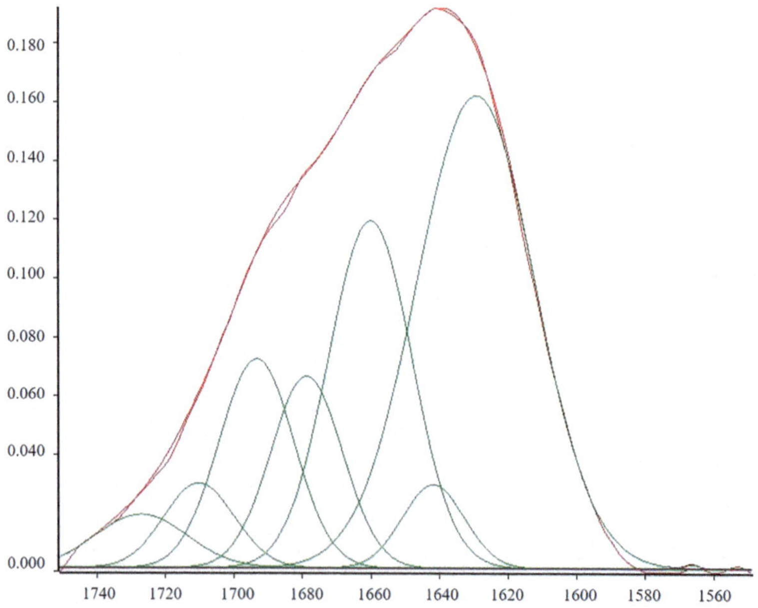

Fig. 6.99 Intermediate infrared spectrum detection fitting curve of ethanol extracting peanut protein concentrate

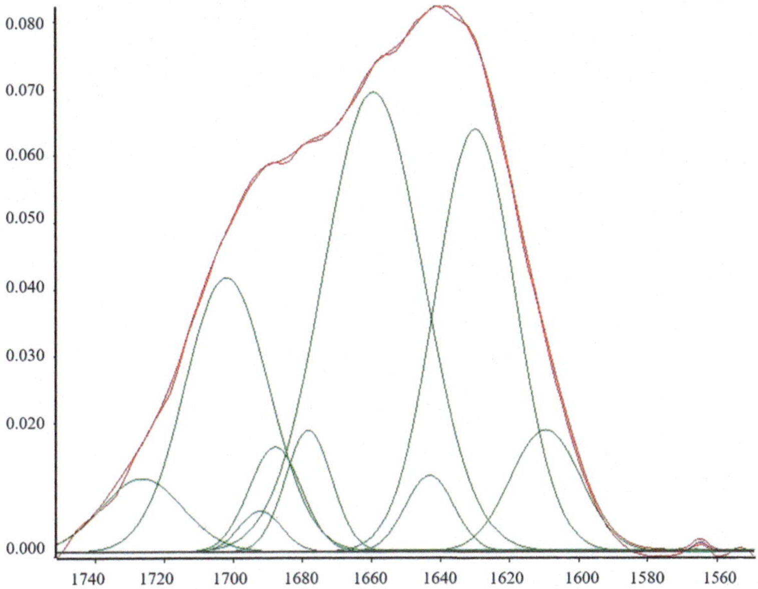

Fig. 6.100 Intermediate infrared spectrum detection fitting curve of PPC modified through compound physical method

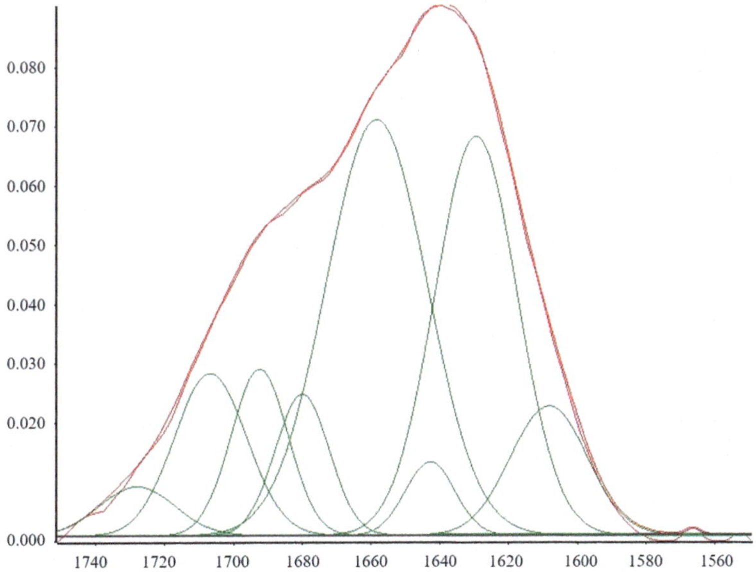

Fig. 6.101 Intermediate infrared spectrum detection fitting curve of PPC modified through restriction enzymolysis

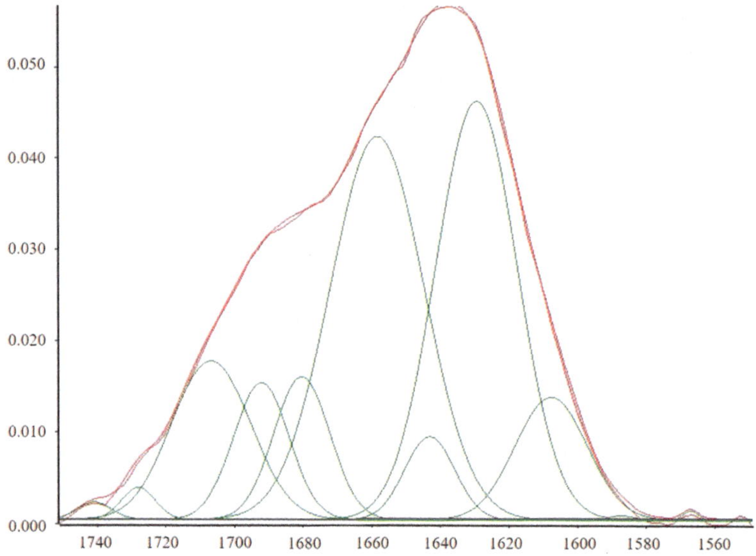

Fig. 6.102 Intermediate infrared spectrum detection fitting curve of PPC modified through high-pressure homogenization

there was little change in the structures of protein, and only a small proportion of β-sheet and β-turn structures were transformed into free coil structures which may be formed by the small molecular weight peptide fragments formed because high-pressure homogenization broke the peptide chains.

3.2.2 Evaluation of Functional Characteristics of Protein

1. Solubility and emulsibility of protein

The changes in the solubility and emulsification of peanut protein during the ethanol extraction and compound physical, restriction enzymolysis, and high-pressure homogenization modification are shown in Fig. 6.103. The experiment results showed that, in the ethanol extraction and modification process, the solubility of protein decreased only in the ethanol extraction step, and the solubility of peanut protein concentrate was improved effectively in all the compound physical modification, restriction enzymolysis modification, and high-pressure homogenization modification.

Emulsibility could usually be improved with the increase of solubility, and the protein with good solubility could spread and adsorb on the interface between the oil and water phases quickly, which could reduce the interfacial tension and enhance the emulsibility of protein. The gelation of highly hydrated interface protein membrane enhanced its surface viscosity and hardness, and thus the emulsion system was stabilized and the emulsibility was improved. The ethanol

Table 6.35 Changes of secondary structures of peanut protein powder, ethanol extracting peanut protein concentrate, and peanut protein concentrate through various modification treatments

Sample	Peak position/ cm^{-1}	Corresponding structure	Content of secondary structure/%
Raw material peanut protein powder	1633.94	β-sheet	50.7711
	1646.61	Free coil	0
	1654.03	α-helix	0
	1680.00	β-sheet	40.4431
	1693.91	β-turn	8.7858
Ethanol extracting peanut protein concentrate	1629.30	β-sheet	46.7638
	1641.66	Free coil	4.3834
	1660.21	β-turn	24.3099
	1678.76	β-sheet	11.5242
	1693.29	β-turn	13.0187
Peanut protein concentrate modified through compound physical method	1629.61	β-sheet	35.3149
	1642.90	Free coil	3.8449
	1659.29	β-turn	47.5556
	1678.14	β-sheet	5.8438
	1687.73	β-turn	5.6608
	1692.06	β-turn	1.7801
Peanut protein concentrate modified through restriction enzymolysis	1629.30	β-sheet	61.4416
	1642.59	Free coil	6.8177
	1680.00	β-sheet	14.2702
	1692.37	β-turn	17.4706
Peanut protein concentrate modified through high-pressure homogenization	1629.30	β-sheet	74.6097
	1636.72	β-sheet	0
	1642.90	Free coil	9.0500
	1692.06	β-turn	16.3403

extraction and various modification processes could improve the emulsification of peanut protein in different degrees, and the effects of ethanol extraction and compound physical modification were more significant. Different from the modification processes, the ethanol extraction reduced the solubility of protein and also improved its emulsification, which may be because the ethanol caused the change of protein conformation and reconstruction of molecular structure, especially the breakage of disulfide bonds would result in the increase of surface hydrophobicity groups, and thus the emulsification of protein would be improved. The improvement of emulsibility by the compound physical modification should be mainly caused by the ultrasonic treatment, the cavitation of ultrasonic wave could loosen the structures of protein molecules, so that the hydrophobic polypeptide part spread toward the lipid and the polar polypeptide part was directed toward the water phase, and thus the emulsibility of protein was enhanced. The improvement of emulsibility by the restriction enzymolysis modification was limited. Because the

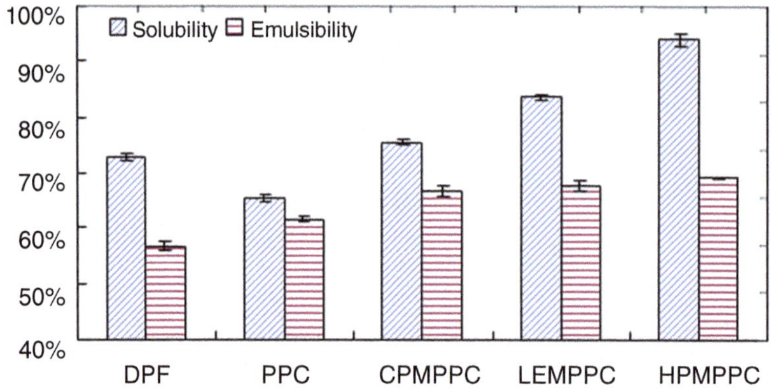

Fig. 6.103 Changes of solubility and emulsibility of peanut protein in the ethanol extraction and modification processes. Note: DPF, PPC, CPMPPC, LEMPPC, and HPMPPC mean raw material peanut protein powder, ethanol extracting peanut protein concentrate, peanut protein concentrate modified through compound physical method, peanut protein concentrate modified through restriction enzymolysis, and peanut protein concentrate modified through high-pressure homogenization, respectively, and they have the same meanings both in Figs. 6.102 and 6.103

enzymolysis process conditions were mainly optimized for the improvement of solubility, the improvement of emulsibility required a relatively larger enzyme dosage and longer reaction time. Some research showed that the emulsibility could be increased by nearly four times by using the papain to carry out enzymolysis for soybean protein concentrate.

The improvement of emulsibility by the high-pressure homogenization modification was also limited. This could be explained by the small change in the secondary structure of peanut protein concentrate before and after homogenization. Therefore, the spreading direction of hydrophobic polypeptide part could not be greatly changed by the homogenization, and thus it was difficult to enhance the emulsibility of protein significantly.

2. Water absorption and oil absorption of protein

The changes in water absorption and oil absorption of peanut protein in the ethanol extraction, compound physical modification, restriction enzymolysis modification, and high-pressure homogenization modification processes are shown in Fig. 6.104. The experiment results showed that the water absorption of protein was reduced significantly after the high-pressure homogenization modification, while it was improved by the ethanol extraction, compound physical modification, and restriction enzymolysis modification at a certain degree in the ethanol extraction and modification processes; the oil absorption of protein did not change in the restriction enzymolysis modification process, while it was improved by the ethanol extraction, compound physical modification, and high-pressure homogenization modification in different degrees.

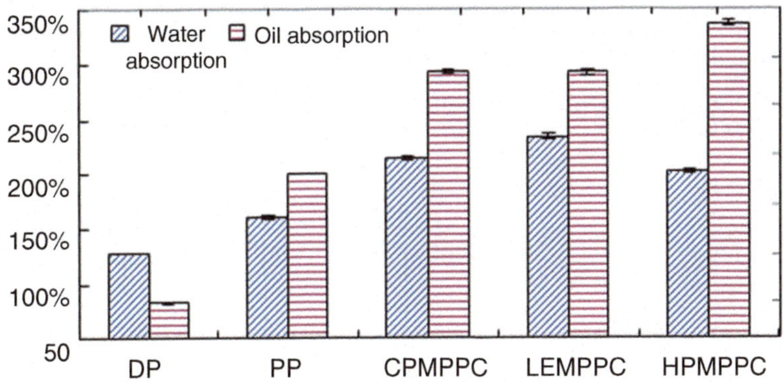

Fig. 6.104 Changes of water absorption and oil absorption of peanut protein in ethanol extraction and modification processes

After the ethanol extraction of raw material peanut protein, the change of conformation made the charged groups and charges on the surface of molecular change, the ethanol extraction made the protein aggregate, the space between the proteins which could hold the water molecule might increase, and thus the water absorption of protein would increase, while the hydrophobic groups exposed due to breakage of disulfide bonds made the oil absorption of protein increase. The compound physical modification made the tight structure of protein loose to a certain extent, the holding space inside the protein molecule was increased, the hydrophobic peptide chains spread to a certain extent, and thus the water absorption and oil absorption of protein were further enhanced. The restriction enzymolysis made the peptide bonds on the protein peptide chains be broken largely, a large number of hydrophilic groups -NH_2 and -COO^- were generated, and thus the water absorption of protein was enhanced. Although the space between the protein molecules decreased, there was little effect on the small water molecule. Enzymolysis could not only made the hydrophobic groups hidden internally exposed but also made the protein molecular weight decrease significantly, and thus the space which could hold oil molecules originally inside the molecule was destroyed, which might be the reason for no change of oil absorption of protein before and after enzymolysis.

The impacts of high-pressure homogenization treatment on the water absorption and oil absorption of protein were completely opposite, and the decrease of water absorption might be because the protein structure was dissolved into strips so that the small water molecules lost their holding space. While the slight increase of oil absorption might be because the further dissociation and opening of side chain buried inside the globular molecules of protein made the lipophilicity of protein enhance, the net framework formed by the strip structure of protein could also hold and support the molecular due to the large volume of oil molecule.

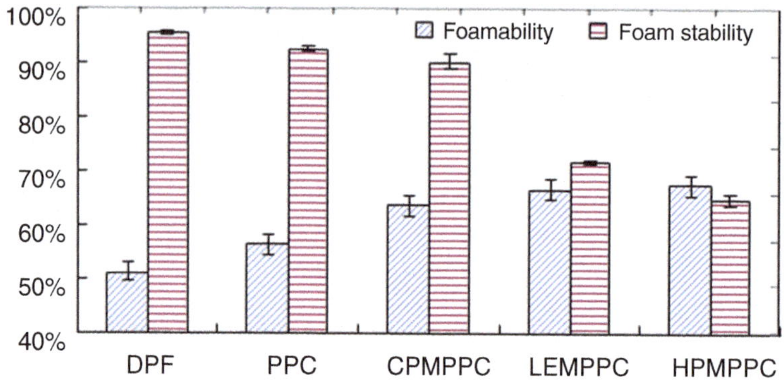

Fig. 6.105 Changes of foamability and foam stability of peanut protein in ethanol extraction and modification processes

3. Foamability and foam stability of protein

The changes of foamability and foam stability of peanut protein in the ethanol extraction and compound physical modification, restriction enzymolysis modification, and high-pressure homogenization modification processes are shown in Fig. 6.105. The experiment results showed that the foamability of protein increased with the decrease of foam stability in the ethanol extraction and modification processes. After the ethanol extraction of protein was carried out, some lipids and alcohol soluble substances which inhibited the formation of foam were removed (Wagner and Sorgentini 1996), so the foamability of peanut protein concentrate increased.

The compound physical modification allowed the hydrophobic peptide chain of protein to spread to a certain extent, and thus the foamability was increased. The decrease of foam stability might be because the decrease of the stability of the structure after the protein was extended resulted in the decrease of stability of protein membrane between the interfaces. When the enzymolysis degree was restricted, the number of hydrogen bonds in the protein molecular increased, and thus the foamability was enhanced to a certain degree. However, due to the significant decrease of the average molecular weight, the stability of the protein membrane between the interfaces would inevitably decrease and thus the foam stability would decrease (Were et al. 1997). The foamability was not increased by the high-pressure homogenization modification basically, because there were small changes of secondary structure of protein caused by homogenization treatment, and the hydrophobic peptide chains in the molecules were not further spread like that after the compound physical modification. The further decrease of foam stability was the same as that after the restriction enzymolysis, which was caused by the continued decrease of the average molecular weight.

4. Viscosity of protein

The factors that affected the apparent viscosity of protein solution included the apparent diameter of protein molecule and strength of interaction between protein molecules or interaction between protein molecules and water molecules. The

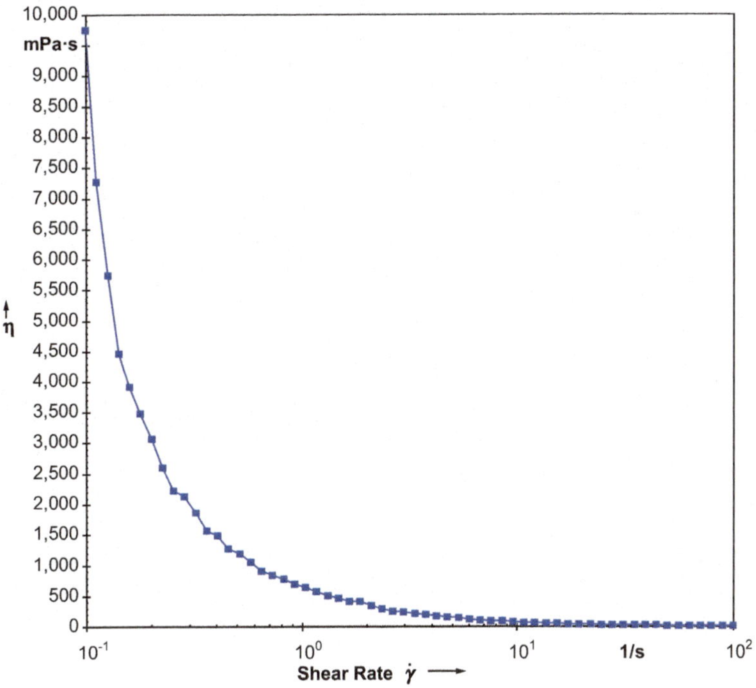

Fig. 6.106 Viscosity curves of peanut protein powder

viscosity curves of raw material protein powder, ethanol extracting peanut protein concentrate, peanut protein concentrate modified through compound physical method, peanut protein concentrate modified through restriction enzymolysis, and peanut protein concentrate modified through high-pressure homogenization are shown in Figs. 6.106, 6.107 6.108, 6.109, and 6.110. The experiment results showed that the viscosity value of peanut protein powder showed a waving trend of increased-decreased-increased-decreased through ethanol extraction, compound physical modification, restriction enzymolysis modification, and high-pressure homogenization modification processes. The decrease of viscosity of protein solution after ethanol extraction might be because the ethanol solution induced the changes of protein conformation, part of disulfide bonds and hydrogen bonds in the protein molecule were opened, this change was favorable to the increase of probability of collision between the peptide chains, and thus the viscosity of solution was increased. The compound physical modification caused the frequent collision, friction, and extrusion between the protein molecules and changed the acting forces between the protein molecules; the mechanical shear force, cavitation effect, and thermal denaturation caused by high-speed stirring and ultrasonic modification changed the structure of ethanol extracting peanut protein concentrate, made the protein molecules become orderly, and weakened the Brownian movement of molecules, and thus the solution viscosity was reduced (Feng and Xiong 2007).

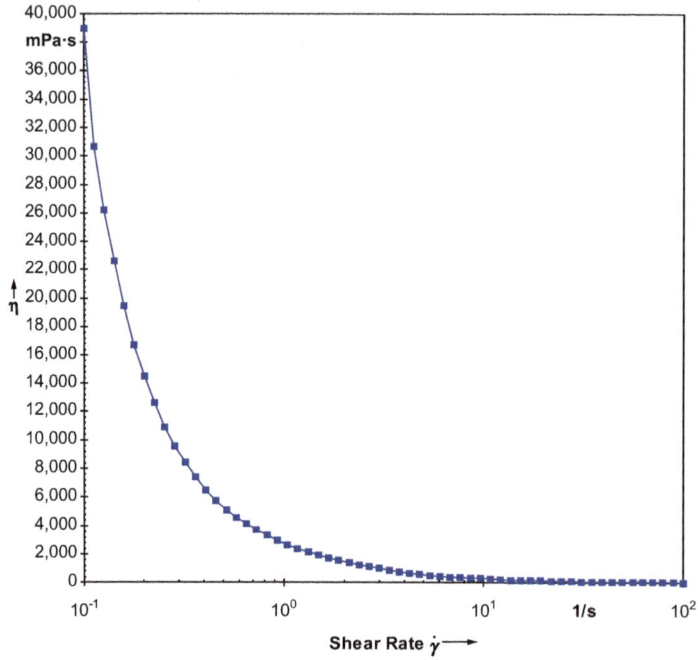

Fig. 6.107 Viscosity curves of ethanol extracting peanut protein concentrate

The improvement of viscosity of protein solution by the restriction enzymolysis was significant. Enzymolysis made the protein peptide chain shorten, the degree of freedom of short peptide chain with low steric hindrance in the water was large, and the capacity of short peptide chain to entangle with each other to form a net structure was weaker than that of long peptide chain. At the same time, the hydrophilic groups -NH_2 and -COO^- were formed due to the breakage of peptide bonds, which increased the degree of solvation of protein. Therefore, the viscosity of the product solution after enzymolysis should be lower than that before enzymolysis, and the higher the degree of hydrolysis, the lower the viscosity of product. But not all the enzymolysis could meet this rule. The viscosity of the product generated when the enzymolysis was carried out by the papain for the soybean protein would be enhanced with the increase of degree of hydrolysis (Zhong Fang 2001). In this experiment, the enzymolysis of papain for peanut protein was also the same, which was because the polypeptide chains in the hydrolysis products with strong interactions formed the polymers with large apparent diameter, or the hydrolysis products had extremely strong hydratability, and thus the viscosity of protein solution was increased. The viscosity of protein solution after high-pressure homogenization modification decreased greatly, which might be because the high-pressure homogeneous treatment destroyed the polymers with large apparent diameter formed between the polypeptide chains after enzymolysis, and thus the viscosity of protein solution was decreased. In addition, the reason for the decrease in the viscosity of product solution by the high-pressure homogenization modification was similar to that by the compound physical

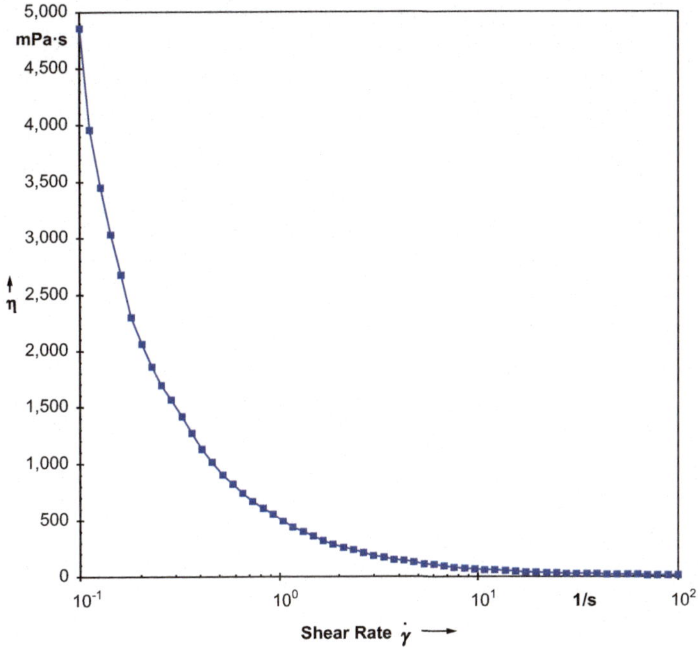

Fig. 6.108 Viscosity curves of PPC modified through compound physical method

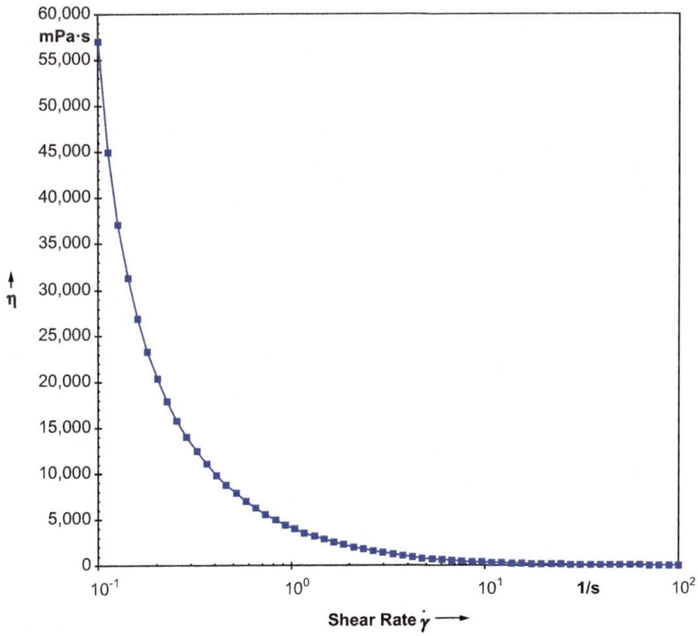

Fig. 6.109 Viscosity curves of PPC modified through restriction enzymolysis

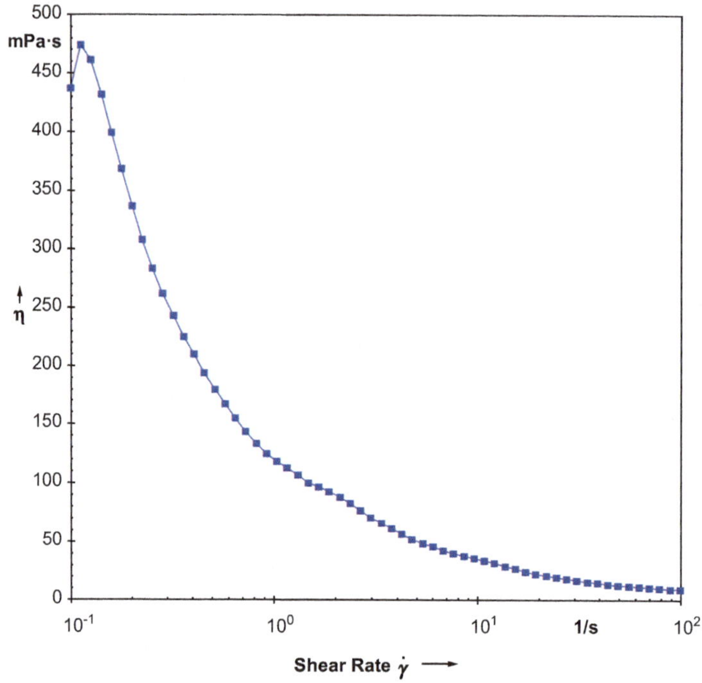

Fig. 6.110 Viscosity curves of PPC modified through high-pressure homogenization

method, the mechanical shear force, cavitation effect, and thermal denaturation caused by the high-pressure homogenization modification that would also weaken the Brownian movement of molecule and thus further reduce the viscosity of solution.

The research on the structural characteristics and functional properties of protein not only provided a theoretical basis for the preparation of functional protein bases by using the ethanol extracting peanut protein concentrate but also provided a solid basis for expanding the application of peanut protein in the food industry.

References

Achouri A, Zhang W, Shiying X. Enzymatic hydrolysis of soy protein isolate and effect of succinylation on the functional properties of resulting protein hydrolysates. Food Res Int. 1998;31(9):617–23.

Aminlari M, Ferrier LK, Nelson AI. Protein dispersibility of spray-dried whole soybean milk base: effect of processing variables. J Food Sci. 1977;42(4):985–8.

Amin NAS, Anggoro DD. Optimization of direct conversion of methane to liquid fuels over Cu loaded W/ZSM-5 catalyst. Fuel. 2004;83(4-5):487–94.

Arnfield SD, Murray ED, Ismond MAH. Effect of salt on the thermal stability of storage proteins from fababean(ViciaFaba) [J]. J Food Sci. 1986;51:371–7.

Arnfield SD, Murray ED, Ismond MAH. Dependence of thermal properties as well as network microstructure and rheology on protein concentration for ovalbumin and vicilin [J]. J Texture Stud. 1990a;21:191–212.

Arnfield SD, Murray ED, Ismond MAH. Influence of salts on the microstructural and rheological properties of heat-induced protein networks from ovalbumin and vicilin [J]. J Agric Food Chem. 1990b;38:1335–43.

Babajimopoulos M, Damodaran S, Rizvi SS, et al. Effects of various anions on the rheological and gelling behavior of soy proteins: thermodynamic observations [J]. J Agric Food Chem. 1983;31 (6):1270–6.

Batista AP, Portugal CAM, Sousa I, Crespo JG, Raymundo A. Accessing gelling ability of vegetable proteins using rheological and fluorescence techniques. Int J Biol Macromol. 2005;36(3):135–43.

Byler DM, Susi H. Examination of the secondary structure of proteins by deconvolved FTIR spectra [J]. Biopolymers. 1986;25(3):469–87.

Daubert CR, Foegeding EA. Rheological principles for food analysis. In: Nielsen SS, editor. Chemical analysis of foods [M]. Boston: Jones and Bartlett; 1998.

Feng LL, Xiong J. The study on viscosity and dynamic modulus of soy protein concentrate [J]. Food Ferment Ind. 2007;33(12):36–9.

Friedman M, Grosjean OK, Zahnley JC. Inactivation of soya bean trypsin inhibitors by thiols [J]. J Sci Food Agric. 1982;33:165–72.

Fukushima D, Van Buren J. Mechanisms of protein insolubilization during the drying of soy milk. role of disulfide and hydrophobic bonds[J]. Cereal Chem. 1970;47:687–96.

Huang G. Researches on Zein about extraction by ultrasound , modification and medicinal release performance[D]. Guangzhou: South China University of Technology; 2004.

Jackson M, Mantsch HH. The use and misuse of FT-IR spectroscopy in the determination of protein structure [J]. Crit Rev Biochem Mol Biol. 1995;30:95–120.

Jambrak AR, Mason TJ, Lelas V, Herceg Z, Herceg IL. Effect of ultrasound treatment on solubility and foaming properties of whey protein suspensions. J Food Eng. 2008;86(2):281–7.

Jian J, Weiyuan L. Polymer structure performance and testing. Beijing: Chemical Industry Press; 2003.

Karren G, Carrasquillo M. Structure-guided encapsulation of model proteins into biocompatible polymers [D]: Dissertation of university of Puerto Rico; 2001.

Krimm S, Bandekar J. Vibrational analysis of peptides polypeptides, and proteins. V. Normal vibrations of beta -turns [J]. Biopolymers. 1980;19(1):1–3.

Lakemond C. Gelation of soy glycinin; influence of pH and ionic strength on network structure in relation to protein conformation. Food Hydrocoll. 2003;17(3):365–77.

Liompart MP, Lorenzo RA, Cela R, et al. Evaluation of supercritical fluid extraction, microwave~assisted extraction and sonication in the determination of some phenolic compounds fron various soil matrices [J]. J Chromatogr A. 2007;74(1-2):243–51.

Ma T. Study on preparation and solubility of peanut protein concentrate[D]. Beijing: Chinese Academy of Agricultural Sciences; 2009.

MacRitchie F. Mechanical degradation of gluten proteins during high-speed mixing of doughs. J Polym Sci Polym Symp. 1975;49(1):85–90.

Mason TJL, Paniwnyk JP. The uses of ultrasound in food technology [J]. Ultrasound Sonochem. 1996;3:253–60.

Matsumura Y, Mori T. Gelation. In: Hall GM, editor. Methods of testing protein functionality: Blackie Academic and Professional; 1996. p. 76–109.

Mecham DK. Changes in flour protein during mixing [J]. Cereal Sci Today. 1968;13:371–3.

Meng G-T, Ma C-Y. Thermal properties of Phaseolus angularis (red bean) globulin. Food Chem. 2001;73(4):453–60.

Nakai S, Li CE. Hydrophobic interact ions in food systems. Boca Raton: CRC Press; 1988. p. 87–104.

Ng PKW, Xu C, Bushuk W. Model of glutenin structure based on farinograph and electrophoretic results [J]. Cereal Chem. 1992;68:321–2.

Nishinari K, Kohyama K, Zhang Y, Kitamura K, Sugimoto T, Saio K, Kawamura Y. Rheological study on the effect of the A5 subunit on the gelation characteristics of soybean proteins. Agric Biol Chem. 1991;55(2):351–5.

Petruccelli S, Anon MC. Relationship between the method of obtention and structural and functional properties of soy protein isolates. 2. Surface properties [J]. J Agric Food Chem. 1994;42(10):2170–6.
Renkema JMS. Relations between rheological properties and network structure of soy protein gels. Food Hydrocoll. 2004;18(1):39–47.
Rooney JA. Other non-linear acoustic phenomena. In: Suslick KS, editor. Ultrasound: its chemical, physical, and biological effects [M]. New York: VCH. p. 1988.
Sanchez AC, Burgos J. Factors affecting the gelation properties of hydrolyzed sunflower proteins. J Food Sci. 1997;62(2):284–8.
Sharma V, Srinivas VR, Surolia A. Cloning and sequencing of winged bean ( ) basic agglutinin (WBA I): presence of second glycosylation site and its implications in quaternary structure. FEBS Lett. 1996;389(3):289–92.
Singh NK, Donovan GR, Batey IL, et al. Use of sonication and size-exclusion HPLC in the study of wheat flour protein. I. Dissolution of total protein in unreduced form [J]. Cereal Chem. 1990;67:150–61.
Singh H, MacRitchie F. Use of sonication to probe wheat gluten structure. Cereal Chem J. 2001; 78(5):526–9.
Susi H, Byler DM. Resolution-enhanced Fourier transform infrared spectroscopy of enzymes [J]. Methods Enzymol. 1986;130:290–311.
Suslick KS. Sonochemistry. Science. 1990;247(4949):1439–45.
Tanaka K, Bushuk W. Changes in flour protein during dough mixing. III Analytical results and mechanisms [J]. Cereal Chem. 1973;50:605–12.
Tao WS, Li W, Jiang YM. The basis of protein molecular [M]. Beijing: Higher Education Press; 1995.
Wagner JR, Sorgentini DA. Thermal and electrophoretic behavior, hydrophobicity, and some functional properties of acid-treated soy isolates. J Agric Food Chem. 1996;44(7):1881–9.
Wang Z, et al. Food chemistry [M]. Beijing: China Light Industry Press; 1991.
Wei YN, Chen YF, Chen GH, et al. Study on subunit composition and molecular weight of peanut proteins [J]. J Guangxi Norm Univ. 1998;(01):73–80.
Were L, Hettiarachchy NS, Kalapathy U. Modified soy proteins with improved foaming and water hydration properties. J Food Sci. 1997;62(4):821–4.
Wu HW, Wang Q, Ma TZ, et al. Comparative studies on the functional properties of various protein concentrates of peanut protein [J]. Food Res Int. 2009;42:343–8.
Yamada T. Dissociation-association behavior of arachin between dimeric and monomeric forms [J]. Agric Biol Chem. 1979;43(12):2549–56.
Yang XQ, Chen Z, Zhao MM. Separation and characterization of peanut proteins [J]. J Chin Cereals Oils Assoc. 2001;16(5):25–8.
Yang SF, Xu K, Zhang ZS. Study on the mechanism of high shear and high-pressure and its application on food industry [J]. Mach Cereals Oil Food Process. 2002;4:33–5.
Yufei H. Physical modification of alcohol leaching soy protein concentrate [D]. Jiangsu: Jiangnan University; 1993.
Zhang T. Study on preparation of chickpea protein isolates and their functional properties [D]. Wuxi: Jiangnan University; 2006.
Zhang T, Jiang B, Wang Z. Gelation properties of chickpea protein isolates [J]. Food Sci. 2006;27(8):108–13.
Zhang C. Peanut protein extrusion texturisation technics and mechanisms [D]. Beijing: Chinese Academy of Agricultural Sciences; 2007.
Zheng B, Matsumura Y, Mori T. Relationship between the thermal denaturation and gelling properties of legumin from broad beans [J]. Biosci Biotechnol Biochem. 1993;57:1087–90.
Zhong Fang. Preparation of bean curd flower and velocity coagulation mechanism of soybean protein [D]. Wuxi: Jiangnan University; 2001.
Zhou H-Y, Liu C-Z. Microwave-assisted extraction of solanesol from tobacco leaves. J Chromatogr A. 2006;1129(1):135–9.

Chapter 7
Improvement of Gelation of Peanut Protein Isolate

Peanut protein isolate (PPI) has a variety of functional characteristics such as emulsibility, water-binding capacity, gelation, solubility, foamability, and film-forming property, so it can be used in meat products, aquatic products, baked food, dairy products, beverages, ice cream, candy, and other food treatment areas. Among them, as one of the major functional characteristics of PPI, gelation can make the protein have high viscosity, plasticity, and springiness, so that the gel formed by protein can be used as the carrier for water, flavor agents, sugar, and other compounds. Over the years, the study in this field has been the focus of food treatment research, and a majority of scholars have attached much attention to it. In order to further improve the gelation of PPI and expand its applications in the field of food treatment, modification method is often used for improvement. Physical modification (microwave, ultrahigh pressure, and other technologies) is simple, safe, and fast with low treatment costs, and it is able to maintain the nutritional value of protein to the largest extent, so it has gained popularity from the majority of scholars. There are a large number of study reports on the physical modification of peanut protein, but the study progress is slow due to the complexity of the globular structure of peanut protein. Although the functional characteristics of peanut protein can be improved in different degrees, the specific physical modification for gelation and its mechanism and other problems have not been described clearly, and thus the use of gelation of peanut protein has not been really industrialized. Therefore, it is urgent to deeply study the new technology, new methods, and related basic theories of the physical modification of peanut protein. High-pressure technology has been widely used in the modification studies of food protein (Messens et al. 1997; Wang et al. 2008; Apichartsrangkoon 2003; Molina et al. 2001, 2002; Puppo et al. 2004, 2005; Hongkang Zhang et al. 2003). Studies showed that the emulsibility, solubility, gelation, and other functional characteristics of soy protein, soy protein components, whey protein, and other protein might change due to high pressure. There are few studies on the ultrahigh-pressure modification of the functional characteristics of peanut protein, and the change laws of functional characteristics after modification and its structure-function

Q. Wang, *Peanut Processing Characteristics and Quality Evaluation*,
DOI 10.1007/978-981-10-6175-2_7

relationship are not clear, and there is no ultrahigh-pressure modification technology suitable for deep treatment. Therefore, the author's study team took the ultrahigh-pressure technology as the supporting point to modify the PPI and reveal the change law of its gelation and its structure-function relationship, so as to provide theoretical guidance to the deep treatment and utilization of peanut protein.

1 Ultrahigh-Pressure Modification

In the meat industry, especially in the production of ham, in order to reduce the cost while maintaining the sensory and nutritional quality of ham, a certain amount of soy protein can be added, which mainly makes use of the gelation, water-binding capacity, oil-binding capacity, and other functional characteristics of soy protein. Compared with soy protein, peanut protein has more advantages: firstly, peanut protein is a kind of plant protein with high nutritional value, so its nutritional value is similar to that of animal protein, and it does not contain cholesterol; secondly, peanut protein contains a large number of essential amino acids that can be absorbed by human bodies easily and has a seductive fragrance and white color. However, there is still a certain gap in the gelation, water-binding capacity, oil-binding capacity, and other functional characteristics between natural peanut protein and soybean protein, so the applications of natural peanut protein in meat products are limited. As we all know, ultrahigh-pressure technology is a physical process, the non-covalent bond of food ingredients will be destroyed or formed, and the enzymes, proteins, and other biological macromolecules will lose their activity and have denaturation, and the color, flavor, and taste deterioration of food and nutrient loss and other disadvantages caused by traditional hot treatment after the high-pressure treatment of food materials will be avoided (Murchie et al. 2005). There are also reports on increasing the solubility and emulsification of PPI by ultrahigh-pressure modification (Zong Wei and Chen Yi-ping 2007, 2008). In order to obtain high-gelatinized PPI, the author's team studied the change law of gelation of PPI under different ultrahigh-pressure treatment conditions, and the optimal ultrahigh-pressure treatment technology was obtained by response surface design on this basis, so as to provide a certain theoretical basis for the applications of PPI in the applications of ham and sausage.

Add deionized water to peanut protein powder at the ratio of 1:10, adjust the PH value to be 9.0 using 0.2 M NaOH solution, stir it for 2 h for extraction at room temperature (25 °C), and then centrifuge it for 10 min at 4200 r/min; adjust the PH value of supernate to about 4.5, centrifuge it again, remove the supernate, stir it evenly after adding a small amount of water, and then freeze and dry it for use. The protein content of sample was $86.46 \pm 0.09\%$ (protein conversion factor was 5.46), ash was $3.35 \pm 0.25\%$, water content was $4.12 \pm 0.13\%$, crude fat content was $0.60 \pm 0.08\%$, and total sugar content was $5.47 \pm 0.10\%$.

1.1 Single-Factor Test

The high-pressure treatment method has been improved on the basis of the report of Cong-Gui Chen et al. (2006). After preparing the protein sample into a certain mass concentration (w/v) and stirring it evenly, put it into the polypropylene vacuum bag, evacuate the bubbles (prevent bag breakage during high-pressure treatment), and then heat-seal the package under vacuum conditions. Check the sealing status of the high-pressure equipment. Place the vacuum bag with samples into the sample container of high-pressure equipment, set the pressure and pressure-maintaining time (pressure fluctuation range of ± 10 MPa) from the computer, take it out after treatment, and then freeze and dry it for use. Prepare the solution with a mass concentration of 14% from the frozen and dried PPI using deionized water, heat it for 1 h at 95 °C, quickly cool it after taking it out, and determine the gel hardness, springiness, and cohesiveness using TA-TX2i texture instrument (probe diameter of 12 mm) after placing it for 24 h at 4 °C. Slightly modify it by the methods used by Pinterits and Arntfield (2008). Operating mode: TPA; pretest speed, 2.0 mm/s; test speed, 0.8 mm/s; pressing distance, 50%; posttest speed, 0.8 mm/s; data acquisition rate, 200/s. Gel hardness (g) = peak force in the first compression (Force 2); springiness = the ratio of the time taken from starting the second compression to ending the compression and the time taken from starting the first compression to ending the compression (time diff 4:5/time diff 1:2); cohesiveness = ratio of the areas under the positive peaks in the first and second compressions.

1. Pressure and time

The effects of different pressures and times on the curdlan hardness, springiness, and cohesiveness of PPI are shown in Figs. 7.1 and 7.2. It was shown from Fig. 7.1 that the hardness of curdlan formed by PPI increased first and then decreased with the increase of pressure. When the pressure was 100 MPa, the maximum hardness of gel formed by PPI was 172.52 g. Compared with other pressure levels, there was a significant difference ($p < 0.05$); when the pressure exceeded 100 Mpa, the hardness decreased sharply; it was shown from the impact of different pressures on gel springiness, within the range from 50 to 120 Mpa, that the gel springiness was high and larger than 0.7, but the change was not significant ($p > 0.05$), and the gel springiness decreased when the pressure exceeded 120 Mpa. It was shown from the figure that with the increase of pressure, the cohesiveness first increased and then decreased slightly and its value changed from 0.33 to 0.38, which was not significant ($p > 0.05$). The natural structures of most proteins were generally stabilized by the interaction of some non-covalent bonds. The high-pressure treatment might destroy the non-covalent bond balance of the protein, which would force the original structure of the protein to extend, and hydrogen bond, hydrophobic bond, ionic bond, and other non-covalent bonds changed, thus leading to the change in the properties of protein (Angsupanich et al. 1999). It is the reason why ultrahigh pressure can change the PPI curdlan. Another study showed that with the increase of pressure, the more complete the protein denaturation was, the more dense and

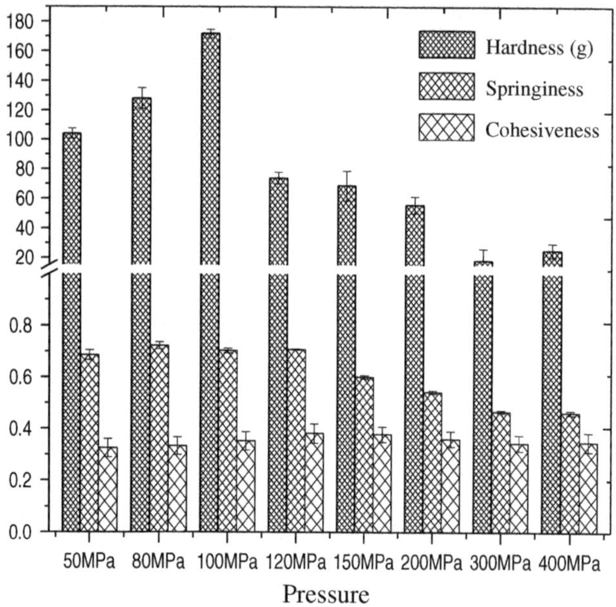

Fig. 7.1 Impact of pressure on the PPI curdlan

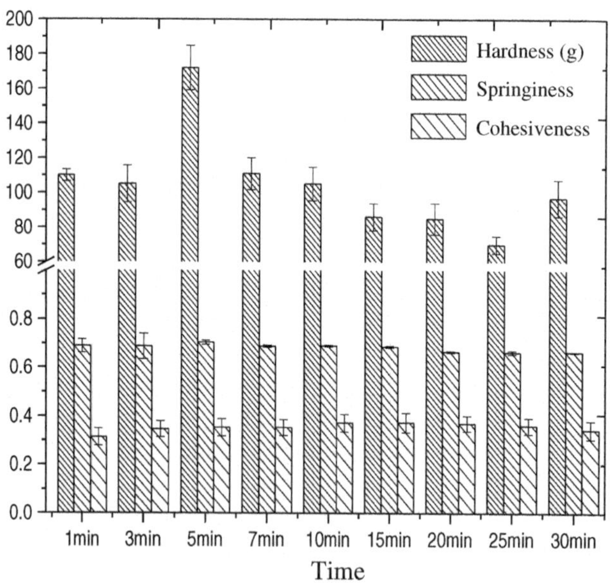

Fig. 7.2 Impact of treatment time on the PPI curdlan

fine the gel network formed by it, which might lead to the increase of gel hardness; however, after the pressure increased to a certain extent, the protein aggregation might occur, and its gelation would decrease. In this study, when the pressure was greater than 100 MPa, with the increase in pressure, protein aggregation might gradually increase, resulting in the significant decrease of hardness of curdlan formed.

It was shown from Fig. 7.2 that the effect of different treatment time on gel hardness was significant ($p < 0.05$), and the gel hardness value was highest after it had been processed for 5 min, and the hardness of PPI significantly reduced over time. With the gradual increase of time, the springiness and cohesiveness PPI gel first increased and then decreased slightly; the springiness reached the maximum value after it had been processed for 5 min, being 0.704; the cohesiveness reached the maximum value after it had been processed for 15 min, being 0.374. The ultrahigh-pressure treatment can enhance the aggregation of protein subunits; however, after the protein polymerization reached a certain extent, the effect on increasing protein polymerization was not obvious after continuing to increase the treatment time, and the dense and orderly gel network might be damaged and aggregation might be caused, and thus its gel hardness would significantly be reduced.

2. Protein concentration and pH

The effects of different protein concentrations and pH on the curdlan of PPI are shown in Figs. 7.3 and 7.4. It was shown in Fig. 7.3 that the gel hardness increased first and then decreased with the increase of the concentration of peanut protein, the gel hardness reached the maximum value when the concentration was 5%, and the gel hardness gradually decreased with the gradual increase of concentration; it was shown from the figure that with the gradual increase of concentration, the springiness and cohesiveness of PPI first increased and then decreased. The springiness reached the maximum value when the protein concentration was 10%, being 0.796; the cohesiveness reached the maximum value when the protein concentration was 15%, being 0.402; the protein concentration continued to increase to 20%; and cohesiveness dropped to 0.372. Under a certain pressure, with the increase of protein concentration, the nonbinding liquid of protein solution with high concentration decreased, and the strength of gel formed was higher than that of protein solution with a low concentration; when the concentration was very large, the gel was difficult to form a dense and orderly network structure, so the strength would be reduced (Briscoe et al. 2002; Gosal and Ross-Murphy 2005).

It was shown from Fig. 7.4 that the pH value had a great effect on the strength and springiness of the gel. The gel hardness reached the maximum value when the PH was 4.2, being 172.52 g, decreased to 10.56 g when pH was 6.8, and then increased to 46.33 g when pH was 8.3; however, it decreased to 7.67 g when pH increased to 10.3; the gel springiness was good when pH was 4.2, decreased to 0.60 when pH increased to 6.8, reached the maximum value of 0.78 when pH was 8.3, and decreased to 0.59 when pH was 10.3. The change trend of cohesiveness was completely different from that of hardness and springiness. With the gradual

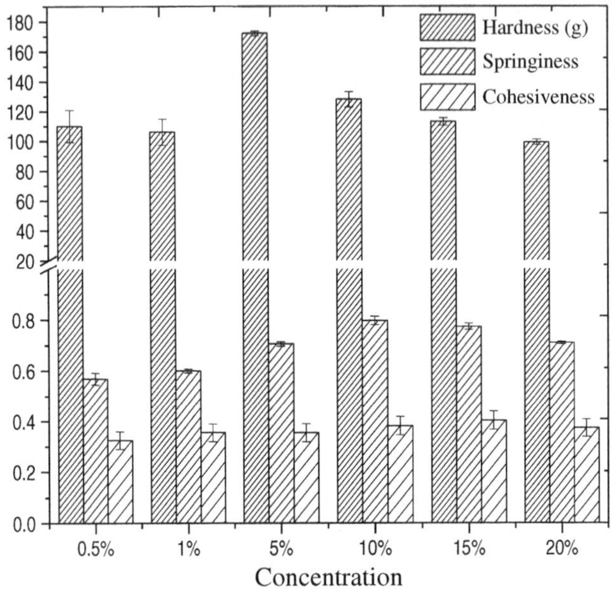

Fig. 7.3 Impact of protein concentration on the PPI curdlan

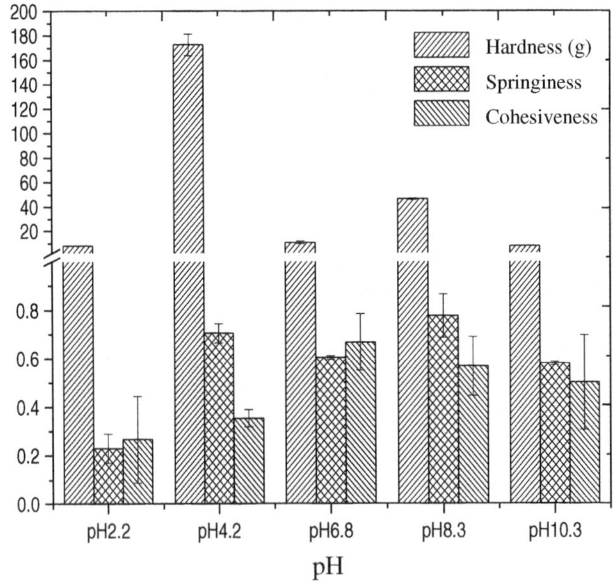

Fig. 7.4 Impact of pH on the PPI curdlan

increase of pH value, the cohesiveness of PPI increased first and then decreased and reached the maximum value at neutral pH conditions, being 0.667. It was shown from the impacts of pH on the hardness, springiness, and cohesiveness of PPI gel that it was more favorable for the formation of curdlan when it was in the vicinity of the isoelectric point, because when pH was close to the isoelectric point, the attractive force was greater than repulsive force, the protein conformation was compact, and a compact gel network was formed; when pH was at the non-isoelectric point, the force between the molecules was repulsive, the spatial conformation was relatively loose, and the structural strength of formed gel network was low. It was shown from the experimental results that the curdlan formed by PPI had good hardness and springiness at the same time when the pH value was 4.2. Therefore, optimization would be conducted when the pH value was 4.2 in the latter processes.

From the above single-factor results, the impact of ultrahigh pressure on the springiness and cohesiveness of PPI curdlan was not large, being within the variation ranges of 0.23–0.78 and 0.26–0.67, respectively. Ultrahigh pressure had a large impact on its hardness, being within the variation range of 18–172.52 g, so this book will optimize the formation process of PPI curdlan by taking gel hardness as an indicator.

1.2 Box-Behnken Combination Design Test

According to the results of single-factor test, Box-Behnken combination design was conducted to pressure, time, and protein by taking gel hardness as the evaluation index (Table 7.1) (Xu 1997) to further evaluate the interaction of pressure, time, and protein concentration on the gel hardness and optimize its process. Test was conducted according to Box-Behnken combination design program; see Table 7.2 for test combination and results.

According to the results of Table 7.2, the regression coefficients were calculated to establish the mathematical regression model between PPI curdlan hardness and the three factors of pressure, time, and concentration.

X_1, X_2, and X_3 represent the pressure, time, and concentration, respectively. The variance analysis of test results is shown in Table 7.3.

The variance analysis significance test results showed that the regression coefficient of the model was $P = 0.000918$ ($p < 0.01$), indicating that the regression was significant, and $R^2 = 0.9771$ in the model, indicating that the model was in good agreement with the actual test, and there was a significant linear relationship between independent variable and response value, and it could be predicted using the theory of PPI curdlan test after ultrahigh-pressure treatment. After removing the nonsignificant items at the significance level of $p = 0.05$, the optimized equation is

Table 7.1 Box-Behnken combination design test level

Canonical variate Z	Pressure (X_1)/MPa	Time (X_2) min)	Concentration (X_3)/%
Upper level (1)	200	10	10
Zero level (0)	100	5	5
Lower level (−1)	50	3	1

Table 7.2 Experimental design and results

Treatment number	X_1 pressure (MPa)	X_2 time (min)	X_3 concentration (%)	Hardness (g)
1	−1(50)	−1(3)	0(5)	84.60
2	−1(50)	1(10)	0(5)	95.95
3	1(200)	−1(3)	0(5)	35.12
4	1(200)	1(10)	0(5)	62.26
5	0(100)	−1(3)	−1(1)	93.56
6	0(100)	−1(3)	1(10)	141.43
7	0(100)	1(10)	−1(1)	101.28
8	0(100)	1(10)	1(10)	126.43
9	−1(50)	0(5)	−1(1)	104.70
10	1(200)	0(5)	−1(1)	34.61
11	−1(50)	0(5)	1(10)	131.94
12	1(200)	0(5)	1(10)	71.79
13	0(100)	0(5)	0(15)	173.20
14	0(100)	0(5)	0(5)	169.24
15	0(100)	0(5)	0(5)	176.45

$$\begin{aligned} Y = {} & 172.9633 - 26.67625X_1 + 17.18X_3 - 66.69792X_1X_1 \\ & - 36.782928X_2X_2 - 20.5054X_3X_3 \end{aligned} \tag{7.1}$$

1. Interaction Effect Between Test Factors

The response surfaces of interactions of treatment time and protein concentration, treatment pressure and protein concentration, as well as pressure and time on the hardness of PPI curdlan are shown in Figs. 7.5, 7.6, and 7.7. It was shown from the figures that when the pressure, time, and concentration were low, the hardness of PPI curdlan increased with the increase of pressure, time, and concentration; when the pressure, time, and concentration were high, the hardness of PPI curdlan decreased with the increase of pressure, time, and concentration. When the pressure, time, and concentration were close to the central values, the hardness of curdlan reached the maximum value; when the factors deviated from the central values, the farther the deviation distance was, the lower the hardness of curdlan was, and the factors have secondary effect on the response value.

Table 7.3 Regression analysis results

Source	DF	SS	MS	F	Pr > F	Significance analysis
X_1	1	5692.979	5692.979	49.57911	0.000892	**
X_2	1	121.758	121.758	1.060368	0.350344	
X_3	1	2361.219	2361.219	20.56343	0.006199	**
X_1*X_1	1	16425.64	16425.64	143.0479	0.0001	**
X_1*X_2	1	62.33102	62.33102	0.542829	0.494358	
X_1*X_3	1	24.7009	24.7009	0.215116	0.662278	
X_2*X_2	1	4995.629	4995.629	43.50602	0.001204	**
X_2*X_3	1	129.0496	129.0496	1.123869	0.337589	
X_3*X_3	1	1552.512	1552.512	13.52055	0.014338	*
Model	9	29180.95	3242.328	F1 = 28.23684	0.000918	**
Error	5	574.1308	114.8262			
(Lack of fit)	3	548.0547	182.6849	F2 = 14.01169	0.067348	not significant
(Pure error)	2	26.07607	13.03803			
Total	14	29755.08				

*Represents significance at the level of 0.05, and **represents the significance at the level of 0.01

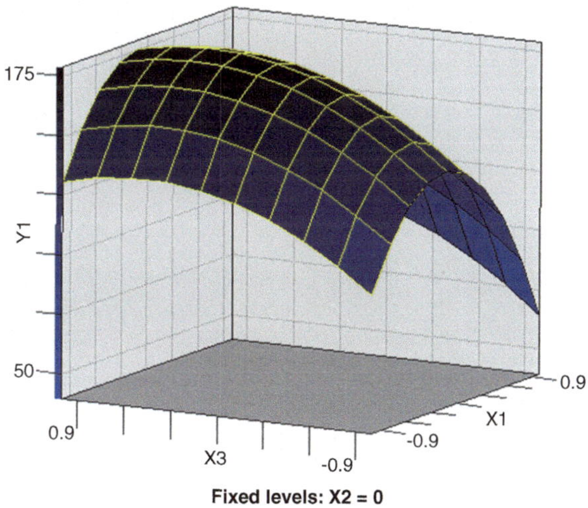

Fig. 7.5 Interaction of pressure and protein concentration on the hardness of PPI curdlan

2. Importance of factors

Analyze the importance of factors by contribution rate method:

$$\delta = \begin{cases} 0 & \text{If}\, F \leq 1 \\ 1 - 1/F & \text{If}\, F \leq 1 \end{cases} \tag{7.2}$$

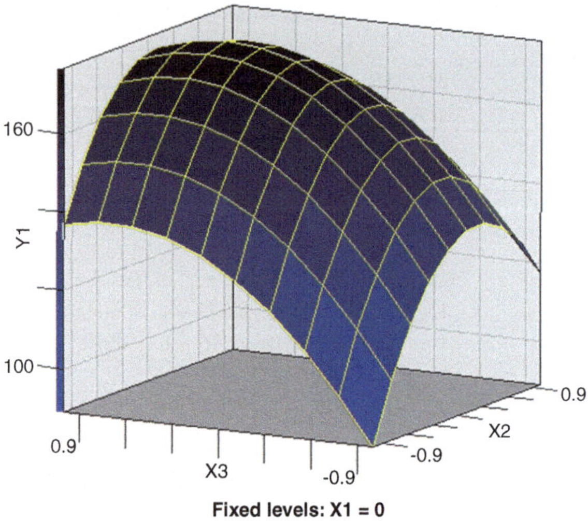

Fig. 7.6 Interaction of time and protein concentration on the hardness of PPI curdlan

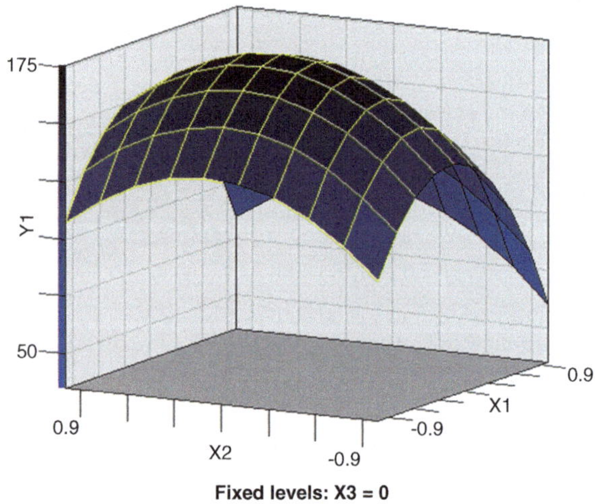

Fig. 7.7 Interaction of pressure and time on the hardness of PPI curdlan

Assume that the mean square ratio of significance test of regression coefficients are $F(j)$, $F(ij)$, and $F(jj)$.

The contribution rate of each factor to the index was expressed as

$$\Delta_j = \delta_j + \frac{1}{2} \sum_{\substack{i=1 \\ i=j}}^{m} \delta_{ij} + \delta_{jj}, j = 1, 2, \ldots\ldots, m \tag{7.3}$$

where δ_j represented the standard conversion value of the F value of the j-th factor, δ_{ij} represented the standard conversion value of the F value of interaction between the i-th factor and the j-th factor, δ_{jj} represented the standard conversion value of the F value of the quadratic term of the j-th factor, and Δ_j represented the contribution size of the j-th factor to the extracted results.

According to the mean square ratio F in Table 7.3, it was shown that the contribution rates of pressure, time, and protein concentration to the hardness of gel were $\Delta_1 = 1.9728$, $\Delta_2 = 1.0891$, and $\Delta_3 = 1.9325$, respectively, because $\Delta_1 > \Delta_3 > \Delta_2$, the sizes of effects on the hardness of gel, were as follows: pressure> concentration of PPI > time.

3. Ultrahigh-Pressure Conditions

According to the mathematical analysis of regression model, the optimum technological parameters of PPI curdlan after ultrahigh-pressure treatment were pressure of 115 MPa, time of 5 min, and PPI concentration of 3.11%. The hardness of PPI gel was 176.96 g under this process condition. In order to further test the reliability of the response surface analysis method, the above optimal conditions were used for the ultrahigh-pressure treatment. The measured hardness of PPI gel was 174.37 g, and its relative error was small compared to the theoretical value, being 1.46%. Therefore, the ultrahigh-pressure condition parameters obtained by analysis and optimization using response surface were accurate and reliable, and they could be used for practical operation.

2 Physicochemical Characteristics

PPI is a kind of plant protein featuring rich sources, low price, high quality, and complete nutrition. Although natural PPI has a variety of functional characteristics, it cannot fully meet the actual production needs during the actual applications. In addition, when using PPI in the food industry as a food additive, it often needs to use several functional characteristics of protein. For example, during the treatment to add PPI to ham, not only the gelation of PPI is used, but also good water-binding capacity and oil-binding capacity of PPI are required. It is necessary to further study whether the changes in the water-binding capacity, oil-binding capacity, grain size, amino acid type, and content of PPI can be obtained properly after gaining the PPI with good gelation by ultrahigh-pressure physical method. Because of the action of a large number of hydrogen bonds, hydrophobic bonds, van der Waals force, ionic bonds, and coordination bonds in peanut protein molecules, the emulsification and solubility of peanut protein can be improved properly under a certain pressure. After making clear the change law of PPI gel characteristics during the ultrahigh-pressure treatment, this book has systematically studied other physicochemical characteristics of PPI, to provide a theoretical support to the actual applications of PPI.

2.1 *Water-Binding Capacity and Oil-Binding Capacity*

The water-binding capacity of protein is an indicator to evaluate the ability of protein products to absorb water. The water-binding capacity of protein plays a major role in the texture of the food, especially meat and baked dough; insoluble protein may cause swelling and result in the changes in volume after absorbing water. In addition, the rheological characteristics of the system may be affected (Zhou 1999). The oil-binding capacity of protein refers to the capacity of protein to absorb oil, and it plays a very important role in the formula and treatment of meat products, dairy products, sandwich biscuit, and other foods. For example, in chopped meat products, it can reduce the weight loss during cooking and help to maintain the stability of the shape (Maruyama et al. 1998).

The change laws of water-binding capacity and oil-binding capacity of PPI after ultrahigh-pressure treatment at different pressures, times, and protein concentrations are shown in Figs. 7.8, 7.9, and 7.10. It was shown from Fig. 7.8 that the water-binding capacity and oil-binding capacity of PPI after pressure treatment significantly increased ($p < 0.05$). With the increase of pressure, the water-binding capacity and oil-binding capacity increased gradually; under different pressure (50–200 MPa) treatment conditions, the oil-binding capacities of PPI increased by 45.98%, 43.65%, 50.01%, 50.31%, and 59.66%, respectively, and the water-binding capacities increased by 1.48%, 2.09%, 7.33%, 9.51%, and 29.46%, respectively. It was shown from the figure that the improvement of oil-binding capacity was more obvious than that of water-binding capacity. The water-binding capacity and oil-binding capacity of protein solution with a certain concentration gradually increased with the increase of pressure after ultrahigh-pressure treatment. This was because the high-pressure treatment might destroy the non-covalent bond balance of the protein, force globular protein molecules to extend, and expose the disulfide group, hydrophobic group, and other functional groups in molecular chain originally, and thus the water-binding capacity and oil-binding capacity of protein would gradually increase.

Figure 7.9 showed that with the extension of ultrahigh-pressure treatment time, the water-binding capacity and oil-binding capacity of PPI increased first and then decreased. The water-binding capacity reached the maximum value when the treatment time was 5 min, being 2.40 g/g; the oil-binding capacity reached the maximum value when the treatment time was 3 min, being 3.73 mL/g. Oil-binding capacity can play a role in both promoting and controlling fat absorption. In many food and biological systems, there is an interaction between proteins and lipids, which is not due to covalent bonds but the hydrophobic interaction between the nonpolar aliphatic chains of lipids and nonpolar regions of proteins. The energy of interaction between protein and lipid is lipid, especially the length of its aliphatic chain and hydrophobicity function of the protein. If the amino acids that constitute protein have a high average hydrophobicity, such protein can often strongly interact with lipids to form lipid-protein complex, which may change the structure of protein. The milk film formed when heating soybean milk is a kind of

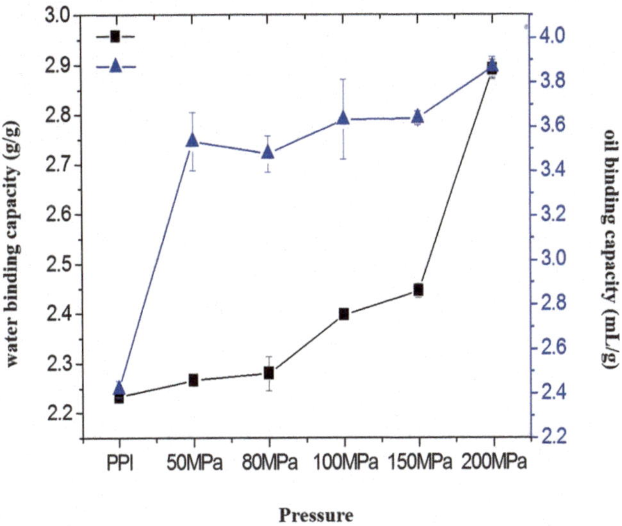

Fig. 7.8 Impact of different pressures on water-binding capacity and oil-binding capacity. ■ Water binding capacity; ▲ Oil binding capacity

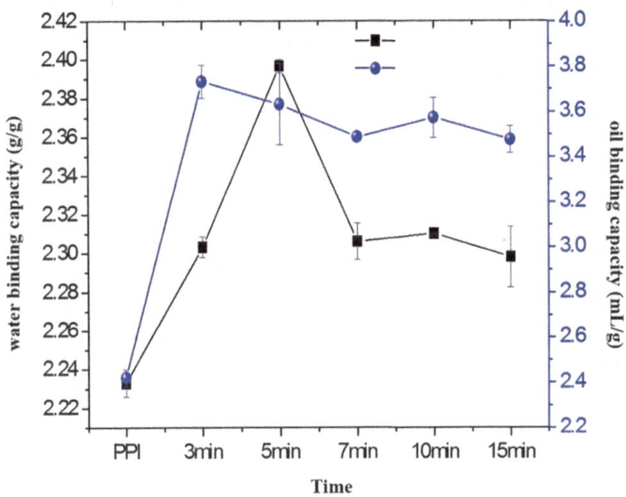

Fig. 7.9 Impact of different times on water-binding capacity and oil-binding capacity. ■ Water binding capacity; ● Oil binding capacity

lipid-protein complex. In this study, after a certain time of treatment, when the surface hydrophobicity index of protein increased, the side chain in the molecule was dissociated and opened, thus increasing the water-binding capacity and oil-binding capacity of protein; when the treatment time was prolonged, the protein was denatured, the structure became dense, and insoluble proteins increased, thus gradually decreasing the water-binding capacity and oil-binding capacity.

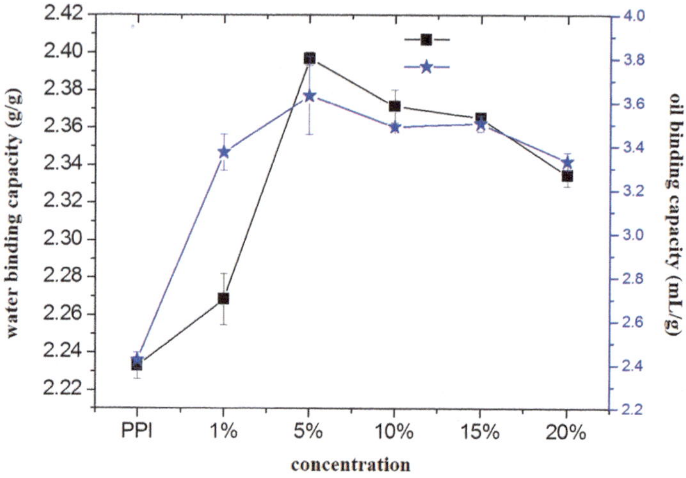

Fig. 7.10 Impact of different concentrations on water-binding capacity and oil-binding capacity. ■ Water binding capacity; ★ Oil binding capacity

It was shown from Fig. 7.10 that the water-binding capacity and oil-binding capacity of PPI at different concentrations changed significantly after high-pressure treatment and first increased and then decreased. The water-binding capacity and oil-binding capacity reached the maximum values when the concentration was 5%, being 2.40 g/g and 3.63 mL/g, respectively. The action of absorption of fat by protein was another form of emulsification. Protein could form an emulsion and a gel matrix when being added to meat, to prevent the fat from moving to the surface. Therefore, it played a role in promoting fat absorption or combination, so as to reduce the loss of fat and juice during the treatment of the meat and maintain the stability of shape. The oil-binding capacity increased with the increase of protein, small particles of low-density protein powder were able to absorb or retain a large number of oil compared with high-density protein powder, and high-pressure homogenization could increase the phase interface, thereby enhancing the protein-lipid interaction degree. The protein part in protein solution was unfolded under the action of ultrahigh pressure and thus caused the exposure of hydrophobic groups, thereby enhancing the water-binding capacity and oil-binding capacity. When the concentration of the protein solution was too high, the hydrophobic group could not be fully unfolded under the ultrahigh-pressure treatment, and then the protein was denatured and its structure became dense, resulting in a decrease in water-binding capacity and oil-binding capacity (Shi Yanguo and Sun Bingyu 2005).

In order to compare the functional characteristics of commercial soybean protein isolates, five kinds of soy protein isolates that were mainly produced and sold at home and abroad were bought. Among them, SPI emulsion type 1 and SPI injection type 1 were domestic commercial soybean protein isolates, and SPI emulsion type 2, SPI injection type 2, and SPI dissolution type were commercial soy protein isolates from Japan. The functional characteristics of each SPI were measured and compared with that of PPI under optimal ultrahigh pressure (115 MPa, 5 min,

Table 7.4 Comparison of functional characteristics of PPI and SPI

Type of protein	Oil-binding capacity (g/g)	Water-binding capacity (ml/g)	Gel hardness (g)	Springiness
SPI emulsion type 1	2.3563 ± 0.03^{b}	3.4027 ± 0.118^{f}	82.08	0.9686
SPI injection type 1	2.3259 ± 0.02^{c}	6.2657 ± 0.47^{e}	97.59	0.9744
SPI emulsion type 2	2.2268 ± 0.04^{d}	5.7216 ± 0.175^{d}	38.99	0.7603
SPI injection type 2	2.1593 ± 0.01^{e}	7.0698 ± 0.04^{c}	45.64	0.9674
SPI dissolution type	2.6790 ± 0.05^{f}	1.8755 ± 0.01^{b}	no gel	–
PPI after ultrahigh-pressure treatment	3.6318 ± 0.01^{a}	2.3965 ± 0.01^{a}	174.37	0.7332
PPI without treatment	2.421 ± 0.01^{b}	2.233 ± 0.02^{a}	115.81	0.7484

The same letter means that the difference is not significant ($p > 0.05$), and different letters mean that the difference is significant ($p < 0.01$)

3.11%) and PPI without ultrahigh-pressure treatment. The results are shown in Table 7.4. It was shown from the figure that the water-binding capacities of five commercial soybean isolates were significantly higher than that of PPI ($p < 0.01$) except SPI dissolution type, including PPI after ultrahigh-pressure treatment and PPI without ultrahigh-pressure treatment. There was no significant difference in the water-binding capacities of two PPIs ($p > 0.05$). However, the oil-binding capacity of PPI after high-pressure treatment was not only higher than that of natural PPI ($p < 0.01$) but also the commercial soy protein isolates ($p < 0.01$). There was no significant difference between the PPI without high-pressure treatment and domestic SPI emulsion type 1 ($p > 0.05$), and it is very significantly different from other types of SPI ($p < 0.01$). After comparison of the hardness of gel formed by various proteins, the hardness of curdlan formed by PPI after ultrahigh-pressure treatment was significantly higher than that of commercial soy protein isolate and PPI without high-pressure treatment ($p < 0.01$). It was shown from the figure that a certain ultrahigh-pressure treatment could improve the oil-binding capacity and gelation of PPI, and it was significantly higher than that of commercial soy protein isolate ($p < 0.01$), but its water-binding capacity was far less than that of commercial soy protein isolate ($p < 0.01$).

2.2 *Molecular Weight Distribution*

Figure 7.11 showed the molecular weight distribution diagrams of the PPI not treated at high pressure and the samples that had been treated for 5 min at 50, 100, and 200 MPa, respectively. The diagrams were obtained at multi-angle laser light scattering instrument through analysis. Table 7.5 is the result obtained from the diagrams in Fig. 7.11 by analysis using the software in the instrument. It was shown from Fig. 7.11 and Table 7.5 that there were significant changes in the distribution of PPI molecular weight after treatment under different pressures. There were mainly two peaks in the untreated protein: the peaks appeared at

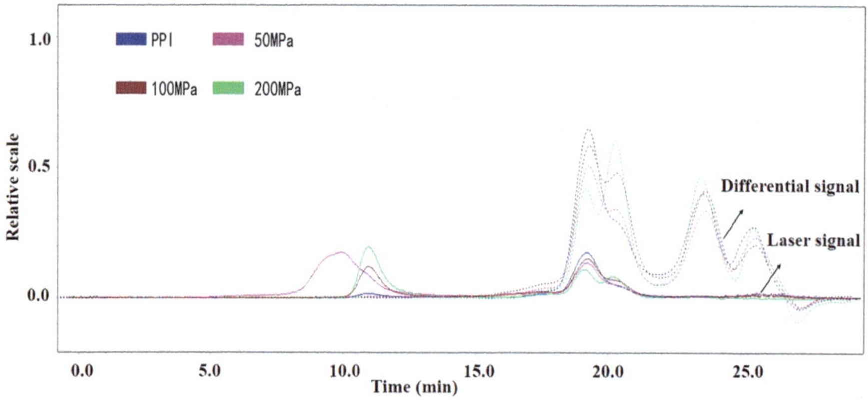

Fig. 7.11 Molecular weight distribution diagram of PPI after treatment at different pressures

Table 7.5 Molecular weight distribution result of PPI at different pressures

Sample	First peak (component)		Second peak (component)		Third peak (component)	
	Mw (Da)	Proportion (%)	Mw (Da)	Proportion (%)	Mw (Da)	Proportion (%)
Contrast	$3.648e^5$	70.87	$2.327e^5$	29.13	–	–
50 MPa	$3.762e^5$	60.44	$3.529e^5$	38.30	$6.292e^7$	1.27
100 MPa	$3.439e^5$	54.32	$2.133e^5$	45.68	–	–
200 MPa	$3.528e^5$	36.08	$1.861e^5$	62.95	$3.665e^7$	0.98

18.083–20.013 min and 20.073–21.585 min, respectively; the molecular weights were $3.648e^5$ Da component (I) and $2.327e^5$ Da component (II), respectively; and the contents were 70.87 % and 29.13%, respectively. After the high-pressure treatment, the component (I) was gradually decomposed. Under the pressure of 50 MPa, 100 MPa, and 200 MPa, it was shown that the peak appearance time and molecular weight of each sample component were basically the same as that of the samples without pressure treatment. However, the proportions of the components in the entire protein changed significantly. Component (I), namely, the main peak with large molecular weight, decreased obviously, and the content decreased to 36.08% at 200 MPa, which was half of that before treatment; component (II) showed a significant increase trend, and the components with larger molecular weight appeared at 50 MPa and 200 MPa, but the content was small, being about 1%. The above results showed that the pressure broke the weak sub-key bonds between the polypeptide chains in PPI, resulting in a decomposition of the components with a large molecular weight and the content of the components with a small molecular weight increased. However, under a certain pressure, a small amount of polymer might be formed in the protein at a certain pressure.

Figure 7.12 showed the molecular weight distribution diagrams of the PPI not treated at high pressure and the samples that had been treated for 3 min, 5 min, and 15 min at 100 MPa, respectively. The diagrams were obtained at multi-angle laser

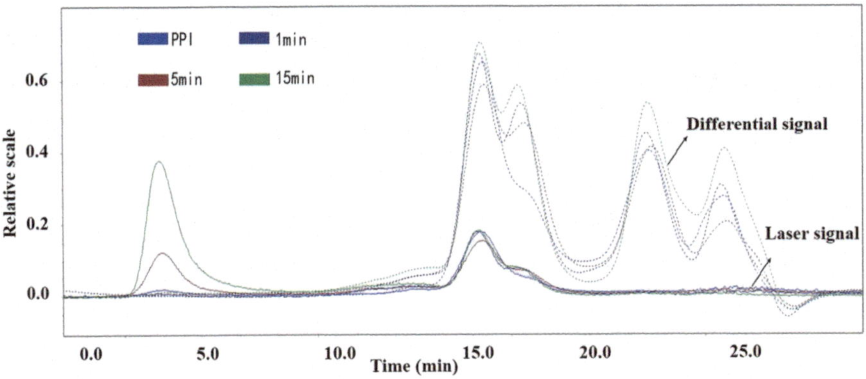

Fig. 7.12 Molecular weight distribution diagram of PPI after treatment for different times

light scattering instrument through analysis. Table 7.6 showed the result obtained from the diagrams in Fig. 7.12 by analysis using the software in the instrument. It was shown from Fig. 7.12 and Table 7.6 that there were significant changes in the distribution of PPI molecular weight after treatment for different times under a certain pressure. The peaks of samples appeared at 18.032–20.013 min and 19.735–21.715 min. Compared with the samples without pressure treatment, most of the components with a large molecular weight (the peak appeared at 18.032–20.013 min) were decomposed, and the contents of the components with a small molecular weight (peak appeared at 19.735–21.715 min) increased obviously. When the time reached 15 min, new polymer appeared in the sample, the peak appeared at 10.044–12.113 min, the molecular weight was $5.667e^7$ Da, and the content was 1.31%. With the increase of treatment time, the components with a large molecular weight were decomposed first quickly and then slowed down. At 5 min, the content of components with a large molecular weight was 54.32%, which was higher than that of samples at 3 min (52.54%) and 15 min (49.96%); the content of components with a small molecular weight was 45.68%, which was lower than that of samples at 3 min (47.46%) and 15 min (48.72%). Compared with the pressure effect (result in Table 7.5), the effect of time was significantly weaker than that of pressure under the same external environment.

Figure 7.13 showed the molecular weight distribution diagrams of the PPI not treated at high pressure and the PPIs with a concentration of 1% (w/v), 5% (w/v), and 20% (w/v), respectively, after treatment at 100 MPa. The diagrams were obtained at multi-angle laser light scattering instrument through analysis. Table 7.7 showed the result obtained from the diagrams in Fig. 7.13 by analysis using the software in the instrument. It was shown from Fig. 7.13 that the peaks of samples appeared at 18.083–20.013 min and 19.738–21.715 min. It was shown from Table 7.7 that compared with the PPI not treated at high pressure, the content of the component with a large molecular weight (the peak appeared at 18.083–20.013 min) decreased sharply first and then gradually increased with the increase of concentration; the content of the component with a small molecular weight (the peak appeared at 19.738–21.715 min) increased sharply first and then

Table 7.6 Molecular weight distribution result of PPI after treatment for different times

Sample	First peak (component)		Second peak (component)		Third peak (component)	
	Mw (Da)	Proportion (%)	Mw (Da)	Proportion (%)	Mw (Da)	Proportion (%)
Contrast	$3.648e^5$	70.87	$2.327e^5$	29.13	–	–
3 min	$1.023e^5$	52.54	$1.412e^5$	47.46	–	–
5 min	$3.439e^5$	54.32	$2.133e^5$	45.68	–	–
15 min	$3.439e^5$	49.96	$2.169e^5$	48.72	$5.667e^7$	1.31

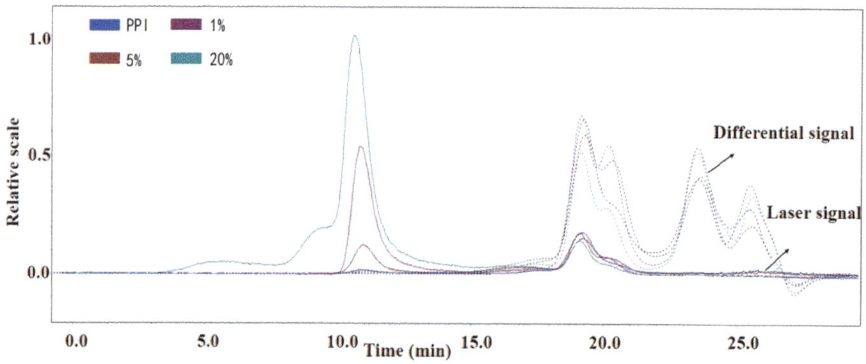

Fig. 7.13 Molecular weight distribution diagram of PPI with different concentrations after ultrahigh-pressure treatment

Table 7.7 Molecular weight distribution result of PPI with different concentrations after ultrahigh-pressure treatment

Sample	First peak (component)		Second peak (component)	
	Mw (Da)	Proportion (%)	Mw (Da)	Proportion (%)
Contrast	$3.648e^5$	70.87	$2.327e^5$	29.13
1%	$3.286e^5$	52.18	$1.723e^5$	47.82
5%	$3.439e^5$	54.32	$2.133e^5$	45.68
20%	$3.491e^5$	65.41	$1.883e^5$	34.59

gradually decreased with the increase of concentration. When the concentration was 1–5%, the molecular weight distribution was in a relatively stable state without obvious changes, and the content change of the two components was only 2.14%; when the concentration was greater than 5%, the content change of the two components increased to 11.09%. The result above was significantly different from the result of the impact of pressure and time on the molecular weight distribution of PPI. This indicated that during the ultrahigh-pressure processing, the impacts of different treatment conditions on the molecular weight distribution

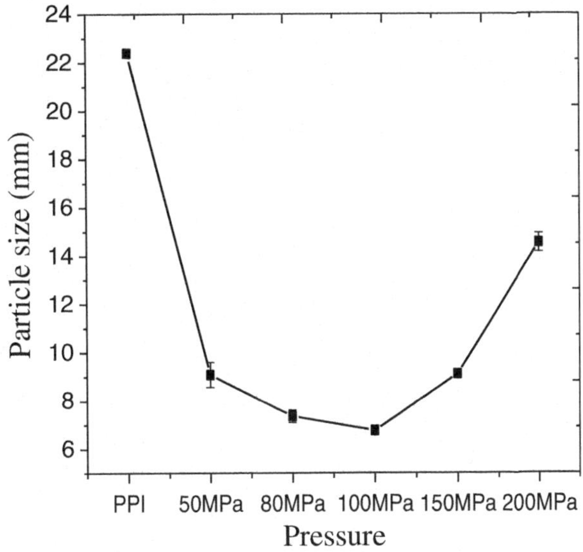

Fig. 7.14 Impact of different pressures on the particle size of PPI

of PPI were completely different, which resulted in completely different of texture characteristics after the formation of curdlan (Figs. 7.1, 7.2, and 7.3). It was known after further analysis that when the contents of the two components in the sample accounted for 50%, respectively, the texture of the curdlan formed was good, and the large polymers in the sample could not improve the texture characteristics of the curdlan formed.

2.3 *Particle Size*

The particle sizes of the samples are shown in Figs. 7.14, 7.15, and 7.16. It was found that the average particle size of PPI after ultrahigh-pressure treatment drastically changed. Figure 7.14 showed the impacts of different pressures on the particle size of PPI. It was shown from the figure that the average particle size of untreated PPI was 22.39 μm, and the average particle sizes after treatment at different pressures were 9.08 μm at 50 MPa, 7.36 μm at 80 MPa, 6.77 μm at 100 MPa, 9.11 μm at 150 MPa, and 14.55 μm at 200 MPa, respectively. The average particle size of PPI greatly decreased after modification by high-pressure treatment and reached the minimum value at 100 MPa. The average particle size decreased by 15.62 μm compared to that before treatment; however, with the further increase of pressure, the average particle size begun to increase, which was consistent with the impact of dynamic ultrahigh-pressure micro-jet technology on the particle size of water-in-oil emulsions in the research of Jafari et al. (2007). The particle size decreased sharply at a low pressure, and it showed an increase trend at a high pressure. There are two different interpretations of this result. One explanation is that after pressure treatment, the final particle size of the substance

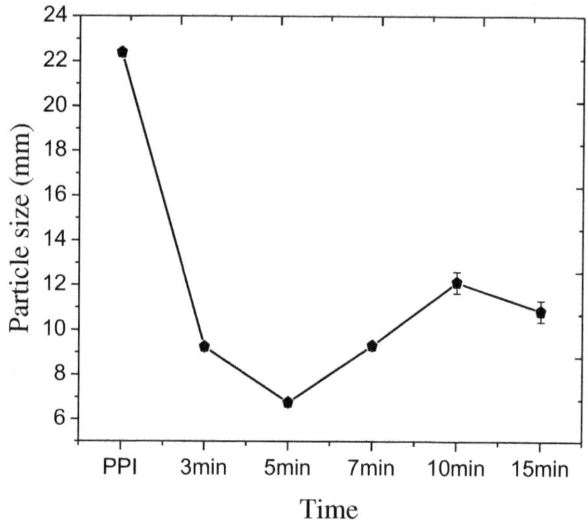

Fig. 7.15 Impact of different times on the particle size of PPI

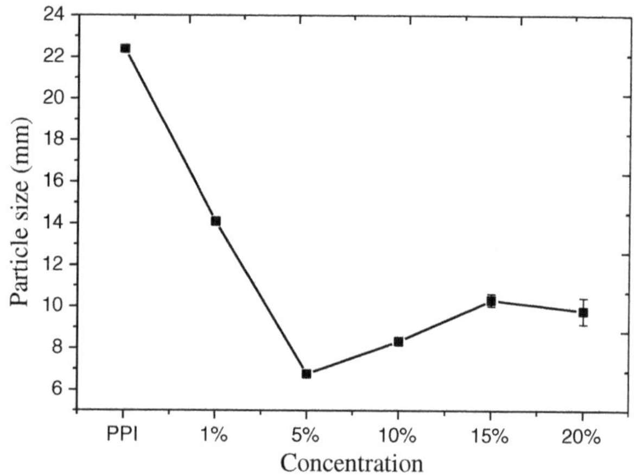

Fig. 7.16 Impact of different concentrations on the particle size of PPI

depends on the competition results of two opposing factors of particle fragmentation and particle reaggregation. When the protein polymer is broken into smaller polymers, a new surface is formed, and the new polymers tend to reaggregate due to the thermodynamical instability. If reaggregation dominates, the particle size will increase. The other explanation is that this result is caused by "overpressure" (Kolb et al. 2001; Desrumaux and Mareand 2002). When the pressure increases, the water will be compressed, and its temperature will increase. It is generally believed that under the adiabatic conditions, the temperature at 20 °C will increase by 3 °C after applying 100 MPa each time; therefore, the possibility of reaggregation will be raised, and thus the particle size will be increased.

Figure 7.15 showed the changes of the particle size of PPI after treatment for different times. It was shown from the figure that with the extension of the treatment time, the particles were broken and the particle size decreased rapidly and reached the minimum value at 5 min; with the continuous extension of treatment time, some tiny protein particles recombined to form new polymers, and thus the particle size increased. Figure 7.16 showed the impacts of different protein concentrations on the particle size of PPI. It was shown from the figure that among the protein solutions with different concentrations, after a certain high-pressure treatment, the protein particles became smaller; the average particle size reached the minimum value when the concentration was 5%, being 6.77 μm; when the concentration continued to increase, the PPI with a small particle size formed a new polymer and thus led to reaggregation within the limited solution space scope, and the particle size slowly increased.

2.4 *Microscopic Morphology*

1. SEM scanning of ultra-high pressure PPI samples

The electron microscopic scanning diagrams of the samples are shown in Fig. 7.17, which are obtained in the magnification of 10,000 times. A in the figure is PPI not treated at ultrahigh pressure; B–F are PPI after treatment at 50, 80, 100, 150, and 200 MPa, respectively; G–J are the samples of PPIs with a concentration of 1%, 10%, 15%, and 20% (w/v), respectively, after treatment for 5 min at 100 MPa; and K–N are the samples of PPIs after treatment for 3, 7, 10, and 15 min, respectively, at 100 MPa.

As shown in Fig. 7.17a, the surface of the original sample of untreated PPI was visibly viscous and smooth, and the protein particles were clustered together with compact structure and free of obvious pores; after the treatment at the pressure of 50 MPa (Fig. 7.17b), the polymer of PPI was broken, the polymer surface was not smooth and the viscosity became smaller, and there were some small pores. When the pressure reached to 80 MPa (Fig. 7.17c), due to the intensified degree of crushing, protein polymer was decomposed into countless irregular small particles under high-pressure mechanical force, and when they would combine together closely, the pores become large, but the surface was not visibly viscous. When the pressure increased to 100 MPa (Fig. 7.17d), the surface of the sample became slightly viscous; its viscosity was between the values in Fig. 7.17b, c. A small amount of spherical particles were combined together, with visible uniform pores. With the further increase of pressure, the peanut globulin particles should be further broken into smaller particles under high pressure conventionally; however, this change trend was not intensified, the size of peanut protein particles became large, and most of them were in the form of sheets. This could be confirmed using the electron microscope scanning diagrams after treatment at 150 MPa and 200 MPa, respectively (Fig. 7.17e, f). The spherical particles in the sample of Fig. 7.17e almost completely disappeared, and the surface was very viscous and in the form of sheets, but there were a small number of larger visible pores, while the spherical

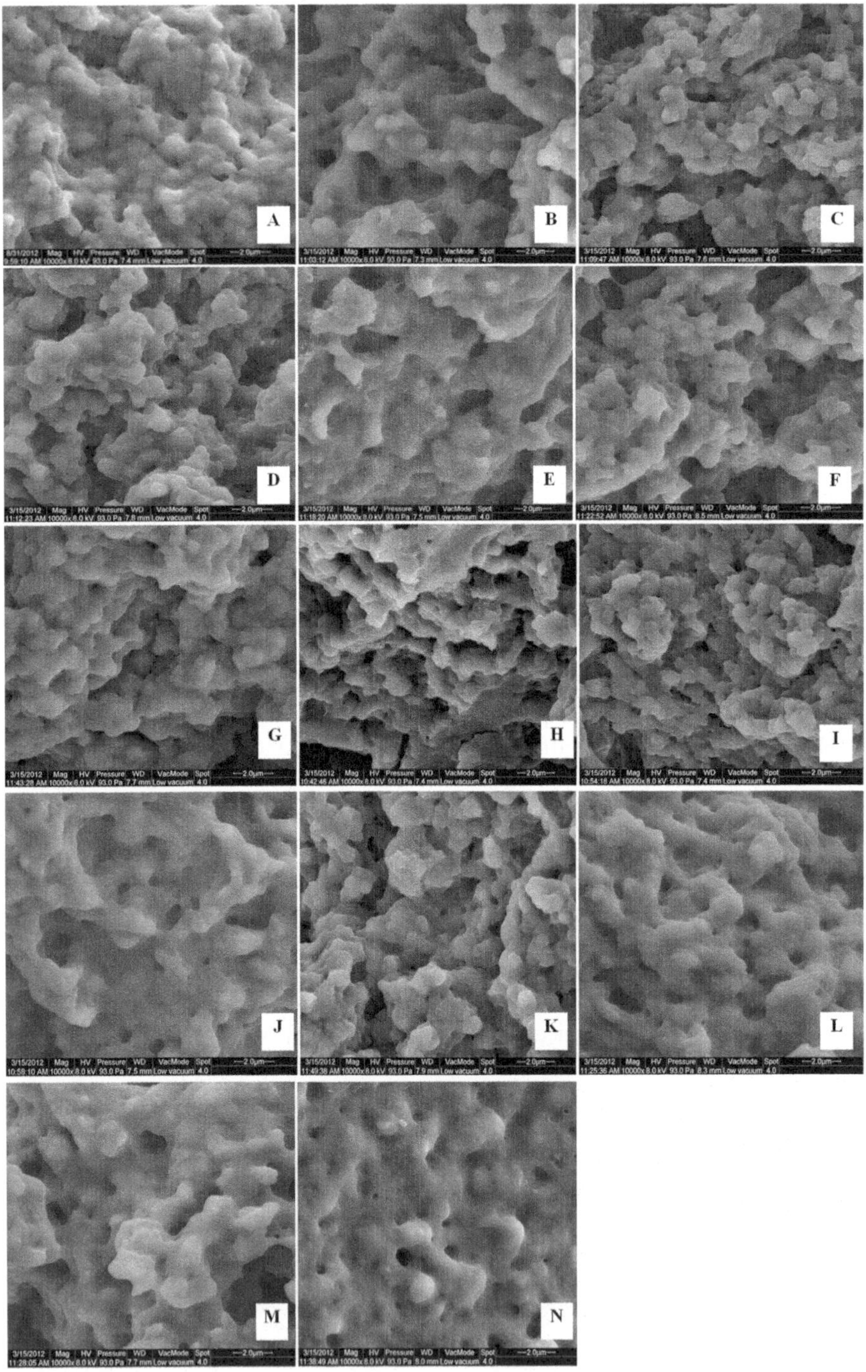

Fig. 7.17 SEM scanning diagrams of PPI samples before and after ultrahigh-pressure treatment (10,000×)

particles in the sample of Fig. 7.17f were incomplete with large pores. The possible explanation is that on the one hand, the large particles continued to rupture; on the other hand, small particles were aggregated into large particles, and the aggregation was predominant, so the particle size became large at large treatment pressures (>100 MPa). The above results showed that the pressure could break the peanut globule protein particles, making its structure become loose; with the increase of treatment pressure, the degree of crushing was intensified and became the most severe at 100 MPa. At this time, the particle size of peanut globule was the smallest, and the surface was uneven; after continuing to apply pressure, the specific surface area of particle increased, the aggregation became dominant, and the particle size increased, which was very consistent with the particle size change of the peanut protein at different pressures measured in Fig. 7.11.

It was shown from Fig. 7.17g–j that there were spherical particles on the surface of PPI with a low concentration (Fig. 7.17g); with the increase of concentration, the spherical particles in the protein disappeared, and protein particles were broken into flaky substances at the concentration of 10% (Fig. 7.17h); however, after continuing to increase the concentration to 15% (Fig. 7.17i), clustered substances appeared at the surface of protein; after the concentration reached 20% (Fig. 7.17j), the sample became viscous visibly without obvious pores. This was consistent with the results of particle size change of PPI at different concentrations in Fig. 7.12. Figure 7.17k–n were the electron microscope scanning diagrams of samples after different times. It was shown from the figures that at a certain pressure (100 MPa), there were obvious spherical particles on the surface of sample within a short time (5 min); however, with the extension of time, the viscosity at the surface of sample increased obviously, and pores decreased obviously.

2. SEM scanning of curdlan samples formed by PPI after ultra-high pressure treatment

Figure 7.18a–h in Fig. 7.18 are the diagrams of curdlan formed by PPI. They were obtained using scanning electron microscopy at 10,000 times. Among them, Fig. 7.18a is the electron microscopy diagram of curdlan formed by PPI not treated at ultrahigh pressure, and Fig. 7.18b–h are the electron microscopy diagrams of curdlans formed by PPI after the treatment at 50 MPa, 100 MPa, 200 MPa, 1% (w/v), 20% (w/v), 3 min, and 15 min, respectively. The microstructures of the gels formed by various samples were significantly different. It was shown from Fig. 7.18a that there were many small pores in the curdlan formed by PPI not treated at high pressure, and protein particles were connected compactly and densely. In the curdlan formed after treatment at 50 MPa, the protein particles were connected together in clusters, and the pores became obviously large (Fig. 7.18b); in the curdlan formed after treatment at 100 MPa, globular protein particles were compactly connected together in clusters or strips (Fig. 7.18c); in the curdlan formed after treatment at 200 MPa, although protein particles were tightly connected together, they did not form the original globular structure (Fig. 7.18d). It was shown from Fig. 7.18e that the parts at the surface of sample were clustered and the structure was not compact with visible large pores. It was shown from Fig. 7.18f

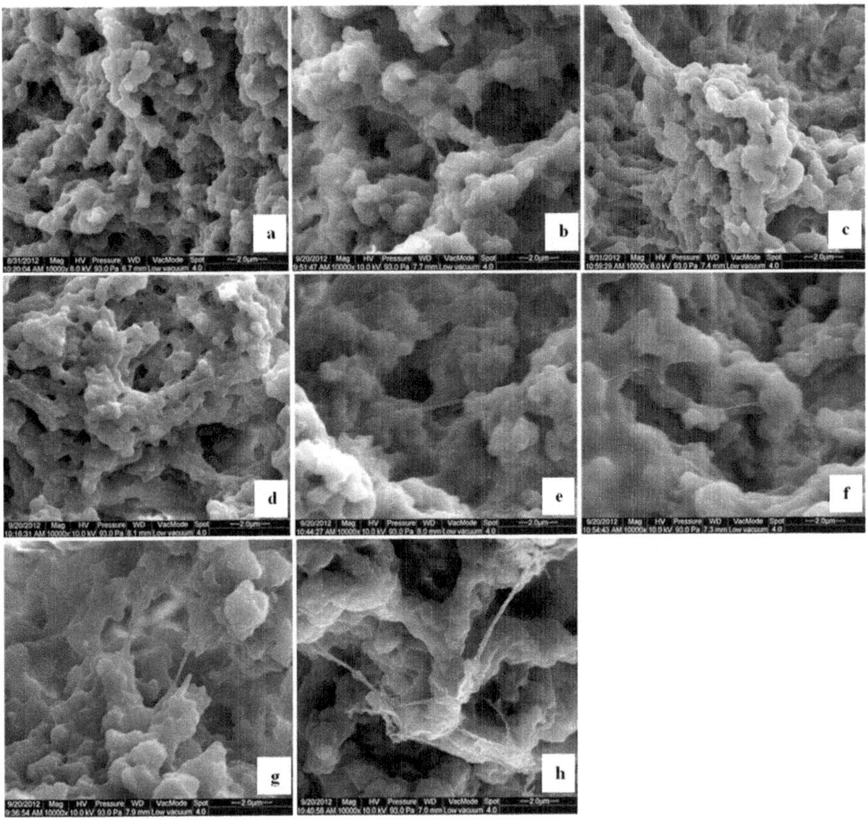

Fig. 7.18 SEM scanning diagrams of curdlan formed by PPI samples before and after ultrahigh-pressure treatment (10,000×)

that the curdlan formed by sample obviously increased, and protein particles formed the cluster structure which was not compact but viscous. For the microstructure of curdlan formed by the sample in Fig. 7.18g, there were a few of small pores at the surface of sample, most of the protein particles formed flakes, and a small part of them formed clusters. For the gel sample in Fig. 7.18h, the visible pores increased obviously, the spherical structure at the surface of sample disappeared, most of the spherical particles formed flake substances, but they did not connect together tightly, so there were many large pores. Compared with the microstructures of the samples before forming curdlan (Fig. 7.16), it was shown that ultrahigh-pressure treatment changed the microstructure of PPI and the microstructures of the curdlan formed by it were significantly different with the changes in pressure, time, and protein, which also led to significantly different texture characteristics of the curdlan formed by the samples (as shown in the results of 2.1.2). After further analysis, it was shown that there were obvious spherical particles at the surface of protein particle sample and a small amount of particles were

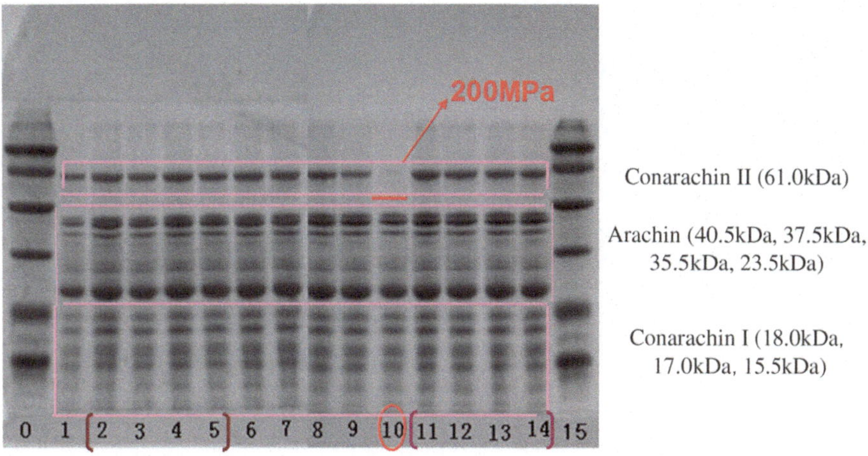

Fig. 7.19 SDS-PAGE analysis of PPI. Note: 0, 15: standard sample; 1: PPI; 2–5: 1%, 10%, 15%, 20%; 6–10: 50, 80, 100, 150, 200 MPa; 11–14: 3, 7, 10, 15 min

combined together to form a uniform pore structure (Fig. 7.18d), and the curdlan formed was in clustered and stripped microstructure (Fig. 7.18c), with good texture characteristics (Fig. 7.1). When the surface of protein appeared viscous or protein particles were closely gathered together, the texture characteristics of the curdlan formed were poor.

2.5 *Protein Components and Bands*

The SDS-PAGE electrophoretograms of PPI before and after ultrahigh-pressure treatment are shown in Fig. 7.19. The subunit bands of PPI were clearly shown in the figure, the molecular weights corresponding to the subunit bands were, respectively, 61.0 kDa, 40.5 kDa, 37.5 kDa, 35.5 kDa, 23.5 kDa, 18.0 kDa, 17.0 kDa, and 15.5 kDa. Among them, 61.0 kDa belonged to conarachin II; 40.5 kDa, 37.5 kDa, 35.5 kDa, and 23.5 kDa belonged to arachin; and 18.0 kDa, 17.0 kDa, and 15.5 kDa belonged to conarachin I (Prakash and Narasinga 1986). The solubilities of protein sample (according to the order in the figure) were 73.88 ± 0.24%, 68.43 ± 0.48%, 68.90 ± 0.25%, 70.09 ± 0.27%, 74.91 ± 0.12%, 65.20 ± 0.23%, 65.45 ± 0.24%, 65.88 ± 0.24%, 65.03 ± 0.24%, 64.35 ± 0.25%, 71.16 ± 0.23%, 70.89 ± 0.10%, 68.24 ± 0.33%, and 67.75 ± 0.08%, respectively.

From the solubilities of samples, the results of Fig. 7.19 were not due to the differences in the solubilities of samples but due entirely to the treatment at different ultrahigh pressures. It was shown after further analysis that compared with that of untreated PPI, the 61.0 kDa subunit bands at different times and in the samples with different protein concentrations did not have any obvious changes, except that the color got darker. With the increase of pressure, the color of 61.0 kDa

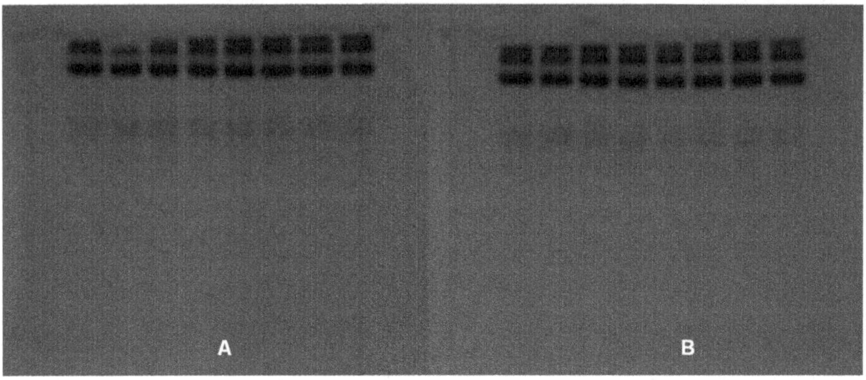

Fig. 7.20 Native PAGE analysis of PPI. Note: (**a**) from *left* to *right*, PPI, PPI2, 50 MPa, 80 MPa, 100 MPa, 150 MPa, 200 MPa, 3 min. (**b**) Samples from *left* to *right*: PPI, 7 min, 10 min, 15 min, 1%, 10%, 15%, 20%

subunit band first got darker and then lighter, showing that PPI first aggregated and then unfolded after ultrahigh treatment. After comparing the electrophoretic bands of PPI before and after ultrahigh-pressure treatment, it was found that there was no significant difference in the subunit bands of conarachin I, showing that ultrahigh-pressure treatment has no impact on conarachin I basically; after the treatment at different pressures, the subunit bands of PPI became darker obviously, showing that ultrahigh pressure caused aggregates in the arachin of PPI. Because β-mercaptoethanol was used in SDS-PAGE electrophoresis to break the disulfide bonds of PPI, the aggregate bonds showed that non-disulfide-covalent cross-linker was formed in PPI due to the effect of pressure. The above results indicated that conarachin II was most sensitive to pressure than arachin and conarachin I.

Figure 7.20 is the Native PAGE of PPI, corresponding to the nonreduction situation in Fig. 7.19. It was shown from the figure that at the top of the separation gel, each sample had three aggregates, being N I, N II, and N III, respectively. Figure 7.19 is the corresponding reduction situation. It was found after comparing Fig. 7.19 with Fig. 7.20 that the bands corresponding to N I, N II, and N III had appeared, showing that disulfide bonds played an important role in the aggregates of N I, N II, and N III. The difference among samples after treatment at different ultrahigh pressures was not reflected obviously in the figure.

2.6 *Thermal Performance*

1. Pressure

DSC is widely used in the research on the thermodynamic and dynamic characteristics of protein denaturation. We can verify whether high-pressure treatment affects the denaturation of PPI through DSC experiment. Table 7.8 shows the

Table 7.8 DSC scanning of PPI samples at different pressures

Sample	T_{d1} (°C)	T_{d2} (°C)	ΔT1/2 (°C)	ΔH (J/g)
Control	93.53 ± 0.27	107.25 ± 0.09	6.35 ± 0.12	8.780 ± 0.58
50 MPa	93.4 ± 0.27	106.84 ± 0.01	6.86 ± 0.52	7.2 ± 0.74
80 MPa	93.01 ± 0.65	107.09 ± 0.09	6.74 ± 0.20	6.99 ± 0.61
100 MPa	92.44 ± 0.31	105.74 ± 0.75	6.4 ± 0.07	6.44 ± 0.32
150 MPa	–	106.73 ± 0.22	6.33 ± 0.15	6.08 ± 0.5
200 MPa	–	107.09 ± 0.01	6.16 ± 0.02	6.05 ± 0.12

analysis results of the impact of different pressures on the thermal characteristics of PPI: T_d is the denaturation temperature; T_d value can reflect the thermal stability of the protein; the higher the T_d, the higher the thermal stability is; and the lower the T_d, the lower the thermal stability is (Wang et al. 2000). ΔH is the enthalpy of the sample; exactly, it is the enthalpy change of the sample, that is, the ΔH before and after the thermal transformation of the sample can reflect the degree of denaturation of the protein, and the reaction of destroying hydrogen bond can produce the endothermic enthalpy, the protein agglomeration and the hydrophobic reaction can produce heat release enthalpy, and the enthalpy size and positive and negative results indicate which reaction is dominant. Half-peak width (ΔT1/2) mainly reflects the synergism of proteins during thermal denaturation.

As shown in Table 7.8, endothermic peak of untreated PPI occurred at 93.53 °C and 107.25 °C. According to the literature, PPI was mainly composed of arachin and conarachin. The research of Du Yin (2012) showed that the denaturation temperatures of arachin and conarachin were 102.27 °C and 87.49°C, respectively. Thus, 93.53 °C and 107.25 °C were the endothermic peaks of conarachin and arachin in PPI, respectively. As can be seen from the table, with the increase in pressure (50–100 MPa), the peak temperature of conarachin gradually decreased; at 150 MPa and 200 MPa, the endothermic peak of PPI was not visible, indicating that after high-pressure treatment, significant denaturation occurred in conarachin (this is basically the same as the SDS-PAGE electrophoretic result in 2.2.5). It was shown from the table that after high-pressure treatment, the endothermic peak of arachin was visible, showing that after the treatment at different pressures, arachin was more stable than conarachin and the temperatures of endothermic peak value changed. At the pressure of 100 MPa, the peak temperature was the lowest, being 105.74 °C and decreasing by 1.51 °C compared with the control. These results indicated that high pressure had different effects on conarachin and arachin. After 150 MPa, the endothermic peak of conarachin disappeared completely, but the endothermic peak of arachin was visible, showing that conarachin was more likely to denature than arachin under the action of pressure.

The study of Arntfield and Murray (1981) showed that the total enthalpy (ΔH) of glycinin and β-conglycinin represented the impact on the proportion of undenatured proteins in the sample or the proteins with ordered structure after pressure treatment. The total enthalpy (ΔH) of PPI after pressure treatment was lower than that of PPI without pressure treatment. With the increase of pressure,

Table 7.9 DSC scanning of PPI sample after treatment for different times

Sample	Td1 (°C)	Td2 (°C)	ΔT1/2 (°C)	ΔH(J/g)
Control	93.53 ± 0.27	107.25 ± 0.09	6.35 ± 0.12	8.780 ± 0.58
3 min	91.88 ± 0.57	106.64 ± 0.1	6.87 ± 0.02	10.14 ± 0.42
5 min	92.44 ± 0.31	105.74 ± 0.75	6.4 ± 0.07	6.44 ± 0.32
7 min	91.90 ± 0.01	106.54 ± 0.15	6.65 ± 0.07	10.18 ± 0.48
10 min	91.55 ± 0.19	106.43 ± 0.08	6.91 ± 0.16	9.61 ± 012
15 min	92.15 ± 0.3	106.56 ± 0.16	6.82 ± 0.21	9.13 ± 0.04

enthalpy decreased gradually, showing that with the increase of pressure, the degree of denaturation of PPI increased gradually, PPI unfolded gradually under the action of pressure, and the three-dimensional structure gradually became disordered, so that PPI was in a gradually unstable state, resulting in the decrease of energy required. The half-peak width of peak value of arachin was related to the transformation and coordination between the natural state and denatured state of protein (Privalov 1982), and it was not affected when treatment is at 50–200 MPa, as shown in Table 7.8. This showed that pressure did not change the coordination of protein denaturation.

2. Time

The DSC scanning results of PPI not treated at high pressure and PPI after treatment for 3–15 min at 100 MPa are shown in Table 7.9. It was shown from the table that the endothermic peaks of arachin and conarachin of PPI were visible after treatment for different times, but peak value temperature decreased compared to that of untreated PPI, showing that PPI did not denature completely after treatment for 15 min at 100 MPa, which was entirely different from the impact of pressure on the thermal characteristics of PPI. It was also seen from the table that after the treatment for different times, the initial denaturation temperature and peak temperature decreased; the degrees of decrease were different due to different treatment pressures. It was shown from the table that the peak values of the conarachins in the samples decreased by 1.65, 1.09, 1.63, 1.98, and 1.38 °C, respectively, relative to untreated PPI, with a decrease range of not more than 2 °C; the peak values of arachins decreased by 0.64, 1.51, 0.71, 0.82, and 0.69 °C, respectively, relative to untreated PPI, with a decrease range of not more than 1 °C, except the value after treatment for 5 min. This showed that the impact on glycinin was large after treatment for 5 min, which could be seen from its enthalpy. It showed after comparing the peak value changes of conarachin and arachin in different samples that at a certain pressure, the impact of different treatment times on conarachin was greater than that of arachin. It was shown from the half-peak widths (ΔT1/2) of the samples that the impact of different treatment times on its half-peak width (ΔT1/2) was small, so we considered that time did not change the coordination of protein denaturation.

Table 7.10 DSC scanning of PPI samples with different concentrations

Sample	T_{d1} (°C)	T_{d2} (°C)	ΔT1/2 (°C)	ΔH (J/g)
Control	93.53 ± 0.27	107.25 ± 0.09	6.35 ± 0.12	8.780 ± 0.58
1%	92.52 ± 0.07	106.71 ± 0.08	6.62 ± 0.25	9.68 ± 0.2
5%	92.44 ± 0.31	105.74 ± 0.75	6.4 ± 0.07	6.44 ± 0.32
10%	91.23 ± 0.14	106.48 ± 0.25	7.0 ± 0.07	9.46 ± 0.21
15%	91.01 ± 0.1	106.64 ± 0.3	6.56 ± 0.03	9.47 ± 0.05
20%	91.84 ± 0.21	106.3 ± 0.23	5.81 ± 0.02	10.01 ± 0.05

3. Protein concentration

The DSC scanning results of PPI not treated at high pressure and PPI with a mass concentration of 1–20% after treatment at 100 MPa are shown in Table 7.10. It was shown from the table that in untreated PPI and PPI with a protein concentration of 1–20% after treatment at 100 MPa, the endothermic peaks of conarachin and arachin were visible, which was consistent with the impact of time on the thermal characteristics of PPI but entirely different from the impact of pressure on the thermal characteristics of PPI, indicating that the impacts of different ultrahigh treatment conditions on the thermal characteristics of PPI were different. This was mainly due to the change in the structure of PPI after ultrahigh-pressure treatment. The change law in its structure was discussed in Article 2.3 of this book. It was shown from the table that, after ultrahigh-pressure treatment, the peak values of the conarachins and arachins in the PPIs with different concentrations decreased, which showed that after ultrahigh-pressure treatment, PPI denatured and thus its thermal stability decreased. Generally, the formation of a unique three-dimensional protein structure was the net result of various repulsive and attractive non-covalent interactions and several disulfide bonds, while denaturation asked to destroy these interactions. The decrease in denaturation temperature and denaturation enthalpy showed that the energy that required these interactions also decreased and the structure of PPI after ultrahigh-pressure treatment was loose. The reason might be because part of the interactions was destroyed or disulfide bonds were exposed to the surface of the protein molecule and some of the disulfide bonds were broken and thus protein molecules partially denatured. From the half-peak widths (ΔT1/2) of different samples in the table, the protein concentration had a slight impact on changing the coordination of protein denaturation.

3 Changes in Advanced Structure

Because the functional characteristics of natural PPI cannot meet the needs of production and products, it is generally used after modification. After ultrahigh-pressure physical modification, the gel properties of PPI were improved and at the same time changed its water-binding capacity, oil-binding capacity, particle size, and microscopic characteristics (the results are shown in Articles 2.1 and 2.2), and the reason of these changes was because protein structure changed due to ultrahigh-pressure treatment. PPI has the characteristics of typical globulin. Studies have shown that the primary structure of protein determines the nutritional characteristics of protein, the high-level structure determines the functional properties of protein, and there is no the report on the impact of ultrahigh-pressure treatment on the primary structure of protein, so it is generally considered that ultrahigh pressure does not have any impact on the primary structure of protein; among the high-level structures, the quaternary structure of protein is the most difficult to obtain. Because the subunits of PPI are complex and it is difficult to separate and purify PPI to obtain individual subunits, there is no report on the study of the quaternary structure of PPI after ultrahigh-pressure treatment. In contrast to the primary and quaternary structures of protein, the study on secondary and tertiary structures of proteins is easier from the means of research or obtaining the detected samples. At present, there are some preliminary reports on the α-helix, β-sheet, and β-turn conformation and surface hydrophobicity in the protein after ultrahigh-pressure treatment, but the correspondence between the structural change and the functional change is not clear. Therefore, the research team has studied the relationship between the change laws of gel characteristics as well as secondary and tertiary structures during the process of ultrahigh-pressure treatment, so as to find out the inherent causes of the change of protein gel characteristics.

3.1 Secondary Structure

According to the energy size at the level after electron transition, the CD spectrum of protein was divided into three wavelength ranges: far-UV spectral region of below 250 nm, the circular dichroism of which was mainly caused by $n \rightarrow \pi^*$ electron transition of peptide bond; near-UV spectral region of 250–300 nm, the circular dichroism of which was mainly caused by $\pi \rightarrow \pi^*$ electron transition of peptide bond of side chain aromatic group; and UV-visible light spectrogram of 300–700 nm, the circular dichroism of which was mainly caused by external chromophore such as the prothetic group of protein (Sreerama and Woody 2004; Hanchang et al. 2007). Far UV CDs could be used to characterize the impact of ultrahigh pressure on the secondary structure of PPI, and the far-UV CD spectra of different samples after ultrahigh-pressure treatment (including different pressures, different times, and different concentrations) are shown in Figs. 7.21, 7.22, and 7.23. Tables 7.11, 7.12, and 7.13 showed the changes of the secondary structures of untreated PPI and PPI after high-pressure treatment obtained through computer Reed's simulation based on the measured CD spectral line. It was shown from Table 7.11 that the conformation of natural PPI contained 9.6% α-helix, 32.1% β-sheet, 10.6% turn, and 47.7% irregular coil, core conformation (α-helix + β-sheet) accounted for 41.7%, random structure (turn + irregular coil) accounted for 58.3%, and at the same time β-folding and irregular coil conformation

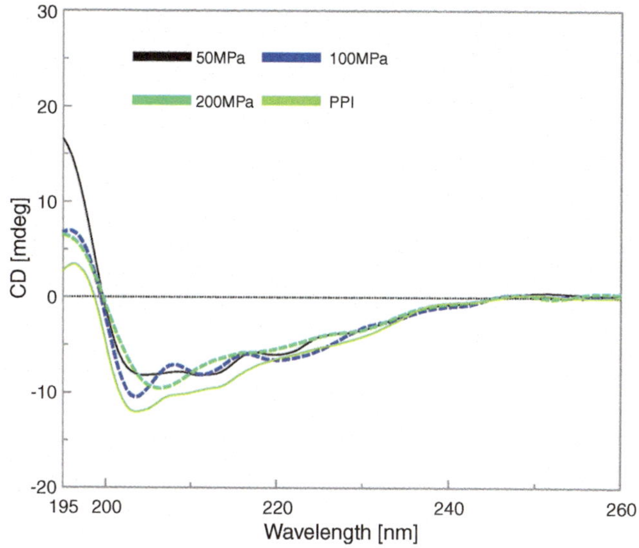

Fig. 7.21 CD spectrogram of PPI at different pressures

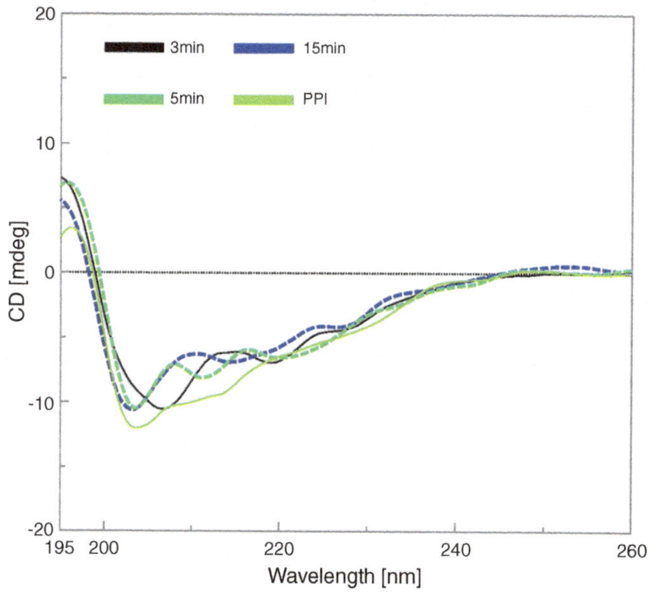

Fig. 7.22 CD spectrogram of PPI at different times

accounted for a large proportion. The CD spectra of protein were generally divided into two wavelength ranges, namely, 178–250 nm for the CD spectra of far-UV region and 250–320 nm for the CD spectra of near-UV region. The CD spectra of

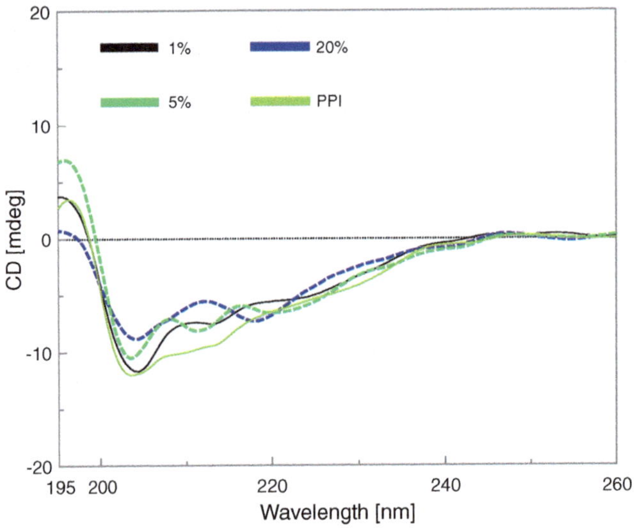

Fig. 7.23 CD spectrogram of PPIs with different concentrations

Table 7.11 Analysis of secondary structure of PPI at different pressures

	Reed's			
	α-Helix (%)	β-Sheet (%)	β-Turn (%)	Irregular coil (%)
PPI	9.6	32.1	10.6	47.7
50 MPa	35.0	0	29.5	35.5
100 MPa	41.4	3.2	15.5	39.9
200 MPa	25.2	15.0	17.9	41.9

Table 7.12 Analysis of secondary structure of PPI at different times

	Reed's			
	α-Helix (%)	β-Sheet (%)	β-Turn (%)	Irregular coil (%)
PPI	9.6	32.1	10.6	47.7
3 min	38.2	0	19.3	42.5
5 min	41.4	3.2	15.5	39.9
15 min	30.2	7.9	14.5	47.4

Table 7.13 Analysis of secondary structure of PPIs with different concentrations

	Reed's			
	α-Helix (%)	β-Sheet (%)	β-Turn (%)	Irregular coil (%)
PPI	9.6	32.1	10.6	47.7
1%	24.7	13.2	14.7	47.5
5%	41.4	3.2	15.5	39.9
20%	5.2	41.9	4.2	48.7

far-UV region reflected the circular dichroism of peptide bonds. In the regular secondary structures of protein or polypeptide, peptide bonds were arranged orderly, and the direction of arrangement decided the division of energy level transition of peptide bond. Therefore, the positions and absorption intensities of CD bands generated by proteins or polypeptides with different secondary structures were different. There was a positive band near 192 nm in α-helix structure, and there were two negative characteristic shoulder peak bands at 222 and 208 nm; there was one negative band at 216 nm in the CD spectrum of β-sheet, and there was one positive band at 185–200 nm; there was one positive CD band at 206 nm in β-turn, and there was negative CD band at the corresponding position of the left helix P2 structure. Therefore, the information on the secondary structures of protein or polypeptide chain could be reflected according to the far-UV CD spectrum of protein or polypeptide measured.

It was shown from Table 7.11 that the secondary conformation of PPI changed significantly after the treatment at different pressures. After the treatment at the pressures of 50, 100, and 200 MPa, PPI changed from mainly β-sheet and irregular coil conformation to α-helix conformation and β-turn conformation. Among them, α-helix conformation and β-turn confirmation significantly increased compared with untreated PPI, and α-helix conformations increased by 25.4%, 31.8%, and 15.6%, respectively, and β-turn confirmations increased by 18.9%, 4.9%, and 7.3%, respectively, while β-sheet conformation and irregular coil significantly decreased. At 50 MPa, there was no β-sheet conformation in PPI. With the increase of pressure, α-helix conformation in PPI first increased and then decreased, and β-sheet conformation first decreased to 0 and then increased with the increase of pressure. The contents of irregular coils decreased to different degrees and reached the minimum value under low-pressure treatment and then increased with the increasing pressure. It showed that α-helix played a major role and the hydrogen bond effect between proteins decreased during the pressure treatment. Turn conformation increased compared with natural PPI. With the increase of pressure, its content first increased and then decreased and reached the minimum value at 100 MPa. Most of β-turns were at the surface of protein molecule, and they could change the direction of polypeptide chain and made it bent, folded back, and reoriented. The change in the content of β-turn structure showed that the shape of PPI molecule might change, protein molecules extended, and the degree of asymmetry increased. It was shown from the table that at 50 MPa, the core conformation was 35%, not containing β-sheet conformation, and the random conformation accounted for 65%; at 100 MPa, the core conformation accounted for 44.6%, and the random conformation accounted for 55.4%; at 200 MPa, the core conformation accounted for 40.2%, and the random conformation accounted for 59.8%. This showed that under low pressure, there was almost no weak hydrogen-bonding interaction generated due to the dipole conversion in PPI. With the increase of pressure, the orderly structure in PPI first increased and then decreased, showing that during the pressure treatment, the protein first unfolded and then recombined due to the instable conformation of unfolded PPI.

The changes in the secondary conformations of PPI samples after different treatment times were shown in Table 7.12. It was shown from the table that at 3 min, the core conformations in the secondary conformations of PPI accounted for 38.2%, without β-sheet conformation, and the random conformations accounted for 61.8%; at 5 min, the core conformations accounted for 44.6%, and the random conformations accounted for 55.4%; at 15 min, the core conformations accounted for 38.1%, and the random conformations accounted for 61.9%. The change mechanism was similar to that of pressure sample.

From Table 7.13, it was shown that the changes in the secondary conformation of PPIs with different concentrations were significantly different. At low concentration (1%), the irregular coil conformations of PPI basically remained unchanged after treatment at100 MPa. They accounted for 47.7% and 47.5% before and after treatment, respectively, and there were changes only among α-helix, β-sheet, and β-turn conformations. When the concentration increased to 5%, the irregular coil decreased by 7.8% compared to that of untreated PPI; when the concentration increased to 20% again, the irregular coil slightly increased by 1% compared to that of untreated PPI. It was shown from the table that when the concentration was 1%, the core conformations accounted for 37.8% and the random conformations accounted for 62.2% in the secondary conformations of PPI; when the concentration was 5%, the core conformations accounted for 44.6%, and the random conformations accounted for 55.4%; when the concentration was 20%, the core conformations accounted for 47.1%, and the random conformations accounted for 52.9%. For the samples with a concentration of 20%, β-sheet and irregular coil conformations accounted for a large proportion in the secondary conformations of the samples after treatment at 100 MPa, and the untreated samples accounted for 90.6% and 79.8% in all the secondary conformations, respectively; α-helix and β-turn conformations accounted for a large proportion in the secondary conformations of the samples with a concentration of 1% or 5% after treatment at 100 MPa, accounted for 72.2% and 81.3% in all the secondary conformations, respectively. The above results showed that with the increase of concentration, the secondary conformations of PPI gradually became orderly and the protein structure extended.

In general, the ultrahigh-pressure treatment completely changed the secondary conformations of PPI, and the changes in the secondary conformations of PPI were completely different under different treatment conditions (including pressure, time, and concentration). Studies showed that high pressure had an impact on the change of the secondary conformation of PPI (Tu Zongcai 2007; Zhang 2001), and thus changes were caused by solubility, emulsification, and other functional characteristics. In this study, ultrahigh-pressure treatment not only changed the secondary conformations of PPI but also caused changes to gelation (see Figs. 7.1, 7.2, and 7.3). The following conclusions can be reached combining the results of the two: appropriate ultrahigh-pressure treatment can increase the relative contents of α-helix and irregular coil conformations in the secondary conformations of PPI, and such changes are advantageous to the formation of PPI.

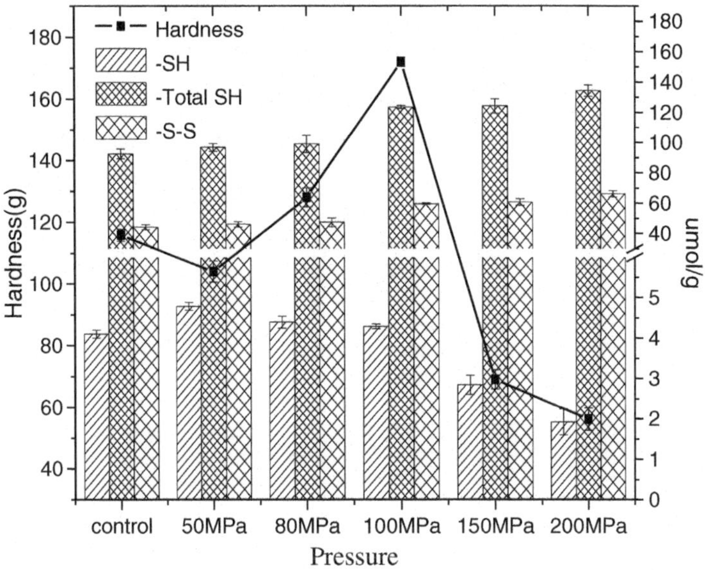

Fig. 7.24 Analysis of sulfhydryl and disulfide bond content in PPI at different pressures

3.2 *Sulfhydryl and Disulfide Bonds*

Sulfhydryl and disulfide bonds are important functional groups of protein, so they play an important role in the functional characteristics of protein. The PPI without ultrahigh treatment and the impacts of different ultrahigh-pressure treatment conditions (different pressure, time, protein concentration) on the contents of sulfhydryl and disulfide bonds and the hardness of curdlan formed were shown in Figs. 7.24, 7.25, and 7.26. The sulfhydryl of protein (including free sulfhydryl and the sulfhydryl hidden in the hydrophobic groups of protein) and total sulfhydryl group (including sulfhydryl and reduced disulfide bonds) were shown in the figure (Wu 2010). It was shown from Fig. 7.24 that the content of sulfhydryl in the PPI not treated at ultrahigh pressure was 4.10 μmol/g pro, which was close to the sulfhydryl content in ovalbumin (4.0 μmol/g pro); higher than skimmed milk (μmol/g pro), flour (2.18 μmol/g pro), and β-lactoglobulin (0.95 μmol/g pro) (Luo Mingjiang et al. 1986); and lower than egg white (50.7 μmol/g pro) (Luo Mingjiang et al. 1986), peanut-concentrated protein (9.09 μmol/g pro) (free sulfhydryl) (Wu 2009), and PPI (12.1 μmol/g pro) (free sulfhydryl). The total sulfhydryl content was 92.53 μmol/g pro, and the disulfide bond content was 44.22 μmol/g pro, which were higher than peanut-concentrated protein (26.22 μmol/g pro) and soy protein (23 μmol/g pro) (Wu 2009) and lower than egg white (79.7 μmol/g pro). Compared with the untreated PPI, with the increase of pressure, the total sulfhydryl content of PPI gradually increased, and sulfhydryl content first increased and then decreased. The sulfhydryl contents in the PPIs after treatment at 150 MPa and 200 MPa decreased

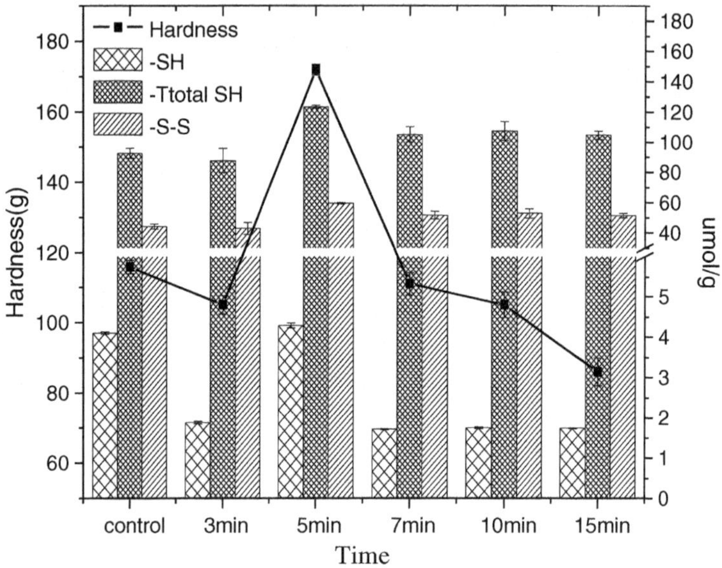

Fig. 7.25 Analysis of sulfhydryl and disulfide bond content in PPI at different times

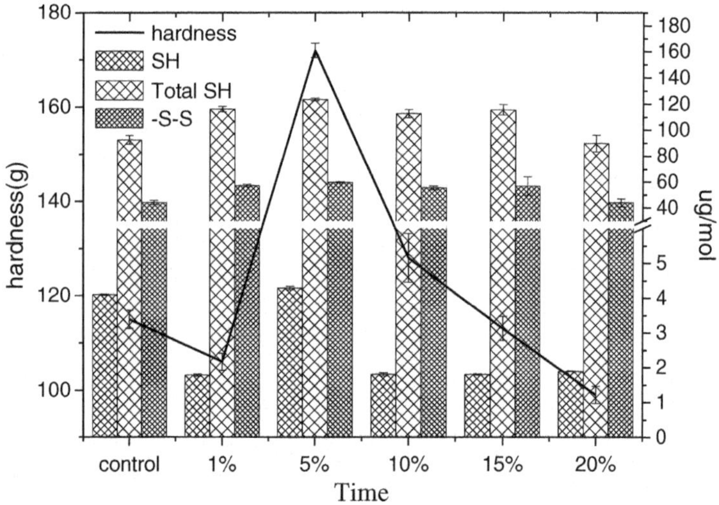

Fig. 7.26 Analysis of sulfhydryl and disulfide bond contents in PPIs with different concentrations

to 2.84 μmol/g pro and 1.92 μmol/g pro, which were all lower than that of untreated samples. This indicated that the pressure caused a significant change in the conformation of PPI, unfolding occurred in PPI, and the sulfhydryls at cysteine residues originally hidden in protein molecules were gradually exposed. However, after the pressure exceeded 100 MPa, the sulfhydryl content in PPI gradually decreased, which might be caused due to -S-S bond combined by exposed sulfhydryls and other groups; therefore, with the increase of pressure, the content of disulfide bond in PPI gradually increased (Wu 2010; Yin Shouwei and Tang Chuanhe 2009). It was shown from the above results that although the pressure contributed to the exposure of the sulfhydryl in PPI, the sulfhydryl exposed in PPI reacted with other groups due to instable structure of PPI treated at a certain pressure. It was also shown from Fig. 7.24 that when the total sulfhydryl content in PPI was 123.39 μmol/g pro, the sulfhydryl content was 4.29 μmol/g pro, and the disulfide bond content was 59.55 μmol/g pro; the PPI gel hardness reached the maximum value.

The above results showed that the three-dimensional structure of peanut changed locally due to ultrahigh-pressure treatment technology. Under the action of ultrahigh pressure, protein molecules might meet the following two cases. First, the protein molecules become loose, the internal disulfide bonds were exposed to the molecular surface, some molecules had more violent activities locally, disulfide bond was broken and reduced to form sulfhydryl, and sulfhydryl content increased. Second, the sulfhydryl groups in protein molecules were also exposed to the molecular surface, and part of them fully combined with the oxygen in the air to form disulfide bonds; in addition, the sulfhydryl groups at the surface of protein molecules were surrounded in the molecules during the protein oxidation and refolding process, so sulfhydryl content decreased. In these two cases, when the pressure was smaller than 100 MPa, the first case was dominant, so that the sulfhydryl content of peanut increased with the increase of pressure; when the pressure was greater than 100 MPa, with the increase of treatment pressure, the second case became the main process of structural change, and thus sulfhydryl content showed a decreasing trend. In general, the sulfhydryl content of PPI after ultrahigh-pressure treatment increased compared with the untreated original sample.

The changes in the sulfhydryl and disulfide bond contents of untreated PPI and PPI that have been treated at 100 MPa for 3–15 min and the hardness after forming curdlan were shown in Fig. 7.25. It was shown from the figure that compared with untreated PPI, the sulfhydryl and total disulfide bond contents of PPI after ultrahigh-pressure treatment first decreased and then increased sharply and the contents of the two reached the maximum values at 5 min, they decreased sharply at 7 min, and the changes in the two slowed down after extending the time. Disulfide bond content first increased and then decreased gradually with the continuous extension of time, and it reached the maximum value at 5 min, which was basically consistent with the change in the hardness of sample after forming curdlan; the hardness of sample reached the maximum value at 5 min. The results showed that the conformation of PPI had completely changed, the protein was de-folded, and more amino acid residues increased. Due to this, the total sulfhydryl content and sulfhydryl in the sample increased sharply after ultrahigh-pressure treatment for 5 min, the active sulfhydryl content decreased significantly, and total sulfhydryl content also decreased after continuing to extend the pressure application time. Its gel hardness was positively correlated with the influence of sulfhydryl content, the hardness reached the maximum value after high-pressure treatment for 5 min, and later the hardness obviously decreased with the extension of time.

The changes in the sulfhydryl and disulfide bond contents of untreated PPI and PPI with a mass concentration of 1–20% after treatment at 100 MPa and the hardness after forming curdlan are shown in Fig. 7.25. It was shown from Fig. 7.26 that under the treatment conditions at 100 MPa, the total sulfhydryl and total disulfide bond contents of PPIs with different mass concentrations first increased and then decreased and the contents of the two in the sample reached the maximum values at the protein concentration of 5%, being 123.39 μmol/g pro and 59.55 μmol/g pro, respectively; the change trend was basically the same as that of hardness after forming curdlan. Compared with the untreated samples, the sulfhydryl content first decreased and then increased sharply with the increase of protein concentration; the sulfhydryl content reached the maximum value when the concentration was 5%, being 4.29 μmol/g pro, it decreased sharply at 7 min, and the sulfhydryl content changed slowly after continuing to increase the concentration, and this change trend was the same as that of the impact of different times on sulfhydryl content of PPI.

Studies showed that some processing methods (such as heating and high pressure) might result in the breakage of sulfhydryl and disulfide bonds and thus caused protein denaturation. Disulfide bond and sulfhydryl were weak secondary bonds to maintain the tertiary structure of protein, and the changes in their contents could reflect the degree of protein denaturation. It was shown from the above results that the contents of disulfide bond and sulfhydryl changed significantly during the ultrahigh-pressure treatment, which had an adverse impact on its functional characteristics such as gelation. In general, the heat gelation separation of PPI could be improved after treatment for 5 min at appropriate high pressure (such as 100 MPa).

3.3 Surface Hydrophobicity

ANS was used as fluorescence probe to measure the surface hydrophobicity indexes of PPI before and after ultrahigh-pressure treatment, and the results are shown in Fig. 7.27. Because the fluorescence spectrum of ANS was very sensitive to environmental changes, it was very sensitive to the conformational change of protein molecules. It was shown from the figure that with the increase of pressure, the surface hydrophobicity indexes of PPI first increased and then decreased, and they all increased obviously compared with that of control ($p < 0.01$). In the protein molecules not treated at ultrahigh pressure, most of the aromatic amino acid molecules that could produce fluorescent were in the protein and surrounded by a variety of nonpolar amino acid residues, so the polarity of the local microenvironment was weaker than that of external aqueous solution of protein molecules, and the surface hydrophobicity was low; in protein denatured after ultrahigh-pressure treatment, the side chain group of the aromatic amino acid molecule was gradually exposed to the aqueous solution, and the polarity of the environment was increased,

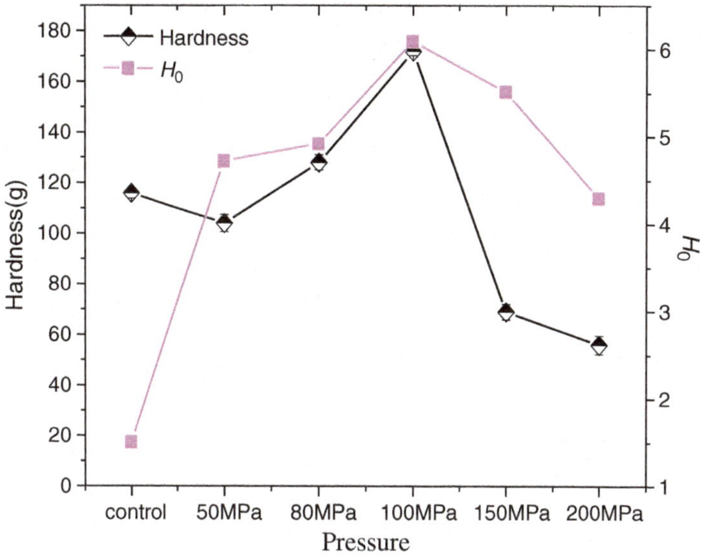

Fig. 7.27 Analysis of surface hydrophobicity indexes of PPI at different pressures

thus resulting in the increase of surface hydrophobic index. At the same time, it was shown from the figure that at 100 MPa, the surface hydrophobic index of PPI reached the maximum value, indicating that more hydrophobic groups in PPI were exposed to the outside and the protein structure was the most loose. After 100 MPa, the surface hydrophobic index of PPI gradually decreased, indicating that the structure of PPI was instable after pressure treatment and the folded protein might accumulate under the condition of increasing pressure. By comparing the changes in the surface hydrophobic index of PPI before and after ultrahigh-pressure treatment, it was shown that ultrahigh-pressure treatment might cause change to the conformations of PPI. And such change to the conformations might change the curdlan hardness of PPI. It was shown from the figure that, after treatment at different pressures, the change trends of surface hydrophobicity index and gel hardness of PPI were basically the same and they all reached the maximum values below 100 MPa. This also indicated that relatively loose protein conformation contributed to gel formation.

The change trends of surface hydrophobic index and gel hardness of untreated PPI and PPIs with different concentrations after treatment at 100 MPa for different times (5–15 min) were shown in Figs. 7.28 and 7.29. It was shown from Figs. 7.28 and 7.29 that the surface hydrophobic index and gel hardness of the samples significantly changed, but the change trends were basically consistent with each other, indicating that at 100 MPa, time and concentration had similar impacts on the exposure of hydrophobic groups in PPI. It was also shown from Figs. 7.28 and 7.29 that when the time was less than 5 min and protein concentration was less than 5%, the amino acid side chain group extended more fully with the extension of time;

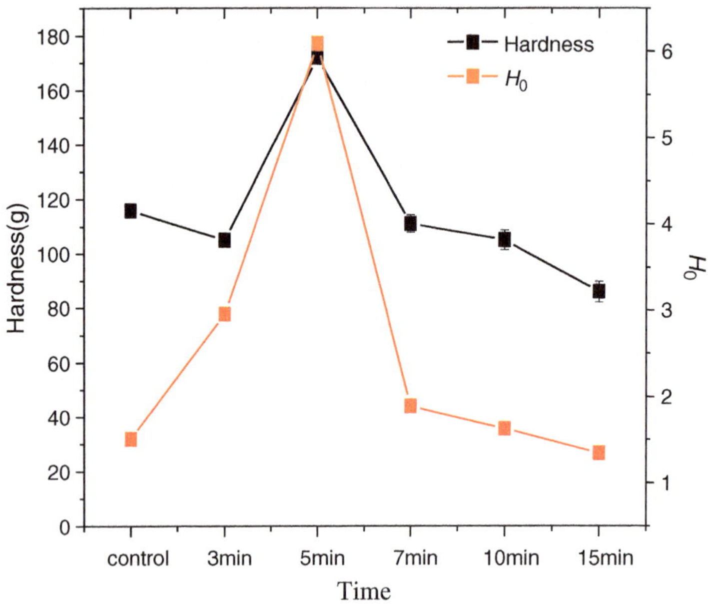

Fig. 7.28 Analysis of surface hydrophobicity indexes of PPI at different times

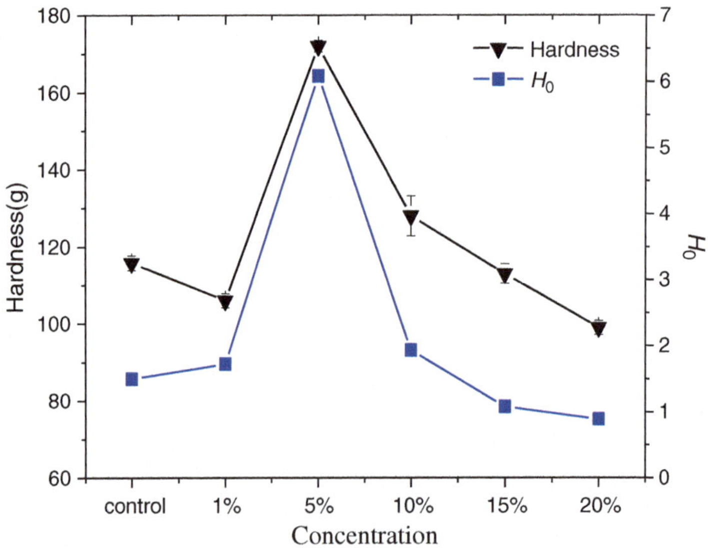

Fig. 7.29 Analysis of surface hydrophobicity of PPIs with different concentrations

more side chains of aromatic amino acid molecules among the protein molecules were exposed, and the hydrophobicity of the samples increased and reached the maximum values at 5 min and 5%, respectively; when the time was greater than

5 min, the expanded proteins reaggregated; when the concentration was greater than 5%, the gaps in aqueous solution became smaller due to the increase of concentration, and protein also reaggregated, resulting in a significant decrease in hydrophobic index.

ANS fluorescence probe method is a classic method to evaluate the surface hydrophobicity of protein, which is a reaction of protein three-dimensional structure in aqueous solution. We can analyze the changes in the three-dimensional structure of protein by observing the changes in the surface hydrophobicity of protein (Molina et al. 2001). Studies showed that hydrophobicity was the primary force to fold proteins into unique three-dimensional structures, and it had a significant impact on the functional characteristics of protein, such as solubility, emulsification, and foamability (Li Ying-qiu and Chen Zheng-xing 2006). After high-pressure treatment, the hydrophobicity of PPI first increased and then decreased, the change trend of which was basically the same as the curdlan formed, indicating that hydrophobicity had a significant impact not only on the solubility, emulsification, foamability, and other functional characteristics of protein but also on its gelation. Proper pressure and treatment time could improve the surface hydrophobicity of protein and improve its gelation.

3.4 Internal Fluorescence Characteristics

The side chain groups of the aromatic amino acid residues in the protein had the characteristic to absorb the incident light from the UV region and emit fluorescence, which could be used to study the changes in the overall spatial conformation of PPI during the ultrahigh-pressure treatment. When the excitation wavelength was 280 nm, the fluorescence spectrum which took tyrosine as the emission group was obtained. The impacts of different pressures, times, and concentrations on the fluorescence spectrum of tyrosine in PPI are shown in Figs. 7.30, 7.31, and 7.32. It was shown from the figures that the maximum fluorescence spectrum of tyrosine moved to the direction of long wave under the conditions of pressure, time, and concentration. With the increase of pressure and PPI concentration, the maximum fluorescence spectrum of tyrosine in PPI increased gradually, showing that the conformation change degree of PPI gradually increased under the treatment conditions of pressure and concentration; when the pressure was 200 MPa, the maximum fluorescence spectrum of tyrosine red shifted to 332 nm from 326.5 nm by 5.5 nm; when the concentration was 20%, the maximum fluorescence spectrum of tyrosine red shifted to 331.5 nm from 326.5 to 5 nm. It was shown that at different pressures and concentrations, the red shift degrees of maximum fluorescence spectrum of tyrosine in PPI were close to each other. This indicated that under these two conditions, the overall spatial conformation change degrees were close and the maximum fluorescence spectrum of tyrosine in protein red shifted, showing that the side chain groups of tyrosine molecules in protein were gradually exposed to aqueous solution, the polarity of the environment gradually increased, and thus

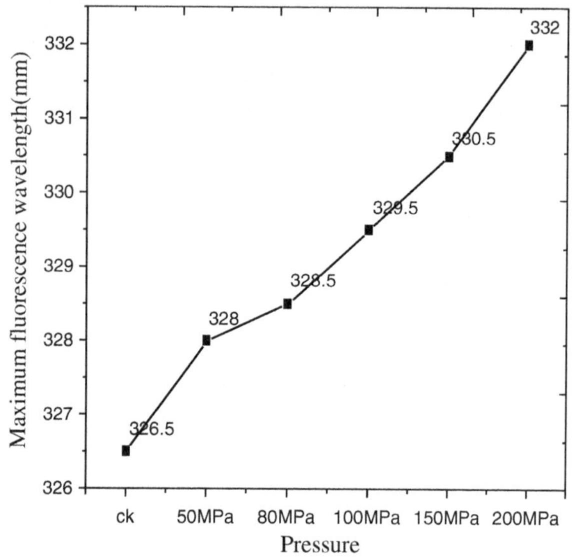

Fig. 7.30 Impact of pressure on the fluorescence wavelength of tyrosine in PPI

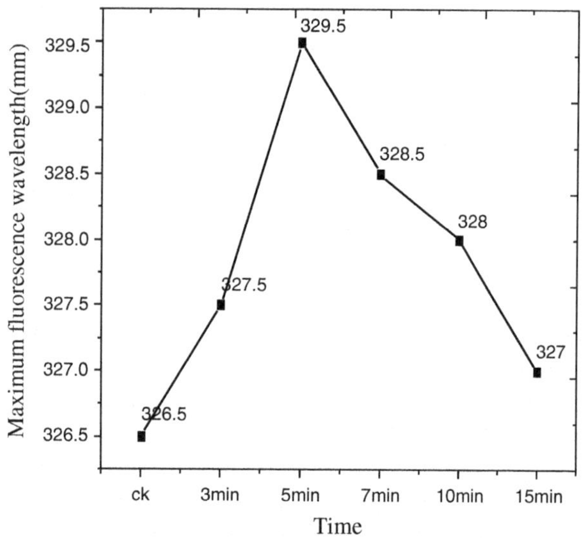

Fig. 7.31 Impact of time on the fluorescence wavelength of tyrosine in PPI

the maximum fluorescence wavelength moved to the direction of long wave. With the increase of time, the maximum fluorescence wavelength of tyrosine in PPI increased first and then decreased, indicating that in a short period of time, the space of PPI could be opened; however, after extending ultrahigh-pressure treatment time, the opened spatial conformation gradually became compact. However, it was shown from Fig. 7.31 that after treatment for 15 min, the maximum fluorescence wavelength of tyrosine in PPI red shifted (1 nm) compared with that of

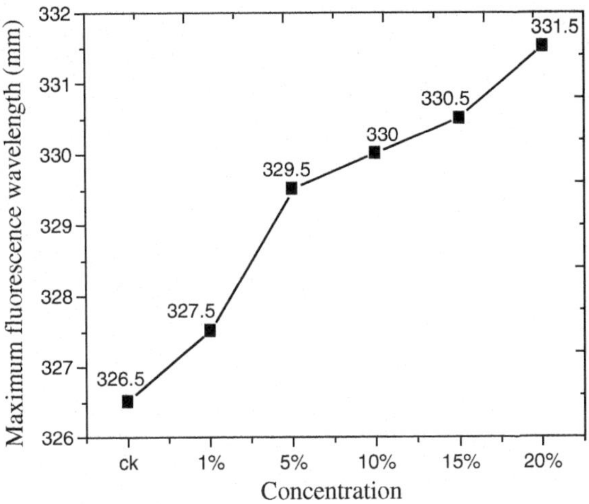

Fig. 7.32 Impact of concentration on the fluorescence wavelength of tyrosine in PPI

untreated PPI, indicating that the spatial conformation of PPI was not recovered to that before treatment.

The fluorescence spectrograms (Figs. 7.33, 7.34, and 7.35) showed the impacts of different pressures, times, and PPI concentrations on the fluorescence wavelength of PPI tryptophan when the excitation wavelength was 295 nm. Generally, the fluorescence of protein was provided by tryptophan residue, tyrosine residue, and phenylalanine residue. When the excitation wavelength was set to 295 nm, the fluorescence spectrum of protein was only provided by tryptophan residue. It was shown from the figure that with the increase of pressure and concentration and the extension of time, the maximum fluorescence wavelength of PPI tryptophan red shifted in different degrees compared with that of untreated PPI, showing that after ultrahigh-pressure treatment, the spatial conformation of PPI was destroyed, and thus the tryptophan residues in internal hydrophobic region were exposed to aqueous solution in different degrees, and PPI had more loose spatial structure than that before treatment. When the pressure was 200 MPa, the maximum fluorescence wavelength of PPI tryptophan red shifted from 329 to 335.5 nm by 6.5 nm; when the concentration was 20%, the maximum fluorescence wavelength PPI tryptophan red shifted from 329 to 336 Nm by 7 nm; under different pressure and concentration conditions, the red shift degrees of the maximum fluorescence wavelength of PPI tryptophan were close to each other, indicating that under these two conditions, the change degrees of overall spatial conformations were close to each other. When the treatment time was 7 min, the maximum fluorescence wavelength of PPI tryptophan red shifted from 329 to 333.5 nm by 4.5 nm; when the treatment time was 15 min, the maximum fluorescence wavelength of PPI tryptophan only red shifted by 1 nm compared with that before treatment, indicating that with the extension of time, the spatial structure of PPI gradually became compact from loose. The above results showed that the spatial conformation of PPI changed in a certain degree during the ultrahigh-pressure treatment.

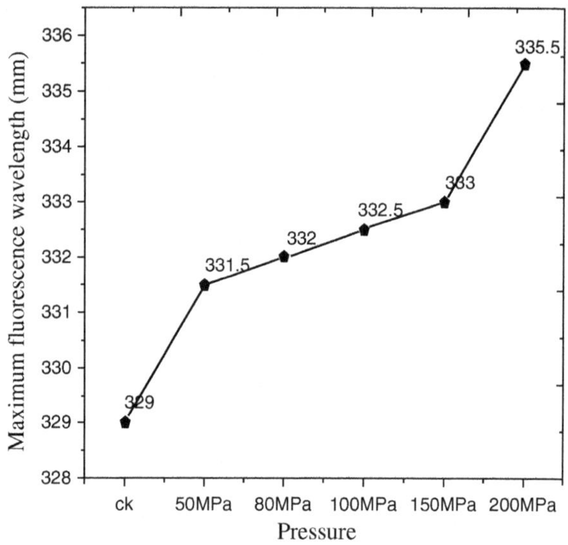

Fig. 7.33 Impact of pressure on the fluorescence wavelength of tyrosine in PPI

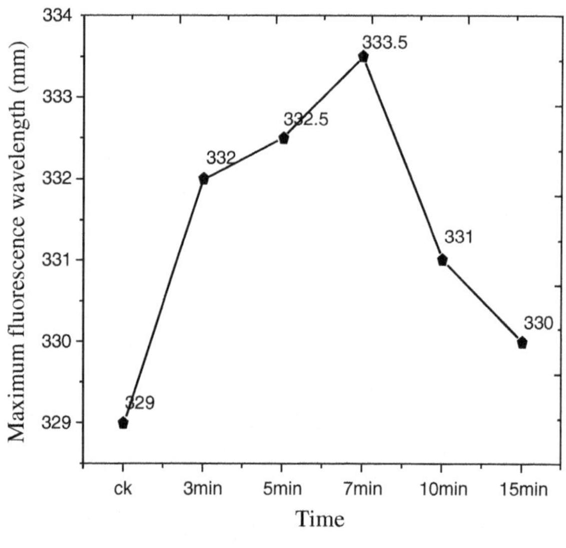

Fig. 7.34 Impact of time on the fluorescence wavelength of tyrosine in PPI

4 Correlation Among Gelation, Physicochemical Characteristics, and Structural Characteristics

Correlation analysis was conducted for the hardness, physicochemical characteristics of curdlan formed by 16 PPI samples, including those untreated and those after ultrahigh-pressure treatment (different pressures between 50 and 200 MPa, different times between 3 and 15 min, and different concentrations between 1% and

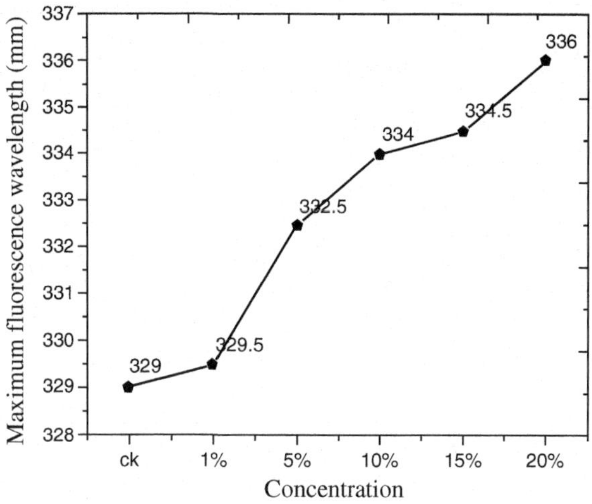

Fig. 7.35 Impact of concentration on the fluorescence wavelength of tyrosine in PPI

Table 7.14 Correlation among the indicators of PPI

Physicochemical characteristics		Structural characteristics	
	Gel hardness		Gel hardness
Oil-binding capacity	−0.035	SH	0.583*
Water-binding capacity	−0.320	Total SH	0.127
Particle size	−0.439	-S-S	0.075
Td1 (conarachin)	0.609**	*H₀*	0.432
Td2 (arachin)	−0.729**	Tyr fluorescence intensity	−0.237
ΔH	−0.213	Tyr fluorescence intensity	−0.132
$T_{1/2}$	0.052	α-Helix content	0.536
Cystine (%)	−0.390	β-Sheet content	−0.345
Hydrophobicity AA (%)	0.335	β-Turn content	−0.056
Polarity AA (%)	−0.311	Irregular coil content	−0.394

Note: df = 15, $\alpha_{0.05} = 0.482$, $\alpha_{0.01} = 0.606$, df = 9, $\alpha_{0.05} = 0.602$, $\alpha_{0.01} = 0.735$, *shows that 0.05 level is significant, **shows that 0.01 level is extremely significant

20%), as well as their tertiary structures. The sample sizes for the correlation analysis of secondary structure were 10, and the results were shown in Table 7.14. It was shown from the table that the correlation coefficient between gel hardness and the denaturation temperature of conarachin among physicochemical characteristics was 0.609, and they were positively correlated with each other significantly, and the correlation coefficient between it and the denaturation temperature of arachin was −0.727, and they were negatively correlated with each other significantly. This further showed that the denaturation of conarachin in PPI was aggravated, which was unfavorable to the increase of gel hardness, whereas arachin was opposite to it. At the same time, it was shown from the table that the correlation coefficient of PPI gel hardness was −0.439, and they were negatively correlated with each other strongly. It was shown that the increase of PPI particle size was not conducive to the increase of gel hardness. Except the above physicochemical indicators, the correlation coefficient between PPI gel hardness and other physicochemical indicators was small with weak correlation.

It was shown from the correlation between PPI gel hardness and structural characteristics that the correlation coefficient between gel hardness and SH content in the sample was 0.583 and they were positively correlated with each other significantly, and the correlation coefficients between it and the surface hydrophobic index and α-helix content in the samples were 0.432 and 0.536, respectively, and they were positively correlated with each other strongly; in addition, the correlation coefficients between it and other structural characteristic indicators were small and the correlations were weak. This further indicated that the changes in SH content, surface hydrophobic index, and α-helix content would directly cause change to the gel hardness of sample, and the gel hardness of sample would increase with the increase of the contents of the above indicators in the sample. The analysis results of correlation between physicochemical characteristics and structural characteristics and gel hardness showed that although a series of changes occurred in the physicochemical characteristics and structural characteristics of PPI after ultrahigh-pressure treatment, there were only six indicators that had strong correlation with gel hardness, including the denaturation temperatures of conarachin and arachin in PPI, particle size, α-helix content, surface hydrophobicity index, and SH content. Among these six indicators, the indicators with the largest correlation were the denaturation temperatures of conarachin and arachin and SH content. This further showed that PPI gel hardness was improved mainly due to the action of the above factors.

Ultrahigh-pressure treatment significantly improved the thermal curdlan properties of PPI, the mechanism of which was mainly understood from the structural changes in the protein. It was found after analyzing the structure of PPI after ultrahigh-pressure treatment (including fluorescence spectra and circular dichroism spectrum) that the molecular structures of natural PPI were compact, the secondary structure was mainly composed of β-sheet (32.1%), the intramolecular hydrogen bond interaction was strong, and its contribution rate was 6.89%, and the secondary structure changed significantly after ultrahigh-pressure treatment, peptide bond was opened, and amino acid conformation changed to form the conformation mainly composed of α-helix. At 100 MPa, its content increased from 9.6% to 41.4%; while β-sheet decreased to 3.2%, the contribution rate of intramolecular hydrogen bond decreased to 3.42%; meanwhile, protein molecules unfolded; the microenvironment polarities of tyrosine and tryptophan increased; the maximum fluorescence wavelengths red shifted to 330.5 nm from 326.5 nm and 333 nm from 329 nm, respectively; a large number of hydrophobic groups were exposed; part of the hydrophobic groups between molecules formed a small amount of uniform aggregates through hydrophobic interaction reaction; and a large number of unreacted hydrophobic sites make the aggregates have very high surface hydrophobicity index. The surface hydrophobicity index of protein increased from 1.50 to 6.09, a large amount of intramolecular free sulfhydryl were exposed, the total sulfhydryl content increased from 92.53 to 123.38 μmol/L, a small amount of disulfide bonds were broken, the adjacent free sulfhydryls form disulfide bonds through oxidation, and the content increased from 44.21 to 59.55 μmol/L. Due to the above structure changes, the PPI gel hardness was enhanced eventually.

References

Angsupanich K, Edde M, Ledward DA. Effects of high pressure on the myofibrillar proteins of cod and turkey muscle[J]. J Agric Food Chem. 1999;47(1):92–9.

Apichartsrangkoon A. Effects of high-pressure on rheological properties of soy protein gels [J]. Food Chem. 2003;80:55–60.

Arntfield SD, Murray ED. The influence of processing parameters on food protein functionality I. Differential scanning calorimetry as an indicator of protein denaturation. Can Inst Food Sci Technol J. 1981;14(4):289–94.

Arnfiled SD, Murray ED, Ismond MAH. Dependence of thermal properties as well as network microstructure and rheology on protein concentration for ovalbumin and vicilin [J]. J Texture Stud. 1990a;21:191–212.

Arnfiled SD, Murray ED, Ismond MAH. Influence of salts on the microstructural and rheological properties of heat-induced protein networks from ovalbumin and vicilin [J]. J Agric Food Chem. 1990b;38:1335–43.

Briscoe B, Luckham PF, Staeritz KU. Pressure induced changes in the gelation of milk protein concentrates [J]. Trends High Pressure Biosci Biotechnol, Proc. 2002;19:445–52.

Cong-Gui Chen, Borjigin Gerelt, Shao-Tong Jiang, Tadayuki Nishiumi, Atsushi Suzuki. Effects of high pressure on pH, water-binding capacity and textural properties of pork muscle gels containing various levels of sodium alginate. Asian Australas J Anim Sci. 2006;19 (11):1658–64.

Desrumaux A, Mareand J. Formation of sunflower oil emulsions stabilized by whey proteins with high-Pressure homogenization(up to 350 MPa): effect of pressure on emulsion characteristics [J]. Int J Food Sci Technol. 2002;37:263–9.

Du Yin. Study on the preparation and gel properties of the peanut protein fractions[D]. Beijing: Chinese Academy of Agricultural Sciences; 2012.

Gosal WS, Ross-Murphy SB. Gelation of whey protein induced by high pressure [J]. Curr Opin Colloid Interface Sci. 2005;34:188–94.

Hanchang H, Zhaofeng J, Zhu Hongji. The methods of protein conformation predicted by uv-circular dichroism [J]. Chemistry, 2007.

Hongkang Zhang, Lite Li, Eizo Tatsumi, Sabir Kotwal. Influence of high pressure on conformational changes of soybean glycinin. Innovative Food Sci Emerg Technol. 2003;4(3):269–75.

Jafari SM, He Y, Bhandari B. Production of sub-micron emulsions by ultrasound and microfluidization techniques [J]. J Food Eng. 2007;82:478–88.

Kolb G, Viardot K, Wagner G, et al. Evaluation of a new high-Pressure dispersion unit(HPN)for emulsification [J]. Chem Eng Technol. 2001;24:293–6.

Li Ying-qiu, Chen Zheng-xing. Effects of high pulsed electric field on hydrophobicity and sulfhydry groups content of soybean protein isolated [J]. Food Chem. 2006;27(05):40–2.

Luo Mingjiang, Luo Chunxia, Wu Ganxiang. Determination of sulfhydryl group and disulfuric chain of protein in food by Ellman's reagent colorimetric method [J]. J Zhengzhou Grain Coll. 1986;23:92–5.

Maruyama N, Katsube T, Wada Y, Moon Hun O, Rosa APBDL, Okuda E, Nakagawa S, Utsumi S. The roles of the N-linked glycans and extension regions of soybean beta-conglycinin in folding, assembly and structural features. Eur J Biochem. 1998;258(2):854–62.

Messens W, Van Camp J, Huyghebaert A. The use of highpressure modify the functionality of food proteins [J]. Trends Food Sci Technol. 1997;8:107–12.

Molina E, Papadopoulou A, Ledward DA. Emulsifying properties of high pressure treated soy protein isolate and 7S and 11S globulins. Food Hydrocoll. 2001;15(3):263–9.

Molina E, Defaye AB, Ledward DA. Soy protein pressure-induced gels. Food Hydrocoll. 2002;16 (6):625–32.

Murchie LW, Cruz-Romero M, Kerry JP, Linton M, Patterson MF, Smiddy M, Kelly AL. High pressure processing of shellfish: a review of microbiological and other quality aspects. Innovative Food Sci Emerg Technol. 2005;6(3):257–70.

Pinterits A, Arntfield SD. Improvement of canola protein gelation properties through enzymatic modification with transglutaminase. LWT Food Sci Technol. 2008;41(1):128–38.

Prakash V, Narasinga RMS. Physicochemical properties of oilseed proteins [J]. Crit Rev Biochem Mol Biol. 1986;20(3):286–98.

Privalov PL. Stability of proteins: proteins which do not presenta single cooperative system [J]. Adv Protein Chem. 1982;35:1–104.

Puppo C, Chapleau N, Speroni F, de Lamballerie-Anton M, Michel F, Añón C, Anton M. Physicochemical modifications of high-pressure-treated soybean protein isolates. J Agric Food Chem. 2004;52(6):1564–71.

Puppo MC, Speroni F, Chapleau N, de Lamballerie M, Añón MC, Anton M. Effect of high-pressure treatment on emulsifying properties of soybean proteins. Food Hydrocoll. 2005;19 (2):289–96.

Shi Yanguo, Sun Bingyu. The effects of ultrasonic wave on oil adsorption of alcohol precipitated soy protein concentrated [J]. Sci Technol Food Ind. 2005;26(10):84–86,89.

Sreerama N, Woody RW. Methods Enzymol [J]. 2004;383:318–51.

Tu Zongcai. Improvement of protein properties through dynamic high pressure micro-fluidization treatment and preliminary study on its mechanism [D]. Jiangxi: Nanchang University; 2007. p. 185–90.

Wang SM, Zhao JF, Huang MX, et al. Study on the mixed symergistic antioxidative activity from tea polyphenols and Vc to emulsion porkfat [J]. Sci Technol Food Ind. 2000;21(10):22–5.

Wang XS, Tang CH, Li BS, et al. Effects of high-pressure treatment on some physicochemical and functional properties of soy protein isolates[J]. Food Hydrocoll. 2008;22:560–7.

Wu HW. Study on preparation and gel-forming mechanism of peanut protein concentration and its application [D]. Beijing: Chinese Academy of Agricultural Science; 2009.

Wu W. Effects of protein oxidation on structure and gel properties of soy protein [D]. Wuxi: Jiangnan University; 2010. p. 25–6.

Xu ZR. Regression analysis and experimental design[M]. Beijing: China Agriculture Press; 1997. p. 102–58.

Yin Shouwei, Tang Chuanhe. Effect of micro-fluidization treatment on conformational and functional properties of kidney bean(Phaseolus vulgaris L.) protein isolates [J]. J South China Univ Technol. 2009;37(10):312–7.

Zhang HK. High pressure effects on biomacromolecules [D]. Beijing: China Agriculture University; 2001. p. 80–3.

Zhou SM. The function of the protein in the bakery product [J]. Cereal Oils. 1999;1:25–6.

Zong Wei, Chen Yi-ping. Effect of ultra high pressure on solubility of isolated peanut protein [J]. Cereals & Oils. 2007;10:16–7.

Zong Wei, Chen Yi-ping. Effect of ultra high pressure on the emulsifying ability of peanut protein isolate [J]. China Oils Fats. 2008;33(3):26–8.

Chapter 8
Gelation Improvement of Peanut Protein Component

Arachin and conarachin are the most abundant and important components in peanuts, and their properties can affect all the functional properties of peanut protein. Therefore, it is of practical significance to optimize the preparation method of peanut protein component and analyze the factors that affect the functional properties of peanut protein components and formation mechanisms for the promotion of healthy development of peanut protein processing industry and increase of utilization rate of peanut protein. The author's research team carried out the isolation and extraction for arachin and conarachin by using the ammonium sulfate precipitation method and freezing precipitation method, determined the process conditions of peanut protein component isolation of two methods, and investigated the external factors (protein concentration, heating temperature, time, pH value, and salt ion strength) that affected the heat-induced gel of arachin and conarachin with the peanut protein component as the research object so as to improve the gelation of peanut protein component and provide a theoretical basis for the practical application in the food industry.

1 Protein Component Isolation Method

At present, the peanut protein components have been mainly prepared by using the ammonium sulfate precipitation method and freezing precipitation method at home and abroad. Based on the salt-out characteristics of protein, the peanut component isolation and extraction by using the ammonium sulfate precipitation method can separate the arachin, conarachin I, and conarachin II. Due to the condition restrictions, the ammonium sulfate precipitation method is only applicable to the small-scale extraction and preparation in the laboratory and not suitable for the industrial production (Chiou 1990). Freezing precipitation method is to isolate the peanut protein components by using the low-temperature cryoprecipitation characteristics of arachin (Basha and Pancholy 1982; Neucere 1969), the degree of protein

Q. Wang, *Peanut Processing Characteristics and Quality Evaluation*,
DOI 10.1007/978-981-10-6175-2_8

denaturation is low, and the operation is simple and convenient with the potential of industrialization, but the purity of protein components isolated is lower than that isolated by using the ammonium sulfate precipitation method. At present, we can only separate the arachin and conarachin preliminarily and then analyze the influencing factors (cryoprecipitation time, protein concentration, and ion strength) in the isolation process. Du Yin (2012) and Wang et al. (2013), and Feng Xiaolong (2014) isolated the arachin and conarachin by using the ammonium sulfate precipitation method and low-temperature cryopreservation method (short for cryopreservation) on the basis, analyzed and determined the components through SDS-PAGE, calculated the purity of the various components (the purity of the component was determined by using the sum of optical density calculated through the corresponding strips of this component to divide the sum of the optical density of all the strips), and thus improved the gelation of peanut protein components on the basis.

1.1 Ammonium Sulfate Precipitation Method

As a protein fractionation method, the ammonium sulfate precipitation method belongs to the solubility classification method, and it is a selective precipitation method. The differences of surface polarity range, the number of charges, and the charge distribution on the protein surface of various kinds of proteins caused the difference in the salting-out order, and thus the protein was fractionated.

1.1.1 Arachin

The arachin could be separated out from the solution by using the ammonium sulfate with the saturation of 0–40%. According to the differences of buffer solutions and saturation degrees of ammonium sulfates used by different scientific research workers, there were the following methods:

1. Tombs method (1965)

   The defatted peanut powder was added to 10% NaCl (w/v) solution at a ratio of 10% (w/v) and stirred for 4–6 h at 25 ± 2 °C. The extracting solution was centrifuged at $4300 \times g$ for 45 min at 25 ± 2 °C. The precipitation with the final saturation of 40% (w/v) was made by adding the solid ammonium sulfate and centrifuged at $6700 \times g$ for 30 min at 25 ± 2 °C. The precipitation was redissolved in 10% NaCl, and the step was repeated three times so as to obtain the pure arachin. The precipitated arachin was redissolved in the distilled water, dialyzed repeatedly, freeze-dried, and stored at 4 °C for later use.
2. Yamada method (1979)

   The defatted peanut powder was dissolved in the 0.01 mol/L phosphate buffer solution (pH 7.9) containing 0.5 mol/L NaCl and stirred and extracted for 1 h;

the precipitate was removed by centrifuging, the solid ammonium sulfate powder was added to the supernatant until its saturation reached 40%, the solution should be stirred slowly in the addition process, and the addition should be completed within 5-10 min. The above mixture was kept in a refrigerator at 4 °C for 3 h. The precipitate was dissolved in a minimum amount of phosphate buffer solution, dialyzed for 24 h with the cooled double distilled water, and freeze-dried to obtain the arachin component; the arachin component was stored at 4 °C for later use.

3. Yang Xiaoquan method (2001)

   The defatted peanut powder was added to the precooled 10 mmol/L phosphate buffer solution (pH 7.9) containing 2 mol/L NaC1, 1 mmol/L phenylmethylsulfonyl fluoride (PMSF), and 10 mmol/L β-mercaptoethanol, dispersed at high speed for 20 min, and stirred and extracted for 3 h at 4 °C. The solution was centrifuged at 16,000 r/min for 30 min by using a high-speed freeze centrifuge. The precipitate was removed, the solid ammonium sulfate powder was added to the supernatant under the effect of the magnetic stirrer until the saturation reached 40%, and then the mixture was placed for 3 h at room temperature (25 °C) and centrifuged at 1280 r/min for 20 min when the precipitate was completely separated; the precipitate was collected, dissolved in the minimum amount of phosphate buffer solution, dialyzed with cooled double distilled water for 24 h, and freeze-dried to obtain the arachin component; the arachin component was stored at 4 °C for later use.

4. Govindaraju method (2006)

   The defatted peanut powder was dissolved in 0.01 mol/L phosphate buffer solution (pH 7.9) containing 0.5 mol/L NaCl at a ratio of l:10 (w/v), stirred and extracted for 1 h at 4 °C, and then centrifuged by using the high-speed freezing centrifuge for 30 min at 10,000 r/min. The solid ammonium sulfate powder was added slowly in the supernatant in the continuous slow stirring process, so that the saturation reached 18 g/100 mL; the mixture was placed for 3 h at 4 °C and then centrifuged for 30 min at 10,000 r/min; the precipitate was resolved in the phosphate buffer solution, the ammonium sulfate was added again until the saturation reached 18 g/100 mL, and then the mixture was centrifuged for 30 min at 10,000 r/min; the precipitate was collected, dissolved in the smallest amount of phosphate buffer solution, dialyzed for 24 h with cooled double distilled water, and then freeze-dried to obtain the crude arachin; the arachin was stored at 4 °C for later use.

Du Yin (2012) optimized and improved the methods researched by predecessors and established a new method to prepare arachin through ammonium sulfate precipitation isolation, as shown in Fig. 8.1.

The above peanut meal total protein extraction liquid was added to the ice bath quickly and cooled to about 4 °C, and the fine-grained solid ammonium sulfate powder was added under the magnetic stirring slowly (the addition was completed within 5–10 min), so that the saturation of ammonium sulfate reached 40%; the mixture was placed for 3 h and centrifuged for 20 min at 10,000 × g to collect the

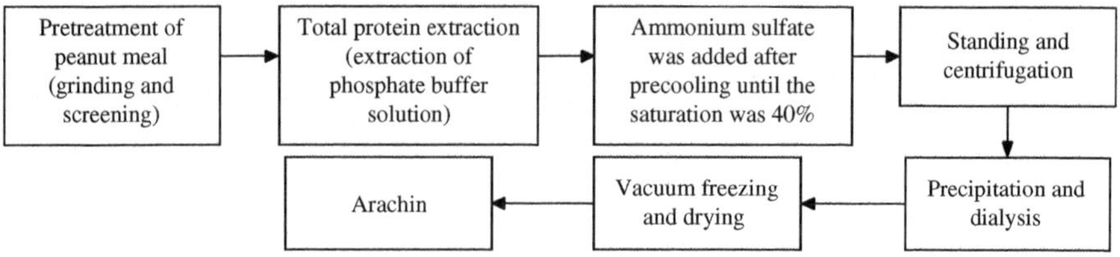

Fig. 8.1 Flow diagram of arachin component preparation through improved ammonium sulfate precipitation method

precipitate. The precipitate was put into a dialysis bag (pre-boiled and pretreated through ethanol) and dialyzed with double distilled water for 24 h (4 °C, slowly stirring), and the water was changed four to five times. The sample after vacuum freezing and drying was the arachin. The arachin was prepared through SDS-PAGE analysis with the purity of 85.53 $\pm$ 0.86% (Fig. 8.3).

1.1.2 Conarachin

1.1.2.1 Conarachin II

Conarachin II was called 7S component in the total peanut protein according to its sedimentation coefficient. Due to the difference of variety, the conarachin II accounted for 15–25% in the total protein, and the main method to isolate the conarachin II was the ammonium sulfate precipitation method (Prakash and Rao 1986). Johnson and Naismith (1953) used the following steps: extract the peanut protein powder with the ammonium sulfate with the saturation of 0.4; continue to add the ammonium sulfate until its saturation reached 0.65, and isolate the precipitated protein; and add the ammonium sulfate in the supernatant until the saturation was 0.85, and isolate the precipitate. Through the ultracentrifugation analysis, most of the conarachin II and a large proportion of conarachin I (2S) were obtained by using this method. Naismith and McDavid (1958) extracted the peanut protein with 10% NaCl solution and precipitated with the ammonium sulfate with the saturation of 0.85. The precipitate was dissolved in phosphate buffer solution at pH 7.0, and the pH was adjusted to 4.7 ($\mu = 0.025$). The supernatant was cooled at 4 °C for precipitation. The precipitation was conducted again by adding alums, which made the ammonium sulfate precipitation with the saturation of 0.65–0.85 obtained by using Johnson method be purified, but there was still conarachin I (Prakash and Narasinga 1986). Li Yin et al. (1998) and Yang Xiaoquan et al. (2001) continued to add the ammonium sulfate to the solution (the ammonium sulfate saturation had been 40 g/100 mL, and the arachin had been removed by centrifuging), so that its saturation reached 65 g/100 mL; the solution was placed for 3 h and centrifuged for 20 min at 10,000 $\times$ g; the precipitate was collected, dialyzed, and freeze-dried to obtain the conarachin II.

1.1.2.2 Conarachin I

1. High-salt buffer solution extraction and ammonium sulfate fractionation precipitation method (Lin Lu 1994)

   The skin of peanut seeds was peeled, the cotyledons were taken and grinded in powder, and then the powder was defatted with n-hexane. After the powder was defatted and dried, it was added in the precooling 10 mmol/L phosphate buffer solution (pH 7.9, containing 2 mol/L NaCl), 10 mmol/L β-mercaptoethanol, and 1 mmol/L phenylmethylsulfonyl fluoride. It was mashed for 10 min at a high speed and extracted overnight at 4 °C. It was centrifuged for 30 min (4 °C) at 20,000 × g. The fractional precipitation was carried out for the supernatant by using ammonium sulfate, and the protein components precipitated by 65–85% (w/v) of ammonium sulfate were put into the dialysis bag and dialyzed for 24 h with redistilled water (4 °C). After the sample was frozen and dried, it is the extracted and isolated crude conarachin I. The crude conarachin I was dissolved in a low-salt extraction buffer solution in a proportion and crossed the Sephadex G-100 column (2.6 cm × 50 cm). The eluent was the same as the extraction buffer solution. The flow rate of peristaltic pump was 25 mL/h, the collection was conducted every 10 min automatically, 280 nm absorption was absorbed, and the required peak sample was collected and dialyzed for 24 h (4 °C) with redistilled water after combination. The sample frozen and dried was the conarachin I.
2. Low-salt buffer solution extraction and heating isolation method (Yang Xiaoquan et al. 1998)

   The defatted powder was added to 50 mmol/L Tris-HCl buffer solution (pH 8.2, containing 0.5 mol/L NaCl), 10 mmol/L β-mercaptoethanol, and 1 mmol/L phenylmethylsulfonyl fluoride. The solution was mixed by vortex vibration for 10 min and extracted overnight at 4 °C. The mixture was centrifuged at 23,000 × g for 20 min (4 °C) and extracted twice; the supernatant was combined, a water bath of 100 °C was conducted for 10 min, an ice batch was carried out for 15 min for cooling, and then the mixture was centrifuged at 10,000 g for 10 min; the supernatant was concentrated with PEG-6000. The sample frozen and dried was the conarachin I.

Du Yin (2012) optimized and improved the methods researched by predecessors and established a new method to prepare conarachin through ammonium sulfate precipitation isolation, as shown in Fig. 8.2.

The above peanut meal total protein extraction liquid was added to the ice bath quickly and cooled to about 4 °C, and the fine-grained solid ammonium sulfate powder was added under the magnetic stirring slowly (the addition was completed within 5–10 min), so that the saturation of ammonium sulfate reached 40%; the precipitate was removed by centrifuging, and the solid ammonium sulfate was added to the supernatant according to the above method so that the saturation reached 65%; the mixture was placed for 3 h and centrifuged for 20 min at 10,000 × g to collect the precipitate; the solid ammonium sulfate was added to

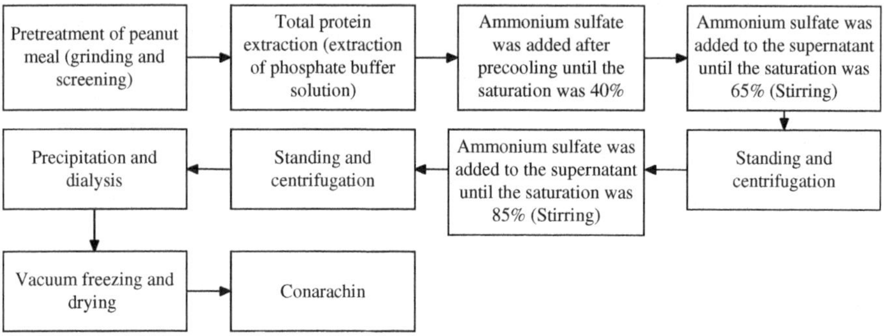

Fig. 8.2 Flow diagram of conarachin component preparation through improved ammonium sulfate precipitation method

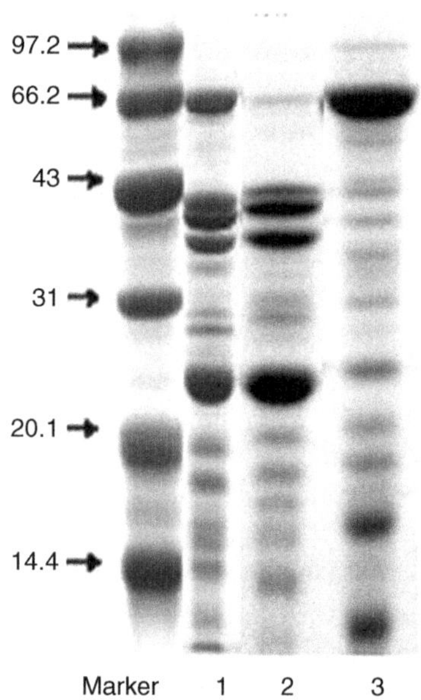

Fig. 8.3 Electrophoretograms of arachin and conarachin prepared by using improved ammonium sulfate precipitation method. Note: *1*, peanut protein isolate; *2*, arachin; *3*, conarachin

the supernatant that the precipitate had been removed until the saturation reached 85%; the mixture was placed for 3 h and centrifuged for 20 min at 40,000 × g to collect the precipitate; the precipitate was put into a dialysis bag (pre-boiled and pretreated through ethanol) and dialyzed with double distilled water for 24 h (4 °C, slowly stirring), and the water was changed four to five times. The sample after vacuum freezing and drying was the conarachin. The conarachin was prepared through SDS-PAGE analysis with the purity of 75.81 ± 1.02% (Fig. 8.3).

The electrophoretograms of arachin and conarachin prepared by using the ammonium sulfate precipitation method are shown in Fig. 8.3.

1.2 Low-Temperature Cryoprecipitation Method

In 1969, Neucere reported the low-temperature cryoprecipitation method for the first time. Peanut grains were grinded and dissolved in the 0.2 mol/L phosphate buffer (pH 7.9); it was stirred for 5 min at 0 °C and centrifuged at 37,000 × g for 30 min (twice) to remove all insoluble substances; after the supernatant dialysis was carried out for 24 h, the low-temperature cryoprecipitation was conducted for 17 h at 2 °C. Then the solution was centrifuged at 3000 × g for 20 min. The precipitate was redissolved in 800 ml of buffer solution and cooled for 17 h at 2 °C again. After the low-temperature centrifugation was completed, the obtained precipitate was the arachin.

The author's research team optimized and improved the process conditions on the basis of predecessors and established a new method to prepare arachin and conarachin through low-temperature cryoprecipitation isolation (Feng Xiaolong (2014) and Wang et al. 2013), as shown below.

1.2.1 Arachin

The peanut protein powder was prepared together with the phosphate buffer solution with a certain pH and ionic strength in proportion; the mixture was stirred for 1 h at room temperature and then centrifuged for 30 min at 8000 r/min. The supernatant was placed in a 2 °C environment for a certain time and then centrifuged at 8000 r/min for 30 min at 2 °C. The precipitate after vacuum freezing and drying was the arachin. The process flow is shown in Fig. 8.4:

Through the optical density analysis, the purity of arachin prepared through the above process was $61.07 \pm 0.15\%$ which was low, so it was necessary to further explore the extraction process.

In the extraction process, the protein content in the supernatant extracted from the peanut protein powder the first time was determined and recorded as M_1; the supernatant extracted the second time was cryoprecipitated overnight, and the protein content in the supernatant after centrifugation was recorded as M_2; the protein (conarachin) content in the cryoprecipitation supernatant was calculated,

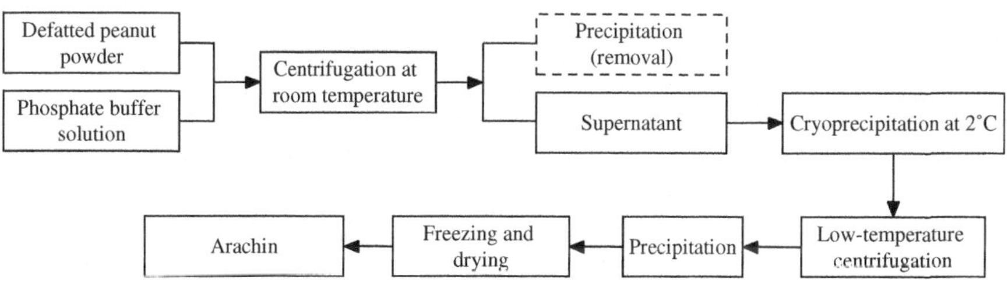

Fig. 8.4 Flow diagram of arachin component preparation through cryoprecipitation method

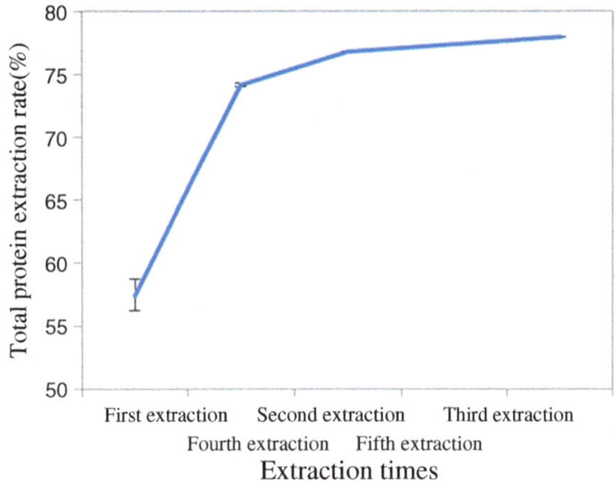

Fig. 8.5 Impact of extraction time on extraction rate of peanut protein

and the precipitate obtained after low-temperature centrifugation was the arachin. Therefore

$$\text{Extraction rate of arachin } (\%) = \frac{M_1 - M_2}{\text{Protein content in peanut}} \times 100$$

1.2.1.1 Extraction Times of Total Peanut Protein

When the solid-liquid ratio was 4:10, the ion strength was 0.2 M, and the pH of extraction solution was 7.5, the extraction times were selected as once, twice and thrice, and four and five times. The optimal extraction time was determined according to the total protein extraction rate of peanut protein. The experiment results are shown in Fig. 8.5.

From Fig. 8.5, it could be seen that the extraction rate of peanut protein increased with the increase of the number of extraction times, there were no significant changes in the extraction rate when the number of extraction times exceeded 2, and the extraction rate of peanut protein only increased by 2.62% during the third extraction; too much extraction would not only increase the cost but also produce a large number of industrial waste water; considering the energy-saving emission reduction of factory, the optimal number of extraction times was 2. On the basis, the second extraction was carried out when the single-factor experiments were carried out for the globulin extraction.

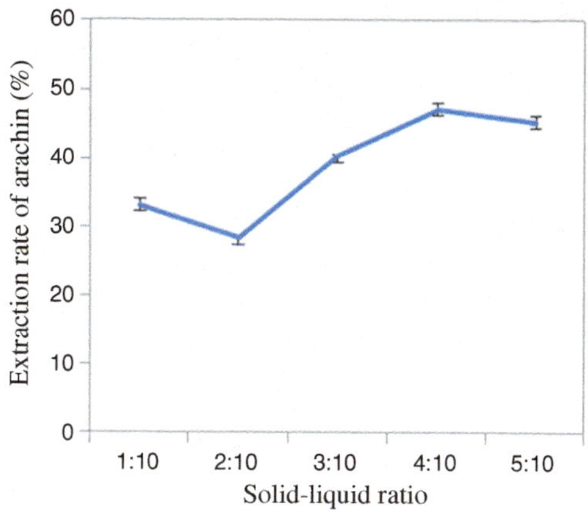

Fig. 8.6 Impact of solid-liquid ratio on extraction rate of peanut protein

1.2.1.2 Single-Factor Experiment of Arachin

1. Solid-liquid ratio

When the ion strength was 0.2 M, the pH of extraction solution was 7.5, and the freezing time was 16 h, the extraction was carried out once, and the optimal solid-liquid ratio was determined according to the extraction rate of arachin. The experiment results are shown in Fig. 8.6.

It could be seen from Fig. 8.6 that the extraction rate of arachin increased gradually with the decrease of solid-liquid ratio but there were no significant changes in the extraction rate of globulin when the solid-liquid ratio was 4: 10 and 5: 10. Therefore, 4: 10 was determined as the optimal solid-liquid ratio.

2. Ion strength

When the solid-liquid ratio was 4: 10, the pH of extraction solution was 7.5, and the cryoprecipitation time was 16 h, the extraction was carried out once, and the optimal ion strength was determined according to the extraction rate of arachin. The experiment results are shown in Fig. 8.7. It could be seen from Fig. 8.7 that the extraction rate of globulin increased gradually with the increase of ion strength but when the ion strength was 0.3 M, there were no significant changes in the extraction rate of globulin and the curve showed a stable trend. Therefore, 0.3 M was determined as the optimal ion strength.

3. pH of solution

When the solid-liquid ratio was 4: 10, the ion strength was 0.2 M, and the cryoprecipitation time was 16 h, the extraction was carried out once, and the optimal pH of solution was determined according to the extraction rate of arachin. The experiment results are shown in Fig. 8.8. It could be seen obviously from Fig. 8.8 that the extraction rate of globulin increased gradually with the increase of

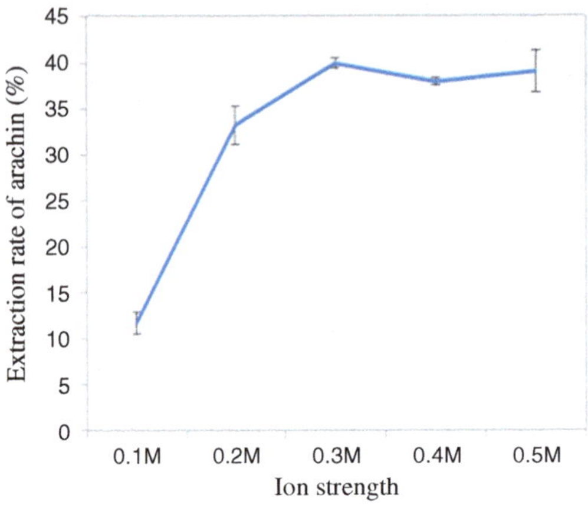

Fig. 8.7 Impact of ion strength on extraction rate of peanut protein

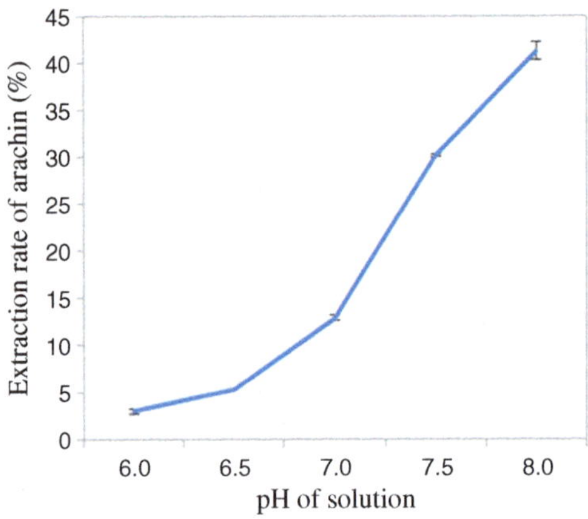

Fig. 8.8 Impact of pH of solution on extraction rate of peanut protein

pH of solution. Due to the low denaturation of protein extracted by using cryoprecipitation method, the pH of solution should not be too high; otherwise the protein would be denatured. In addition, there were some reports on the peanut protein component isolation by using the cryoprecipitation at abroad (Neucere 1969), and the pH of extraction solution should be maintained within 8.0; therefore, 7.5 was selected as the suitable pH.

4. Cryoprecipitation time

When the solid-liquid ratio was 4: 10, the ion strength was 0.2 M, and the pH of solution was 7.5, the extraction was carried out once, and the optimal cryoprecipitation time was determined according to the extraction rate of arachin. The experiment results are shown in Fig. 8.9. It could be seen from Fig. 8.9 that the extraction rate of arachin increased with the extension of cryoprecipitation time. In

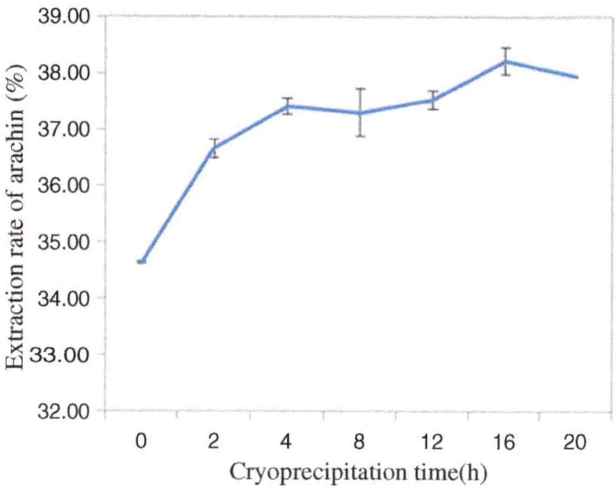

Fig. 8.9 Impact of cryoprecipitation time on extraction rate of peanut protein

the range of 0–4 h, the extraction rate of arachin increased sharply with the increase of cryoprecipitation time. However, when the cryoprecipitation time exceeded 4 h, the extraction rate of arachin increased slowly, and the impact on the extraction rate became small with the extension of cryoprecipitation time, which might be because the cryoprecipitation of arachin reached a dynamic balance. Therefore, 4 h was the most suitable cryoprecipitation time.

1.2.1.3 Cryoprecipitation Extraction Conditions of Arachin

1. Orthogonal optimization experiment

According to the results of single-factor experiment, the optimal parameters were determined as follows: the solid-liquid ratio was 4:10, the pH of solution was 7.5, the ion strength was 0.3 M, and the cryoprecipitation time was 4 h. The orthogonal experiment design is shown in Table 8.1.

The arachin was extracted through the two defatted peanut powder extraction. On the basis of single factor, the L_9 (4^3) four-factor three-level orthogonal test design was carried out. The results were analyzed by using Duncan's new multiple range method, and the significance analysis was carried out for the impacts of various factors. The orthogonal experiment results and range analysis are shown in Table 8.2. If the ranges of factors were large, it indicated that the impacts of ranges on the extraction rate of arachin were significant. The impacts of various factors on the extraction rate of arachin in priority were as follows: D (cryoprecipitation time)> A (solid-liquid ratio) >C (ion strength) >B (pH of solution). The optimal combination was $A_3B_3C_3D_2$, which indicated that the solid-liquid ratio was 5:10, the pH of solution was 7.9, the ion strength was 0.4 M, and the cryoprecipitation time was 4 h.

Table 8.1 Table of orthogonal test design

Level	Factor			
	Solid-liquid ratio	pH value of solution	Ion strength (M)	Cryoprecipitation time (h)
1	3:10	7.1	0.2	0
2	4:10	7.5	0.3	4
3	5:10	7.9	0.4	8

Table 8.2 Table of orthogonal test range analysis

Treatment no.	A Solid-liquid ratio	B pH of solution	C Ion strength	D Cryoprecipitation time	Extraction rate of arachin (%)
1	1	1	1	1	25.34
2	1	2	2	2	44.73
3	1	3	3	3	45.47
4	2	1	2	3	52.29
5	2	2	3	1	36.38
6	2	3	1	2	50.77
7	3	1	3	2	54.71
8	3	2	1	3	52.42
9	3	3	2	1	37.54
K1	115.54	132.34	128.53	99.26	
K2	139.44	133.53	134.56	150.21	
K3	144.67	133.78	136.56	150.18	
k1	38.51	44.11	42.84	33.09	
k2	46.48	44.51	44.85	50.07	
k3	48.22	44.59	45.52	50.06	
R	9.71	0.48	2.68	16.98	
Optimal level	A3	B3	C3	D2	
Priority order	D (cryoprecipitation time) >A (solid-liquid ratio) >C (ion strength) >B (pH of solution)				

2. Experiment optimization and validation

$A_3B_1C_3D_2$ was the extraction process condition with the highest extraction rate in the nine groups in the orthogonal test, and the extraction rate was 54.71%. The confirmatory experiment was carried out once under the $A_3B_1C_3D_2$ and $A_3B_3C_3D_2$ process conditions. The results are shown in Table 8.3. According to the orthogonal experiment range analysis and experimental validation results, $A_3B_1C_3D_2$ was determined as the optimal extraction process parameter of arachin: the solid-liquid ratio was 5:10, the pH of solution was 7.1, the ion strength was 0.4 M, and the cryoprecipitation time was 4 h. The arachin was extracted under the optimal extraction process. After centrifugation, the supernatant dried was the arachin with the purity of 76.40 ± 0.53% (Fig. 8.10).

Table 8.3 Model validation test

Experiment no.	Condition	Extraction rate (%)
1	$A_3B_1C_3D_2$	53.60 ± 1.27
2	$A_3B_3C_3D_2$	50.83 ± 0.72

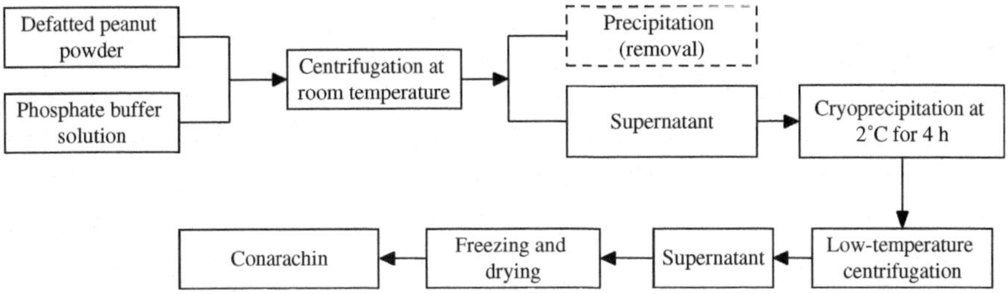

Fig. 8.10 Flow diagram of conarachin component preparation through cryoprecipitation method

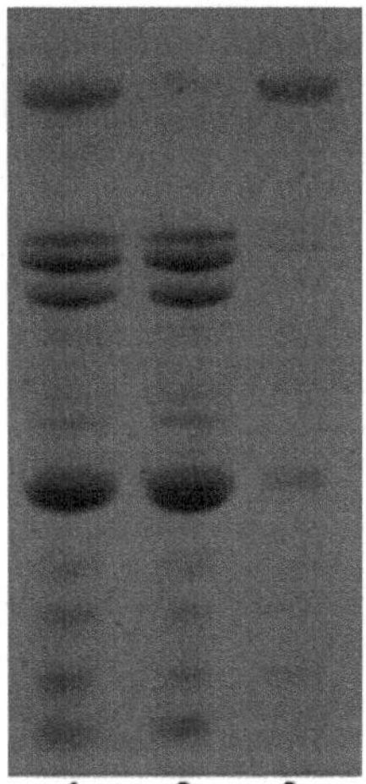

Fig. 8.11 Electrophoretogram of arachin and conarachin preparation through cryoprecipitation method. Note: *1*, peanut protein isolate; *2*, arachin; *3*, conarachin

1.2.2 Conarachin

The peanut powder was prepared together with the phosphate buffer solution with a certain pH and ionic strength in proportion; the mixture was stirred for 1 h at room temperature and then centrifuged for 30 min at 8000 r/min. The supernatant was placed in a 2 °C environment for 4 h and then centrifuged at 8000 r/min for 30 min at 2 °C. The supernatant dried directly through vacuum freezing or acid precipitation was the arachin (the process flow is shown in Fig. 8.10), and the purity was $58.35 \pm 0.35\%$ (Fig. 8.11).

2 Gelation Improvement of Arachin

2.1 *Determination Method of Gelation*

2.1.1 Gel Texture Property

The determination was carried out by using the TA-TX2i texture analyzer (probe type was P0.5R, diameter was 12 mm) at room temperature of 25 °C. The determination results were analyzed by using the TPA-macro tool, and the hardness, springiness, cohesiveness, and chewiness of protein gel were calculated, respectively. The calculation formulas are shown below:

$$\text{Gel hardness}(N) = \text{Peak force during the first compression(Force2)}.$$

$$\text{Gel springiness} = \text{Time spent during the second compression/time spent during the first compression(Time diff } 4:5/\text{Time diff } 1:2).$$

$$\text{Cohesiveness} = \text{Area under positive peak during the second compression/area under positive peak during the first compression(Area } 4:6/\text{Area } 1:3).$$

$$\text{Chewiness} = \text{Hardness} \times \text{cohesiveness which means the work made when the semisolid food is chewed until it can be swallowed.}$$

2.1.2 Water-Binding Capacity of Gel

The gel sample was stored in the refrigerator overnight at 4 °C, and a certain amount of gel pieces were filled in a 50 mL centrifuge tube and then centrifuged at 10,000 g for 15 min at 4 °C. The isolated water was poured out, the centrifuge tube was inverted to drain the water and dried with filter paper, and then the sample was weighed.

Water-binding capacity of gel (%) = $W_2/W_1 \times 100$ (W_1, gel weight before centrifugation/g; W_2, gel weight after centrifugation/g).

2.2 *Gelation Improvement Method*

2.2.1 Arachin Concentration

Figure 8.12 showed the gelation that the arachin formed the thermal gel in different protein concentrations. The arachin could not form gel when the concentration was 6%; when the concentration increased from 8% to 18%, the hardness, chewiness, and water-binding capacity of arachin increased continuously, because the number of molecules in unit volume increased and the interaction between the molecules

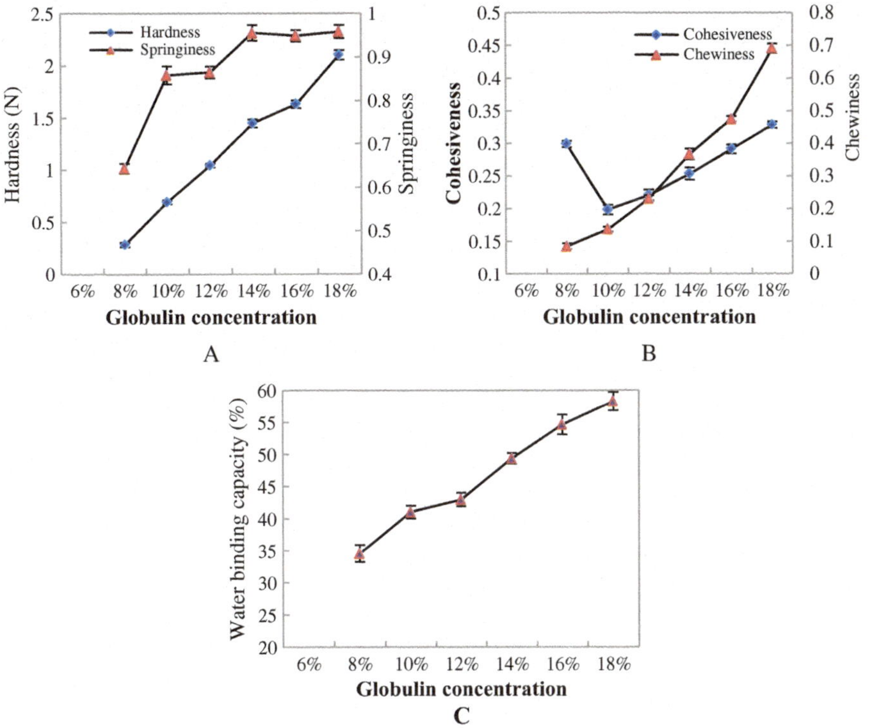

Fig. 8.12 Impact of arachin concentration on protein gelation. Note: (**a**) hardness and springiness; (**b**) cohesiveness and chewiness; (**c**) water-binding capacity

was enhanced with the increase of protein concentration. The springiness was smooth basically when the concentration was 14% and the cohesiveness increased first and then increased.

2.2.2 Phosphate Concentration

Figure 8.13 showed the changes in gelation of arachin in different phosphate buffer solution concentrations. It could be seen that when the phosphate concentration was 0.01 mol/L, the hardness, springiness, and cohesiveness reached the maximum and the changing trend of cohesiveness was not obvious. The water-binding capacity decreased with the increase of phosphate concentration, indicating that the interaction between the low concentration of phosphate solution and protein was obvious and more free water could be combined with better water-binding capacity.

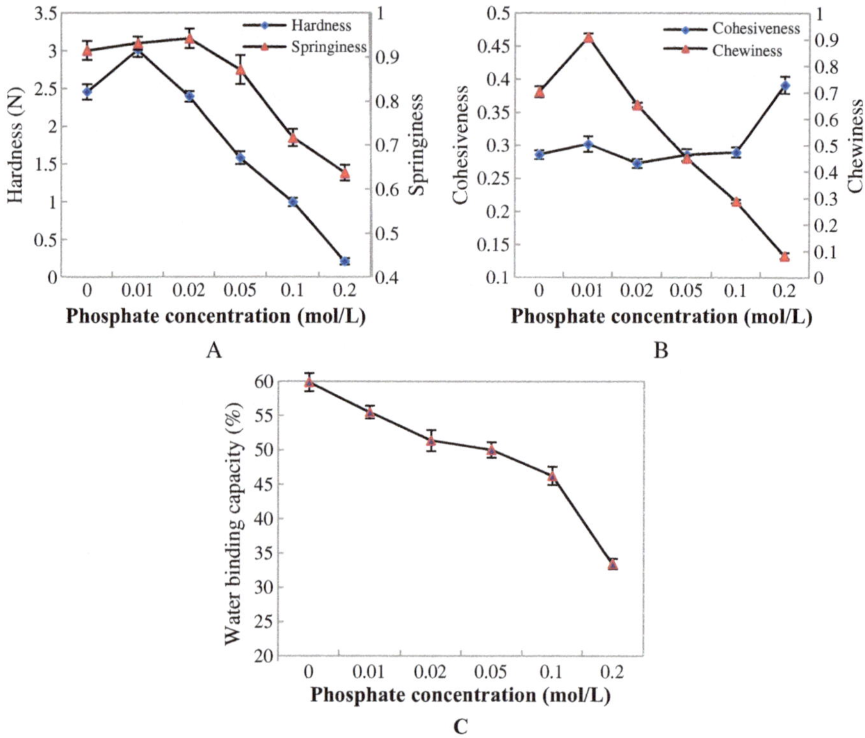

Fig. 8.13 Impact of phosphate concentration on arachin gelation. Note: (**a**) hardness and springiness; (**b**) cohesiveness and chewiness; (**c**) water-binding capacity

2.2.3 pH Value

pH value could change the ionization of protein function groups and double layer thickness and affect the interaction between protein-protein or protein-solvent. Figure 8.14 reflected the changes in gelation of arachin with pH. It could be seen that when pH was below 6, the arachin could not form gel. The change trends of hardness, springiness, and chewiness were basically the same, and they increased sharply when pH was from acidity to alkalinity; when pH was more than 7, their changes were mild. The cohesiveness and water-binding capacity were the opposite, and they were the lowest when pH was neutral.

2.2.4 Heating Temperature

Heat treatment is a common method that allows proteins to form gels. Figure 8.15 showed the changes in the gelation of arachin with the heating temperature. The hardness, springiness, chewiness, and water-binding capacity of arachin gel

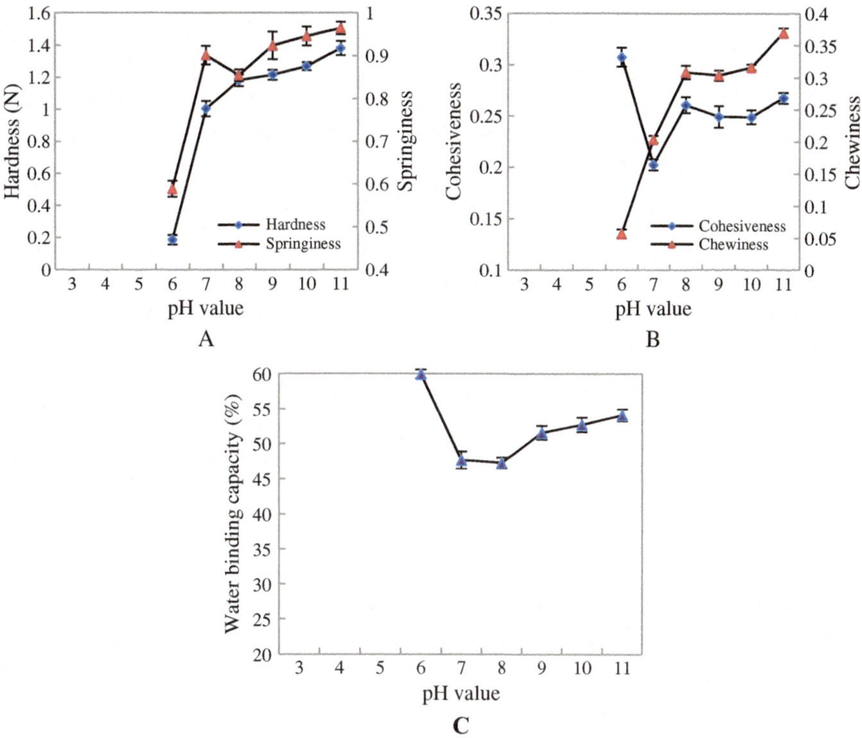

Fig. 8.14 Impact of pH value on arachin gelation. Note: (**a**) hardness and springiness; (**b**) cohesiveness and chewiness; (**c**) water-binding capacity

increased with the increase of temperature. When the temperature was 90–95 °C, these characteristics changed sharply. The results were consistent with the theory of requirement of globulin gelation for protein denaturation. The mechanism was that the cyclic structures in the gel network expanded into spread nonpolar amino acid peptide chains during the thermal denaturation of protein molecule, so that the hydrophobic groups originally buried inside the coil structure were exposed to cause the interaction between protein-protein (Kella and Poola 1985). At the same time, with the enhancement of heating movement of protein molecules, the probability of collision between the molecules increased greatly, which could make the molecules cross-linked through the hydrophobic interaction to increase the gel strength of protein (Molina et al. 2004).

2.2.5 Heating Time

Figure 8.16 showed the changes in gelation of arachin at different heating time. It could be seen that the hardness and springiness of gel increased with the extension

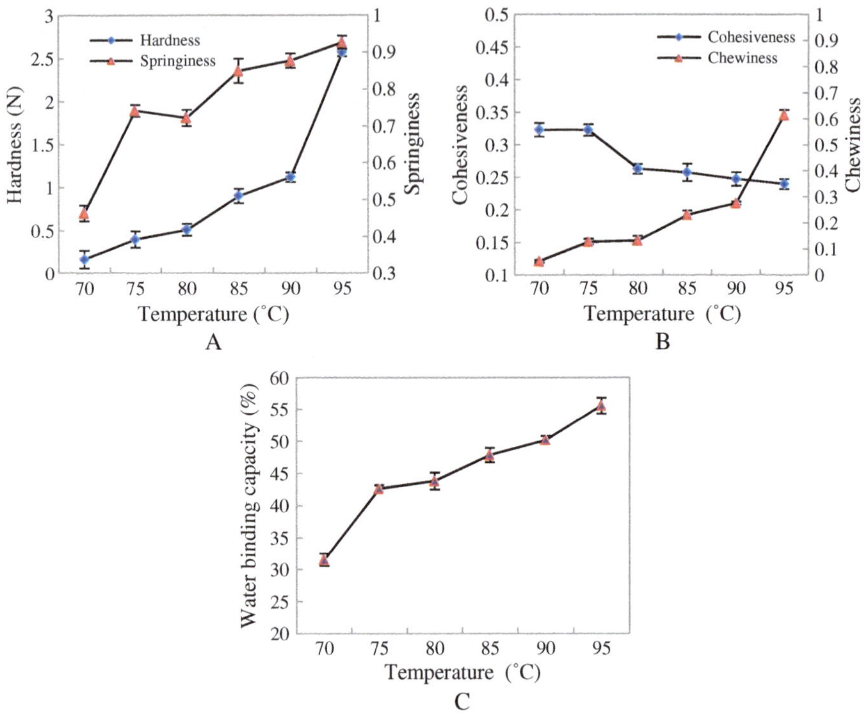

Fig. 8.15 Impact of heating temperature on arachin gelation. Note: (**a**) hardness and springiness; (**b**) cohesiveness and chewiness; (**c**) water-binding capacity.

of heating time at 95 °C but the impact of heating time on the hardness and springiness of gel was only limited to that of the initial 30 min, which indicated that the protein must be fully denatured and spread during heating so as to form an ordered three-dimensional network structure of gel.

2.2.6 Quadratic Orthogonal Rotation Test

Response surface methodology (RSM) is an effective method to optimize the process conditions. It can determine the impacts of factors and their interactions on the investigation indexes (response values) during the process operation and fit a complete quadratic polynomial model through the central combination test. It is excellent in the experiment design and result representation aspects. Du Yin (2012) used the response surface methodology to optimize the main process parameters of gel formation of arachin.

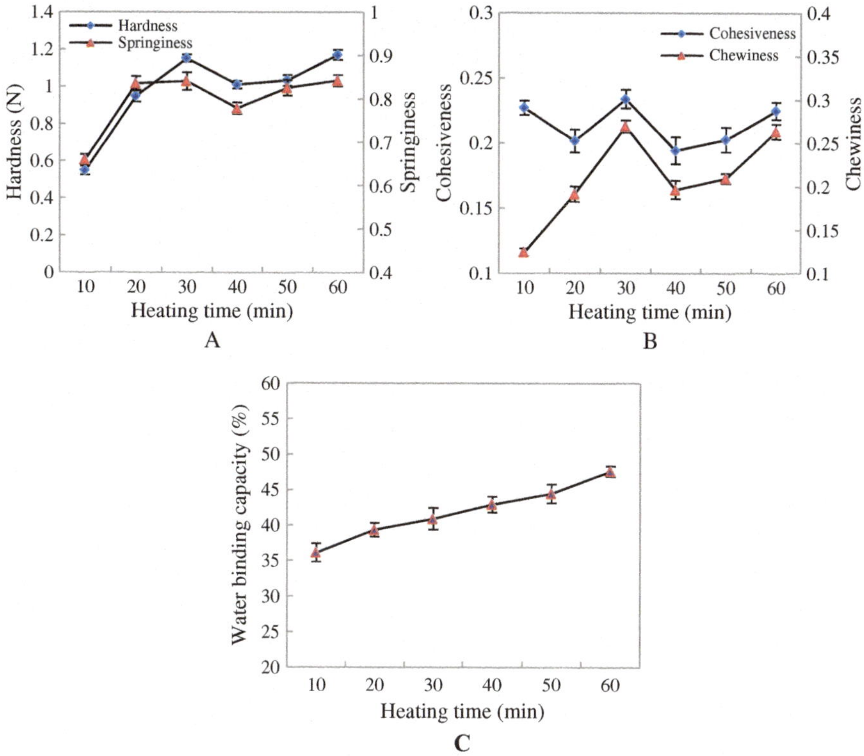

Fig. 8.16 Impact of heating time on arachin gelation. Note: (**a**) hardness and springiness; (**b**) cohesiveness and chewiness; (**c**) water-binding capacity

Table 8.4 Quadratic orthogonal rotation test factor level

Standard variable Z	Heating temperature (°C) X_1	Heating time (min) X_2	pH value X_3
Upper level (1)	95	40	8
Zero level (0)	90	30	7
Lower level (−1)	85	20	6

2.2.6.1 Results of Orthogonal Rotation Combination Test

The Box-Behnken response surface optimization experiment design was carried out with the gel hardness as the index. On the basis of single-factor experiment, the concentration of arachin was set as 14%, and the concentration of phosphate buffer solution was 0.01 mol/L. The three-factor three-level quadratic orthogonal rotation test was designed with the heating temperature (°C) X_1, heating time (min) X_2, and pH X_3 as zero level, with the hardness of arachin (N) as the response value Y. The factor level codes are shown in Table 8.4. The results are shown in Table 8.5. The

Table 8.5 Orthogonal rotation combination test results

Treatment no.	X_1 Temperature (°C)	X_2 Time (min)	X_3 pH	Y Hardness (N)
1	1(95)	1(40)	0(7)	3.20
2	0(90)	−1(20)	1(8)	1.43
3	0(90)	0(30)	0(7)	1.29
4	0(90)	0(30)	0(7)	1.18
5	0(90)	0(30)	0(7)	1.33
6	−1(85)	0(30)	1(8)	0.99
7	0(90)	0(30)	0(7)	1.23
8	−1(85)	1(40)	0(7)	0.79
9	0(90)	1(40)	1(8)	1.53
10	1(95)	−1(20)	0(7)	2.73
11	0(90)	0(30)	0(7)	1.53
12	0(90)	1(40)	−1(6)	0.44
13	−1(85)	0(30)	−1(6)	0.25
14	−1(85)	−1(20)	0(7)	0.54
15	1(95)	0(30)	1(8)	3.08
16	1(95)	0(30)	−1(6)	0.93
17	0(90)	−1(20)	−1(6)	0.35

Note: Tests are carried out in random order

regression analysis and significance test were carried out for the experimental data by using the statistical analysis software Design-Expert 7.0.

2.2.6.2 Variance of Regression Model

The multiple regression fitting was carried out for the test results in Table 8.5 by using the Design-Expert 7.0 software, and the quadratic polynomial regression model equation between the gel hardness of arachin Y and the independent variable temperature, heating time, and pH value was obtained:

$$\begin{aligned} Y = {} & 1.303568 + 0.908916X_1 + .125097X_2 \\ & +0.632747X_3 + 0.080238X_1X_2 + 0.353845X_1X_3 \\ & +0.001393X_2X_3 + 0.434018X_{12} + 0.055309X_{22} \\ & -0.4205X_{32} \end{aligned}$$

The variance analysis test (ANOVA) of the whole and various factors was carried out for the quadratic orthogonal rotation test results of above gel hardness model. The results are shown in Tables 8.6 and 8.7. The model had a high determination coefficient ($R^2 = 0.951714$), indicating that there was a good degree of fitting in the gel hardness regression equation and test results and the model was reliable. $F_1 = 5.6410$, the lack of fit item was not significant, and the equation was not lack of fit; $F_2 = 15.3299 > F_{0.01}$ (9,10) $= 5.05$ and the regression item was

Table 8.6 Variance analysis of regression model

Source of variance	Quadratic sum	Degree of freedom	Mean sum of square	F	P
Regression	11.9379	9	1.3264	$F_2 = 15.3299$	0.0008[a]
Residue	0.6057	7	0.0865		
Lack of fit	0.4899	3	0.1633	$F_1 = 5.6410$	0.0640
Error	0.1158	4	0.0289		
Sum	12.5436	16			
	$R^2 = 0.951714$				

[a]Means the regression item was highly significant

Table 8.7 Variance analysis of various factors

Variable	Quadratic sum	Degree of freedom	Mean sum of square	Ratio F	p value
X_1	6.609027	1	6.609027	76.38231	<0.0001[b]
X_2	0.125193	1	0.125193	1.446894	0.2681
X_3	3.202952	1	3.202952	37.01738	0.0005[b]
X_1^2	0.793145	1	0.793145	9.166591	0.0192[a]
X_2^2	0.012881	1	0.012881	0.148864	0.7111
X_3^2	0.744511	1	0.744511	8.604511	0.0219[a]
X_1X_2	0.025752	1	0.025752	0.297625	0.6023
X_1X_3	0.500826	1	0.500826	5.788185	0.0471[a]
X_2X_3	7.77E-06	1	7.77E-06	8.98E-05	0.9927

[a]Means there was a significant difference
[b]Means there was a very significant difference

highly significant ($P < 0.01$), indicating that the difference between the different treatment methods was highly significant.

From the significant analysis of the regression equation coefficients in Table 8.7, it could be seen that the impact of X_1 (heating temperature) and X_3 (pH) on the gel hardness of arachin reached a highly significant level ($P < 0.01$) and the impact of X_2 heating time (min) on the gel hardness of arachin did not reach a significant level. The impacts of the factors on the gel hardness of arachin in priority were temperature > pH > heating time.

2.2.6.3 Interaction Effect Response Surface

According to the quadratic regression equation, the response surface diagram was established, and the interaction between two factors could be analyzed intuitively. The heating temperature and pH value all had significant quadratic interactions ($p = 0.0471 < 0.05$) (Fig. 8.17). The impact on the gel hardness of arachin was significant with the increase of temperature; the changes of gel hardness of arachin were significant with the increase of pH. The interaction effects between the other factors were not significant.

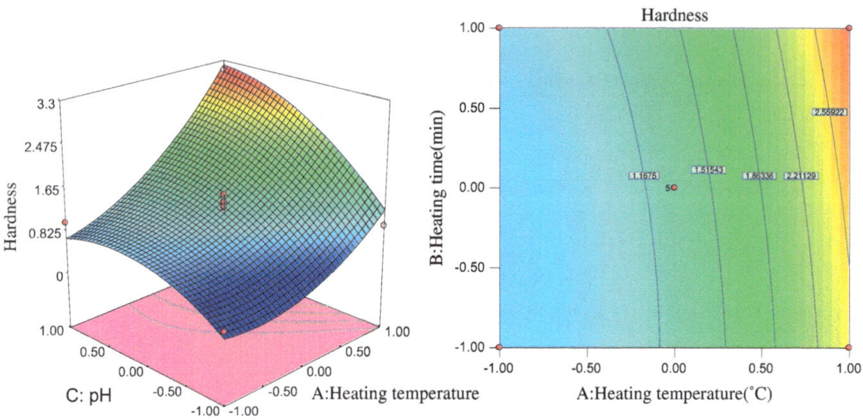

Fig. 8.17 Response surface analysis and contour plot of impact of interaction between the factors on gel hardness of arachin

Table 8.8 Model validation test

Factor	Heating temperature (°C) X_1	Heating time (min) X_2	pH X_3	Hardness Y max
Coded value	0.88	0.96	1.00	3.21
Theoretical value	94.4	39.6	8	3.21
Actual measured value	94	40	8	3.17 ± 0.18

2.2.6.4 Optimal Condition

The optimal conditions of gel hardness of arachin were predicted by using the statistical analysis software Design-Expert, and the highest testing points of gel hardness of arachin were shown below: the heating temperature was 94.4 °C, the heating time was 39.6 min, and the pH was 8.0. Under these conditions, the theoretical value of gel hardness of arachin was 3.21 N. The preparation conditions were adjusted slightly, as shown below: the heating temperature was 94 °C, the heating time was 40 min, and the pH was 8.0. Under these conditions, the confirmatory experiment was carried out, and the gel hardness of arachin was 3.17 ± 0.18 N which was close to the theoretical value (Table 8.8). The results showed that the theoretical prediction value of the gel preparation process parameters of the arachin optimized by using the response surface method was consistent with the actual measured value, the model was constructed correctly, and the process parameters were reliable.

3 Gelation Improvement of Conarachin

3.1 Determination Method of Gelation

3.1.1 Gel Texture Characteristic

The determination was carried out by using the TA-TX2i texture analyzer (probe type was P0.5R, diameter was 12 mm) at room temperature of 25 °C. The determination results were analyzed by using the TPA-macro tool, and the hardness, springiness, cohesiveness, and chewiness of protein gel were calculated, respectively. The calculation formulas are shown below:

$$\text{Gel hardness}(N) = \text{Peak force during the first compression(Force2)}.$$

$$\text{Gel springiness} = \text{Time spent during the second compression/time spent during the first compression(Time diff4 : 5/Time diff1 : 2)}.$$

$$\text{Cohesiveness} = \text{Area under positive peak during the second compression/area under positive peak during the first compression (Area 4 : 6/Area 1 : 3)}.$$

$$\text{Chewiness} = \text{Hardness} \times \text{cohesiveness which means the work made when the semisolid food is chewed until it can be swallowed.}$$

3.1.2 Water-Binding Capacity of Gel

The gel sample was stored in the refrigerator overnight at 4 °C, and a certain amount of gel pieces were filled in a 50 mL centrifuge tube and then centrifuged at 10,000 g for 15 min at 4 °C. The centrifuge tube was inverted to drain the water and dried with filter paper, and then the sample was weighed.

Water-binding capacity of gel (%) = $W_2/W_1 \times 100$ (W_1, gel weight before centrifugation/g; W_2, gel weight after centrifugation/g).

3.2 Gelation Improvement Method

3.2.1 Protein Concentration

The impact of change of protein concentration on the gelation of conarachin is shown in Fig. 8.18. The impact was similar to that on the arachin. The conarachin could not form gel when the concentration was lower than 6%; when the concentration increased from 8% to 18%, the hardness and chewiness increased continuously, because the number of molecules in the same volume increased and the interaction between the molecules was enhanced with the increase of protein

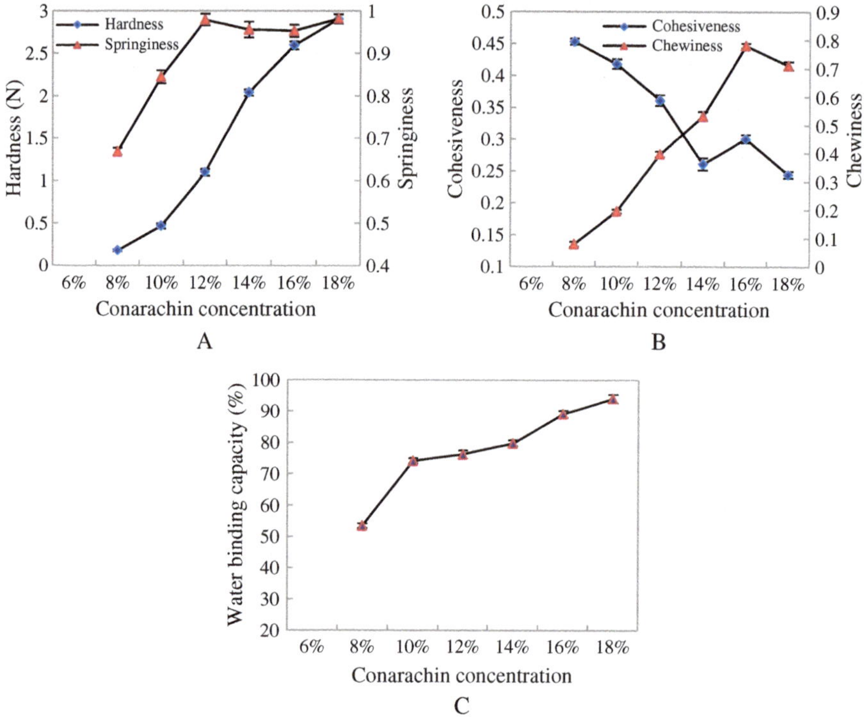

Fig. 8.18 Impact of conarachin concentration on gelation of protein. Note: (**a**) hardness and springiness; (**b**) cohesiveness and chewiness; (**c**) water-binding capacity

concentration. The springiness was smooth basically when the concentration was 12%, the cohesiveness decreased, the chewiness increased first and then increased, and the chewiness was the best when the concentration was 16%. Therefore, 16% was selected as the most suitable conarachin concentration.

3.2.2 Calcium Chloride Ion Concentration

The impact of salt ions on protein gels was complex, and there were salting-in effect and salting-out effect, as shown in Fig. 8.19. It could be seen from the appearance that the tenacity of protein gel surface formed after the addition of calcium ions increased and the hardness was greater than that when calcium salt was not added. With the increase of Ca^{2+} ion concentration, the gel hardness increased. When the Ca^{2+} ion concentration reached 0.2 mol/L, the gel hardness was the largest, and the springiness and chewiness were the best, which might be caused by the salting-in effect; when the Ca^{2+} ion concentration exceeded 0.2 mol/L, the gel hardness, springiness, and chewiness decreased, which might be caused by the salting-out effect, and its internal mechanism needed to be researched further. With the

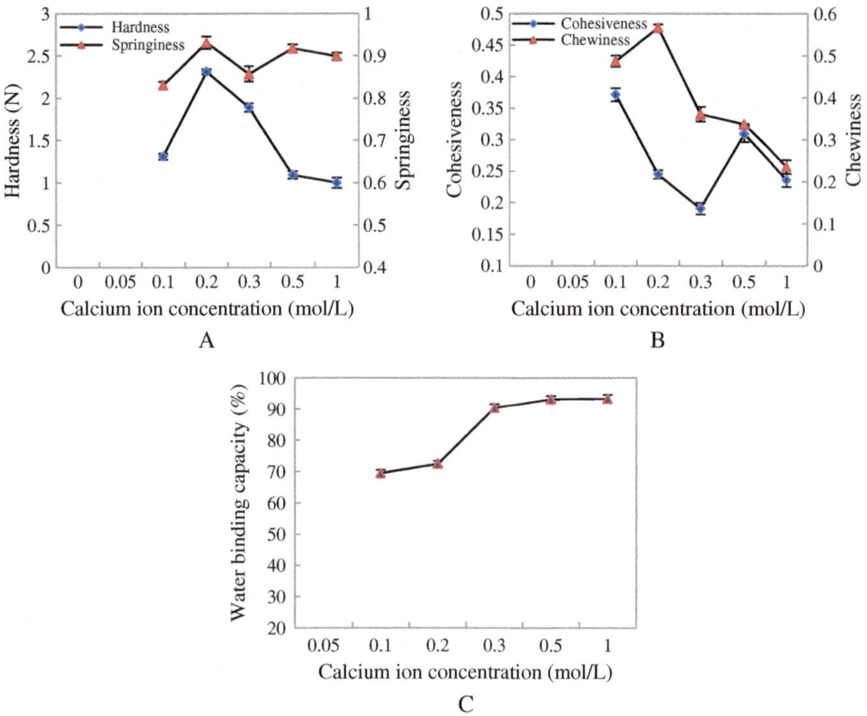

Fig. 8.19 Impact of calcium ion concentration on gelation of conarachin. Note: (**a**) hardness and springiness; (**b**) cohesiveness and chewiness; (**c**) water-binding capacity

increase of Ca^{2+} ion concentration, the water-binding capacity of gel increased continuously and then changed smoothly, which might be because the Ca^{2+} ions made the internal structure of gel become loose and combine with more free water.

3.2.3 pH Value

As shown in Fig. 8.20, pH value was an important factor affecting the formation and texture characteristics of conarachin gel, and there were significant differences in the gelation under acidic conditions and alkaline conditions. The changes in pH value would affect the ionization and net charge values of protein molecules so as to change the attractiveness and repulsive force of protein molecules and the ability to combine protein molecules with water molecules and would also affect the gel formation and maintenance acting force. Figure 8.20 showed that when the pH value was in the acidic range between 3 and 6, the gel hardness of conarachin increased first and then decreased with the increase of pH; when the pH value was 5, the hardness was the highest. When the pH value was in the alkaline range of 7–11, the gel hardness also increased first and then decreased with the increase of

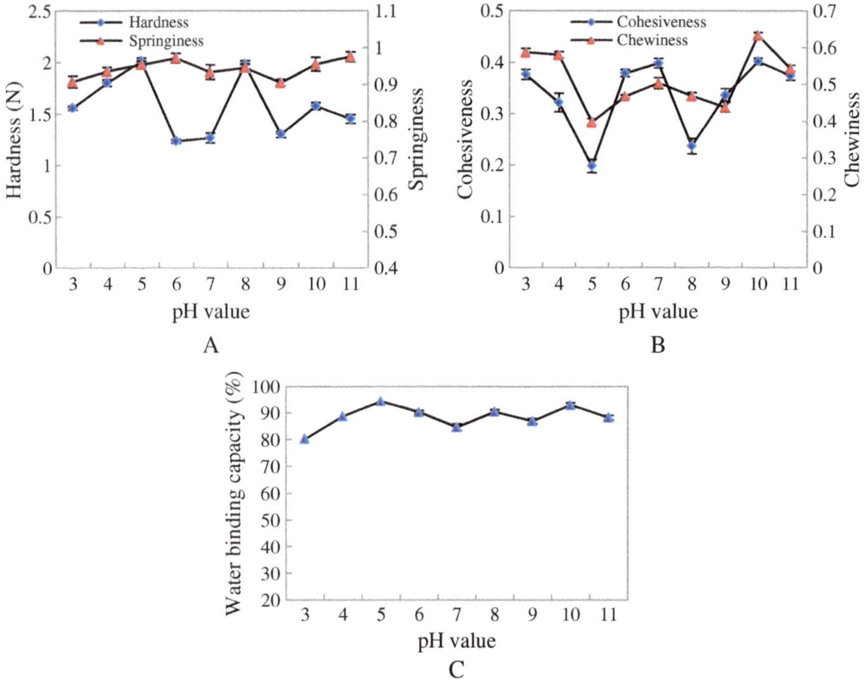

Fig. 8.20 Impact of pH value on gelation of conarachin. Note: (**a**) hardness and springiness; (**b**) cohesiveness and chewiness; (**c**) water-binding capacity

pH; when the pH value was 8, the hardness was the highest. The changes of springiness and water-binding capacity of gel with the pH value were not significant. When the pH was less than 3 or higher than 11, the conarachin could not form gel, because the conarachin had been dissociated partially before heating in the acidic conditions and the protein solution heated existed in the state with floccules. However, under the alkaline conditions, too high pH value might cause the denaturation of protein, resulting in protein aggregation and water precipitation. Therefore, the subsequent optimization experiments should be carried out, respectively, under the acidic conditions and alkaline conditions to research the gelation of conarachin in different solution environments so as to provide a certain theoretical reference for its application in food.

3.2.4 Heating Temperature

As shown in Fig. 8.21, the gel hardness, springiness, chewiness, and water-binding capacity of conarachin increased with the increase of heating temperature. Because the thermal denaturation of conarachin occurred with the increase of temperature, the intramolecular active groups were exposed fully, the molecular were cross-

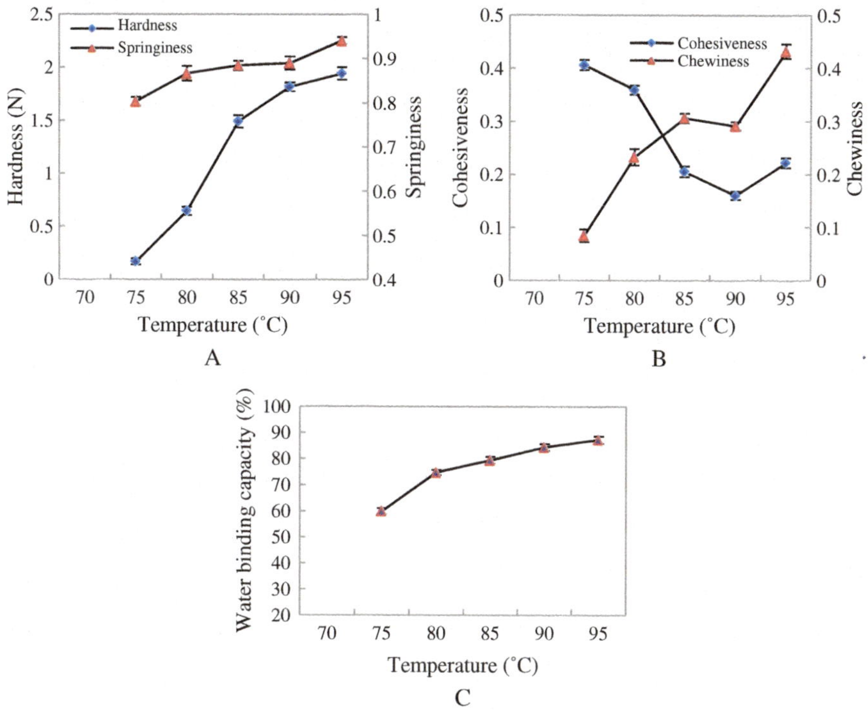

Fig. 8.21 Impact of heating temperature value on gelation of conarachin. Note: (**a**) hardness and springiness; (**b**) cohesiveness and chewiness; (**c**) water-binding capacity

linked with each other, and thus a three-dimensional network structure was formed. The higher the heating temperature, the more sufficient the degree of denaturation of conarachin, the more the exposed intramolecular active groups, and the more perfect the formed three-dimensional network structure. The degree of denaturation of conarachin during the heat treatment was determined by using the differential scanning calorimetry (DSC), and the denaturation temperature (Td) was 87.49 °C (see Fig. 8.22). Therefore, after the conarachin was heated in 90 °C, it had been denatured fully, and there was little change in the hardness, springiness, and water-binding capacity of its protein gel, compared with that heated in 95 °C.

3.2.5 Heating Time

The changes in gelation of conarachin at different heating time were shown in Fig. 8.23. It could be seen that the gel hardness and springiness at 90 °C increased with the extension of heating time but the impact of heating time on the springiness was only limited to that of the initial 30 min, which indicated that the protein must be fully denatured and spread during heating so as to form an ordered three-dimensional network structure of gel. However, if the heating time continued to

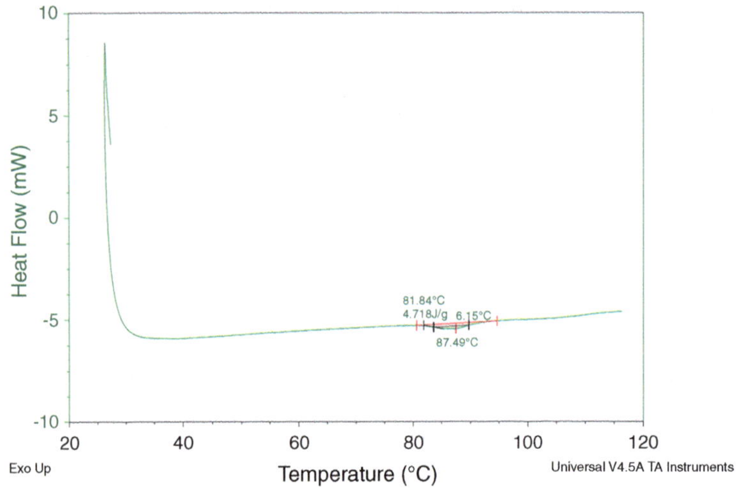

Fig. 8.22 DSC scanning atlas of conarachin component

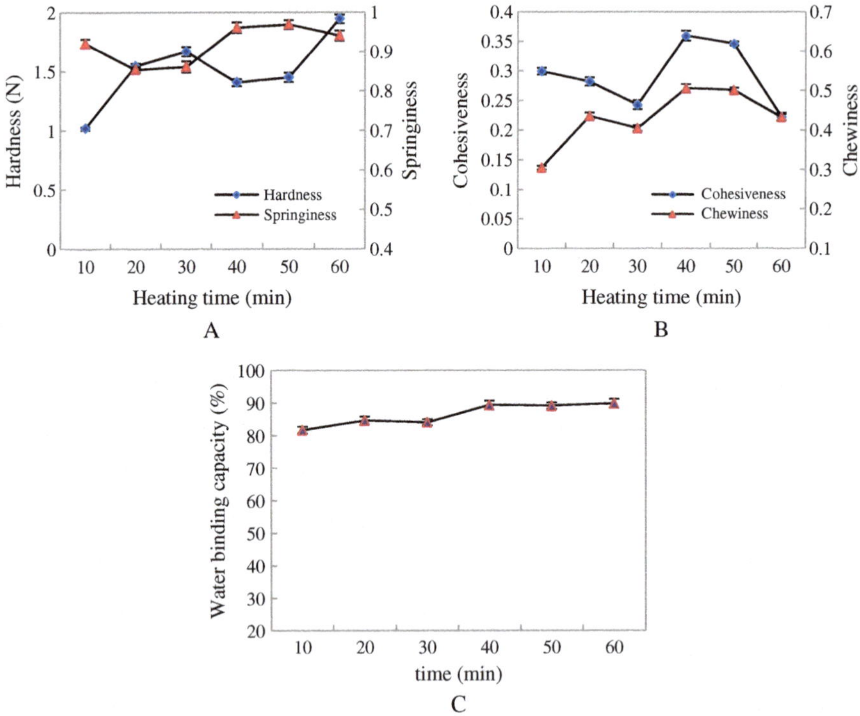

Fig. 8.23 Impact of heating time on gelation of conarachin. Note: (**a**) hardness and springiness; (**b**) cohesiveness and chewiness; (**c**) water-binding capacity

increase, the springiness would decrease, which might be because the long hours of heating made the opportunities of interactions between the protein molecules increase and some active groups would generate chemical reaction, for example, the sulfhydryl was oxidized and formed aggregation, which were bad for protein gelation. It could also be seen from the figure that the impact of heating time on the water-binding capacity was not significant and the cohesiveness and chewiness were the highest at 40 min. Therefore, 40 min was selected as the appropriate heating time finally.

3.2.6 Orthogonal Optimization Test

pH value was an important factor affecting the formation and texture characteristics of conarachin gel, and there were significant differences in the gelation under acidic conditions and alkaline conditions. According to the different change trends of hardness values of conarachin with pH value under acidic and alkaline conditions, the orthogonal tests were carried out under these two conditions, and the hardness value (N) was used as the investigation index. The orthogonal test results were analyzed by using Duncan's new multiple range method.

3.2.6.1 Acidic Condition

According to the above single-factor test results, 16 % was set as the conarachin concentration, 40 min was set as the heating time, three main factors and value ranges were determined, and the L_9 (4^3) orthogonal test design was carried out by using the DPS software (see Table 8.9).

The comparative analysis was carried out for the significance of impacts of various factors by using Duncan's new multiple range method in the test results. The orthogonal test results and range analysis are shown in Table 8.10. From Table 8.10 and Fig. 8.24, it could be seen that the greater the range of the factor, the more significant the impact of the factor on the gel hardness; the impacts of the factors on the conarachin gel hardness in priority were B (calcium chloride concentration)> C (pH value)> A (temperature). The optimal level combination was $A_3B_3C_2$ which meant that the temperature was 95 °C, the calcium chloride concentration was 0.3 mol/L, and the pH value was 5.

Table 8.9 Orthogonal test design table of conarachin gel hardness under acidic conditions

	Factor			
Level	A Temperature (°C)	B Calcium chloride concentration (mol/L)	C pH value	D (null column)
1	85	0.1	4	
2	90	0.2	5	
3	95	0.3	6	

Table 8.10 Range analysis results of orthogonal test under acidic conditions

Treatment no.	A Temperature (°C)	B Calcium chloride concentration (mol/L)	C pH value	D Null column	Hardness (N)
1	1	1	1	1	1.22
2	1	2	2	2	1.69
3	1	3	3	3	1.42
4	2	1	2	3	1.44
5	2	2	3	1	1.80
6	2	3	1	2	1.70
7	3	1	3	2	1.24
8	3	2	1	3	1.66
9	3	3	2	1	2.11
K1	4.32	3.90	4.58		
K2	4.93	5.15	5.23		
K3	5.01	5.22	4.46		
k1	1.44	1.30	1.53		
k2	1.64	1.72	1.74		
k3	1.67	1.74	1.49		
R	0.69	1.32	0.78		
Optimal level	A3	B3	C2		
Order	B Calcium chloride concentration	C pH value	A Temperature		

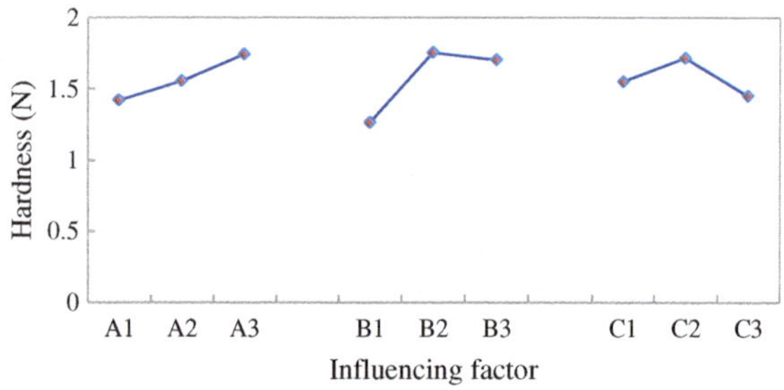

Fig. 8.24 Impacts of various factors on conarachin gel hardness under acidic conditions

3.2.6.2 Alkaline Condition

According to the above single-factor test results, 16 % was set as the conarachin concentration, 40 min was set as the heating time, three main factors and value ranges were determined, and the L9 (4^3) orthogonal test design was carried out by using the DPS software (see Table 8.11).

The comparative analysis was carried out for the significance of impacts of various factors by using Duncan's new multiple range method in the test results.

Table 8.11 Orthogonal test design table of conarachin gel hardness under alkaline conditions

Level	Factor			
	A Temperature (°C)	B Calcium chloride concentration (mol/L)	C pH value	D (null column)
1	85	0.1	7	
2	90	0.2	8	
3	95	0.3	9	

Table 8.12 Range analysis results of orthogonal test under alkaline conditions

Treatment no.	A Temperature	B Calcium chloride concentration (mol/L)	C pH value	D (null column)	Hardness (N)
1	1	1	1	1	1.16
2	1	2	2	2	1.69
3	1	3	3	3	1.45
4	2	1	2	3	1.73
5	2	2	3	1	1.59
6	2	3	1	2	1.47
7	3	1	3	2	1.30
8	3	2	1	3	1.68
9	3	3	2	1	2.01
K1	4.30	4.19	4.32		
K2	4.80	4.97	5.43		
K3	4.99	4.93	4.34		
k1	1.43	1.40	1.44		
k2	1.60	1.66	1.81		
k3	1.66	1.64	1.45		
R	0.69	0.78	1.12		
Optimal level	A3	B2	C2		
Order	C pH value	B Calcium chloride concentration	A Temperature		

The orthogonal test results and range analysis are shown in Table 8.12. From Table 8.12 and Fig. 8.25, it could be seen that the greater the range of the factor, the more significant the impact of the factor on the gel hardness; the impacts of the factors on the conarachin gel hardness in priority were C (pH value)>B (calcium chloride concentration)>A (temperature). The optimal level combination was $A_3B_2C_2$ which meant that the pH value was 8, the calcium chloride concentration was 0.3 mol/L, and the temperature was 95 °C.

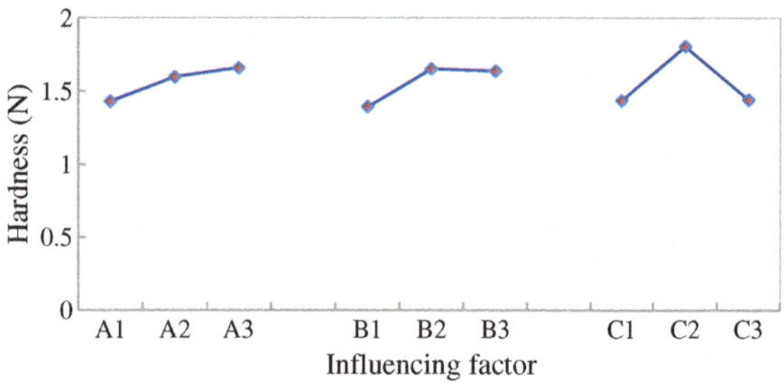

Fig. 8.25 Impacts of various factors on conarachin gel hardness under alkaline conditions

4 Physicochemical Characteristics of Protein Component

4.1 Thermal Characteristic

The thermal differential scanning was carried out for the prepared arachin and conarachin components to determine their degrees of denaturation. The results of initial denaturation temperature (Tm), denaturation temperature (Td), and endothermic enthalpy (ΔH) are shown in Table 8.13. It could be seen from Figs. 8.26 and 8.27 that each of arachin and conarachin had a denaturing peak. T_d value could reflect the thermal stability of protein; the denaturation temperature of arachin was 102.27 °C, and the denaturation temperature of conarachin was 87.49 °C, which could be seen that the denaturation temperature of arachin was higher than that of conarachin and the heat sensitivity of conarachin was stronger than that of arachin. When the arachin and conarachin had the same quality, the arachin might have more ordered spatial conformation and disulfide bonds, hydrophobic bonds, ionic bonds, and other functional groups required to maintain the spatial conformation (Arntfield and Murray 1981). Therefore, the arachin had higher thermal stability than the conarachin. Some research reported that the thermal stability of soybean protein isolate and kidney bean protein isolate could be improved by using the TGase modification treatment (Tang et al. 2006, 2008).

4.2 Protein Content and Amino Acid Composition

4.2.1 Protein Content

The protein contents of raw material peanut meal and peanut protein components prepared were determined by using the Dumas combustion method. The results are shown in Table 8.14. The protein content of arachin was 91.31 ± 0.10%, and the protein content of conarachin was 93.45 ± 0.26%. Both of the protein contents were >90%, so the purity was high.

Table 8.13 DSC thermal characteristics of arachin and conarachin

Protein sample	Initial denaturation temperature (Tm/°C)	Denaturation temperature (Td/°C)	Endothermic enthalpy (ΔH/J.g^{-1})
Arachin	95.68	102.27	4.831
Conarachin	81.84	87.49	4.718

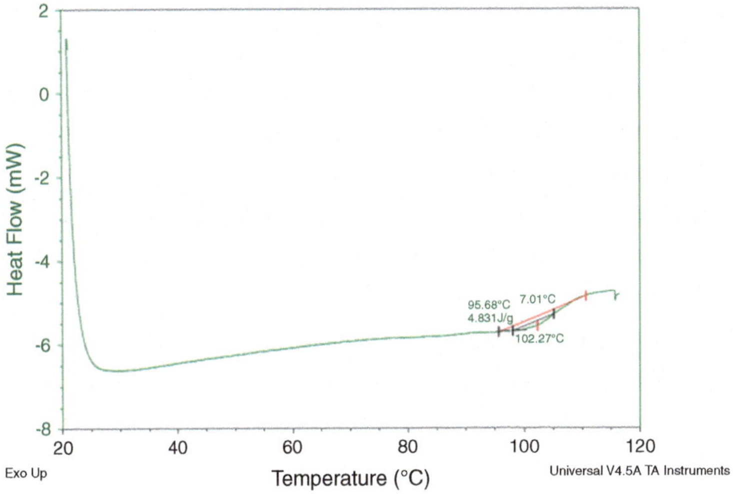

Fig. 8.26 DSC scanning atlas of arachin component

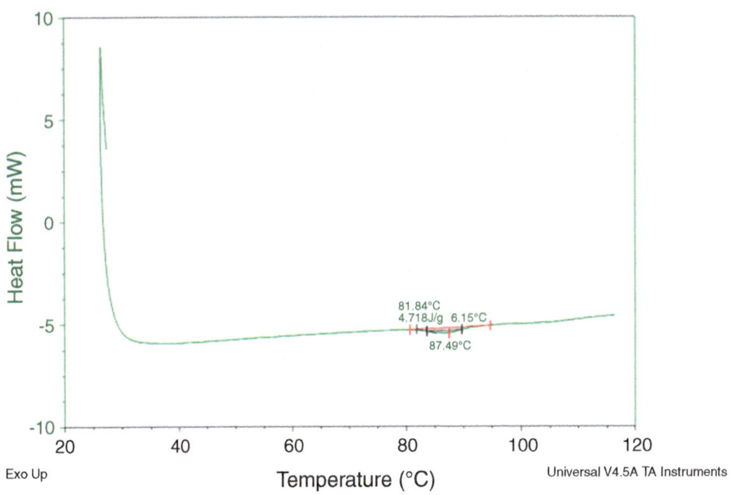

Fig. 8.27 DSC scanning atlas of conarachin component

Table 8.14 Protein content determination results of raw material peanut meal and peanut protein components (wet basis)

Sample	Protein content (%)
Peanut meal	47.31 ± 0.15
Arachin	91.31 ± 0.10
Conarachin	93.45 ± 0.26

4.2.2 Amino Acid Composition and Content of Peanut Protein Component

The amino acid analysis was carried out for these two peanut protein components prepared. The determination results are shown in Table 8.15 and Figs. 8.28 and 8.29. It could be seen from Table 8.15 that the amino acid composition of peanut protein component was very rich and the arachin and conarachin contained 17 kinds of amino acids. These three kinds of amino acids with the highest content were glutamic acid, arginine, and aspartic acid. Aspartic acid could be used to treat heart disease, liver disease, and hypertension, and it also had the function of fatigue prevention and recovery; glutamate was mainly used for the treatment of hepatic coma and improvement of children's intellectual development; arginine was an essential amino acid in the growing period of children. The sum of the contents of the three amino acids accounted for about 47% of the total amino acids. The content level of methionine was low both in these two peanut protein components, being 7.17 mg/g and 16.19 mg/g protein, respectively. It could also be seen from the table that the content of sulfur-containing amino acids (methionine and cystine) in conarachin was significantly higher than that in arachin ($P < 0.05$). The content of methionine was 2.25 times that of arachin, and the content of cystine was 1.31 times that of arachin. In the gel formation process, the spatial structure of protein was mainly maintained by the disulfide bonds. Therefore, it could be preliminarily inferred that the conarachin was more conductive to the gel formation in the heat treatment process from the aspect of amino acid composition. In addition, the content of lysine in conarachin was 1.41 times that in arachin. Hu Xiao researched and reported (2011) that the ε-amino group of lysine residue in polypeptide chain could form the intramolecular and intermolecular ε-(γ-glutamyl)lysine isopeptide covalent bond of protein after being catalyzed by transglutaminase, so that the protein molecules were cross-linked to change the structure, which played an important role in the improvement of gelation, thermal stability, and other functional characteristics. Therefore, it could be seen that it was easier for arachin to be cross-linked through transglutaminase effect to improve the gelation.

Table 8.15 Amino acid composition analysis of main components of peanut protein (unit: mg/g protein)

Chinese name	English name	Arachin	Conarachin	Chinese name	English name	Arachin	Conarachin
Aspartic acid	Asp	114.65	101.15	Leucine	Leu[a]	65.85	54.18
Threonine	Thr[a]	23.72	18.52	Tyrosine	Tyr	51.94	35.90
Serine	Ser	45.78	43.34	Phenylalanine	Phe[a]	59.70	41.00
Glutamic acid	Glu	225.10	213.63	Lysine	Lys[a]	23.52	33.17
Glycine	Gly	42.69	59.66	Histidine	His	21.46	19.67
Alanine	Ala	38.55	27.74	Tryptophan	Trp[a]		
Cystine	Cys	19.58	25.70	Arginine	Arg	116.81	111.76
Valine	Val[a]	39.69	32.40	Proline	Pro	43.60	34.02
Methionine	Met[a]	7.17	16.19	Total amino acid		971.78	894.80
Isoleucine	Ile[a]	31.97	26.77				

[a]Refers to essential amino acid; the tryptophan Trp is not detected because acid hydrolysis is used for sample treatment

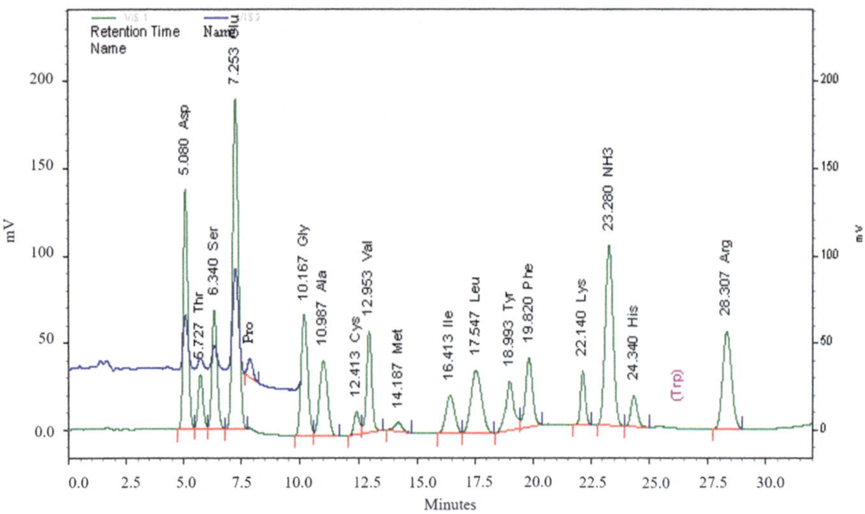

Fig. 8.28 Amino acid analysis atlas of arachin

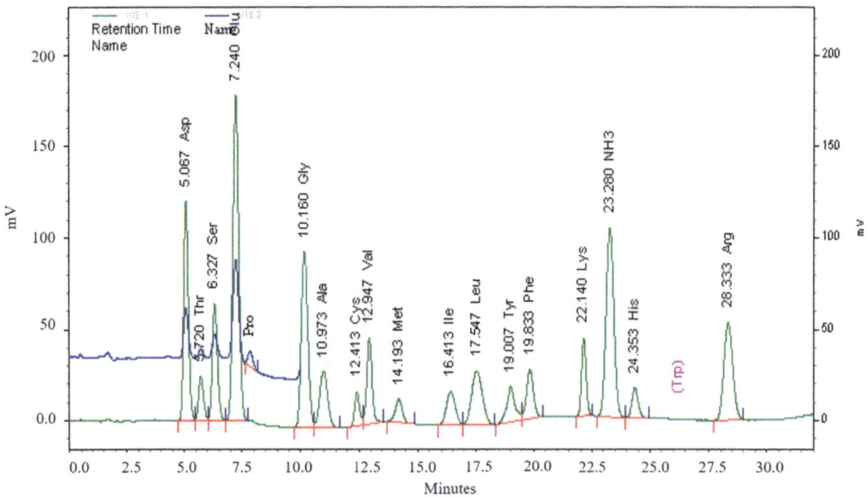

Fig. 8.29 Amino acid analysis atlas of conarachin

References

Arntfield SD, Murray ED. The influence of processing parameters on food protein functionality. I. Differential scanning calorimetry as an indicator of protein denaturation [J]. Can Inst Food Sci Technol J. 1981;14:289–94.

Basha SMM, Pancholy SK. Isolation and characterization of two cryoproteins from Florunner peanut (*Arachis hypogaea* L.) seed [J]. Agric Food Chem. 1982;30:36–41.

Chiou RYY. Effects of heat treatments on peanut arachin and conarachin [J]. J Food Biochem. 1990;14:219–32.

Du Yin. Study on the preparation and gel properties of the peanut protein fractions [D]. Beijing: Chinese Academy of Agricultural Sciences; 2012.

Feng Xiaolong. Preparation, modification and application of peanut protein fractions [D]. Beijing: Chinese Academy of Agricultural Sciences; 2014.

Johnson P, Naismith WEF. High molecular weight systems. The physico-chemical examination of the arachin fraction of the groundnut globulins (Arachis hypogea L.) [J]. Discuss Faraday Soc. 1953;13:98.

Kella NKD, Poola I. Sugars decrease the thermal denaturation and aggregation of arachin [J]. J Pept Protein Res. 1985;26:390–9.

Kella NKD, Poola I. Sugars decrease the thermal denaturation and aggregation of arachin[J]. Chem Biol Drug Des. 2010;26(4):390–9.

Kolb G, Viardot K, Wagner G, et al. Evaluation of a new high-pressure dispersion unit (HPN) for emulsification. Chem Eng Technol. 2001;24:293–6.

Lakemond CMM, Harmen HJ. Gelation of soy glycinin: influence of pH and ionic strength on network structure in relation to protein conformation. Food Hydrocoll. 2003;17:365–77.

Li Yin, Huang Shangzhi, Fu Jiarui. One and two-dimensional polyacrylamide gel electrophoretic analysis of seed polypeptide composition of peanut cultivars [J]. J Integr Plant Biol. 1998;40 (6):534–41.

Lin Lu FJ. 2S protein of peanut seeds [J]. Peanut Sci. 1994;3:1–4.

Molina OSE, Puppo MC, Wagner JR. Relationship between structural changes and functional properties of soybean protein isolates-carrageenan systems [J]. Food Hydrocoll. 2004;18:1045–53.

Naismith WEF, McDavid HMR. Fractionation studies on the proteins of the groundnut (Arachis hypogea L.) [J]. Russ J Appl Chem. 1958;8:605.

Naismith WEF, Mcdavid HMR. Fractionation studies on the proteins of the groundnut (arachis hypogaea)[J]. Int J Adv Res Comp Eng Technol. 2010;8(9):605–10.

Neucere NJ. Isolation α-Arachin,the major peanut globulin. Anal Biochem. 1969;27:15–24.

Prakash V, Narasinga RMS. Physicochemical properties of oilseed proteins [J]. Crit Rev Biochem Mol Biol. 1986;20(3):286–98.

Prakash V, Rao MS. Physicochemical properties of oilseed proteins. CRC Crit Rev Biochem. 1986;20(3):286–301.

Tanaka K, Bushuk W. Changes in flour protein during dough mixing. III analytical results and mechanisms. Cereal Chem. 1973;50:605–12.

Tang CH, Chen Z, Li L, et al. Effects of transglutaminase treatment on the thermal properties of soy protein isolates. Food Res Int. 2006;39(6):704–11.

Tang CH, Sun X, Yin SW, et al. Transglutaminase-induced cross-linking of vicilin-rich kidney protein isolate: influence on the functional properties and in vitro digestibility. Food Res Int. 2008;41(10):941–7.

Tombs MP. An electrophoretic investigation of groundnut proteins: the structure of arachin A and B. J Biochem. 1965;96:119–33.

Wang Qiang, Wang Chunyan, Hu Hui. Structural characterization and structure-activity relationship of ACE inhibitory peptides from peanut [J]. Food Sci. 2013.

Yang Xiaoquan, Chen Zhong, Zhao Mouming. Separation and characterization of peanut proteins [J]. J Chin Cereals Oils. 2001;16(5):25–8.

Yang Xiaoquan, Zhang Shuihua, Li Yin. Separation and characterization of peanut 2S protein [J]. J S China Univ Technol (Nat Sci). 1998;26(4):1–5.

Chapter 9
Preparation of Functional Peanut Oligopeptide and Its Biological Activity

Biopeptides are peptidic compounds that are beneficial or biologically active to the life activities of living organisms (Liu and Cao 2002; Pang Guangchang et al. 2001). Since Brantl et al. (1979) first reported that small peptide with morphine-like activity was found from the small intestine of guinea pig fed with bovine casein zymolyte, a variety of bioactive peptides have been isolated from the animals, plants, and microorganisms (Gill et al. 1996). Among them, peptides that are composed of two to more than ten amino acids with biological activity are known as functional oligopeptides. Functional oligopeptides belong to biologically active peptides. Modern nutrition studies have found that proteins are not completely absorbed in the form of free amino acids after the action of digestive tract enzymes but are absorbed mainly in the form of oligopeptides; the biological meaning of functional oligopeptides is mainly reflected in both a variety of functional activity and completely independent absorption mechanism.

The peanut protein products on the domestic market mainly include peanut protein powder, and it is mainly used as the basic raw material for food processing. Peanut protein powder has such shortcomings as poor functional properties and biological activity, so peanut protein powder is used as the raw material to make functional peanut oligopeptides to improve the functional characteristics of peanut protein, release the functional peptide fragments in protein, fill the blank in peanut protein processing products in China, and effectively improve the conversion rate and utilization rate of peanut processing. This is of great significance to optimize peanut industry structure and product structure, as well as promote the development peanut protein industry and even functional food industry.

At present, China's soybean protein peptide and milk peptide have achieved industrialization, the production mostly adopts directional enzymatic hydrolysis and controllable proteolysis, microbial fermentation, and other technologies. However, there is a lack of detailed research and optimization for oligopeptide production process and its parameters in China, so that most of the peptide products are crude peptides with low purity and low degree of hydrolysis, and functional

Q. Wang, *Peanut Processing Characteristics and Quality Evaluation*,
DOI 10.1007/978-981-10-6175-2_9

oligopeptide products with high purity and high degree of hydrolysis are lacked; in addition, the raw material selection for oligopeptides with corresponding functional activities remains to be established. Zhang and Wang (2007) optimized compound enzyme method to prepare functional peanut oligopeptides. Wang et al. (2013) isolated and purified peanut oligopeptides and analyzed the structure-function relationship between peanut oligopeptide and its structure. On this basis, Wang et al. (2013) simulated the molecular conformation of peanut oligopeptides with the function to reduce blood pressure using Discovery Studio 3.5 software. These studies provide a theoretical basis for the production, processing, and application of functional peanut oligopeptides in China.

1 Composition of Peanut Oligopeptides

1.1 Basic Components

Functional peanut oligopeptides refer to the peptide matters with relative molecular mass of below 5000 Da that are made from peanut meal or peanut protein by enzymatic hydrolysis or microbial fermentation method. Refer to GB/T 22492–2008 (soybean peptide powder). The sensory quality of peanut peptide powder should meet the requirements of Table 9.1.

Soybean peptides are divided into three levels in GB/T 22492–2008. As the products with high-added value among peanut peptides, the physicochemical indicators of functional peanut oligopeptide should meet the level I standards of soybean peptides (Table 9.2).

1.2 Amino Acid Composition

The analysis results of amino acid showed that the free amino acid content of functional peanut oligopeptides was 66.4 mg/g (Fig. 9.1). It was found that although the hydrolytic degree of compound enzymatic hydrolysis was large, the products were mainly oligopeptides and the content of free amino acids was small.

1.3 Molecular Weight Distribution

The molecular weight distribution of functional peanut oligopeptides was generally analyzed using high-performance liquid chromatography (HPLC). The basic

Table 9.1 Sensory quality of peanut peptide powder

Item	Quality requirements	Item	Quality requirements
Fineness	Pass the sieve with a pore diameter of 0.250 mm by 100%	Taste and smell	Have the special taste and smell of the product, without other smell
Color	White, faint yellow, yellow impurities no visible foreign matter	Impurities	Without visible foreign matters

Table 9.2 Physicochemical indicators of functional peanut oligopeptides

Item	Peanut oligopeptides
Crude protein (calculated by dry basis, $N \times 6.25$)/%	≧90.0
Peptide content (calculated by dry basis)/%	≧80.0
≧Relative molecular mass of 80% peptide fragments/Da	≦2000
Ash (calculated by dry basis)/%	≦6.5
Moisture/%	≦7.0
Crude fat (dry basis)/%	≦1.0
Urease activity	Negative

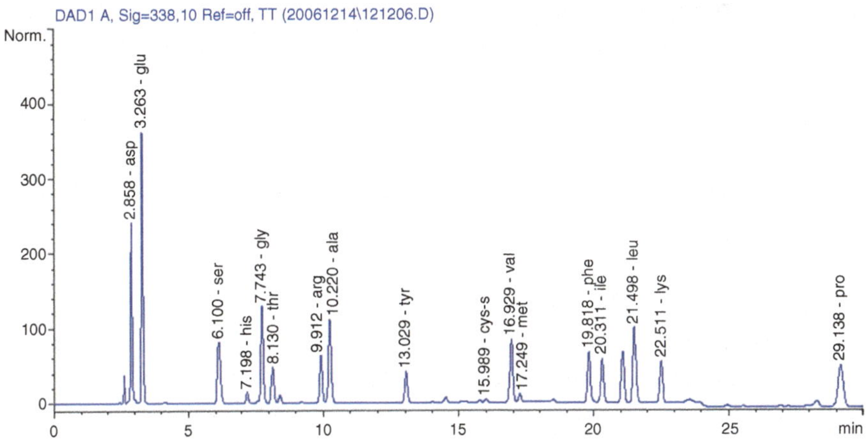

Fig. 9.1 Amino acid composition of functional peanut oligopeptides

conditions are as follows: chromatographic column, TSKgel2000 SWXL 300 × 7.8 mm; mobile phase, acetonitrile/water/trifluoroacetic acid, 20/80/0.1 (V/V); testing wavelength, 220 nm; flow rate, 0.5 ml/min; testing time, 30 min; and column temperature, room temperature; the molecular weight markers were cytochrome C (M_W12500), bacitracin (M_W1450), aminoethanoic acid-aminoethanoic acid-tyrosine-arginine (M_W451), and aminoethanoic acid-aminoethanoic acid-aminoethanoic acid (M_W189).

When making relative molecular mass standard curve, mobile phase should be used to prepare the above peptide standard solutions with different relative molecular masses at mass concentration of 1 mg/mL. After mixing it with a certain proportion and filtering it using the organic phase membrane with a pore size of 0.2–0.5 μm, samples were introduced, and then the chromatogram of standard samples was obtained. The retention time was plotted using the logarithm of the relative molecular mass, or linear regression was made to obtain the relative molecular mass correction curves and their equations.

Zhang and Wang (2007) analyzed the molecular weight distribution of functional peanut oligopeptides with cold-pressed peanut meal as raw materials. The regression equation of the standard sample was $\lg Mw = 7.18–0.239T$, and the correlation coefficient was $R^2 = 0.9925$, indicating that the relative molecular weight logarithm was well correlated with the elution times of standard samples, and the molecular weight distribution of functional peanut oligopeptides could be tested accurately (Table 9.3). Figure 9.2 was the high-performance liquid chromatogram of volume exclusion of functional peanut oligopeptides. It was shown from the figure that the relative molecular weights of functional peanut oligopeptides were distributed between 126 and 11,197 Da, mainly between 126 and 949 Da.

2 Testing Method of Oligopeptide

2.1 *Improvement of Soluble Nitrogen Determination Method*

2.1.1 Improvement of Lowry Method

Lowry method means that protein can react with Folin-phenol reagent under alkaline conditions to form copper-protein compound, and then this compound reduces phosphomolybdic acid-phosphotungstic acid reagent producing a dark blue of the molybdenum blue and tungsten blue compound, and such dark blue compound has the maximum absorption peak at 745–750 nm, and the color depth (absorption value) is proportional to the protein concentration (Wang Jiazheng 2002), and the content of protein can be calculated according to the light absorption value at 750 nm. However, the protein concentration and the linear range of absorption value are narrow, being 5–100 μg/ml, because when the protein concentration is greater than 100 μg/ml, the combination rate of protein and copper will decrease with the increase of protein concentration. Therefore, it is required to adjust the protein concentration in the sample to this concentration range during the actual determination; otherwise, a large deviation may be caused to the result. Coakley and James (1978) established a double-reciprocal equation for this case:

Table 9.3 Molecular weight distribution of functional peanut oligopeptides

Retention times of peak values (min)	Relevant molecular weight distribution scope (Da)	Proportion of peak area (%)
13.100–14.933	11,197–4083	2.94
14.933–16.500	4083–1723	3.12
16.500–17.583	1723–949	4.15
17.583–19.050	949–423	20.18
19.050–20.123	423–235	26.49
20.123–21.250	227–126	34.64

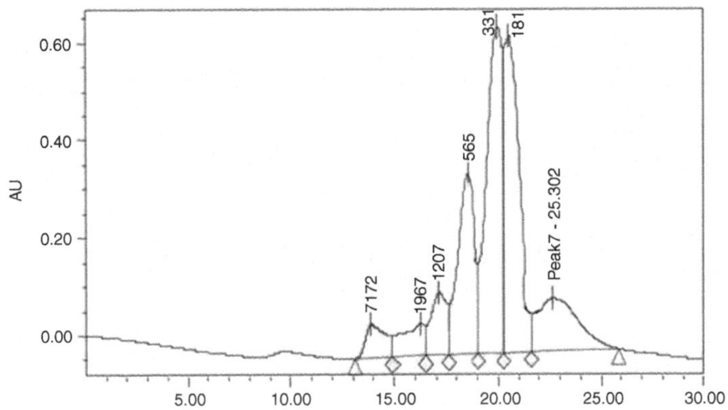

Fig. 9.2 Molecular weight distribution of functional peanut oligopeptide (Standard curve of relative molecular weight: lgMw = 7.18–0.239T, $R = 0.9925$)

$$\frac{1}{A_{750}} = \frac{\Phi}{[P]} + \chi \tag{9.1}$$

This equation can be used to accurately calculate the protein concentration in the sample within the protein concentration range of 0.1–10 mg/ml. The constants can be calculated by two-point method:

$$\Phi = \frac{y_1 - y_2}{x_1 - x_2} \tag{9.2}$$

where $y_1 = \frac{1}{A_{750(1)}}$, $y_2 = \frac{1}{A_{750(2)}}$, $x_1 = \frac{1}{[P]_1}$, and $x_2 = \frac{1}{[P]_2}$
where $[P]$, bovine serum albumin (BSA) concentration, and A_{750}, absorbance of BSA solution with the corresponding concentration at 750 nm:

$$\chi = \bar{y} - \Phi\bar{x} \tag{9.3}$$

where $\bar{y} = \dfrac{y_1 - y_2}{2}$ and $\bar{x} = \dfrac{x_1 - x_2}{2}$

Setting $[P]_1 = 0.1$ mg/ml and $[P]_2 = 5$ mg/ml, we obtained $A_{750\,(1)} = 0.254$, $A_{750\,(2)} = 2.728$ by determination.

And $y_1 = 3.937$, $y_2 = 0.3666$, and $\Phi = 0.3643$:

$$\bar{y} = 2.1518, \bar{x} = 5.1, \chi = 0.2939.$$

When the protein content in the sample was within the range of 0.1–10 mg/ml:

$$\frac{1}{A_{750}} = \frac{0.3643}{[P]} + 0.2939 \tag{9.4}$$

When the protein content in sample was within the range of 5–100 μg/ml/ml, a standard curve was established (Fig. 9.3):

Therefore, through the segmentation function, the protein concentration could be calculated when the protein content in the sample was within the range of 5–10 mg/ml. When the absorbance was less than 0.254, the linear equation was used; when the absorbance was more than 0.254, the double-reciprocal equation was adopted.

2.1.2 Verification of Improved Lowry Method

The linearity and double-reciprocal equations were used to calculate the protein content of BSA solution when the concentration was within the range of 120–200 mg/ml. The results are shown in Table 9.4. The recovery rate was always about 100%. It was shown that the double-reciprocal equation could be applied to the determination of protein concentration in the sample with a protein content of 0.1–10 mg/ml.

2.2 *Yield of Oligopeptides*

In the preparation of oligopeptides, yield was one of the most important evaluation indicators. The yield of oligopeptides was usually determined by trichloroacetic acid precipitation method. That is, first mix the trichloroacetic acid with a concentration of 20% with aqueous solution in 1:1, precipitate macromolecular protein, obtain the supernatant by centrifugation, and then determine soluble protein nitrogen and calculate the yield.

The determination methods of soluble nitrogen usually include Coomassie brilliant blue method, biuret method, Folin-phenol method, and so on. Coomassie brilliant blue method is the most accurate method in the determination of soluble nitrogen. The principle is that Coomassie brilliant blue G-250 can combine with protein in the acidic solution, and thus the maximum absorption peak position of the dye will become 595 nm from 465 nm, and the color of the solution becomes blue from brownish black. The absorbance value determined at 595 nm is proportional to

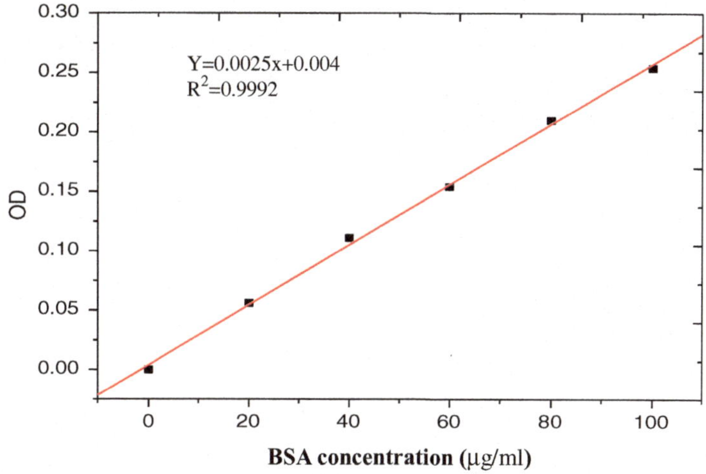

Fig. 9.3 Standard curve by Lowry method

Table 9.4 Verification of improved Lowry method

Actual concentration (mg/ml)	Linear equation (mg/ml)	Recovery rate (%)	Double-reciprocal equation (mg/ml)	Recovery rate (%)
120	120.4 ± 3.61	100.33	122.1 ± 1.32	101.75
140	135.2 ± 2.77	96.57	138.5 ± 2.84	98.93
160	150.8 ± 2.34	94.25	158.3 ± 3.17	98.94
180	168.2 ± 3.19	93.4	183.61 ± 1.71	102.00
200	186.4 ± 1.95	93.20	198.6 ± 2.43	99.30

the protein concentration. However, Coomassie brilliant blue is not suitable for the determination of small-molecule peptide, because the small-molecule peptide cannot be effectively combined with Coomassie brilliant blue G-250 effectively, thus resulting in greater error. The principle of the biuret method is that the biuret is combined with the copper ions in the alkaline solution to form a purple complex. The peptide bonds of protein and polypeptide are similar with the structure of biuret, and they can form purple complex with Cu^{2+}, and its maximum light absorption is at 540 nm. Its color depth is proportional to the protein concentration. But the determination accuracy of biuret method is poor, so it does not apply to accurate quantitative experiments. The principle of Folin-phenol method is to form copper-protein compound using the reaction of protein with Folin-phenol reagent under alkaline condition, and then this compound can reduce phosphomolybdic acid-phosphotungstic acid reagent to produce the dark blue molybdenum blue and tungsten blue compound. Such dark blue compound has the maximum absorption peak at 745–750 nm, the color depth (absorbance) is proportional to the protein concentration, and the protein content can be calculated according to the light absorption value at 750 nm. The method can be used to accurately determine small-molecule proteins and polypeptides and is more suitable for the determination of polypeptide yield.

2.3 Degree of Hydrolysis

DH (degree of hydrolysis) represents the degree or percentage of peptide bond breakage in protein during the hydrolysis process. The mathematical expression is:

$$DH = \frac{h}{h_{hot}} \times 100\% \quad (9.5)$$

where h is the number of peptide bonds broken, h_{tot} is the total number of peptide bonds in the protein, and h and h_{tot} are often expressed in mmol/g. For a particular protein, h_{tot} is a constant, and it can be calculated according to the components of amino acids in the protein. h_{tot} of peanut protein is about 7.13 mmol/g.

In the hydrolysis process, when the peptide bond is broken, new –COOH and –NH will be formed, so the number of hydrolyzed peptide bonds can be determined by measuring the number of newly formed terminal –COOH or –NH groups after hydrolysis.

Commonly used methods for the determination of degree of hydrolysis include pH-stat method, trinitrobenzene sulfonic acid method (TNBS), o-phthalaldehyde method (OPA), hydrated ninhydrin method, and formol titration method. Among them, pH-stat method is used to determine DH using the protons released during titration and hydrolysis process, and it is usually applied only in the hydrolysis of neutral or alkaline environment. OPA, TNBS, and hydrated ninhydrin and formol titration method are all measured by chemical reaction of free amino groups. Among them, the determinations of DH with OPA and TNBS are highly correlated with high accuracy but long time. Hydrated ninhydrin method is generally used for the determination of DH in the low degree of hydrolysis. It is usually possible to select a reasonable DH determination method based on the nature of the hydrolysate, hydrolysis environment, and laboratory conditions.

2.4 Average Peptide Chain Length (PCL)

The PCL of the hydrolysate is usually calculated based on the degree of hydrolysis. If the average chain length of the protein is PCL_0, the formula is established after the protein was hydrolyzed into n peptide molecules:

$$DH = \frac{n-1}{PCL_0 - 1} \times 100\% \tag{9.6}$$

Meanwhile, $PCL_0 = n \times PCL$; the following formula is obtained after putting it into the above equation:

$$DH = \frac{1}{PCL} - \frac{\frac{1}{PCL_0}}{1 - \frac{1}{PCL_0} \times 100\%} \tag{9.7}$$

Because the molecular weight of peanut protein is large, so $\frac{1}{PCL_0} = 0$.

Therefore, $DH = \frac{1}{PCL} \times 100\%$, and thus the following formula is obtained:

$$PCL = \frac{1}{DH} \tag{9.8}$$

3 Preparation Process and Technology

Biological enzyme method was used to hydrolyze the protein under mild conditions, and the generated polypeptide had a very high nutritional value. The modern nutrition study showed (Grimble and Silk 1990; Sheng 1993; Dun and Chen 2004; Zhang and Feng 2004): The oligopeptides with a molecular weight of less than 1000 Da could be easily absorbed by the human body and they had a strong functional activity. This required the selected enzymolysis process to improve the degree of hydrolysis of the protein as far as possible under the premise of ensuring the yield of oligopeptides during the preparation of oligopeptides.

At present, in the process of preparing oligopeptides using proteolysis (Zhang 2002; Guanhong 2005; Mo Qi 1996; Wu and Ding 2002), the enzymolysis time was 360–960 min, the yield of oligopeptides was about 75%, and the degree of hydrolysis of the system was about 16 %. This showed that it was a major problem to be solved how to obtain the oligopeptides with high degree of hydrolysis and yield within a short time. In order to solve the problems of low degree of hydrolysis and yield of peptide products, we prepared peanut oligopeptides using cold-pressed peanut protein powder as the raw material by enzymatic hydrolysis process and strived to prepare the peanut oligopeptides with high yield and degree of hydrolysis through the study on various influencing factors in the enzymatic hydrolysis process and optimization of the process parameters of peanut protein by enzymatic hydrolysis, so as to provide theoretical basis for the industrial production of functional oligopeptides of peanut.

3.1 Enzymolysis Conditions

3.1.1 Single Protease

Enzyme type is one of the most important influencing factors in the enzymatic hydrolysis process. In this experiment, five peanut proteins by protease hydrolysis were selected; the results are shown in Fig. 9.4. Under the recommended operation conditions of various proteases, the TCA-NSI of 1,398 neutral protease, FM, Protamex, N120p-hydrolyzed peanut protein was between 45 and 53% and DH was between 9.4 and 12.1%, while the TCA-NSI and DH of Alcalase-hydrolyzed peanut protein was 63.80 ± 0.47% and 15.01 ± 0.88%, respectively, which were significantly higher than other proteases ($p < 0.01$). Therefore, Alcalase was selected as the tool enzyme of hydrolyzed peanut protein for the following experiment.

3.1.2 Single-Factor Test of Alcalase-Hydrolyzed Peanut Protein

3.1.2.1 Dosage of Enzyme

In the experiment, the dosage was further improved on the basis of 304 U/g (dosage of enzyme), to study the impact of the dosage of enzyme on TCA-NSI and DH (Fig. 9.5). The results showed that within the concentration range of 304–2428 U/g, the TCA-NSI and DH of enzymatic hydrolysate increased rapidly with the increase of dosage of enzyme. When the dosage of enzyme reached 3642 U/g, the TCA-NSI and DH reached 71.63 ± 0.16% and 19.74 ± 0.27%, respectively, with an increase of 10.86% and 11.45%, respectively, compared with that when the dosage of enzyme was 304 U/g. When the dosage of enzyme was within the range of 3642–6070 U/g concentration, TCA-NSI basically no longer changed, DH increase was also not obvious, only being 0.85%. In the early stage of enzymatic hydrolysis, macromolecule proteins rapidly decreased, and small-molecule peptide rapidly increased; meanwhile, peptide was further hydrolyzed to the peptide molecules with lower molecular weight as the substrate of protease, and both of them had competitive combination with protease during the hydrolysis process. The initial rate of system reaction would increase with the increase of dosage of enzyme, which made most of large molecular proteins degraded into small-molecule peptides within a short time; when large molecular proteins decreased to a certain extent, regardless of the concentration of protease of the system, the interaction probability between it and large molecular proteins would greatly decreased, and at this time, the main substrate of protease was peptide. Therefore, when the dosage of enzyme reached 3642 U/g, the further increase of dosage of enzyme did not have significant impact on improving the yield of peptide. Although the protease would still have an impact on peptide, DH of the system rose very slowly, because the action sites of protease had been greatly reduced. Thus, when the dosage of enzyme

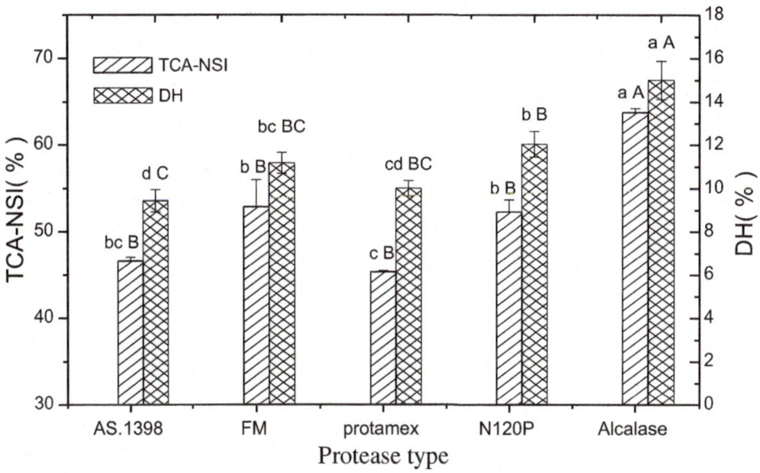

Fig. 9.4 Comparison of effects of enzymatic effects of different proteases (*Note*: Lowercase letters mean 0.05 significance level; capital letters mean 0.01 significance level)

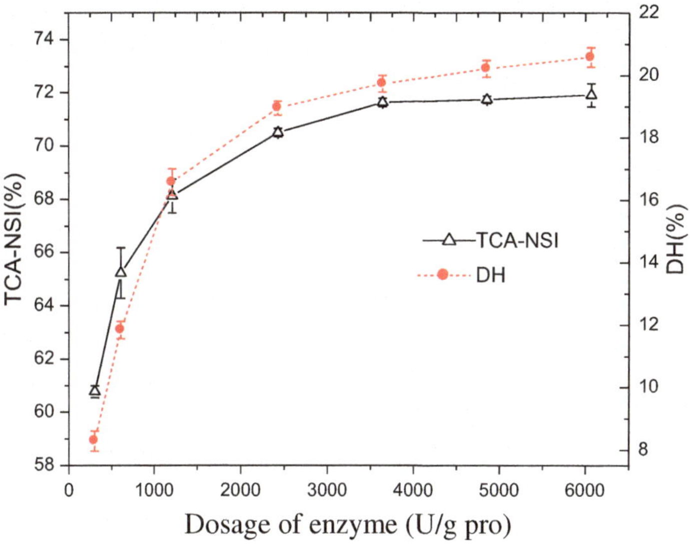

Fig. 9.5 Impact of Alcalase dosage on TCA-NSI and DH

reached 3642 U/g, the yield of oligopeptides of the system is basically unchanged, and the increase range of DH was very small with the further increase of dosage of enzyme, so the appropriate dosage of enzyme was about 3642 U/g.

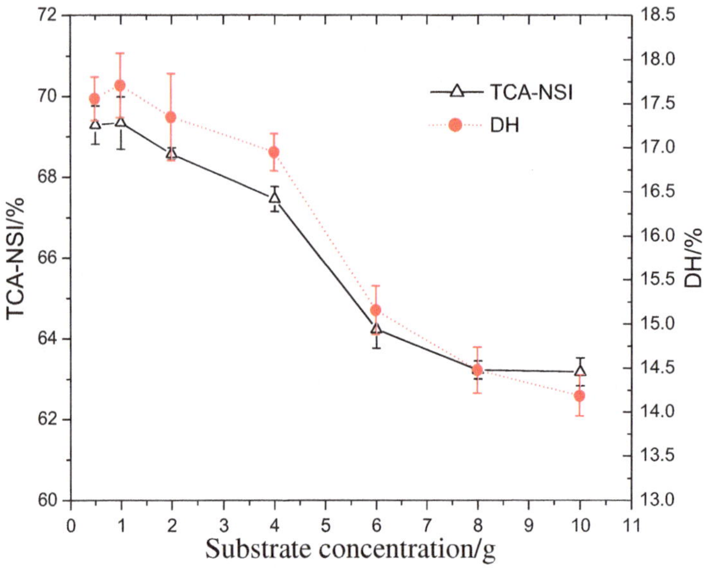

Fig. 9.6 Impact of substrate concentration on TCA-NSI and DH

3.1.2.2 Substrate Concentration

It was shown from Fig. 9.6 that when the substrate concentration was within the range of 1–10%, there was a substrate inhibition in peanut protein-Alcalase system, that is, TCA-NSI and DH decreased with the increase of substrate concentration; the substrate inhibition disappeared when substrate concentration further decreased to 0.5%, and TCA-NSI and DH all decreased compared with substrate concentration of 1%. When the substrate concentration was 4%, TCA-NSI and DH were, respectively, 67.46 ± 0.31% and 16.95 ± 0.21%, which were lower than that when the substrate concentration was 2% by 1.11% and 0.4%, respectively, and higher than that when the substrate concentration was 6% by 3.23% and 1.79%. It was shown that when the substrate concentration increased from 4% to 6%, the decrease ranges of TCA-NSI and DH were large. In theory, if the system substrate concentration was too large, the effective water concentration would be too low (Tello 1994), thus reducing the diffusion and movement of substrate and protease and inhibiting the hydrolysis; when the substrate concentration was too low, the collision probability of protease and active substrate would be reduced, thus inhibiting the hydrolysis reaction. The experiment results showed that when the substrate concentration was 0.5%, the collision probability of protease and substrate was reduced, the impact of which was larger than that of high effective moisture concentration and reflected in the inhibition of low substrate concentration to hydrolysis; within the concentration range of 1–10%, it was shown by the inhibition of high substrate concentration to hydrolysis reaction, that is, when the substrate concentration was 1%, TCA-NSI and DH reached the maximum values. However, during the actual production, too low protein concentration in raw materials would result in low absolute yield of product (Deng 1981). Under a condition of a certain amount of processing, it was necessary to increase the output of products by repeated production, which would increase the energy consumption in the production process and reduce equipment utilization, so too low substrate concentration should not be used in the actual production. The substrate concentration of 4% was selected as the optimum substrate concentration in the final experiment.

3.1.2.3 pH Value

The impact of pH on the enzymatic reaction was mainly reflected in the impact on the enzyme activity (Munilla-moran and Saborido-rey 1996). The changes of TCA-NSI and DH when the pH was within the range of 6–9 were analyzed. The results are shown in Fig. 9.7. With the increase of pH value, TCA-NSI and DH rapidly increased; when the system pH reached 8.0, TCA-NSI and DH reached $8.61 \pm 1.23\%$ and $16.36 \pm 0.33\%$, respectively; when the system pH reached 9.0, TCA-NSI and DH decreased compared with that when the pH was 8.0, with a decrease range of 1.59% and 0.84%, respectively. It was shown that Alcalase was more suitable for alkaline environment, so pH 8.0–8.5 was selected as the optimum pH for Alcalase.

3.1.2.4 Reaction Temperature

The changes of TCA-NSI and DH within the temperature range from 40 to 75 °C were analyzed. The results showed that system TCA-NSI and DH increased gradually when the temperature rose from 40 to 55 °C, and system TCA-NSI and DH reached the maximum values at 55 °C, which was $67.70 \pm 0.58\%$ and $15.96 \pm 0.22\%$, respectively. After the temperature exceeded 55 °C, system TCA-NSI and DH began to decrease and then began to decrease rapidly after the temperature exceeded 60 °C. The hydrolysis temperature was closely related to the stability of protease molecules (Wang and Ng 2001), because the peptide bond of protease molecule had a specific spatial structure. If the reaction temperature exceeded a certain limit, the dissociation secondary bond might be caused easily, and thus protein would lose catalytic activity fully or partially. However, if the reaction temperature was too low, the degree of molecular movement of the system would be greatly reduced, and thus the collision probability of protease and substrate would be reduced. Therefore, the action temperature suitable for Alcalase was about 55 °C (Fig. 9.8).

3.1.2.5 Hydrolysis Time

The changes of system TCA-NSI and DH under the impact of different hydrolysis times within the range of 0–240 min were analyzed in this experiment. It was shown

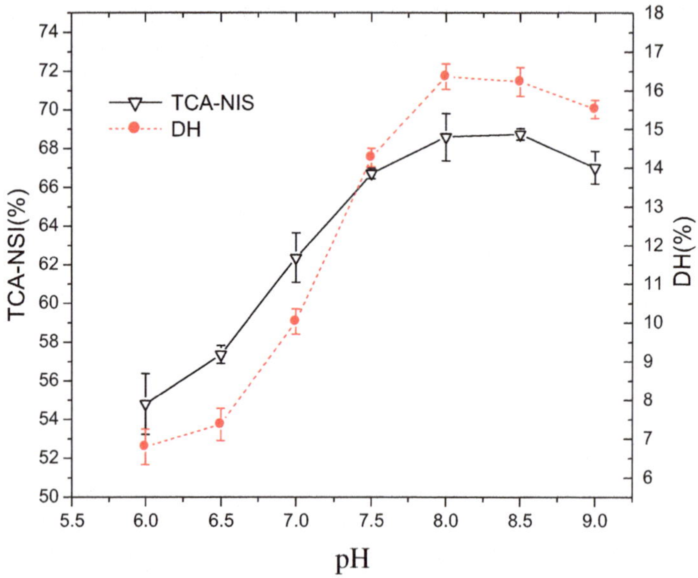

Fig. 9.7 Impact of pH on TCA-NSI and DH

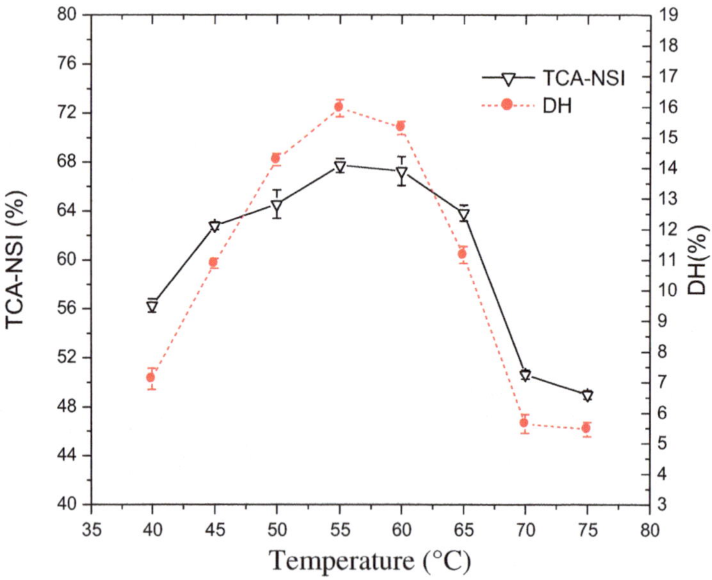

Fig. 9.8 Impact of reaction temperature on TCA-NSI and DH

from Fig. 9.9 that 120 min before hydrolysis, TCA-NSI and DH increased; after hydrolysis for 120 min, TCA-NSI and DH reached 69.69 ± 0.47% and 15.81 ± 0.26%, respectively; after hydrolysis for 180 min, although DH continued

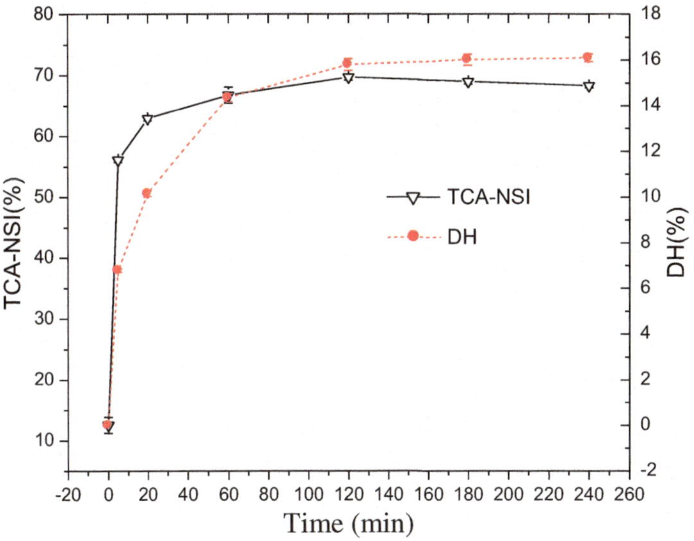

Fig. 9.9 Impact of hydrolysis time on TCA-NSI and DH

to rise, the increase range was only 0.21%, while TCA-NSI showed a decrease trend. This may be due to the fact that Alcalase itself is a complex protease consisting of endonucleases and enzymes. During the hydrolysis process, the activity of enzyme might gradually appear with the gradual decrease of endonuclease sites, and some peptide molecules of the system were hydrolyzed to amino acids under the action of enzyme in the hydrolysis system. In the later stage of hydrolysis, the hydrolysis rate of endonucleases of the system was greater than that of enzyme. Therefore, after 180 min, the peptide extraction rate of the system slightly decreased. It was shown that when the hydrolysis time was 120 min, the extraction rate of peptide of the system reached the maximum value. At this time, the increase range of the degree of hydrolysis was very small with the extension of hydrolysis time. Therefore, it was determined that the optimum hydrolysis time was about 120 min.

3.1.3 Orthogonal Rotation Combination Experiment of Alcalation Hydrolyzed Peanut Protein

The results of TCA-NSI and DH measured through 23 treatment combination experiments of four-factor quadratic rotation combination are shown in Table 9.5. SPSS working platform was used to analyze and process the results, establish the mathematical models of TCA-NSI and DH with various factors, obtain the optimum process conditions by taking TCA-NSI as the first indicator and using the analysis of oligopeptide yield model, and determine the degree of hydrolysis of the system using DH model under the optimal conditions.

Table 9.5 Results of orthogonal rotation experiment

Processing number	Matrix design				TCA-NSI/%			Degree of hydrolysis/%		
	X1	X2	X3	X4	Experiment value	Fitted value	Fitted error	Experiment value	Fitted value	Fitted error
1	1	1	1	1	62.79	61.37	1.42	11.39	12.04	−0.65
2	1	1	−1	−1	48.80	49.68	−0.88	5.16	4.98	0.18
3	1	−1	1	−1	65.45	62.54	2.91	10.02	9.54	0.48
4	1	−1	−1	1	58.80	58.07	0.73	10.53	11.36	−0.83
5	−1	1	1	−1	71.97	70.57	1.40	11.61	11.62	−0.01
6	−1	1	−1	1	66.57	66.10	0.47	13.81	13.44	0.37
7	−1	−1	1	1	75.65	78.96	−3.31	17.21	18.00	−0.79
8	−1	−1	−1	−1	65.90	67.27	−1.37	10.01	10.94	−0.93
9	−1.682	0	0	0	72.20	70.36	1.84	12.59	12.39	0.20
10	1.682	0	0	0	46.50	48.82	−2.32	5.49	5.63	−0.14
11	0	−1.682	0	0	80.21	79.27	0.94	15.52	14.89	0.63
12	0	1.682	0	0	70.13	71.24	−1.11	11.06	11.63	−0.57
13	0	0	−1.682	0	64.56	63.77	0.79	12.17	12.81	−0.64
14	0	0	1.682	0	76.08	77.35	−1.27	16.43	17.21	−0.78
15	0	0	0	−1.682	60.60	61.65	−1.05	8.59	9.07	−0.48
16	0	0	0	1.682	68.29	67.72	0.57	17.05	16.54	0.51
17	0	0	0	0	75.16	75.26	−0.10	16.14	15.01	1.13
18	0	0	0	0	74.47	75.26	−0.79	14.52	15.01	−0.49
19	0	0	0	0	76.08	75.26	0.82	15.76	15.01	0.75
20	0	0	0	0	73.56	75.26	−1.70	13.91	15.01	−1.10
21	0	0	0	0	78.15	75.26	2.89	14.52	15.01	−0.49
22	0	0	0	0	74.47	75.26	−0.79	15.29	15.01	0.28
23	0	0	0	0	75.16	75.26	−0.10	14.91	15.01	−0.10

3.1.3.1 Impact of Enzymatic Hydrolysis on TCA-NSI

SPSS data processing system was used to fit the experimental data by standard polynomial regression method, thus establishing the quadratic polynomial Formula 9.9:

$$\begin{aligned} Y = \; & 75.22 - 6.40X1 - 2.39X2 + 4.04X3 + 1.80X4 \\ & -5.54X12 + 0.06X22 - 1.66X32 - 3.74X42 \\ & -0.60X1X2 + 0.34X1X3 + 0.19X1X4 + 0.19X2X3 \\ & +0.34X2X4 - 0.60X3X4 \end{aligned} \tag{9.9}$$

After excluding the nonsignificant items at $\alpha = 0.05$ significance level, the simplified regression equation is Formula 9.10:

$$Y = 75.22 - 6.4X1 - 2.39X2 + 4.04X3 + 1.8X4 - 5.54X12 - 1.66X32 - 3.74X42 \tag{9.10}$$

To test the significance of regression equation, F-test should be conducted to the regression equation. The results are shown in Table 9.6. From $F_{0.05}$ (2, 6) = 5.14 < F_1 = 5.641< $F_{0.01}$ (2, 6) = 10.92. It was shown that the item in lack of fit was significant at the level of $\alpha = 0.05$, but it was nonsignificant at the level of $\alpha = 0.01$, indicating that there were factors in lack of fit and this might come from the high-order interaction between the factors. From F_2 = 25.106> $F_{0.01}$ (9, 15) = 5.56, it was shown that the regression equation itself was significant at the level of $\alpha = 0.01$, and the maximum fitting error between fitting value and experiment value was 5.0%, indicating that the regression equation and the experiment were fitted properly.

After the model was established, further analysis was made to the characteristics of the model, and the regression equation model (9.9) was rewritten as a matrix (Eq. 9.1) form:

$$[y] = 107.42 + \begin{bmatrix} x_1 \\ x_2 \\ x_3 \\ x_4 \end{bmatrix}^T \begin{bmatrix} -6.40 \\ -2.39 \\ 4.04 \\ 1.80 \end{bmatrix} + \begin{bmatrix} x_1 \\ x_2 \\ x_3 \\ x_4 \end{bmatrix}^T A \begin{bmatrix} x_1 \\ x_2 \\ x_3 \\ x_4 \end{bmatrix} \tag{9.11}$$

where

$$A = \begin{bmatrix} -11.08 & -0.60 & 0.34 & 0.19 \\ -0.60 & 0.12 & 0.19 & 0.34 \\ 0.34 & 0.19 & -3.32 & -0.60 \\ 0.19 & 0.34 & -0.60 & -7.48 \end{bmatrix}$$

To obtain the stationary points of Formula 9.11, set:

Table 9.6 Variance analysis of TCA-NSI test results

Source of variation	Quadratic sum	Degree of freedom	Mean square	Ratio F	Significance level p
X_1	560.2580	1	560.2580	118.10791	0.00000
X_2	77.9261	1	77.9261	16.42759	0.00367
X_3	222.8255	1	222.8255	46.97381	0.00013
X_4	44.3946	1	44.3946	9.35883	0.01560
X_1^2	480.9496	1	480.9496	101.38891	0.00001
X_2^2	0.2693	1	0.2693	0.05678	0.81766
X_3^2	40.4939	1	40.4939	8.53650	0.01924
X_4^2	216.1033	1	216.1033	45.55671	0.00015
X_1X_2	5.8182	1	5.8182	1.22653	0.30027
X_1X_3	1.8831	1	1.8831	0.39698	0.54622
X_1X_4	0.5586	1	0.5586	0.11775	0.74033
X_2X_3	0.5586	1	0.5586	0.11775	0.74033
X_2X_4	1.8831	1	1.8831	0.39698	0.54622
X_3X_4	5.8182	1	5.8182	1.22653	0.30027
Regression	1667.2850	14	119.0918	$F_2 = 25.106$	
Residue	37.9489	8	4.7436		
Lack of fit	24.7737	2	12.3869	$F_1 = 5.641$	
Error	13.1751	6	2.1959		
Sum	1705.2241	22			

$$\frac{\partial y}{\partial x} = \left(\frac{\partial y}{\partial x_1} \cdots \frac{\partial y}{\partial x_4}\right)^T = 0 \tag{9.12}$$

The stationary points can be obtained from Formula 9.12: $x_1 = -0.578$, $x_2 = -1.682$, $x_3 = 1.197$, and $x_4 = 0.241$.

Judge whether the stationary points were saddle points, and determine the properties of positive and negative of Matrix A, and establish Formula 9.13:

$$|A - \lambda I| = 0 \tag{9.13}$$

After solving the opposite sign of λ value, it was shown that Matrix A was negative, so this stationary point was saddle point, and y only had the maximum value.

From this, the optimal combination when TCA-NSI was the maximum included: hydrolysis temperature, 53 °C; substrate concentration, 1%; dosage of enzyme, 3637 U/g; and hydrolysis time, 105 min, which were basically consistent with single-factor test results. Under this condition, the yield of oligopeptide measured in the test under this condition was 83.8%.

From the results of the model analysis, it was found that the enzymatic hydrolysis efficiency decreased gradually with the increase of the protein concentration within the selected range. However, the increase of the protein concentration in the production could reduce the energy consumption and improve the utilization rate of the equipment. Therefore, a high substrate concentration could be selected according to the model during the actual production. The single-factor test showed that 4% was the optimum substrate concentration, so 4% was chosen as the optimum substrate concentration. Combined with the regression model, it was found that the TCA-NSI of the system was 80.35% when the substrate concentration was 4% and other factors were under the best conditions.

3.1.3.2 Impact of Enzyme Hydrolysis on DH

SPSS data processing system was used to fit the test data by standard polynomial regression method, thus establishing the quadratic polynomial formula 9.14:

$$\begin{aligned} Y = \; & 15.01 - 2.01X1 - 0.97X2 + 1.31X3 + 2.22X4 \\ & -2.12X12 - 0.62X22 - 0.26X32 - 0.78X42 \\ & -0.14X1X2 + 0.05X1X3 + 0.17X1X4 + 0.17X2X3 \\ & +0.05X2X4 - 0.14X3X4 \end{aligned} \tag{9.14}$$

After excluding the nonsignificant items at $\alpha = 0.05$ significance level, the simplified regression equation is Formula 9.15:

$$\begin{aligned} Y = \; & 15.01 - 2.01X1 - 0.97X2 + 1.31X3 + 2.22X4 - 2.12X12 \\ & - 0.62X22 - 0.78X42 \end{aligned} \tag{9.15}$$

To test the significance of regression equation, F-test should be conducted to the regression equation. The results are shown in Table 9.7. From $F_1 = 8.785 < F_{0.05}(2, 6) = 5.14$, it was shown that the item in lack of fit was not significant at the level of $\alpha = 0.05$, indicating that there were no factors in lack of fit. From $F_2 = 24.64 > F_{0.01}$ (9, 15) = 5.56, it was shown that the regression equation itself was significant at the level of $\alpha = 0.01$ and the maximum fitting error between fitting value and experiment value was 9.29%, indicating that the regression equation and the experiment were fitted properly.

After substituting the optimal combination, namely, $x_1 = -0.578$, $x_2 = -0.238$, $x_3 = 1.197$, and $x_4 = 0.241$, into DH regression model, it was found that the system DH acting under this condition was DH = 17.72 %, PCL = 5.64, and Mw = 638.4.

3.1.4 Screening of Compound Enzymatic Proteases

Enzyme type is one of the most important factors in the enzymatic process. In this test, Alcalase was used to hydrolyze peanut protein, and then four kinds of proteases were used to continue enzymolysis for peanut protein respectively. The results are shown in Fig. 9.10. Under the recommended conditions of various proteases, the

Table 9.7 Variance analysis of DH test results

Source of variation	Quadratic sum	Degree of freedom	Mean square	Ratio F	Significance level p
X_1	55.2975	1	55.2975	76.88987	0.00002
X_2	12.9540	1	12.9540	18.01224	0.00282
X_3	23.4207	1	23.4207	32.56591	0.00045
X_4	67.5275	1	67.5275	93.89534	0.00001
X_1^2	70.6569	1	70.6569	98.24665	0.00001
X_2^2	5.6395	1	5.6395	7.84157	0.02319
X_3^2	0.8864	1	0.8864	1.23257	0.29916
X_4^2	9.2521	1	9.2521	12.86475	0.00712
X_1X_2	0.3025	1	0.3025	0.42062	0.53480
X_1X_3	0.0324	1	0.0324	0.04505	0.83722
X_1X_4	0.4422	1	0.4422	0.61490	0.45554
X_2X_3	0.4422	1	0.4422	0.61490	0.45554
X_2X_4	0.0324	1	0.0324	0.04505	0.83722
X_3X_4	0.3025	1	0.3025	0.42062	0.53480
Regression	248.1178	14	17.7227	$F_2 = 24.643$	
Residue	5.7534	8	0.7192		
Lack of fit	2.1355	2	1.0677	$F_1 = 1.771$	
Error	3.6179	6	0.6030		
Sum	253.8704	22			

DH of peanut proteins which were Alcalase hydrolyzed by N120P and FM were significantly higher than that of other two proteases, being 20.96 ± 0.33% and 20.08 ± 0.37%, respectively, and there were no significant difference between them ($p > 0.05$); the TCA-NSI value of peanut protein Alcalase hydrolyzed by N120P reached 84.6 ± 0.74%, which increased by 6.45% compared with that hydrolyzed by Alcalase alone and was significantly higher than that of other protease ($P < 0.05$). It was shown that in the enzymatic hydrolysate of peanut protein Alcalase hydrolyzed by N120P, DH and the yield of oligopeptide were all high; therefore, N120P was selected as the tool enzyme for Alcalase hydrolyze peanut protein in the following study.

3.1.5 Single-Factor Test of N120P-Hydrolyzed Peanut Protein

3.1.5.1 Dosage of Enzyme

The test analyzed the changes of the system TCA-NSI and DH when the N120P concentration was within the range of 300–3600 U/g (Fig. 9.11). The results showed that with the increase of the dosage of N120P, the TCA-NSI and DH in enzymatic hydrolysate gradually rose. When the dosage of N120P was smaller than 1800 U/g, the increase speeds of TCA-NSI and DH were quick; when the dosage of

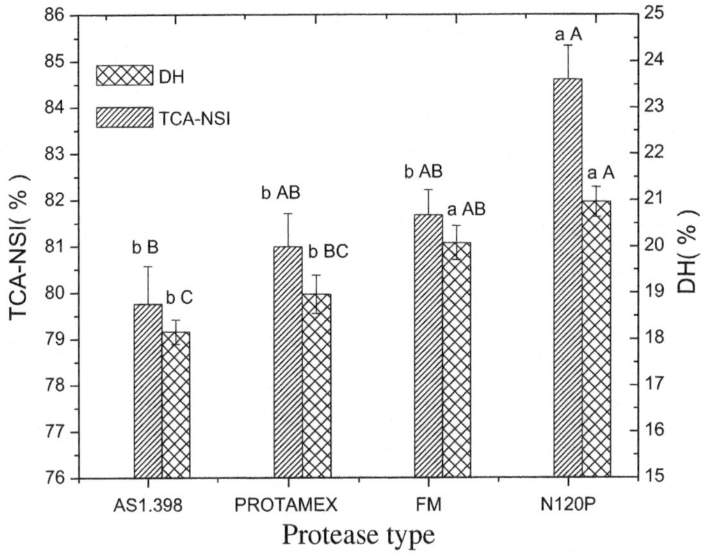

Fig. 9.10 Screening of compound enzymolysis protease

N120P reached 1800 U/g, TCA-NSI and DH were $54.44 \pm 0.42\%$ and $13.89 \pm 0.44\%$, respectively; and when the dosage of enzyme was within the range of 1800–3600 U/g, the increase ranges of TCA-NSI and DH were only 1.17% and 0.707%. According to enzyme kinetics, too large or too small dosage of enzyme was unfavorable to hydrolysis, and too large dosage of enzyme could not achieve the hydrolysis effect but might cause the waste of resources. So the appropriate dosage of enzyme should be 1800 U/g.

3.1.5.2 pH Value

The impact of pH value on enzymatic reaction was mainly reflected in enzyme activity. The test analyzed the changes of the system TCA-NSI and DH when pH value was within the range of 5–8. It was shown in Fig. 9.12 that TCA-NSI and DH gradually increased with the increase of pH value and TCA-NSI and DH reached the maximum values when pH value was 6.0, being $51.94 \pm 0.42\%$ and $13.83 \pm 0.34\%$, respectively; when the pH value was greater than 6.0, TCA-NSI and DH began to decrease. This showed that too acidic or alkaline system would affect the activity of enzyme; N120P was more suitable for neutral to acidic conditions, so pH 6.0 was selected as the optimal pH.

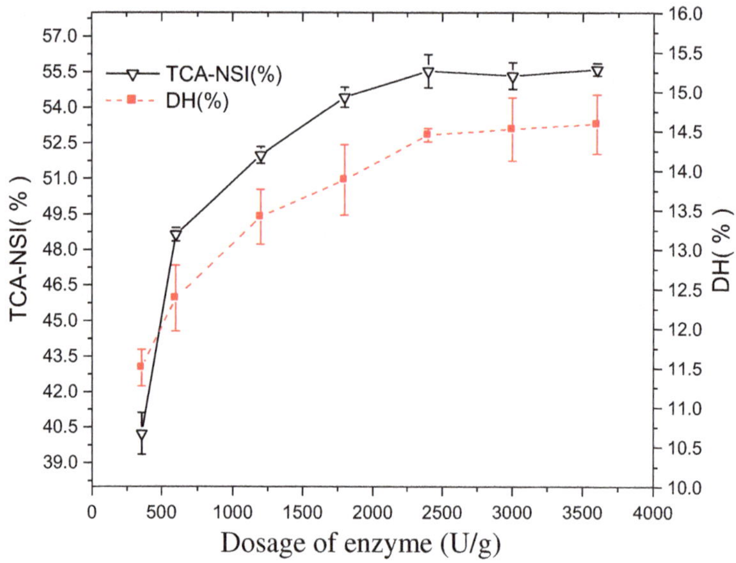

Fig. 9.11 Impact of dosage of N120P on TCA-NSI and DH

3.1.5.3 Reaction Temperature

The activity of the enzyme depends largely on the temperature used. This test analyzed the changes in TCA-NSI and DH over the temperature range from 40 to 70 °C. As shown in Fig. 9.13, the system TCA-NSI and DH increased greatly when the temperature rose from 40 to 55 °C, and system TCA-NSI and DH reached the maximum values at 55–60 °C, being 53.29 + 1.48 % and 13.97 + 0.56 %, respectively; system TCA-NSI and DH began to decrease after the temperature reached 60 °C. This was because the peptide bond of protease molecule had a specific spatial structure. If the reaction temperature exceeded a certain limit, the molecule would absorb too much energy and thus cause the secondary bonds to disintegrate or lose fully or partially its catalytic activity (Wang and Gu 1999). Therefore, the temperature suitable for N120P might be between 5 and 60 °C.

3.1.5.4 Reaction Time

This test analyzed the impacts of different enzymolysis times within the range of 0–240 min on system TCA-NSI and DH. It was shown from Fig. 9.14 that 20 min before enzymolysis, TCA-NSI and DH increased rapidly; however, after 20 min, the increase speed began to slow down; TCA-NSI and DH reached 48.99 ± 1.34 % and 12.8 ± 0.37%, respectively, at 60 min. However, when the enzymolysis time was within the range of 60–240 min, the increase ranges of TCA-NSI and DH were

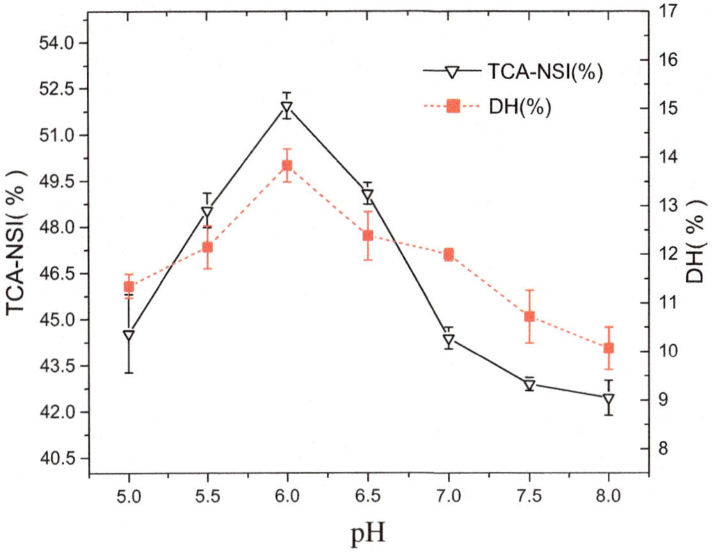

Fig. 9.12 Impact of pH on TCA-NSI and DH

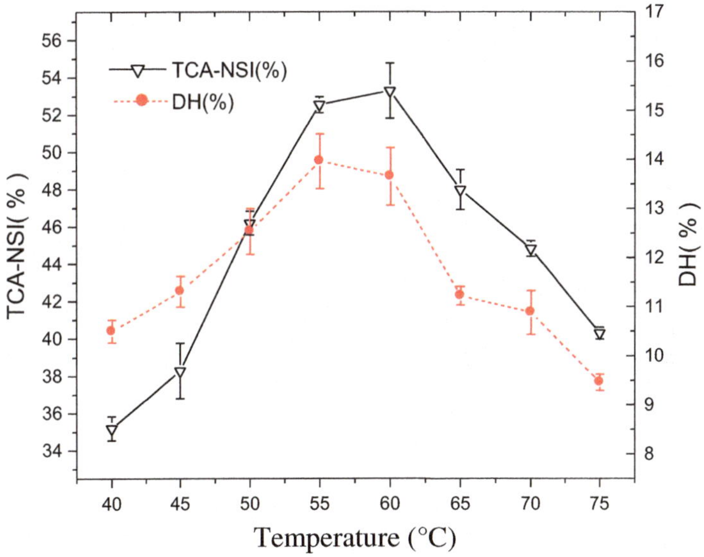

Fig. 9.13 Impact of temperature on TCA-NSI and DH

only 2.92% and 1.21%. The increase speed was very slow, so the suitable time was about 60 min.

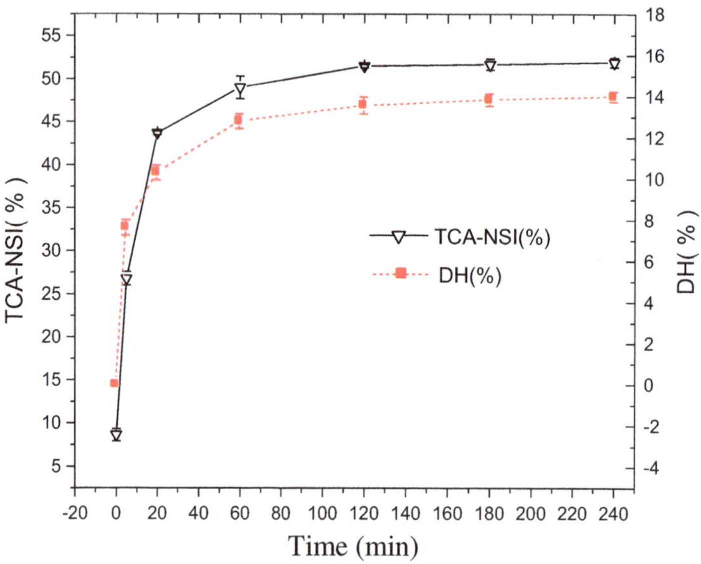

Fig. 9.14 Impact of time on TCA-NSI and DH

3.1.5.5 Uniform Design of N120P-Hydrolyzed Peanut Protein

The results of the TCA-NSI measured by 11 treatment combinations with uniform design are shown in Table 9.8. SPSS working platform was used to analyze and process the results and establish the mathematical model for the relationship between TCA-NSI and various factors, and the optimum process conditions were obtained by analyzing the model.

Quadratic polynomial stepwise regression analysis was conducted to the test results. Take the multiple correlation coefficient R as the indicator and semiautomatically delete the nonsignificant items in the equation, thus establishing the following regression model finally:

$$Y = 67.852 + 2.450X2 + 4.480X3 - 0.057X12 - 0.143X22 - 0.273X32 + 0.045X1X3 - 0.172X2X3$$

The correlation coefficients of regression model: $R^2 = 0.98269$, $P = 0.0328$, equation Durbin-Watson statistics $d = 1.76122687$, indicating that the residuals are independent of each other. The maximum fitting error between fitted value and test value was 2.77%, indicating that the fitting between regression equation and test was good, and the regression model established in this test was credible.

The intuitive judgment based on the three-dimensional response diagram of all factors is shown in Fig. 9.15. The optimal combination when the regression model TCA-NSI was highest was $X_1 = 2.81124$, $X_2 = 4.31006$, and $X_3 = 7.08699$, that is, hydrolysis time was 65 min, hydrolysis temperature was 57 °C, and the enzyme-

Table 9.8 Results of uniform design test

Processing number	Time (min)	Temperature (°C)	Dosage of enzyme (U/g)	TCA-NSI(%)		
				Actual value	Fitted value	Fitting error
1	5	6	11	84.00	83.27	0.73
2	7	1	4	81.23	84.21	−2.98
3	10	9	2	77.79	78.39	−0.60
4	9	3	10	85.40	85.13	0.27
5	6	11	5	83.77	82.69	1.08
6	4	4	1	78.47	78.54	−0.07
7	8	8	8	84.00	83.66	0.34
8	3	10	9	81.23	83.48	−2.25
9	2	2	7	88.21	88.32	−0.12
10	1	7	3	85.17	83.10	2.07
11	11	5	6	85.40	83.87	1.53

substrate ratio was 2061 U/g, which was basically consistent with that of single-factor test results. The TCA-NSI measured under this condition was 89.01%, which was 10.86% higher than that under the single action of Alcalase.

3.1.5.6 Determination of DH of Peanut Protein Compound Enzymatic Hydrolysate

The compound enzymolysis time of peanut protein under the optimal conditions was 170 min; the degree of hydrolysis DH of the system was 23.76 ± 0.93%, which was 6.24% higher than that of single enzymolysis by Alcalase; PCL was 4.21 ± 0.16, which was1.5 lower than that of single enzymolysis by Alcalase; and the average molecular weight of the system was 481.1. It was shown that compound enzymolysis could greatly increased the yield of peanut functional oligopeptides and significantly decreased the length of peptides in peanut functional oligopeptide system ($p < 0.01$).

3.2 *Enzymolysis Kinetics*

The enzymatic preparation of oligopeptides refers to optimizing the hydrolysis parameters, determining the relationship between the degree of hydrolysis of the products and molecular weight distribution, and thus obtaining the oligopeptides with the target molecular weight distribution through the study on protein enzymolysis conditions, degree of hydrolysis and the molecular weight distribution

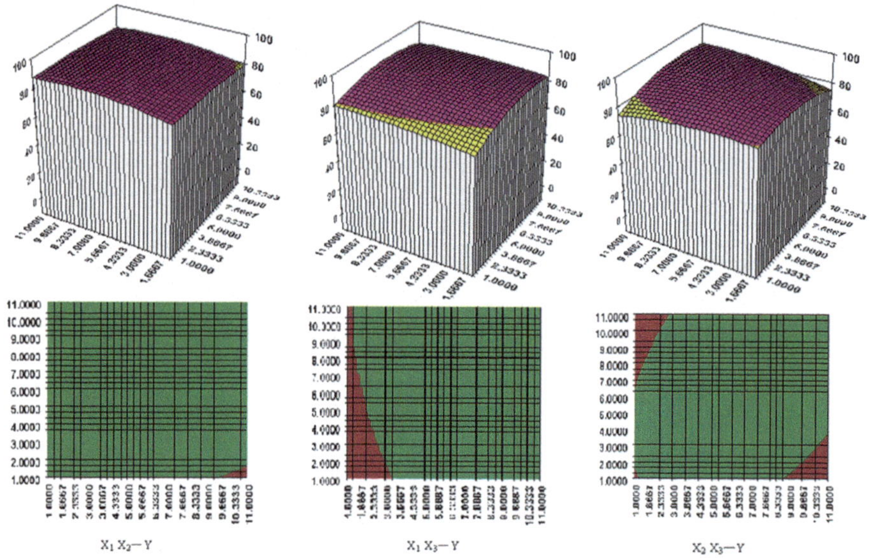

Fig. 9.15 Response surface diagram of interaction of factors on compound enzymolysis

of hydrolysis products, and other indicators. The enzymolysis of protein is a complex process. Protein may contain different types of peptide bonds, and proteases may have different degrees of hydrolysis on different peptide bonds. Due to this, sometimes it may be difficult to clearly describe and explain the protein hydrolysis process. Therefore, it is of practical significance for the deep understanding of the protein hydrolysis process to derive the kinetic relationship formula for describing the enzymatic hydrolysis process law of protein, calculate the kinetics parameters of the enzyme, and establish the experimental model that is consistent with the experimental data from the reaction mechanism.

A large amount of data, derivation, operation, and repeated verification were required to establish the polymer substrate hydrolysis kinetics model that could reflect the hydrolysis process, and the whole process needs to take a lot of time; the model established only relying on the fitting of several sets of data could not reflect the entire hydrolysis process. In response to this problem, Marquez and Fernandez (1993) argued that for the hydrolysis reaction with the same mechanism, the model established before could be used to combine with the actual data for constructing the kinetic curve of hydrolysis process and thus obtaining some information in the hydrolysis process.

3.2.1 Kinetic Model of Enzymolysis

3.2.1.1 Kinetic Model of Single Enzymolysis

The enzymatic hydrolysis reaction of protein should meet the double-substrate sequence reaction mechanism, and it was composed of the following steps. In the aqueous solution, the second step was to limit the speed, and the reverse process was negligible:

$$E + S \underset{K-1}{\overset{K_1}{\rightleftarrows}} ES \tag{9.16}$$

$$ES + \mathrm{H_2O} \xrightarrow{K_2} E + P - \mathrm{OH} + H - P \tag{9.17}$$

$$ES + S \underset{K-3}{\overset{K_3}{\rightleftarrows}} ES_2 \tag{9.18}$$

Therefore, the derivative equation with the following reaction rate could be reached. The reaction rate was represented by R, and S_0 represented the initial substrate concentration:

$$R = S_0 \frac{\mathrm{d}(DH)}{\mathrm{d}(t)} = K_2|ES| \tag{9.19}$$

In the enzymatic reaction process, the reaction formula of the protease being passivated due to the inhibition of substrates and products was as follows, where E_a represented the active protease and E_i represented the passivated protease:

$$E + ES \xrightarrow{K_d} E_a + E_i + S \tag{9.20}$$

The kinetic equation of the above reaction process was

$$-\frac{\mathrm{d}e}{\mathrm{d}t} = K_d|E||ES| \tag{9.21}$$

The following formula was established after dividing Eq. 9.19 by Eq. 9.21:

$$-S_0 \frac{\mathrm{d(DH)}}{\mathrm{d}e} = \frac{K_2}{K_d|E|} \tag{9.22}$$

There were two existence forms of protease in the hydrolysis system: one was free state E, the other was complex state ES, and total protease should be the sum of these two parts. Therefore:

$$e = |E| + |ES| \tag{9.23}$$

When in the equilibrium and stabilization phase, the produced ES complex was equal to the decomposed ES complex, thus establishing the following equation:

$$K_1|E||S| = (K_1 + K_2)|ES| \tag{9.24}$$

$$|ES| = \frac{K_1|E||S|}{K_1 + K_2} = \frac{|E||S|}{K_m} \tag{9.25}$$

$$K_m = \frac{K_1 + K_2}{K_1} \tag{9.26}$$

After substituting Eq. 9.25 into Eq. 9.23, and setting $|S| = S_0$, the following equation was established:

$$|E| = \frac{K_m e}{K_m + S_0} \tag{9.27}$$

K_m was Michaelis-Menten constant, and its physical meaning was the substrate concentration in mol/L when the enzyme reaction rate reached half the maximum reaction rate. K_m could only be used to determine the specificity of enzyme and natural substrate, $\frac{1}{Km}$ could approximately represent the size of the enzyme affinity for the substrate, and the greater the $\frac{1}{K_m}$ was, the larger the affinity was because the greater the $\frac{1}{K_m}$ was, the smaller the substrate concentration required to achieve half of the maximum reaction rate was. Obviously, when r optimal substrate was reached, the enzyme affinity was the largest, while K_m was the smallest. The K_m values of enzyme preparations measured at present were all small, so $K_m < < S_0$ was set, and Eq. 9.27 could be simplified as:

$$|E| = \frac{K_m e}{S_0} \tag{9.28}$$

After substituting Eq. 9.28 into Eq. 9.22, the following equation was established:

$$-\frac{\mathrm{d(DH)}}{\mathrm{d}e} = \frac{K_2}{K_d K_m} * \frac{1}{e} \tag{9.29}$$

After definite integration for Eq. 9.29, the upper and lower limits of the degree of hydrolysis were DH and 0, respectively, while the upper limit of total enzyme amount e was e, its lower limit was e_0; therefore, its integral equation was:

$$\int_{e0}^{e} \frac{\mathrm{d}e}{e} = -\frac{K_d K_m}{K_2} \int_0^{\mathrm{DH}} \mathrm{d(DH)} \tag{9.30}$$

The following equation was established by integration of the both sides of Eq. 9.30:

$$\ln e - \ln e_0 = -\frac{K_d K_m}{k_2}(\mathrm{DH}) \tag{9.31}$$

After simplifying the equation, the following formula was established:

$$e = e_0\left[-\frac{K_d K_m}{K_2}(\mathrm{DH})\right] \tag{9.32}$$

Through arranging Eqs. 9.19, 9.25, 9.28, and 9.32, the following formula was established:

$$R = K_2 e_0 \exp\left[-\frac{K_d K_m}{K_2}(\mathrm{DH})\right] \tag{9.33}$$

After merging Eq. 9.19 with Eq. 9.33, the following formula was established:

$$\frac{\mathrm{d(DH)}}{\mathrm{d}t} = \frac{K_2 e_0}{S_0}\exp\left[-\frac{K_d K_m}{K_2}(\mathrm{DH})\right] \tag{9.34}$$

Based on the previous study results, the kinetic model of the single enzyme hydrolysis process of protein was:

$$\frac{\mathrm{d(DH)}}{\mathrm{d}t} = a\exp[-b*(\mathrm{DH})] \tag{9.35}$$

From Eqs. 9.34 and 9.35, the following equations were established:

$$a = \frac{K_2 e_0}{S_0} \tag{9.36}$$

$$b = \frac{K_d K_m}{K_2} \tag{9.37}$$

Therefore, the kinetic model of single enzyme hydrolysis process of protein was based on Eqs. 9.33, 9.36, and 9.37:

$$R = aS_0\exp[-b(\mathrm{DH})] \tag{9.38}$$

From Formula 9.36, it was shown that the size of a was related to the initial substrate concentration and the initial protease concentration of the enzymolysis system and it decreased with the increase of the initial substrate concentration and increased with the increase of the initial protease concentration. Because K_2 was related to hydrolysis temperature, the size changed with the change in hydrolysis temperature. From Formula 9.37, the size of *b* was not related to the initial protein concentration and the initial protease concentration but related to the hydrolysis temperature. In the constant temperature hydrolysis reaction, the size of *b* should be a constant; the size of *a* was only related to the initial substrate concentration of the enzymolysis system and the initial protease concentration. From Eq. 9.38, the larger the value of *a* was, the greater the reaction rate was; when a was less than 0 and *R* was negative, enzymatic reaction could not be carried out.

3.2.1.2 Kinetic Model of Compound Enzymolysis

Compound enzymolysis refers to the process that the second enzyme continues to hydrolyze the substrate after the first enzyme. Although the action conditions were different, at this time, the first enzyme in the system still had a weak hydrolysis on the substrate, and it played a certain role in promotion relative to the second enzyme; on the other hand, some sites of the second enzyme had been cut by the first enzyme, so it reduced part of the activity of the second enzyme. Therefore, when making definite integration for Formula 9.29, the lower limit of total enzyme amount e was set to be $e_0 - c$, and thus the following integral equation was established:

$$\int_{e_0-c}^{e} \frac{\mathrm{de}}{e} = -\frac{K_d K_m}{K_2} \int_0^{\mathrm{DH}} d(\mathrm{DH}) \tag{9.39}$$

The following formula was established after integration of the both sides:

$$\ln e - \ln e_0 - c = -\frac{K_d K_m}{k_2} (\mathrm{DH}) \tag{9.40}$$

After making simplification:

$$e = (e_0 - c)\left[-\frac{K_d K_m}{K_2} (\mathrm{DH})\right] \tag{9.41}$$

By arranging Eqs. 9.19, 9.25, 9.28, and 9.32, the following equation was established:

$$R = K_2 (e_0 - c) \exp\left[-\frac{K_d K_m}{K_2} (\mathrm{DH})\right] \tag{9.42}$$

After merging Eq. 9.19 with Eq. 9.42, the following equation was established:

$$\frac{\mathrm{d(DH)}}{\mathrm{d}t} = \frac{K_2(e_0 - c)}{S_0}\exp\left[-\frac{K_d K_m}{K_2}(\mathrm{DH})\right] \tag{9.43}$$

The following equation was established based on Eqs. 9.42 and 9.43:

$$a = \frac{K_2 e_0}{S_0} - \frac{K_2 c}{S_0} \; b = \frac{K_d K_m}{K_2} \tag{9.44}$$

The c value in Eq. 9.44 was produced due to the action of the first enzyme on the substrate, so $c = c_0 S_0$, and the following equation was established:

$$a = \frac{K_2 e_0}{S_0} - K_2 c_0 \tag{9.45}$$

3.2.2 Kinetics Model Parameters

The nonlinear regression model was used to fit the sets of data with different substrate concentrations and different enzyme concentrations, derive the kinetics model parameters during the enzymolysis process, effectively establish the kinetics model of enzymolysis process, and explain the phenomenon of the enzymolysis process based on this. Zhang Yuhao and Wang Qiang (2007) derived the kinetics parameters of the hydrolysis process of main tool enzyme Alcalase and auxiliary tool enzyme N120P during the preparation of peanut oligopeptide and established the corresponding kinetics model. The derivation results and process are as follows.

Nonlinear regression model was used to fit the sets of data with different substrate concentrations and different enzyme concentrations and derive the kinetics model parameters of Alcalase enzymolysis peanut protein based on this. The fitting effect was good when using logarithm model. The details are shown in Table 9.9.

Logarithm model is represented as $y = c - d\ln(x + f)$, and the following formula is established after differentiation:

$$\mathrm{d}x = -\frac{\exp\left(\frac{c-y}{d}\right)}{d}\mathrm{d}y \tag{9.46}$$

$$\text{So} \quad \frac{\mathrm{d}y}{\mathrm{d}x} = -d \times \exp\left(-\frac{c}{d}\right) \times \exp\left(\frac{y}{d}\right) \tag{9.47}$$

In contrast with Eq. 9.35, the following formula is established:

$$a = -d \times \exp\left(-\frac{c}{d}\right) \tag{9.48}$$

Table 9.9 Fitting result of logarithm model to Alcalase

Sample	Equation	Related coefficient
Substrate concentration 4%	DH = 1.62146 + 2.99959* ln(*t* + 0.59445)	0.9867
Substrate concentration 6%	DH = 0.09613 + 2.99061* ln(*t* + 0.99643)	0.9874
Substrate concentration 8%	DH = − 0.37737 + 2.97314* ln(*t* + 1.17604)	0.9858
Substrate concentration 10%	DH = − 1.11463 + 2.95763* ln(*t* + 1.52124)	0.9887
Dosage of enzyme 606	DH = − 3.9214 + 2.95251* ln(*t* + 3.99737)	0.9902
Dosage of enzyme 1214	DH = − 0.43601 + 3.00082* ln(*t* + 1.20441)	0.9809
Dosage of enzyme 3642	DH = 2.84359 + 3.00759* ln(*t* + 0.38989)	0.9983

$$b = -\frac{1}{d} \tag{9.49}$$

The kinetics parameters a and b corresponding to different initial substrate concentrations and initial protease concentrations in the process of Alcalase enzymolysis peanut protein were obtained based on Table 9.9 and Eqs. 9.48 and 9.49, as shown in Table 9.10:

As can be seen from the experimental results in Table 9.10, the kinetics parameter a decreased with the increase of initial substrate concentration and increased with the increase of initial Alcalase concentration. There was a small difference in kinetics parameter b, and it was close to a constant and fluctuated up and down near its average value of 0.3352 under different initial substrate concentrations and initial protease concentrations. Therefore, in a constant temperature hydrolysis reaction, b could be regarded as a constant, which was consistent with the conclusion of the model derivation.

From Eq. 9.36, it was known that the value was the ratio between initial protease concentration and initial substrate concentration during the constant temperature hydrolysis reaction, that is, E_0/S_0 was in a linear relationship. Therefore, relevant data were extracted from Table 9.10 for linear fitting, as shown in Fig. 9.16.

The linear relationship formula $y = 254.65x$, $R^2 = 0.9903$ obtained from Fig. 9.16 was consistent with Eq. 9.36, so the equation corresponding to the relationship curve of a value and E_0/S_0 value was:

$$a = 254.65\frac{E_0}{S_0} \tag{9.50}$$

In contrast with Eqs. 9.36 and 9.50, $K_2 = 254.65$ (min^{-1}) was established.

After substituting a and b values into Eq. 9.38, the kinetics model of Alcalase enzymolysis peanut protein was established:

Table 9.10 Kinetics parameters of Alcalase enzymolysis

S_0 (g/L)	E_0 (g/L)	E_0/S_0 (g/g)	a (min^{-1})	*b*
100	0.8	0.008	2.0290	0.3381
80	0.8	0.010	2.6187	0.3363
60	0.8	0.013	3.0880	0.3344
40	0.8	0.020	5.1502	0.3334
40	0.2	0.005	0.7823	0.3387
40	0.4	0.010	2.5950	0.3332
40	1.2	0.030	7.7416	0.3325

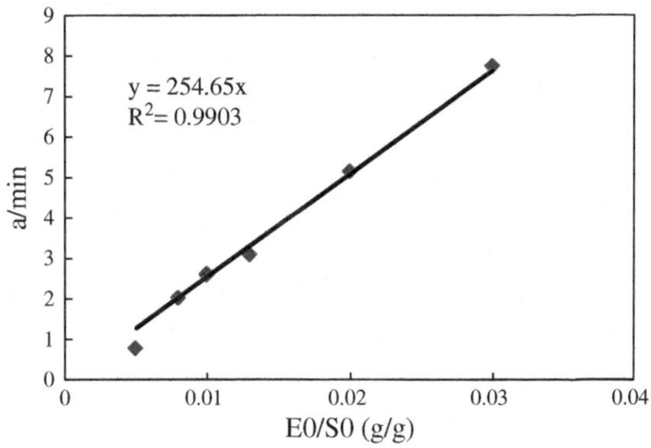

Fig. 9.16 Change trend of a value of Alcalase with different E_0/S_0 values

$$R = 254.65E_0\exp[-0.3352(\text{DH})] \tag{9.51}$$

It can be seen from Eq. 9.51 that the initial reaction rate of Alcalase enzymolysis peanut protein was only related to the initial Alcalase concentration and increased with the increase of initial Alcalase concentration. As the reaction progressed, the reaction rate decreased with the increase of degree of hydrolysis. After substituting the substrate concentration of 4% and the fitting result of logarithm model of Alcalase dosage of 2428 U/g into Eq. 9.51, the fitting curve of hydrolysis rate *R* and reaction time *t* were obtained.

It was shown from Fig. 9.17 that hydrolysis rate constantly decreased as enzymatic hydrolysis progressed, the reaction rate was only 0.733 at 120 min, and the reaction had been carried out very slowly.

Logarithm model was used to fit the data of Alcalase enzymolysis peanut protein by N120P, as shown in Table 9.11.

Based on Table 9.11 and Eqs. 9.41 and 9.42, the enzymolysis kinetics parameters a and b of Alcalase enzymolysis peanut protein by N120P were derived, as shown in Table 9.12.

From the results of Table 9.12, it was found that the change trend of enzymolysis kinetics parameters of Alcalase enzymolysis peanut protein by N120P was similar with that of Alcalase enzymolysis peanut protein. Among them are a decreased with the increase of initial substrate concentration and increased with the increase of

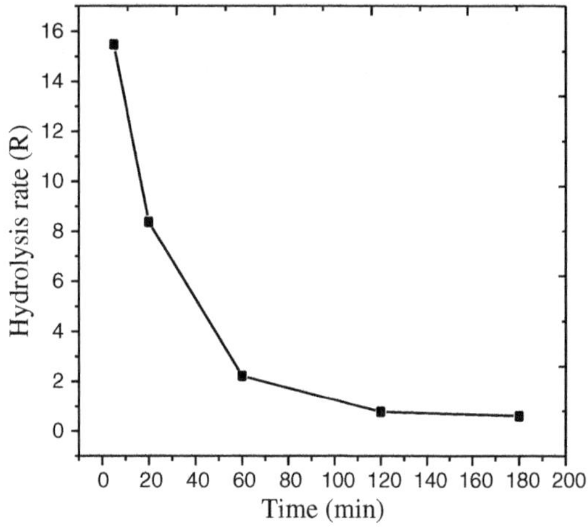

Fig. 9.17 Change of reaction range during Alcalase hydrolysis process

Table 9.11 Fitting results of logarithm model to N120P

Sample	Equation	Related coefficient
Substrate concentration 4%	$DH = -9.03114 + 3.78836^* \ln(t + 10.71919)$	0.9989
Substrate concentration 6%	$DH = -10.164 + 3.88376^* \ln(t + 13.45518)$	0.9989
Substrate concentration 8%	$DH = -10.06973 + 3.71547^* \ln(t + 14.64098)$	0.9964
Substrate concentration 10%	$DH = -11.04904 + 3.8058^* \ln(t + 11.04904)$	0.9944
Dosage of enzyme 600	$DH = -12.06741 + 3.86007^* \ln(t + 22.10365)$	0.9957
Dosage of enzyme 2400	$DH = -10.53295 + 3.86974^* \ln(t + 14.83674)$	0.9955
Dosage of enzyme 3600	$DH = -8.15434 + 3.7117^* \ln(t + 0.4125)$	0.9980

initial N120P concentration. The kinetics parameter b fluctuated up and down near its average value of 0.2629.

The linear fitting result of its a value and E_0/S_0 is shown in Fig. 9.18. The linear relationship formula $y = 9.46563x + 0.14519$, $R^2 = 0.9794$ obtained from Fig. 9.18 was consistent with Eq. 9.45, so the equation corresponding to the relationship curve of a value and E_0/S_0 value was:

$$a = 9.46563\frac{e_0}{S_0} + 0.14519 \tag{9.52}$$

In contrast to Eqs. 9.45 and 9.52, $K_2 = 9.46563$ (min^{-1}) was established.

After substituting the values of a and b into Eq. 9.38, the kinetics model of Alcalase enzymolysis peanut protein by N120P was established:

$$R = (9.46563E_0 + 0.14519)\exp[-0.2629(\text{DH})] \tag{9.53}$$

Table 9.12 Enzymolysis kinetics parameters by N120P

S_0(g/L)	E_0(g/L)	E_0/S_0(g/g)	a (min^{-1})	*B*
100	0.8	0.008	0.2087	0.2628
80	0.8	0.10	0.2472	0.2691
60	0.8	0.013	0.2836	0.2575
40	0.8	0.20	0.3492	0.2640
40	0.2	0.005	0.1694	0.2591
40	0.4	0.010	0.2544	0.2584
40	1.2	0.030	0.4125	0.2694

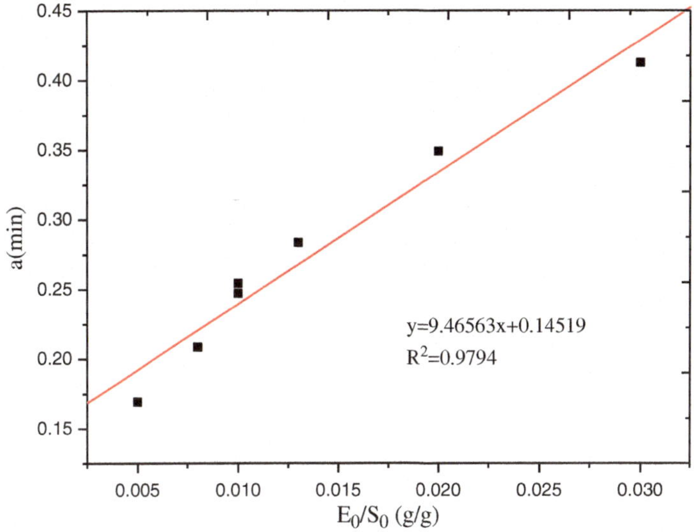

Fig. 9.18 Change trend of a value of N120P with different E_0/S_0 values

From Eqs. 9.64–9.53, it was shown that the hydrolysis reaction was still carried out when N120P was not added. At this time, the hydrolysis rate was very slow. With the addition of N120P, the reaction rate decreased with the increase of hydrolysis degree. After substituting the logarithm model fitting results when the substrate concentration was 4% and N120P dosage was 2400 U/g into Eq. 9.53, the fitting curve of hydrolysis rate *R* and reaction time *t* was obtained (Fig. 9.19).

It was shown in Fig. 9.19 that hydrolysis rate constantly decreased as enzymatic hydrolysis progressed, the reaction rate was only 1.1265 at 65 min, and the reaction proceeded very slowly.

3.2.3 Inactivation Constant of Protease

The activity of protease gradually decreased during the hydrolysis process, and the derivation and determination of the inactivation constant were helpful to explain the

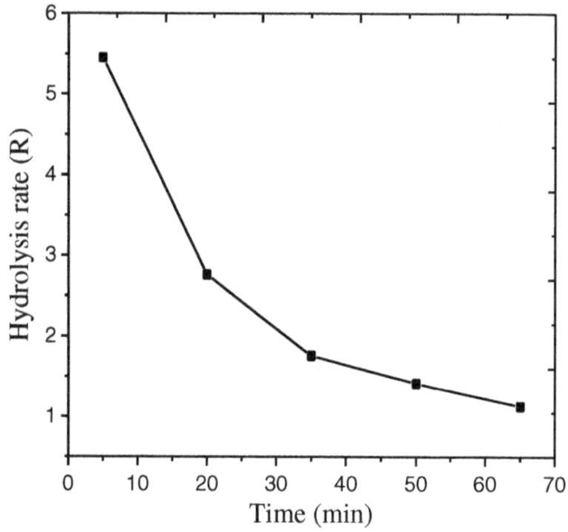

Fig. 9.19 Change of reaction rate during N120P hydrolysis process

role and inactivation of the protease in the hydrolysis process. Zhang Yuhao and Wang Qiang (2007) derived the kinetics parameters of the hydrolysis process of main tool enzyme Alcalase and auxiliary tool enzyme N120P during the preparation of peanut oligopeptide and established the corresponding kinetics model. The derivation results and process were as follows.

The following equation was established based on Eqs. 9.22, 9.26, and 9.29:

$$-\frac{\mathrm{d}e}{\mathrm{d}t} = K_d K_M \frac{e^2}{S_0} = K_4 \frac{e^2}{S_0} \tag{9.54}$$

where

$$K_4 = K_d K_M \tag{9.55}$$

K_4 was the Alcalase inactivation constant for the hydrolysis reaction.

After multiplying Eq. 9.37 with Eq. 9.38 and then merging with Eq. 9.55, the following equation was established:

$$ab = \frac{K_4 e_0}{S_0} \tag{9.56}$$

The *ab* value and the linear fitting results of E_0/S_0 were shown in Fig. 9.20.

The linear relationship formula $y = 84.777x$, $R^2 = 0.9908$ obtained from Fig. 9.20 was consistent with Eq. 9.36, so the equation corresponding to the relationship curve of *ab* value and E_0/S_0 value was:

$$ab = 84.777 \frac{E_0}{S_0} \tag{9.57}$$

Fig. 9.20 Change trend of *ab* value of Alcalase with different E_0/S_0 values

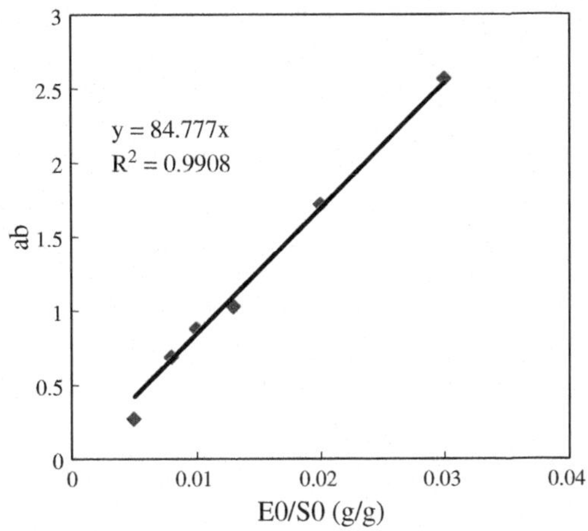

In contrast with Eqs. 9.56 and 9.57, the kinetic constant of Alcalase inactivation was $K_4 = 84.777\ (\text{min}^{-1})$.

Multiplying Eqs. 9.45 and 9.38 and merging with Eq. 9.55, the following equation was established:

$$ab = \frac{K_4 e_0}{S_0} - K_4 c_0 \tag{9.58}$$

The linear fitting results of *ab* value and E_0/S_0 are shown in Fig. 9.21. The linear relationship formula $y = 2.60511x + 0.03675$, $R^2 = 0.98438$ obtained from Fig. 9.21 was consistent with Eq. 9.58, so the equation corresponding to the relationship curve of *ab* value and E_0/S_0 value was:

$$ab = 2.60511\frac{E_0}{S_0} + 0.03675 \tag{9.59}$$

In contrast with Eqs. 9.58 and 9.59, the kinetics constant of N120P inactivation was $K_4 = 2.60511\ (\text{min}^{-1})$.

3.2.4 Kinetics Model

The calculation results of enzyme kinetics model were compared with the actual hydrolysis results, to verify the practical application of the kinetics model. Therefore, the fitting between the hydrolysis degree curve of the system and the corresponding kinetics model calculation results was measured under a certain hydrolysis condition. Zhang Yuhao and Wang Qiang (2007) verified the kinetics model of the hydrolysis process of main tool enzyme Alcalase and auxiliary tool enzyme N120P during the preparation of peanut oligopeptide. The specific

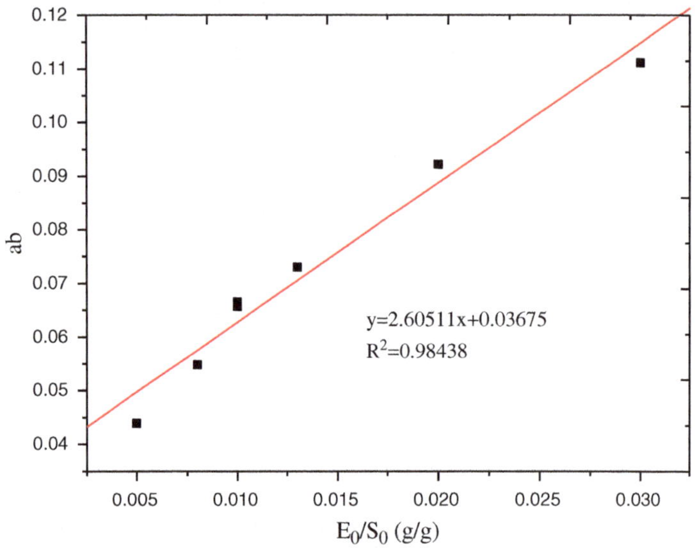

Fig. 9.21 Change trend of *ab* value of N120P with different E_0/S_0 values

conditions were as follows: initial substrate concentration of 4%, Alcalase dosage of 2428 U/g substrate, pH 8.0, reaction temperature of 53 °C, and reaction time of 105 min; later N120P continued to act on Alcalase, with dosage of 2400 U/g substrate, pH 6.0, reaction temperature of 57 °C, and reaction time of 65 min. As shown in Fig. 9.22, the model was well fitted and the error of most of the points was less than 5%, so the established kinetics model was credible.

3.3 *Refining Classification*

3.3.1 Desalination

During peanut protein hydrolysis process, it is required to adjust or maintain pH stability through acid-base regulation, so that peanut peptides contain a certain percentage of salt, while too high salt content will have a negative impact on product applications. Therefore, product quality should be improved by desalination process. The traditional desalting methods include ion exchange resin desalination method and macroporous resin method. The ion exchange resin method needs to exchange resin through anion and cation respectively, so that the desalination process is more cumbersome and the peptide recovery rate is low. This is because when the hydrolysate passes through the cation resin, the solution system is acidic, so that a large number of peptide molecules have positive charges in the solution, thus resulting in a large loss of peptide, and the result is the same when it passes through the anion resin. Macroporous resin method needs absorption and

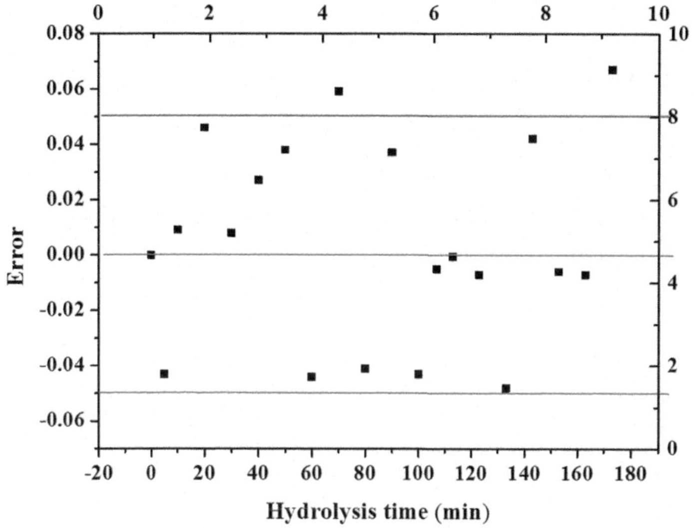

Fig. 9.22 Fitting of experimental hydrolysis degree and theoretical hydrolysis degree

elution process, so that the operation is tedious, the yield of peptide is low, and organic solvents need to be used during elution, and therefore it is not easy to achieve industrialization by this method.

Zhang Yuhao and Wang Qiang (2007) studied the desalting method of peanut peptides, established the mixed bed desalination method of anion and cation resins based on the traditional desalination method, and compared the processes of mixing bed, traditional ion exchange method, and macroporous resin method. The experimental results (Fig. 9.23) showed that during the peanut peptide desalination using mixed bed, both the desalination rate and oligopeptide yield were high. This was because during the desalination using traditional anion and cation resins, the entire solution system was acidic when hydrolysate passes through the cation resin, so that a large number of peptide molecules with positive charges were absorbed on the resin in the solution, and the result was the same when it passed through the anion resin. The other peptide molecules would be adsorbed on the pore size of the resin, so a great oligopeptide loss might be caused during desalination using anion and cation resins respectively. The pH of hydrolysate was stable during the desalination when using mixed bed, thus greatly reducing the loss of peptide during desalination; in addition, the interaction between the anions and cations could reduce the adsorption of resins to oligopeptide. It was shown that the desalination using mixed bed had obvious advantages in terms of oligopeptide yield compared with the other two methods.

Zhang Yuhao and Wang Qiang (2007) also optimized the factors in the desalination process using mixed bed. The results showed that the optimum conditions for the anion and cation exchange resin mixed bed desalination method were as follows: column passing speed of hydrolysate was five to ten times of column

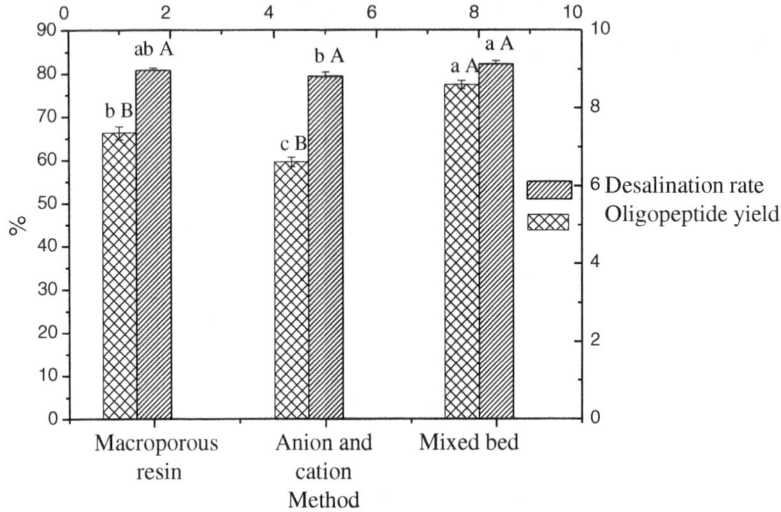

Fig. 9.23 Comparison of desalination method

volume/h, the pH value of hydrolysate was 4.5, and the ratio between anion and cation resins was 3:2. Under this condition, the desalination rate of peanut oligopeptide and protein recovery rate were all more than 80%.

3.3.2 Ultrafiltration

Peanut protein hydrolysates also contain some large molecular weight proteins, proteases, organic colloids, and other substances, which may limit their applications in the field of food, so it is required to conduct ultrafiltration classification for peanut oligopeptide. Ultrafiltration is the system to take the energy difference between the two sides of the membrane and realize heterogeneous system separation of the components in the solution due to the difference in particle size, with no interphase changes, low energy consumption, easy operation, small footprint, and other advantages. It is applied to the classification of protein hydrolysate, to realize the separation of polypeptides with different molecular weights. One of the major problems with ultrafiltration separation of protein hydrolysates is that membrane flux decreased with the extension of running time due to the impact of concentration polarization, membrane pore clogging, gel layer formation, and other factors. Therefore, it is very important for the long, safe, and stable running of ultrafiltration system as well as the increase of product yield to grasp and perform appropriate operating parameters.

3.3.2.1 Impact of Operation Pressure on Membrane Flux

The impact of operating pressure on ultrafiltration is mainly reflected in the formation of the gel layer on the membrane surface. Under a certain solute concentration and ultrafiltration temperature, appropriate operating pressure can delay the formation and thickening of the gel layer. During the ultrafiltration process, the greater the operating pressure is, the greater the pressure difference on both sides of the membrane is, and the greater the membrane flux is. However, as the ultrafiltration progresses, concentration polarization phenomenon may be caused at the membrane surface. The greater the operating pressure is, the more serious the phenomenon is. With the occurrence of concentration polarization, gel layer will be gradually formed at the membrane surface. The greater the operating pressure is, the thicker the formed gel layer is, the larger the mass transfer resistance is, and the smaller the initial membrane flux increase range is; therefore, when the operating pressure is too large, the thickness and density of gel layer formed during the ultrafiltration will increase correspondingly in addition to increasing energy consumption, thus resulting in reduced efficiency of ultrafiltration and serious ultrafiltration membrane pollution. However, too low ultrafiltration pressure might cause too low initial membrane flux, thus resulting in the low efficiency of the entire ultrafiltration process.

Zhang Yuhao and Wang Qiang (2007) conducted primary ultrafiltration to peanut oligopeptide using the ultrafiltration membrane with a molecular weight cutoff of 5000 Da and then conducted secondary ultrafiltration to the filtered solution using the ultrafiltration membrane with a molecular weight cutoff of 1000 Da. In this study, the impact of operating pressure on ultrafiltration effect was studied and explored (Figs. 9.24 and 9.25). The results showed that the optimum operating pressures for primary ultrafiltration and secondary ultrafiltration were 0.18 MPa and 0.21 MPa, respectively.

3.3.2.2 Impact of Feed Liquid Concentration on Membrane Flux

Protein and polypeptide are amphoteric compounds with strong surface activity, so they may be easily absorbed on the surface of polymer. Therefore, the impact of feed liquid concentration on ultrafiltration is mainly reflected in concentration polarization at the membrane surface and the formation of gel layer. Appropriate feed liquid concentration can allow more small-molecule solutes to pass through under the relatively small mass transfer resistance, thus improving the overall efficiency of ultrafiltration.

Zhang Yuhao and Wang Qiang (2007) studied and explored the impact of the feed liquid concentration on the ultrafiltration effect (Figs. 9.26 and 9.27) during the ultrafiltration process. The results showed that the optimum feed liquid concentrations for primary ultrafiltration and secondary ultrafiltration were 5–7.5% and about 7.5%, respectively.

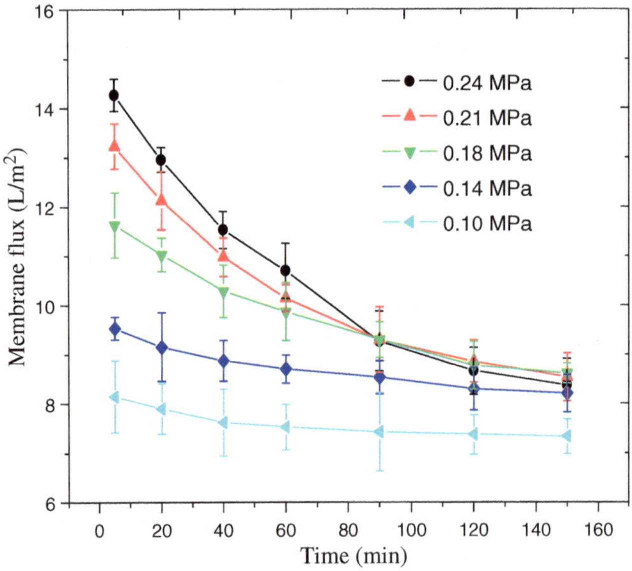

Fig. 9.24 Impact of operating pressure on primary ultrafiltration

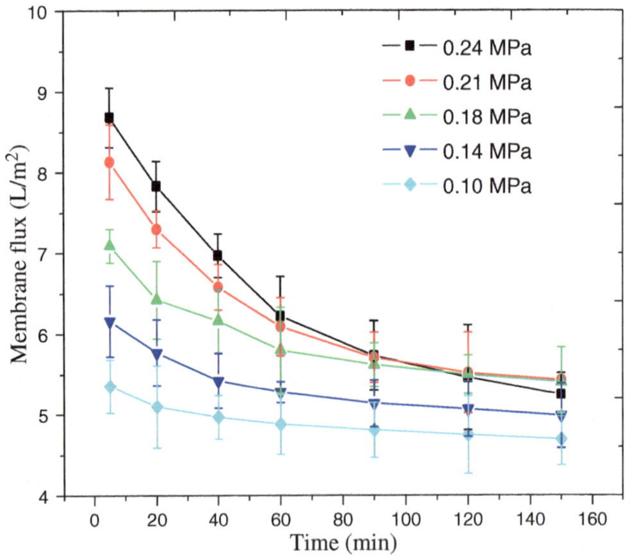

Fig. 9.25 Impact of operating pressure on secondary ultrafiltration

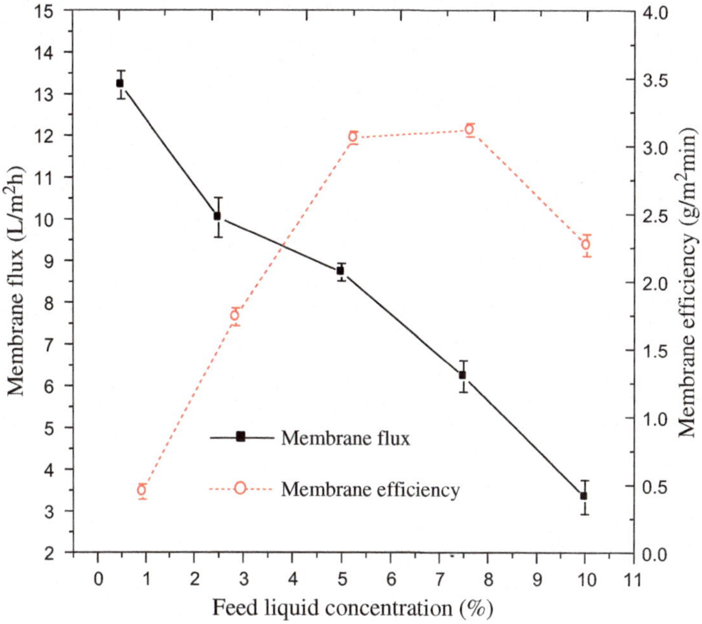

Fig. 9.26 Impact of feed liquid concentration on primary ultrafiltration

3.3.2.3 Impact of Operating Temperature on Membrane Flux

The impact of operating temperature on the ultrafiltration process is mainly reflected in its impact on feed liquid and then the impact on the polarization at the membrane surface. The increase of operating temperature might cause the viscosity of feed liquid to decrease and the diffusion coefficient to increase, which will reduce the impact of concentration polarization, so the membrane flux will increase with the rise of temperature. However, the adsorption of protein macromolecules on the membrane surface will increase with the rise of temperature, which will aggravate the contamination of membrane and rapidly reduce the increase range of membrane flux.

Zhang Yuhao and Wang Qiang (2007) studied and explored the impact of operating temperature on ultrafiltration effect during the ultrafiltration process (Figs. 9.28 and 9.29). The results showed that the optimum temperatures for primary ultrafiltration and secondary ultrafiltration were 35 °C and 40 °C, respectively.

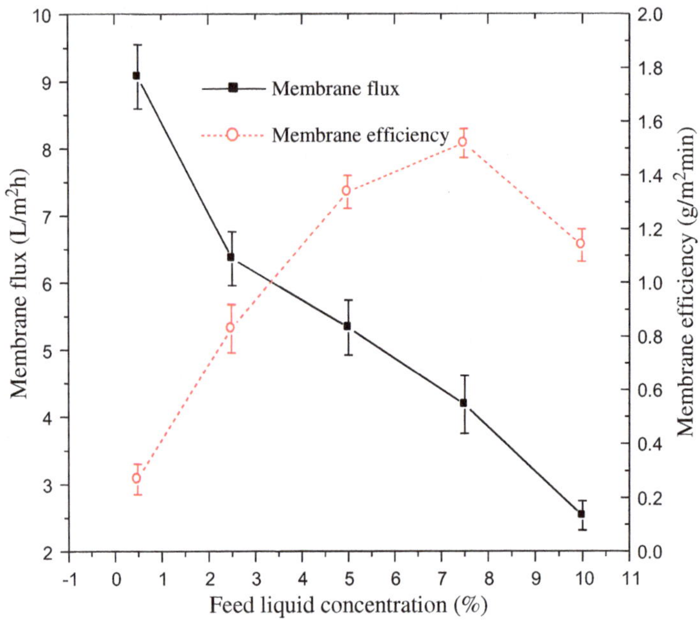

Fig. 9.27 Impact of feed liquid concentration on secondary ultrafiltration

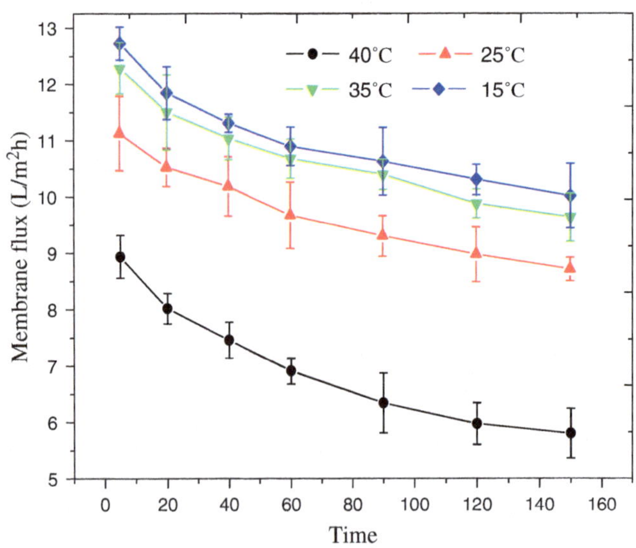

Fig. 9.28 Impact of operating temperature on primary ultrafiltration

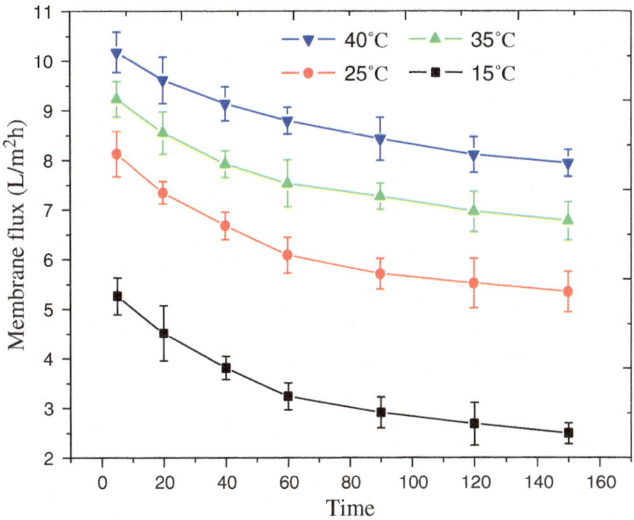

Fig. 9.29 Impact of operating temperature on secondary ultrafiltration

4 Evaluation of Biological Activity

Functional oligopeptides not only have a better absorption mechanism than that of amino acids but also have incomparable functional activity compared with a variety of amino acids. The relevant tests showed that the functional activity of oligopeptides mainly includes antihypertensive, antioxygenation, immune regulation, anti-thrombosis, and so on. In recent years, due to the insecurity of chemical antioxidant additives, the natural food antioxidants with high efficiency and low toxicity have become the current research focus; at the same time, due to the development of free radical life science, the functional foods and drugs with antioxidant function have also attracted the attention of many scholars. Studies (Shen Beiying 1996; Chen et al. 1995; Chen Meizhen et al. 2002) showed that the oligopeptides with molecular weight of about 1000 Da were effective free radical scavengers and antioxidants, because they could remove hydroxyl radicals, superoxide anion radicals, and other free radicals that were very harmful to the human body, prevent the damage to DNA due to excessive free radicals, and significantly prolong the life of rats and mice.

Since Cushman et al. (2002) derived the model of ACE active site according to the chemical structure of ACE substrate and designed the first chemically synthesized ACE inhibitor captopril based on this, synthetic ACE inhibitors have been generally recognized by the medical field. However, synthetic ACE inhibitors often tended to produce side effects in clinical applications, such as cough, taste dysfunction, and rashes (Chen Xiu et al. 2003). Therefore, it has attracted the great concern of the majority of scientific workers to look for ACE inhibitors from natural and safe food sources to prevent and treat high blood pressure, and peanut oligopeptide is a natural blood pressure functional factor that meets this standard.

4.1 Antioxidant Activity

Jamdar et al. (2010) studied the antioxidant effects of peanut polypeptides. Studies showed that 2.0 mg/mL peanut peptide could reach DPPH free radicals by 50%, and the peanut peptide reduction force determination result of 10–40% peanut peptides with a hydrolysis degree of 2.5 mg/mL was between 0.72 and 0.98. It was also verified that peanut peptides had a certain ferrous ion chelating activity. Form this, peanut peptides had a strong action and reducing power to remove DPPH free radicals, but the antioxidant activity of peanut peptides was not compared with that of many antioxidant peptides.

Zhang Yuhao and Wang Qiang (2007) studied the antioxidant activity of desalinated peanut oligopeptide. The results showed that peanut oligopeptide had obvious antioxidant activity, and especially the activity of removing DPPH free radicals was strong. These results were the same as that of the study by Jamdar et al. (2010). However, compared with antioxidant peptides of soybean, the overall antioxidant effect of peanut peptides was not significant (Table 9.13). This might be related to the amino acid composition of peanut protein. In general, peptides with antioxidant activity contained histidine, methionine, and cysteine residues, which typically had a certain antioxidant activity in the free state with the ability to act as a nucleophile. Figure 9.30 listed the graphs of the chemical structures of histidine, methionine, and cysteine, respectively. It was shown from the figure that histidine molecule contained an imidazolyl group, which was both weak acid and weak base, so it was an excellent nucleophile with the ability to capture free radicals. Methionine and cysteine molecules contained sulfur atoms, among which methionine could capture hydroxyl radical and it was oxidized to methionine sulfone, thus playing an role in resisting oxidation and scavenging free radicals; sulfur in cysteine molecules existed in the form of sulfhydryl, and the molecules were polarizable, so it was a good nucleophile. Under the attack of reactive oxygen and other free radicals, sulfhydryl was oxidized to disulfide compound, and cysteine was oxidized to cystine.

Generally, the histidine, methionine, and cysteine contents in peanut protein are lower than that of soybean protein. This may be the reason why the antioxidant effect of peanut oligopeptide is not as good as that of soybean peptide.

4.2 ACE Inhibition and Antihypertensive Activity

Hypertension is an important risk factor for cardiovascular diseases such as heart failure, stroke, coronary heart disease, and myocardial infarction (Kannel 1996).

Table 9.13 Antioxidant activity of peanut oligopeptide

Implementation item	Concentration of peanut peptides (mg/ml)	Activity (%)	Concentration of soybean peptides (mg/ml)	Activity (%)
Superoxide radical	25	33.83 ± 3.31	25	41.70
Hydroxyl radical	15	62.15 ± 1.06	10	59.02
DPPH radical	10	71.60 ± 1.95	10	70.56
Ferrous ion chelating activity	50	40.01 ± 2.06	50	70.19

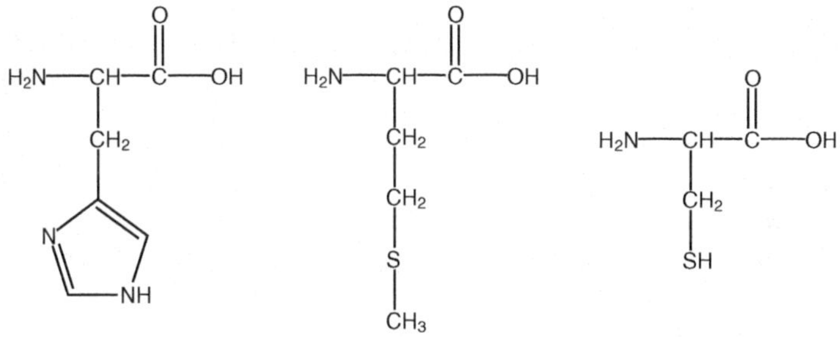

Histidine Methionine Cysteine

Fig. 9.30 Chemical structure graphs of histidine, methionine, and cysteine

Twelve million of people may die due to hypertension in the world. At present, the existing hypertensive patients in China have exceeded 120 million, so it is a very important issue to treat and prevent hypertension disease in today's society (Liu Lisheng 2001). Angiotensin-converting enzyme (ACE) plays an important role in the regulation of blood pressure in the human body; it participates in the regulation of blood pressure through renin-angiotensin system (RSA) and kallikrein-kinin system (KKS). RSA is a pressure increasing system that acts on angiotensin I by ACE and removes His-Leu from its end to generate angiotensin. As the strongest vasoconstrictor known in the RSA system, it acts on small arteries to shrink the vascular smooth muscle and rapidly cause pressure-increasing effect; meanwhile, angiotensin II can result in the increase of sodium storage and blood volume through stimulating aldosterone secretion and direct effect on the kidneys, thus increasing the blood pressure; on the contrary, KKS is a pressure-decreasing system, and it is also known as antihypertensive factor. Among them, kinin is an antihypertensive substance; ACE makes it lose the Phe-Arg and Ser-Pro at C-terminal by acting on soothing kinin in this system, thus degrading kinin into inactivation fragments and causing the elevation of blood pressure (Ondetti and Cushman 1977; Patchett et al. 1980).

Zhang Yuhao and Wang Qiang (2007) determined the ACE inhibition activity of peanut-mixed peptides after enzymolysis. The results showed that peanut oligopeptide ACE inhibited IC_{50} value to be 0.517 mg/ml and the mixed antihypertensive peptide ACE inhibited IC_{50} value to be within the range of 0.126–246.7 mg/ml. It was shown that the inhibition activity of ACE of peanut oligopeptide was high. In recent years, Jamdar et al. (2010), Enyonam Quist et al. (2009), and Guang and Phillips (2009) have demonstrated that the functional peanut enzymolysis with high ACE activity can be prepared by the enzymatic hydrolysis of peanut protein.

4.2.1 Impact of Refining on the Inhibition Activity of ACE of Peanut Oligopeptide

4.2.1.1 Desalination

Oligopeptide is usually prepared by enzymolysis method of protein. During the enzymolysis process of peanut protein, in order to make the protease within the optimum pH range, it is required to maintain its hydrolysis rate as far as possible and add some alkali or acid to adjust the pH value of the hydrolysis system. Therefore, there is a certain amount of salt in the protein hydrolysate obtained finally, which may affect the purity and application scope of product and bring a negative impact on its functional activity, thus limiting its application in the food field. Therefore, desalination treatment should be conducted to peanut oligopeptide.

The desalination method of oligopeptide includes ion exchange method and macroporous resin method and nanofiltration treatment method. Nanofiltration is usually used as the pretreatment of ion exchange or macroporous resin to partially desalt polypeptides, so as to relieve the pressure of ion exchange resin or macroporous resin, and extend its saturation time. However, both macroporous resin treatment and ion exchange treatment may cause a certain degree of loss of oligopeptide. For example, ion exchange resin may cause the loss of oligopeptide rich in polar amino acids, and macroporous resin may cause the loss of oligopeptide rich in hydrophobic amino acid. Therefore, it is required to conduct a comprehensive assessment to the impact of desalination on the activity of oligopeptide.

Zhang Yuhao and Wang Qiang (2007) established the desalination method using mixed bed and evaluated the impact of desalination process on peanut oligopeptide ACE I activity. Studies showed that peanut oligopeptide ACE inhibited IC_{50} value to be 0.711 mg/ml before desalination and 0.517 mg/ml after desalination. It showed that the inhibition activity of peanut oligopeptide ACE increased after desalination, and the analysis of amino acid of peanut oligopeptide before and after desalination showed that the proportions of hydrophobic amino acids before and after desalination were 22.46% and 28.44%, respectively, indicating that desalination by mixed bed had enrichment effect on ACE I peptide to some extent. In addition, the purity of peanut peptides greatly increased after desalination, and the content of oligopeptide in the unit volume also increased, which was the cause why the ACE inhibition activity of peanut oligopeptide increased after desalination. It was found that the desalination by mixed bed was favorable to the preparation and enrichment of peanut ACE I oligopeptide.

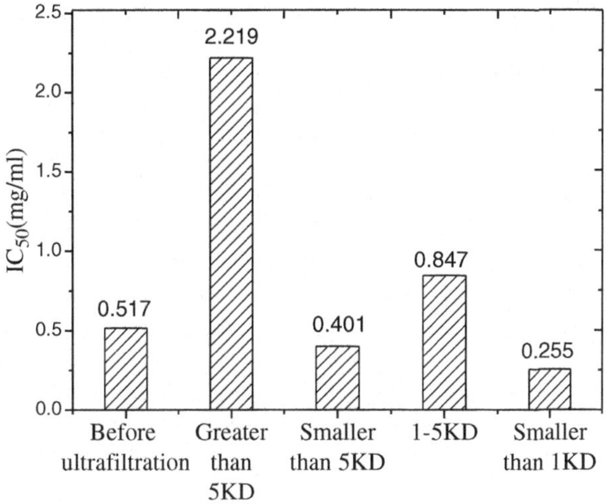

Fig. 9.31 Impact of ultrafiltration on the ACE inhibition activity of peanut oligopeptide

4.2.1.2 Ultrafiltration

Studies showed that most of the antihypertensive peptides were oligopeptides with a molecular weight of less than 1000 Da, so the peptide fragments with low molecular weight could be effectively enriched through ultrafiltration treatment. The studies of Zhang Yuhao and Wang Qiang (2007) showed that the ACE inhibition effect of the component with the molecular weight of less than 1000 Da in peanut oligopeptide after ultrafiltration separation was the best (Fig. 9.31), while the ACE inhibition activity of the component with the molecular weight of more than 5000 Da was the worst. The results showed that the ACE inhibition activity of peanut oligopeptide was closely related to its molecular weight, and ACE inhibition activity increased with the decrease of molecular weight, which was consistent with the previous study conclusions. Among them, the ACE inhibition activity of oligopeptide with the molecular weight of less than 1000 Da was especially prominent. Its IC_{50} could reach 0.255 mg/mL after ultrafiltration.

4.2.2 ACE Inhibition Activity Mechanism

4.2.2.1 ACE Inhibition Kinetics

Enzyme inhibition kinetics is the inhibition mode to establish enzyme kinetics model by investigating the relationship between the changes in substrate concentration and reaction rate under different inhibitor concentrations and thus judge the inhibition mode of the inhibitor. Enzyme inhibition mode is divided into irreversible inhibition and reversible inhibition; studies show that the inhibition of ACE inhibitory peptide to ACE is reversible. Reversible inhibition is divided into three types:

Competitive inhibition: Inhibitor and substrate compete for the combination site of enzyme, thus affecting the normal combination of substrate and enzyme. The structure of most of the competitive inhibitors is similar with that of substrate, so it can combine with the active site of enzyme and form reversible EI compound with enzyme. However, EI compound cannot be decomposed into products, and thus the rate of enzyme reaction is reduced.

Noncompetitive inhibition: It is characterized by the fact that both the substrate and the inhibitor combined with the enzyme at the same time, both of which have no competitive effect. After the enzyme combines with the inhibitor, it can combine with substrate; after the enzyme combines with the substrate, it can combine with the inhibitor. However, the intermediate ternary compound cannot be further decomposed into products, resulting in reduced enzyme activity.

Uncompetitive inhibition: The enzyme can combine with the inhibitor only after combining with the substrate.

Zhang Yuhao and Wang Qiang (2007) studied the inhibitory kinetics of peanut oligopeptide with a molecular weight of less than 1000 Da. From Lineweaver-Burk (Fig. 9.32) model, it was found that the inhibition model of peanut oligopeptide to ACE was competitive inhibition, that is, peanut oligopeptide competed against substrate and combined with ACE active site to form a certain balance, thereby reducing the response rate of ACE to substrate. ACE inhibitory peptides in most of the reports were the competitive inhibitory peptides of ACE, and some peptides were the noncompetitive and uncompetitive inhibitors of ACE. For example, Saito et al. (1994) isolated ACE noncompetitive inhibitory peptides, while Nakagomi et al. (1998) prepared ACE uncompetitive inhibitory peptides from serum.

4.2.2.2 ACE Inhibitor Type

The premise for ACE inhibitory peptide to play a role in the human body after oral administration is that ACE inhibitory peptide can resist digestive enzyme hydrolysis and then be absorbed by the intestine in a complete form and go into the blood circulation and finally reach the target organ. Therefore, the ACE inhibition activity measured after simulating the digestive tract environment for the treatment of ACE inhibitory peptide is more close to the inhibition activity in human body. There are many reports on digestion test in vitro, and the difference in different study reports is mainly focused on the digestive enzyme species used, hydrolysis conditions, and product recovery and analysis methods. The best pH for digesting protein in the stomach by protease is 1–2; in the duodenum, trypsin, chymotrypsin, elastase, and carboxypeptidases A and B can continue to hydrolyze the protein in the best condition of pH 7–8. In the case of eating, the time for half emptying the food in the stomach is 0.5–3 h, the residence time in the duodenum and jejunum is 2–2.75 h, and the time in the ileum is 5–7 h.

Fig. 9.32 Lineweaver-Burk model inhibited by peanut oligopeptide ACE

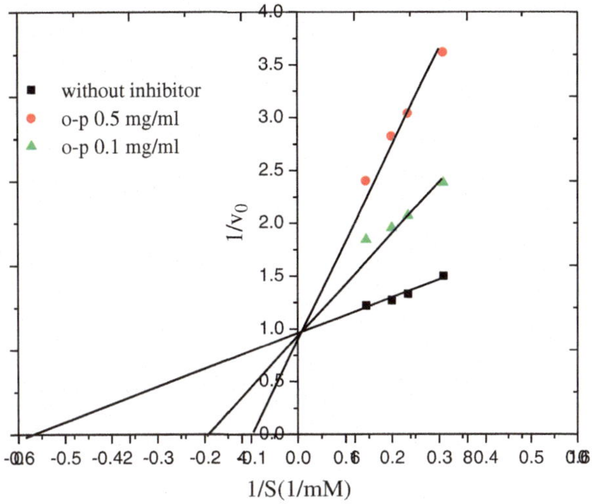

In order to investigate the resistance of protein hydrolysate to gastrointestinal digestive enzymes and the possibility of generating new ACE inhibitory peptides, further digestion treatment should be conducted to the peanut oligopeptide with the molecular weight of less than 1000 Da under the condition of simulating the physiological action of gastrointestinal digestion, to determine the changes in the ACE inhibition IC_{50} value of hydrolysate before and after digestion enzyme treatment and thus evaluate the digestion stability of hydrolysates. ACE inhibitory peptides can be divided into three types according to their digestion stability:

True inhibitor type: After in vitro intestinal enzyme simulation experiments, oligopeptide ACE inhibition activity is not affected.

Substrate type: After in vitro intestinal enzyme simulation experiments, oligopeptide ACE inhibition activity will decrease obviously.

Pro-drug type: After in vitro intestinal enzyme simulation experiments, oligopeptide ACE inhibition activity will increase obviously.

Zhang Yuhao and Wang Qiang (2007) studied the inhibition type of peanut oligopeptide with the molecular weight of less than 1000 Da. The results showed that peanut oligopeptide remained active after treatment with gastrointestinal digestive enzymes, indicating that peanut oligopeptide had a certain resistance to the hydrolysis of digestive enzyme. Therefore, peanut oligopeptide belonged to true ACE inhibitory peptide with true inhibition effect. This might be because peanut oligopeptide was less likely to be hydrolyzed by other proteases due to its low molecular weight. Therefore, the peanut oligopeptide with a molecular weight of less than 1000 Da could be absorbed into the blood circulation through the gastrointestinal tract in vivo, and functional peanut oligopeptide might play a role in lowering blood pressure after oral administration (Fig. 9.33).

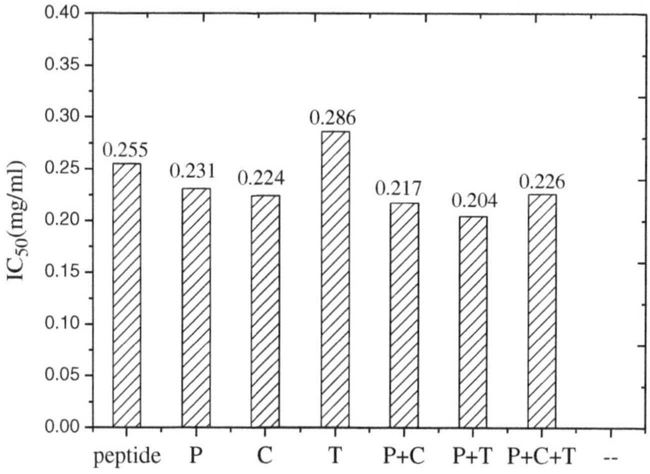

Fig. 9.33 Analysis of inhibition activity stability of peanut oligopeptide ACE (*Note*: *P* pepsase, *C* chymotrypsin, *T* trypsin)

4.2.3 Relationship Between Protein Amino Acid Composition and ACE Inhibitory Activity of Its Hydrolysates

At present, the studies on the ACE inhibitory peptides from food protein source are mainly focused on the optimization of enzymatic conditions, the isolation and purification of ACE inhibitory peptides, the structural characterization, and the in vivo activity test, while there are a small number of studies on the relationship between protein amino acid composition characteristics and ACE inhibitory activity. Zhang Yuhao and Wang Qiang (2007) collected the ACE inhibition IC_{50} value of some amino acid components from food protein source and their hydrolysates (Table 9.14) (Guanhong 2005; Wu and Ding 2002; Byun and Kim 2001), established the model between the amount of different types of amino acids and ACE inhibition IC_{50} value, and thus analyzed the relationship between amino acid species and protein hydrolysate ACE inhibition activity.

The scatter diagram of various amino acid contents and ACE inhibition activity is shown in Fig. 9.34. The regularity of aromatic amino acids was obvious; the regularity of hydrophobic amino acids and branched chain amino acids was relatively obvious, and only rice protein data were different from them; the alkaline amino acids, acidic amino acids, and amino acids without charges at side chains had large randomness in the scatter diagram, without any law; therefore, nonlinear fitting was conducted to hydrophobic amino acids, branched chain amino acids, aromatic amino acids, and ACE inhibition activity. Because rice protein data might affect the fitting of the regression function between hydrophobic amino acids and branched chain amino acids with ACE inhibition activity, this sample was removed when fitted. The results are shown in Figs. 9.35 and 9.36.

Table 9.14 Protein amino acid composition and its ACE inhibition activity

Amino acid	Peanut protein	Mung bean protein	Rice protein	Zein	Casein	Bovine serum albumin	Soybean protein	Collagen of *Gadus*
Asp	11.97	11.83	8.52	4.49	6.21	11.02	12.8	3.59
Glu	22.78	19.79	19.78	22.31	21.6	17.5	20.47	5.94
Ser	4.66	5.67	5.21	4.31	5.18	4.43	5.60	5.74
His	2.39	3.05	2.47	1.15	2.7	3.74	2.77	1.75
Gly	4.37	3.31	4.09	1.26	1.66	1.81	4.57	39.89
Thr	2.28	3.05	3.55	2.66	3.91	5.92	4.4	2.49
Ala	4.05	3.71	4.3	8.23	1.59	6.3	4.66	9.41
Arg	13.26	7.62	10.23	1.72	5.33	6.03	7.91	5.66
Tyr	4.05	2.97	5.74	5.1	5.36	5.45	3.43	0
Cys	0.68	0.26	0.8	0.8	0.21	6.38	1.45	0
Val	4.66	4.61	5.54	3.86	5.35	6.35	5.25	1.78
Met	1.3	1.26	2.76	2	3.37	0.9	1.38	1.43
Phe	5.79	6.45	5.84	6.85	5.16	6.71	5.41	1.36
Ile	4.01	3.54	3.8	4.15	3.95	2.76	4.97	1.2
Leu	7.07	8.28	8.54	17.3	8.47	12.05	7.81	1.99
Lys	3.04	7.12	3.25	0	6.8	12.98	6.98	1.96
Pro	3.63	4.51	4.87	9.19	8.76	4.85	6.01	9.38
Trp	0.87	1.21	1.58	1.01	1.41	0.61	0.93	0
Hyp	0	0	0	0	0	0	0	6.44
Hydrophobic amino acid[a]	35.43	36.54	42.97	57.69	43.42	45.98	39.85	26.55
Polar amino acid without charges at its side chain[b]	7.62	8.98	9.56	7.77	9.3	16.73	11.45	14.67
Basic amino acid[c]	18.69	17.79	15.95	2.87	14.83	22.75	17.66	9.37
Acidic amino acid[d]	34.75	31.62	28.3	26.8	27.81	28.52	33.27	9.53
Aromatic amino acid[e]	10.71	10.63	13.16	12.96	11.93	12.77	9.77	1.36

(continued)

Table 9.14 (continued)

Amino acid	Peanut protein	Mung bean protein	Rice protein	Zein	Casein	Bovine serum albumin	Soybean protein	Collagen of *Gadus*
Branched chain amino acid[f]	15.74	16.43	17.88	25.31	17.77	21.16	18.03	4.97
IC_{50}	0.71	0.62	0.14	0.27	0.46	0.39	0.54	0.84
$\lg\left(\frac{1}{IC50}\right)$	0.149	0.21	0.85	0.57	0.34	0.41	0.27	0.08

[a]Includes Ala, Ile, Leu, Met, Phe, Pro, Trp, Try, and Val
[b]Includes Ser, Thr, Cys, and Hyp
[c]Includes Lys, His, and Arg
[d]Includes Glu and Asp
[e]Includes Tyr, Phe, and Trp
[f]Includes Leu, Ile, and Val

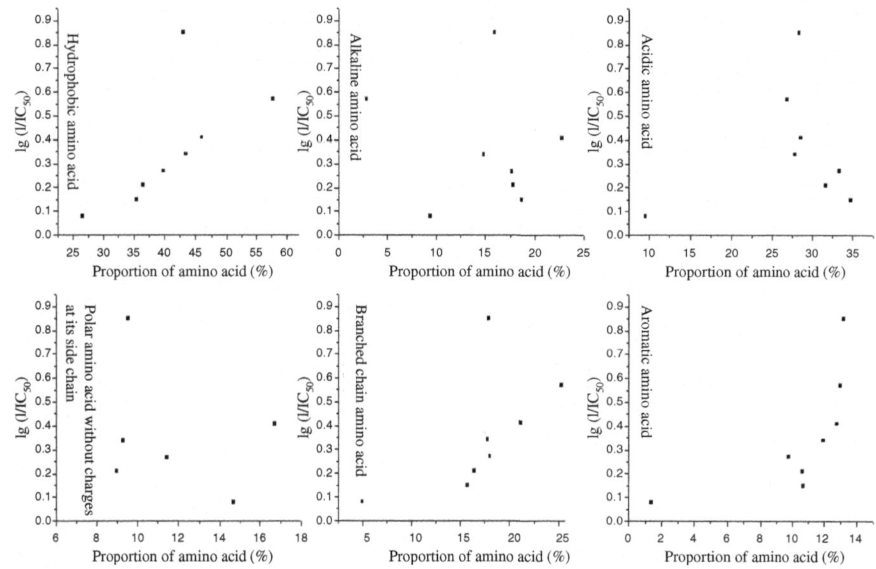

Fig. 9.34 Scatter diagram of amino acid content – ACE inhibition activity

The three fitting models of hydrophobic amino acid content and ACE inhibition activity of protein hydrolysate are shown in Fig. 9.35:

$$\text{Logistic model is} : y = -\frac{0.58}{1 + \left(\frac{x}{43.43}\right)^{7.11}} + 0.64 R^2 = 0.9945 \tag{9.60}$$

$$\text{ExpGro1 model is} : y = -2.52 + 2.2 e^{\frac{x}{168.38}} R^2 = 0.9737 \tag{9.61}$$

$$\text{HyperbolaMod model is} : y = \frac{x}{-3.12x + 277.48} R^2 = 0.9188 \tag{9.62}$$

where x – proportion of hydrophobic amino acids in the protein:

$$y\text{—}\lg\left(\frac{1}{IC_{50}}\right) \tag{9.63}$$

The three fitting models of aromatic chain amino acid content and ACE inhibition activity of protein hydrolysate are shown in Fig. 9.36:

$$\text{ExpGro2 model is} : y = -0.06 + 9.3 \times 10^{-24} e^{\frac{x}{0.25}} + 0.02 e^{\frac{x}{4.98}} R^2 = 0.9653 \tag{9.64}$$

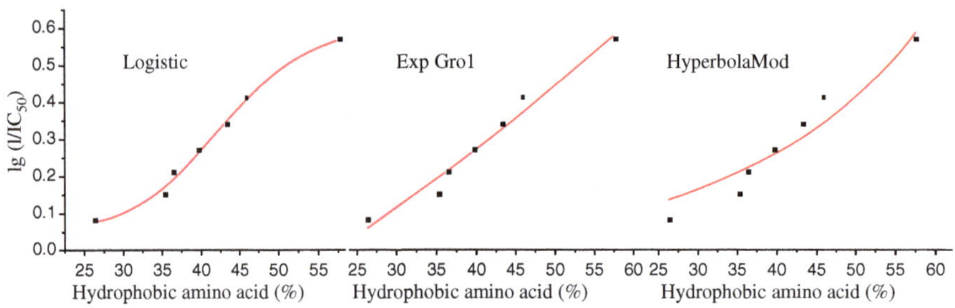

Fig. 9.35 Hydrophobic amino acid content – ACE inhibition activity model

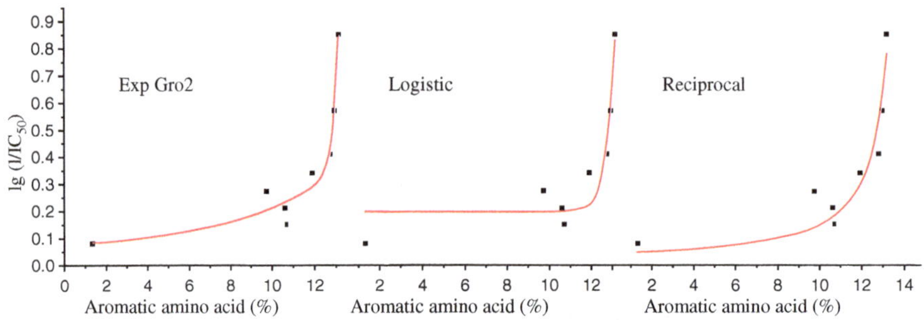

Fig. 9.36 Aromatic amino acid content – ACE inhibition activity model

$$\text{Logistic model is} : y = -\frac{3452.32}{1 + \left(\frac{x}{17.13}\right)^{32.6}} + 3452.5R^2 = 0.9177 \quad (9.65)$$

$$\text{Reciprocal model is} : y = \frac{x}{24.01 - 1.73x}R^2 = 0.9107 \quad (9.66)$$

where x – proportion of aromatic chain amino acids in the protein:

$$y\text{—lg}\left(\frac{1}{IC_{50}}\right) \quad (9.67)$$

The three fitting models of branched chain amino acid content and ACE inhibition activity of protein hydrolysate are shown in Fig. 9.37:

$$\text{Logistic model is} : y = -\frac{0.577}{1 + \left(\frac{x}{19.57}\right)^{6.54}} + 0.65R^2 = 0.9600 \quad (9.68)$$

$$\text{Boltzmann model is} : y = -\frac{0.551}{1 + e^{\frac{x-19.24}{2.76}}} + 0.62R^2 = 0.9569 \quad (9.69)$$

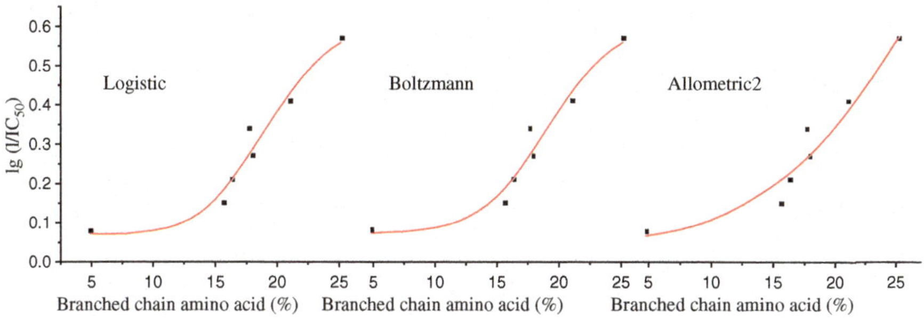

Fig. 9.37 Branched chain amino acid content – ACE inhibition activity model

$$\text{Allomentric 2 model is}: y = 0.06 + 0.00012x^{2.59} R^2 = 0.9385 \quad (9.70)$$

where x – proportion of branched chain amino acids in the protein:

$$y\text{—lg}\left(\frac{1}{\text{IC50}}\right) \quad (9.71)$$

The three kinds of model fitting of Logistic, ExpGro1, and HyperbolaMod were conducted to hydrophobic amino acid content and ACE inhibition activity; the correlation coefficients of the three models established were greater than 0.9, indicating that the correlation of models was good. The trends reflected in the three models were basically consistent with each other, that is, the ACE inhibition activity of hydrolysate increased with the increase of hydrophobic amino acid content, indicating that the higher the hydrophobic amino acid content in the protein was, the higher the ACE inhibition activity of its hydrolysate was. The three kinds of model fitting of ExpGro2, Logistic, and Reciprocal were conducted to aromatic amino acid content and ACE inhibition activity. The model showed that high aromatic amino acid content was helpful to generating the oligopeptide with high ACE inhibition activity; however, different from hydrophobic amino acid, the ACE inhibition activity of hydrolysate would obviously increase with the increase of its content only when aromatic amino acid content was greater than 10%. The three kinds of model fitting of Logistic, Boltzmann, and Allometric 2 were conducted to branched chain amino acid content and ACE inhibition activity. The information reflected by the model was basically similar with that of hydrophobic amino acid, that is, branched chain amino acid content was positively correlated with the ACE inhibition activity of the hydrolysate.

The scatter diagram and nonlinear regression model showed that polar amino acid without charges at its side chain, basic amino acid, and acidic amino acid contents in the protein did not have any obvious impact on the ACE inhibition activity of protein hydrolysate; the increase of hydrophobic amino acid content, aromatic amino acid content, and branched chain amino acid content was helpful to increase the ACE inhibition activity of hydrolysate, that is, the protein with such amino acid composition characteristics might generate a large amount of ACE inhibitory peptide with a high activity. Ruiz et al. (2004) considered that the impact of the content of Pro and Leu and their positions at oligopeptide C-terminal on ACE inhibition activity was large; Stevens et al. (2002) and Rohrbach et al. (1981) considered that when Pro was at the third position at C-terminal, the ACE inhibition activity of oligopeptide was high. Jung et al. (2006) considered that high-activity antihypertensive peptides contained a large amount of hydrophobic amino acids and aromatic amino acids, but the position of amino acid had a large impact on the ACE inhibition activity of oligopeptide. The ACE inhibition activity was stronger when the hydrophobic amino acids and aromatic amino acids of oligopeptide were at C-terminal. Other studies (Cheung et al. 1980) showed that when each position of tripeptide at C-terminal contained hydrophobic amino acid residues (aromatic or branched amino acid residues), it had a strong affinity with ACE; when there were aromatic amino acids and Pro at C-terminal of dipeptide or tripeptide and there were branched chain amino acid at N-terminal, it was more suitable for combination with ACE in a competitive inhibition manner. Thus, the conclusions of this study were consistent with previous studies. It was inferred from previous studies that although there is possibility of generating ACE inhibitory peptide from the amino acid composition characteristics of protein to a certain extent, the inner order of amino acids of protein also strongly affected the activity of ACE inhibitory peptides in addition to the composition of amino acids of protein, because ACE inhibitory peptide did not have a fixed structure. With the development of bioinformatics, more and more protein amino acid sequences have been elucidated, so it is helpful to select the protein material for the enzymatic preparation of ACE inhibitory peptides by combining the protein amino acid composition and sequence.

4.3 Antihypertensive Activity In Vivo

Based on the study of ACE inhibition activity in vitro, the hypotensive activity of oligopeptide usually requires further validation of their in vivo effects by animal experiments. Spontaneously hypertensive rats (SHR) are usually used as experimental animals, and the experimental process is usually divided into single dosing and multiple dosing experiments.

4.3.1 Antihypertensive Effect of Single Dosing

Zhang Yuhao and Wang Qiang (2007) studied the antihypertensive effect of single dosing for the peanut peptides with a molecular weight of less than 1000 Da. The results (Fig. 9.38) showed that after gavage to SHR for peanut oligopeptide and captopril with different dosages, the change trend of SBP of SHR was the group of peanut oligopeptide with low dosage remained antihypertensive activity within 1–7 h after gavage ($p < 0.05$), and the systolic blood pressure (SBP) of SHR

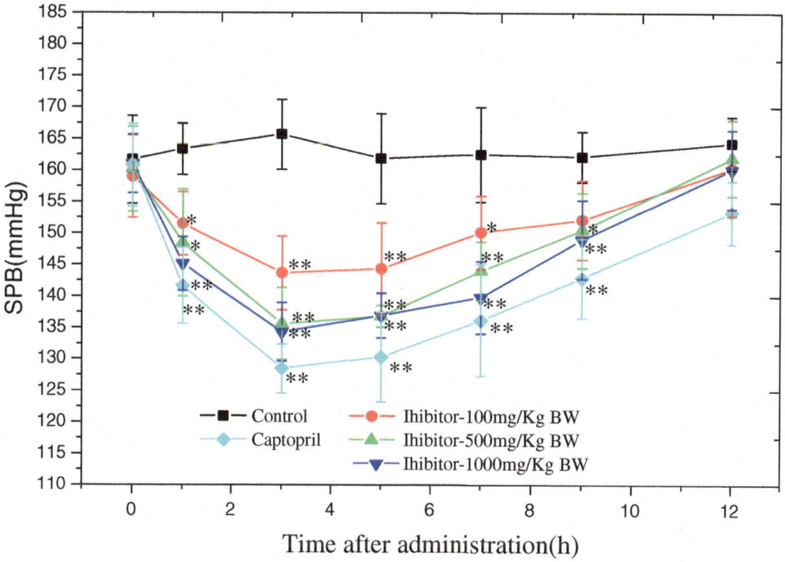

Fig. 9.38 Impact of single dosing on SBP of SHR

significantly decreased ($p < 0.05$) within 1–9 h after gavage in the group with medium and high dosages. The maximum decrease ranges of the three dosage groups all appeared 3 h after gavage, being 15.3 mmHg, 24.5 mmHg, and 26.8 mmHg, respectively. When the dosage of oligopeptide to SHR gavage was 500–1000 mg/ml, the maximum decrease range of SBP was 14–23 mmHg.

The studies on the changes in SHR heart rate after oral administration of peanut oligopeptide (Fig. 9.39) showed that there was no significant difference ($p > 0.05$) in the heart rate between all experimental groups and the control groups within 12 h and there was no significant difference ($p > 0.05$) between the experimental groups, indicating that peanut oligopeptide did not cause adverse impact on the circulatory system of SHR. Sigmoidal model (Fig. 9.40) was established using the maximum decrease range of SHR tail artery systolic pressure and oligopeptide dosage in the groups with different dosages, and the following equation was established:

$$Y = \frac{42.206}{1 + e^{\frac{x+0.43325}{0.5046}}} - 25.654, R^2 = 0.9955 \tag{9.72}$$

This showed that the maximum decrease range of SHR tail artery systolic pressure tended to a fixed value with the increase of dosage of peanut peptide, rather than increasing with the increase of dosage of peanut peptide, so hypotension phenomenon would not be caused due to dosage deviation in the actual applications. The study results showed that peanut antihypertensive peptides had hypotensive effect on hypertensive rats or patients and the hypotensive effect was smooth,

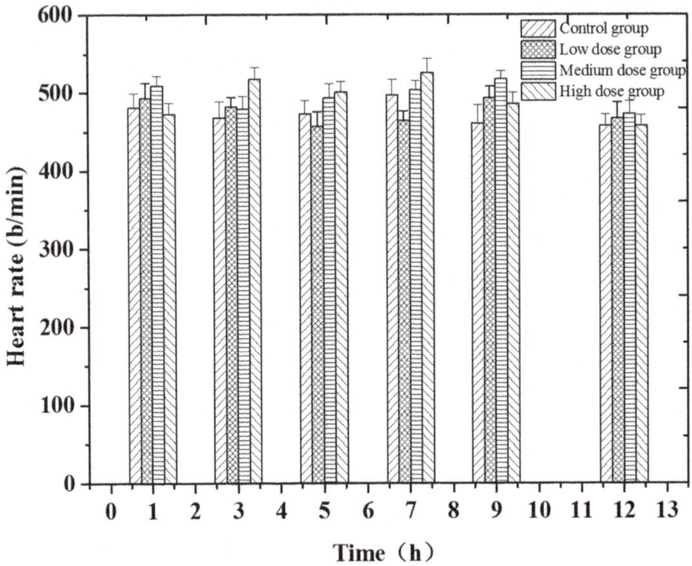

Fig. 9.39 Impact of peanut oligopeptide on SHR heart rate

Fig. 9.40 Dose-effect sigmoidal model of peanut oligopeptide

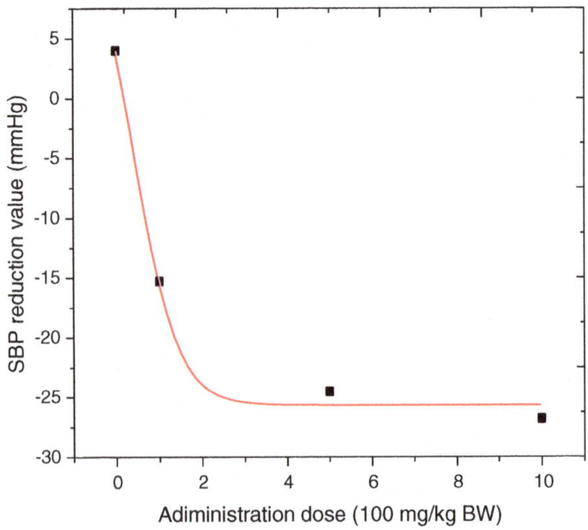

but it did not have any impact on the ones with normal blood pressure; therefore, it was safe without any side effects.

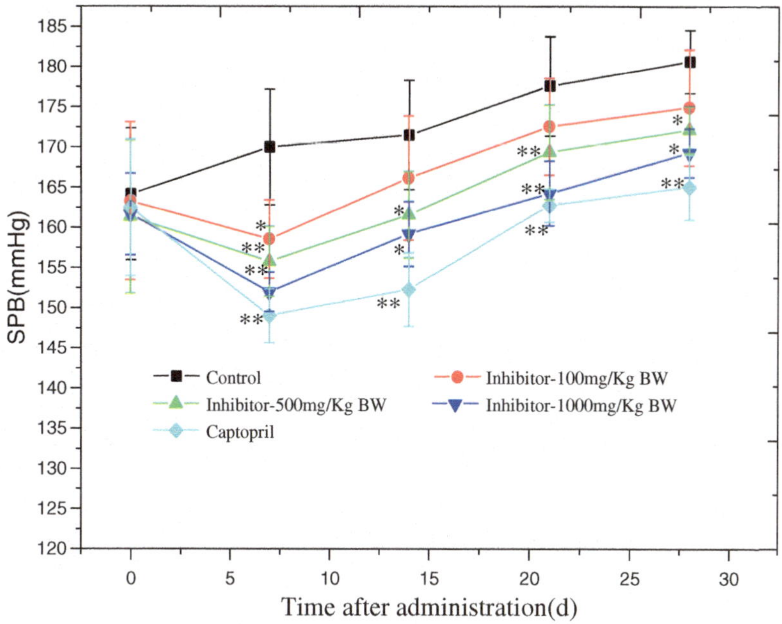

Fig. 9.41 Impact of repeated administrations of peanut oligopeptide on SBP of SHR

4.3.2 Antihypertensive Effect After Repeated Administrations

Zhang Yuhao and Wang Qiang (2007) studied the antihypertensive effect of peanut peptide with a molecular weight of less than 1000 Da after repeated administrations (Fig. 9.41). The results showed that the gavage administration of peanut oligopeptide with different dosages had different degrees of antihypertensive effects on SHR. The SBP of SHR in the three oligopeptide groups decreased by 11.4–18 mmHg 1 week after administration, with significant difference ($p < 0.05$) compared with that of blank group, and later the SBP of the dosage groups rose with the extension of time, because its SBP would rise with the increase of rat age of SHR. Within 4 weeks, the SBPs of medium- and high-dosage groups were significantly lower than the blank group ($p < 0.05$), indicating that long-term blood pressure reduction would be achieved after administration of the peanut oligopeptide with a dosage of 500 mg/kg BW and above.

Studies showed that small peptides and even oligopeptides above tetrapeptide were easily absorbed into the blood circulation in a complete form, the peptide vector genes of mammalian and human had been cloned and expressed, and it was verified that there were two peptide carriers in vivo. After oral administration of yogurt containing ACE inhibitory peptide Ile-Pro-Pro and Val-Pro-Pro, SHR blood pressure was significantly reduced, and at the same time the ACE activity of aorta abdominalis was significantly inhibited, and these two kinds of peptides were tested in the aorta abdominalis.

Table 9.15 Determination of ACE activity in the serum of SHR after repeated administration ($p = 0.05$)

Sample	ACE activity of plasma blood (AU)	Sample	ACE activity of plasma blood (AU)
Blank	30.58 ± 1.96	500 mg/kg BW	24.06 ± 3.05*
100 mg/kg BW	25.71 ± 2.41	1000 mg/kg BW	22.69 ± 1.36*
		Catropril	17.01 ± 2.62**

The studies of Matsui et al. (1999) demonstrated that ACE inhibitory peptide Val-Tyr (IC_{50} = 26 μM) could be fully absorbed in the blood of persons with normal blood pressure and hypertensive patients. Vermeirssen et al. (2002) found that the ACE inhibitory peptide Ala-Leu-Pro-Met-His-Ile-Arg could completely be transported from mucosal layer to the other side from β-lactoglobulin using the study on Caco-2 single cell layer. Zhang Yuhao and Wang Qiang (2007) determined the ACE activity in the serum of SHR after repeated administration, and the results (Table 9.15) showed that the ACE activities in the serum of SHR in the groups with medium and high dosages were significantly lowered than that of the blank group ($p < 0.05$). It proved that peanut peptides could be completely absorbed into the blood circulation and thus produce hypotensive activity.

The study results of Zhang Yuhao and Wang Qiang (2007) showed that in both single administration experiment and multiple administration experiment, the antihypertensive effect of captopril was better than that of peanut oligopeptide. However, it is well known that chemically synthesized antihypertensive drugs have obvious side effects such as the toxic and side effects of the kidneys and other side effects such as hypotension and dry cough. The study results showed that peanut oligopeptide was ACE inhibitor from natural and safe food sources. Therefore, peanut oligopeptide can be used as a new functional food for the prevention and control of hypertension.

5 Structural Characterizations and Structure-Function Relationship of ACE Inhibitory Peptide

The separation and purification of oligopeptide are the processes to enrich, purify, and identify the oligopeptide by using gel chromatography, ion exchange chromatography, HPLC, and HPLC-MS and taking a specific functional activity as the evaluation indicator. Gel filtration chromatography is a chromatographic method for achieving the separation effect according to the different molecular weights of the separated substances. The filler used in the gel filtration chromatography is a material having a three-dimensional network structure with a uniform diameter in bead-like particles. When a mixture solution containing different molecular weights is passing through a gel column, the buffer solution and small and large molecular

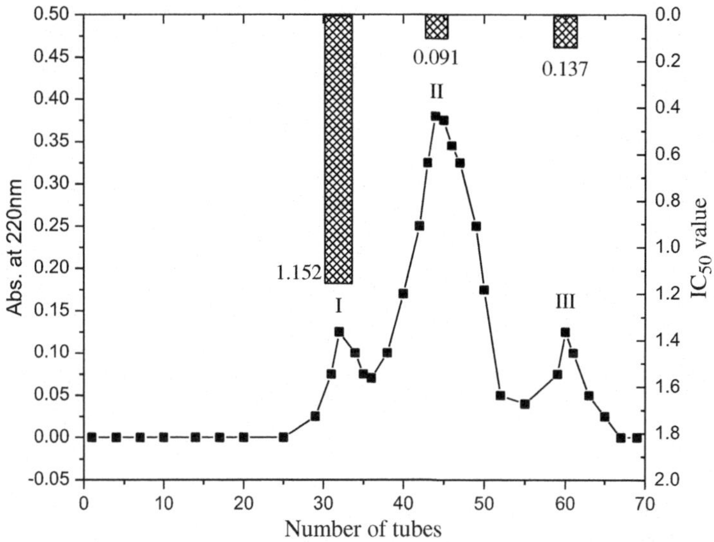

Fig. 9.42 Gel filtration separation of peanut oligopeptide

compounds are able to freely spread and penetrate in the sieve pores, while macromolecular substances are blocked outside the sieve pores. Therefore, the substances with high molecular weight flows in the gap of filler particles, so they can be eluted earlier than the substances with low molecular weight.

5.1 Gel Filtration Separation

The author's study group (2007) conducted gel chromatography separation to peanut ACE I peptides using Sephadex G-15. The results showed that the peanut oligopeptide with a molecular weight of less than 1000 Da could be separated into three peaks (Fig. 9.42), among which the hypotensive activity of peak 2 (PP-II) was the strongest and its IC_{50} value could reach 0.091 mg/ml.

5.2 Separation of PP-II by Semi-preparative RP-HPLC

RP-HPLC principle is to separate the samples according to different molecular polarities of components. Due to such advantages as high resolution, fast analysis speed, and high product recovery, it has become one of the most effective means for the separation and preparation and analysis of polypeptide substances. Peptide component PP-II automatically collected 27 major components after separation by semi-preparative RP-HPLC, as shown in Fig. 9.43. After repetitive sample

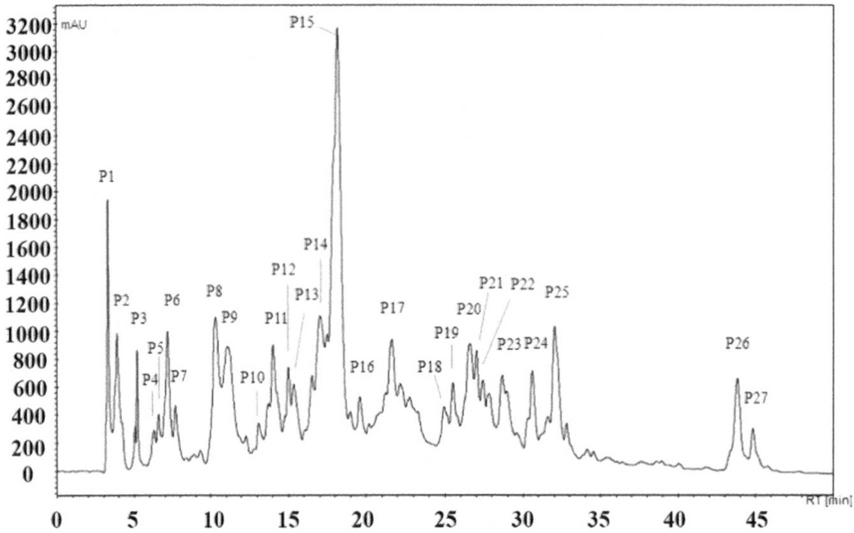

Fig. 9.43 PP-II graph of peanut peptides separated and purified by semi-preparative RP-HPLC

introduction, the peptide concentration of 0.010 mg/mL was prepared by merging the same components obtained each time and then freezing and drying it, and then dissolving it into the distilled water, to determine the ACE inhibition activity of components, as shown in Fig. 9.44. It was shown that the activity of component 8 (recorded as P8) was strong, and the ACE inhibition range reached $85.77 \pm 3.95\%$, followed by P3 and P11, the ACE inhibition ranges of which were $66.79 \pm 2.00\%$ and $66.15 \pm 2.11\%$, respectively. Therefore, component P8 was selected for structural identification. Chromatographic column, Varian C18 (21.2×150 mm); column temperature, 25 °C; flow rate, 10 mL/min; detection wavelength, 280 nm; sample introduction sample, 1 mL; elution condition, 0–50 min, 95%–70% A, 5–30% B (A, water +0.05% TFA; B, acetonitrile +0.05% TFA).

5.3 *Purity Identification of P8 After Purification*

To examine the RP-HPLC separation effect described above, the author's study group conducted purity identification to P8 component. It was shown from Fig. 9.45 that there was a single peak in the RP-HPLC graph of P8, indicating that P8 component might contain only one peptide fragment, and the next step could be conducted for structural analysis. Chromatographic column, C_{18} column (4.6×250 mm); flow rate, 0.5 mL/min; detection wavelength, 280 nm; sample introduction sample, 20 μL; column temperature, 30 °C; elution condition,

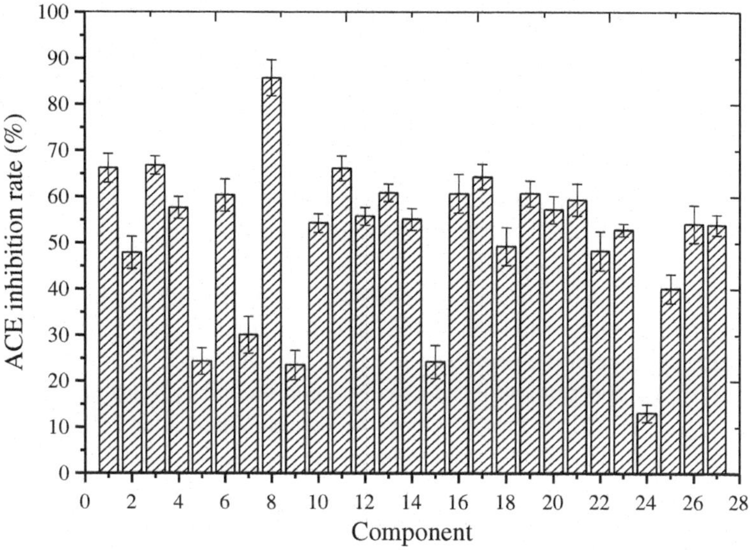

Fig. 9.44 ACE inhibition rates of PP-II components of peanut peptides separated and purified by semi-preparative RP-HPLC

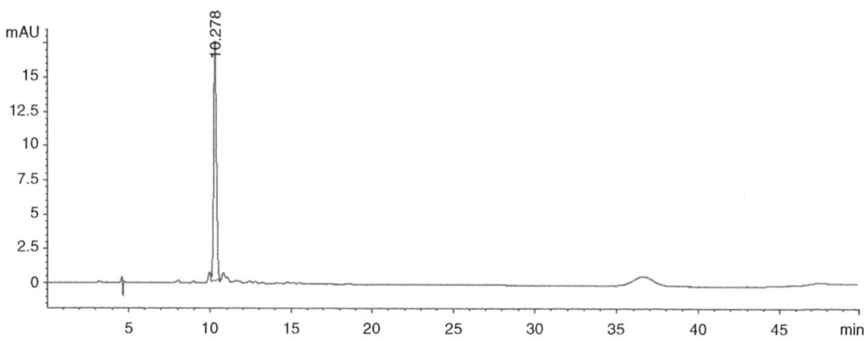

Fig. 9.45 RP-HPLC analysis graph of P8 after purification

0–20 min, 95–70 % A, 5–30 % B (A, water +0.05% TFA; B, acetonitrile +0.05% TFA).

5.4 Analysis of Amino Acid Sequence of ACE Inhibitory Peptide Using MALDI-TOF-TOF

The traditional sequencing methods for peptide and protein analysis include enzymolysis method at C-terminal, Edman degradation method, and DNA translation method. These methods have some shortcomings: for example, as the standard method for determining peptide and protein sequences, N-terminal Edman method (also known as PTH method) has such shortcomings as high requirement for the purity of sample, large consumption amount of sample, and slow sequencing speed; due to incorrect identification to modified amino acid residues, it cannot conduct sequencing for peptide chain with N-terminal protection; it is difficult to find the ideal chemical probe by C-terminal chemical degradation sequencing method (Liao Dan-kui et al. 2006). Mass spectrometry features high sensitivity, accuracy as well as easy operation, and fast speed. With the appearance of FAB, ESI, MALDI, and other soft ionization methods, its applications in biological field have developed rapidly. MALDI-TOF-TOF can be used directly to determine the amino acid sequence of the peptide. The parent ions of peptide fragment generated at primary mass spectrum will selectively enter secondary mass spectrum, the peptide fragments will be broken along the peptide chain after collision against the inert gas, and the mass difference value of the peptide fragment obtained will be used for deducing the peptide sequence.

The tandem mass spectrometry data obtained can be used to analyze peptide sequence from the beginning using software, or it can combine with amino acid analysis to obtain the sequence results by manual analysis. The breakage of peptide follows a certain rule, and the general peptide breakage mode and fragment type are shown in Fig. 9.46 (Guo Hongying 2009). N-terminal series of fragment ions (b series) and C-terminal series of fragment ions (y series) will be obtained after the breakage of peptide bond. These fragments can further form dehydrated and deaminized ions.

MALDI-TOF-TOF analysis (4700 Proteomics Analyzer mass spectrometer) was conducted to peanut oligopeptide component P8, and secondary mass spectrum analysis was conducted to the fragmented ions of parent ions to obtain MS/MS graph (Fig. 9.47). In MS graph, the peak (m/z = 810.8) was molecular ion peak $[M + 2H]^{2+}$, and the molecular weight of component P8 was 808.8 (theoretical relative molecular weight was 808.7). In the analysis using software, the amino acid sequence of P8 was Lys-Leu-Tyr-Met-Arg-Pro. IC_{50} value was 0.0052 mg/mL (6.42 μM) in the determination of ACE inhibition activity of component P8 (Table 9.16).

5.5 Verification of ACE Inhibition Activity of Synthetic Oligopeptide KLYMRP

In order to further verify the functional activity of peanut hexapeptide Lys-Leu-Tyr-Met-Arg-Pro (KLYMRP), the oligopeptide with the same amino acid sequence was synthesized by solid-phase synthesis method, and ACE inhibition IC_{50} value was 0.0038 mg/mL (4.69 μM) in the determination of activity of synthetic peptide. Experiments showed that the oligopeptide with such amino acid sequence had

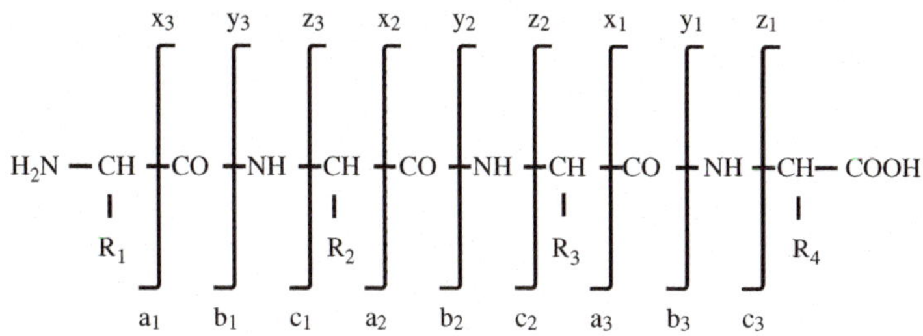

Fig. 9.46 Peptide skeleton breakage and naming of its fragment ions

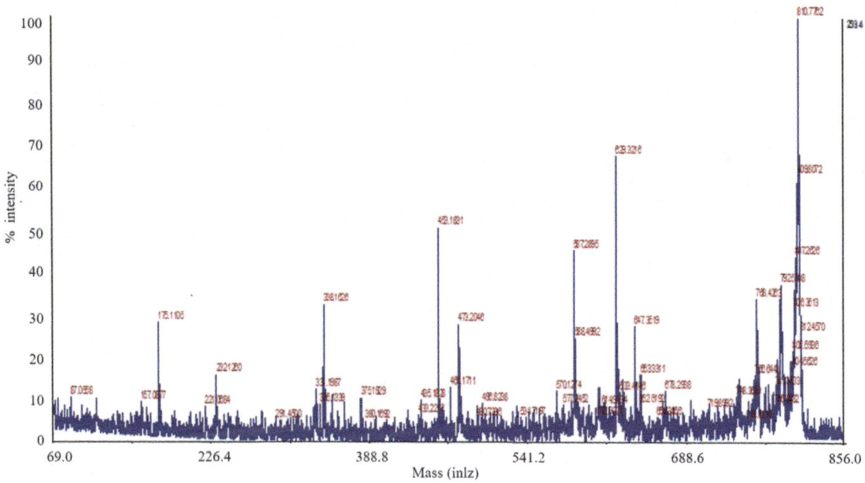

Fig. 9.47 MALDI MS/MS graph of oligopeptide P8

Table 9.16 Amino acid sequence and activity of peanut ACE inhibitory peptide

Peanut ACE inhibitory peptide P8	Analysis results
Theoretical relative molecular weight	807.7
Measured mass spectrum mass	808.8
Amino acid sequence	Lys-Leu-Tyr-Met-Arg-Pro
IC_{50}	0.0052 mg/mL (6.42 μM)

in vitro ACE inhibition activity, and the activity of synthetic peptide was higher than that of oligopeptide KLYMRP from peanut protein source.

5.6 Structure-Activity Relationship of Oligopeptide KLYMRP

The most stable molecular conformation of high-activity ACE inhibitory peptide P8 was specified by computer simulation method (Discovery Studio, Accelrys). It would have different activities due to the transition of the tertiary structure of oligopeptide based on the conformation when it played a role in antihypertensive activity after docking with ACE.

5.6.1 Conformation Optimization of Oligopeptide KLYMRP

Molecular dynamics has been widely used in the field of life sciences, such as the study on the mechanism of protein folding, study on the mechanism of enzyme catalysis reaction, study on the movement of function-related proteins, and study on the large-scale conformational changes of biological macromolecules, so it has become one of the essential methods in the study on theoretical biology.

Oligopeptide chain initial structure was constructed according to the amino acid sequence of oligopeptide P8 (KLYMRP). Conformation optimization was conducted through molecular dynamics simulation. The main steps included pretreatment, giving position, adding solvent environment, temperature rise, equilibrium, and sampling. The lowest potential energy conformation was searched and analyzed, that is, the most stable molecular conformation of the oligopeptide would be used for subsequent molecular docking link. Figure 9.48 showed the most stable molecular conformation (pH 7.4, 300 K) of oligopeptide after 200S of kinetics simulation under the position of CHARMm, and the distance from its C-terminal and N-terminal was 9.448 Å.

5.6.2 Analysis of the Interaction Between Oligopeptide KLYMRP and ACE

The interaction between oligopeptide and ACE was analyzed using molecular docking technique. This technology is to place ligand molecules at the position of active site of receptor molecule; evaluate the interaction between the ligand and the receptor in real time according to the principle of geometry complementarity, energy complementarity, and chemical environment complementarity; and then find the best combination mode between the two molecules. Angiotensin-converting enzyme (ACE) plays an important role in regulating blood pressure, electrolyte and body fluid balance, cardiovascular system development, and structural remodeling, and its inhibitors have been successfully used as first-line drugs for the treatment of hypertension. Oligopeptide can have different activities due to the transformation of the tertiary structure by using oligopeptide KLYMRP as a

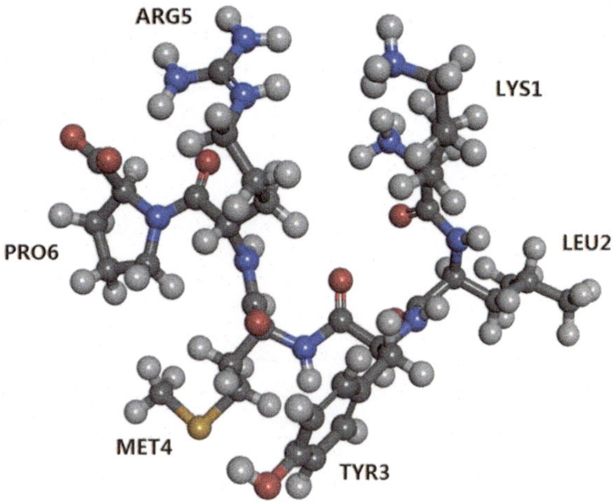

Fig. 9.48 Most stable molecular conformation of oligopeptide KLYMRP

ligand molecule in the active site of the receptor molecule ACE and studying the interaction between this oligopeptide and ACE.

The module of DS_CDOCKER was used for molecular docking. Among them, ACE protein structure came from PDB database (the code was 1UZE), the docking receptor was obtained by removing the ligand (EAL3002) and water molecules in this structure, and the ligand was the oligopeptide molecule KLYMRP after optimization by kinetics model, while the active sites of receptor during docking were obtained using "active pocket search" in DS3.5 software. After the assessment on the docking results using CDocker Energy, CDocker Interaction Energy, LibDockScore, and other scoring functions, the optimal docking compound conformation at pH 7.4 and 300 K was obtained (Fig. 9.49). It was clearly shown from the conformation of this compound that oligopeptide KLYMRP could act well in the "active pocket search" of ACE and then form a variety of interaction with its amino acid residues (Douglas and Sturrock 2012).

Figure 9.50 showed the interaction between oligopeptide KLYMRP and amino acid residues in ACE. There were many hydrogen bond acceptors and donors at the surface of oligopeptide KLYMRP, and these structures were helpful to form stable KLYMRP-ACE compound during the interaction with ACE. The main force formed by KLYMRP and ACE residues is shown in Fig. 9.51. The results showed that KLYMRP and ACE mainly formed three kinds of forces, including hydrogen bond force, electrostatic force, and Pi bond force, and the main ACE amino acid residues that formed these forces included 403-, 123-, and 143-bit glutamic acids, 355- and 517-bit serine, and 124-bit arginine. Glutamic acid is a kind of acidic amino acid with two carboxyls, and it is easily used as the proton receptor to form hydrogen bonds; serine is the amino acid that takes hydroxyl as side chain, and it can be used as both proton donor and proton receptor due to small space occupation; arginine contains a guanidyl; this group has a strong positive charge, so it can be used as a proton donor to form hydrogen bonds. In KLYMRP, LYS1, ARG5, and

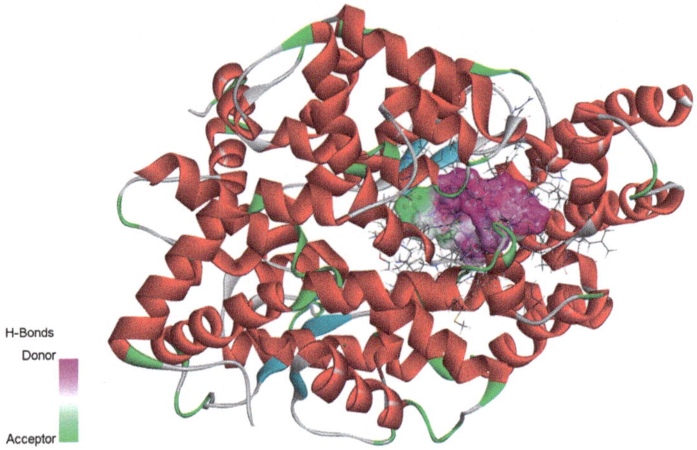

Fig. 9.49 Oligopeptide KLYMRP-ACE compound conformation with the highest scores in CDocker Energy

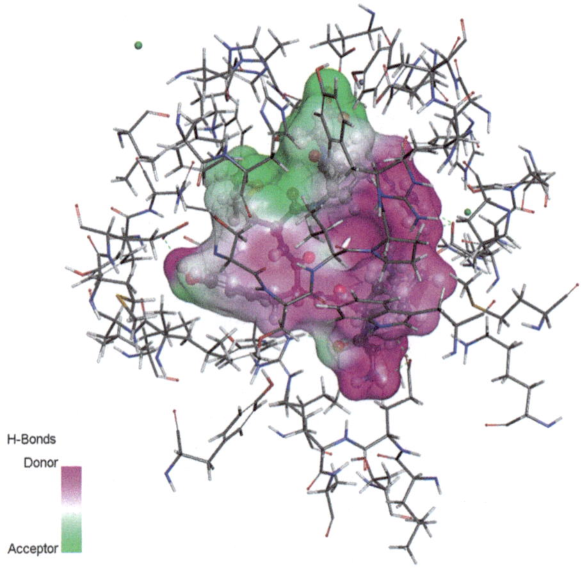

Fig. 9.50 Interaction between oligopeptide KLYMRP and ACE amino acid residues

PRO6 are the main action sites, the polarity and space size of side chain are the main factors to form these action force.

Table 9.17 listed the bond length and force size of oligopeptide KLYMRP and the main interactive groups of ACE. Among them, the glutamic acid residues on ACE and oligopeptide KLYMRP had the strongest interaction force, being −143.2075 kcal/mol; the interaction between arginine at ACE and oligopeptide

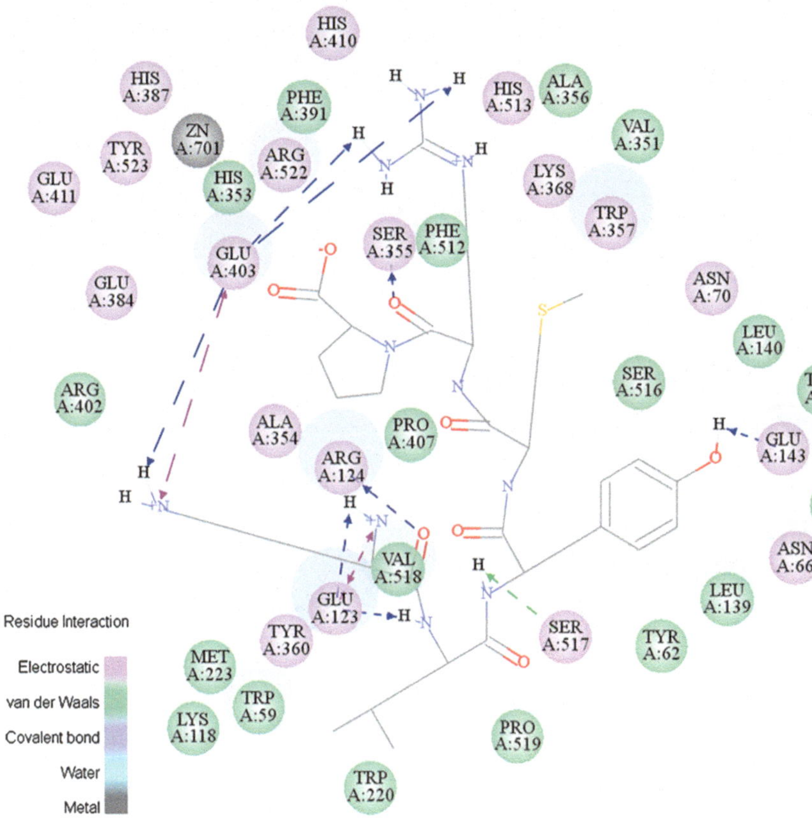

Fig. 9.51 Main interaction forces between oligopeptide KLYMRP and ACE amino acid residues (pH 7.4, 300 K)

KLYMRP was positive, and this may be caused due to the electrostatic repulsion generated by other adjacent groups.

In addition, it was shown from the above table that the metal ions in ACE could strongly interact with oligopeptide KLYMRP, and the distances from Zn ion to O_{119} and O_{120} on KLYMRP were 2.261 Å and 2.314 Å, respectively, which were close to the sum of the covalent radiuses (2.14 Å) of O atom (1.4 Å) and Zn atom (0.74 Å), indicating that the carbonyl at oligopeptide KLYMRP could form a coordination bond with Zn ion (Guo Huiqing et al. 2010). It was also the reason why the interaction between Zn atom and oligopeptide KLYMRP could reach −118.7700 kcal/mol.

It was shown from Fig. 9.52 that Zn ion could form a coordination bond with the carboxyl oxygen atoms (O_{119} and O_{120}) of PRO6 on oligopeptide, and the distance from it to the carboxyl oxygen atoms on 384-bit glutamic acid and 411-bit glutamic acid in ACE was also enough to form the coordination bonds. These three groups and the six coordination bonds formed by Zn ions could stabilize the structure of

Table 9.17 Interaction between oligopeptide KLYMRP and ACE

Action group	Bond length (Å)	Interaction force (kcal/mol)
ACE:ARG124:HH21-O_{24}	2.205	7.7197
ACE:SER355:HG-O_{105}	1.850	−9.6588
H_3-ACE:GLU123:OE2	2.164	−51.1355
H_{26}-ACE:GLU123:OE2	2.434	
H_{45}-ACE:SER517:O	2.498	−4.8263
H_{62}-ACE:GLU143:OE1	1.957	−12.4212
H_{22}-ACE:GLU403:OE1	2.003	−79.6508
H_{100}-ACE:GLU403:OE2	2.049	
H_{103}-ACE:GLU403:OE1	2.013	
H_{103}-ACE:GLU403:OE2	2.146	
ACE:Zn701-O_{119}	2.261	−118.7700
ACE:Zn701-O_{120}	2.314	
ACE:Cl703-O_{113}	16.680	0
ACE:Cl704-O_{35}	4.892	−9.9818

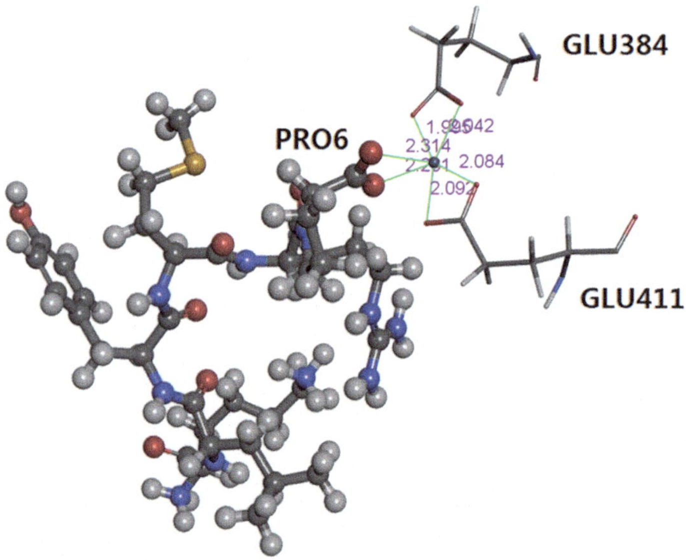

Fig. 9.52 Interaction between Zn ion and amino acid residues

ACE-oligopeptide compound to a large extent, thereby inhibiting ACE activity (Yamamoto et al. 2007).

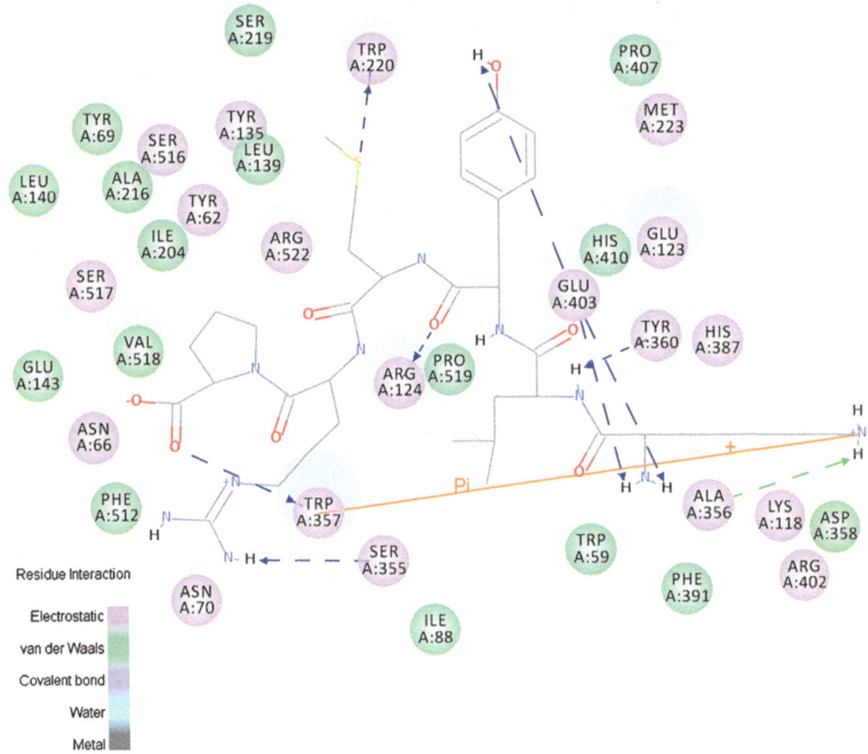

Fig. 9.53 Main interaction force between oligopeptide KLYMRP and ACE amino acid residues (pH 8.4, 300 K)

5.6.3 Stability of Interaction Between Oligopeptide KLYMRP and ACE

As the important factors to impact the interaction between molecules, temperature and pH also have a significant impact on the interaction between oligopeptide KLYMRP and ACE. The DS3.5 molecular simulation software can be used to simulate the molecular docking process at different temperatures and pHs to obtain the optimal docking compound. The interaction between oligopeptide KLYMRP and ACE was analyzed, as shown in Figs. 9.53 and 9.54.

With the increase of pH and temperature, 403-bit glutamic acid and 124-bit arginine on ACE always participated in the interaction between them. Meanwhile, some new groups such as tyrosine, alanine, and aspartic acid could also interact with oligopeptide KLYMRP. Specific bond length and interaction force are shown in Table 9.18. In addition, 357-bit and 220-bit tryptophans in ACE could also form the Pi bond with oligopeptide KLYMRP, and it could play a role in stabilizing their interaction.

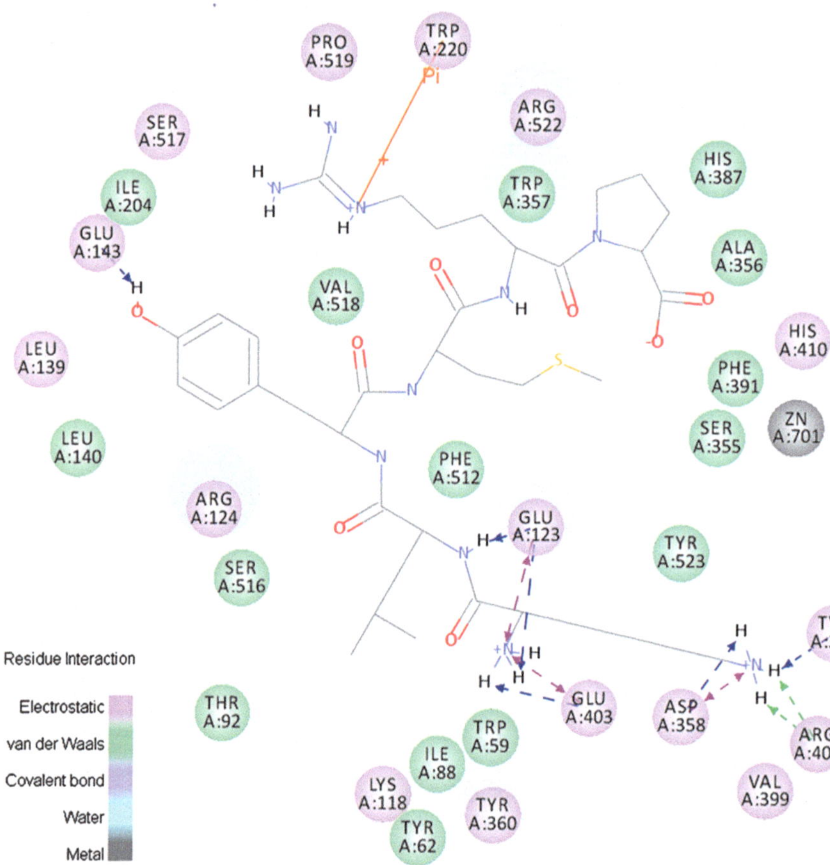

Fig. 9.54 Main interaction force between oligopeptide KLYMRP and ACE amino acid residues (pH 7.4315 K)

In comparison with Tables 9.17 and 9.18, it was shown that the increase in pH and temperature might lead to a decrease in the interaction between the oligopeptide KLYMRP and ACE. As an important residue after the interaction between the two, the interaction force between glutamic acid on ACE and oligopeptide KLYMRP changed from −143.2075 to −28.5356 kcal/mol and −87.8504 kcal/mol, respectively; when pH was 8.4, the interaction force between Zn ion and oligopeptide KLYMRP was positive, indicating that under this condition, there were repulsive forces between Zn ion and some groups on oligopeptide and the interaction was unstable; when the temperature increased, the distance between Zn ion and the sulfur atom on the oligopeptide was 6.070 Å, which was much larger than the sum of the covalent radiuses (2.14 Å) of oxygen atom (1.4 Å) and zinc ion (0.74 Å); the acting force was weak and the interaction was unstable.

Table 9.18 Interaction force between oligopeptide KLYMRP and ACE

Condition	Action group	Bound length (Å)	Interaction force (kcal/mol)
pH 8.4, 300 K	ACE:ARG124:HH21-O_{63}	2.367	−9.9144
	ACE:TRP220:HE1-S_{74}	2.302	−2.0483
	ACE:TRP357:HE1-O_{118}	2.497	−20.3167
	H_2-ACE:GLU403:OE2	2.152	−28.5356
	H_3-ACE:GLU403:OE2	2.434	
	H_{61}-ACE:GLU403:OE1	1.932	
	H_{21}-ACE:ALA356:O	1.816	−12.2489
	H_{25}-ACE:TYR360:OH	2.445	0.0911
	H_{98}-ACE:SER355:OG	2.205	−0.4414
	ACE:Zn701-H_{102}	4.633	23.1370
	ACE:Cl703-H_{98}	19.073	0
	ACE:Cl704-H_{102}	6.614	−2.8376
Ph7.4, 315 K	H_2-ACE:GLU403:OE1	2.488	−38.4605
	H_2-ACE:GLU403:OE2	2.076	
	H_{20}-ACE:ARG402:O	1.830	−7.6691
	H_{21}-ACE:ARG402:O	2.450	
	H_{21}-ACE:TYR394:OH	2.179	−8.6251
	H_{22}-ACE:ASP358:OD2	2.052	−24.4351
	H_3-ACE:GLU123:OE1	2.491	−37.4369
	H_{26}-ACE:GLU123:OE1	1.965	
	H_{62}-ACE:GLU143:OE2	1.870	−11.9530
	ACE:Zn701-H_{78}	3.788	−5.7339
	ACE:Zn701-S_{75}	6.070	
	ACE:Cl703-H_{59}	21.182	0
	ACE:Cl704-H_{94}	4.908	−9.0554

References

Brantl V, Teschemacher H, Henschen A, et al. Novel opioid peptides derived from casein (beta-casomorphins). I. Isolation from bovine casein peptone[J]. Hoppe Seyler's Z Physiol Chem. 1979;360(9):1211.

Byun HG, Kim SK. Purification and characterization of angiotensin I converting enzyme (ACE) inhibitory peptides from Alaska Pollack (Theragra chalcogramma) skin[J]. Process Biochem. 2001;36(12):1155–1162.

Chen Meizhen, Yu Jie, Guo Huimin. Study on the scavenging activity on hydroxyl radical of enzymic hydrolysates of soy protein isolates [J]. Food Sci. 2002;23(1):43–6.

Chen Xiu, Chen Weizhou, Zeng Guiyun. Cardiovascular pharmacology. 3rd Edition[M] ed. People's Health Publishing House: Beijing; 2003. p. 352–3.

Chen HM, Muramoto K, Yamauchi F. Structural analysis of antioxidative peptides from soybean. β-conglycinin [J]. J Agric Food Chem. 1995;43(3):574–8.

Cheung HS, Wang FL, Ondetti MA, et al. Binding of peptide substrates and inhibitors of angiotensin-converting enzyme. Importance of the COOH-terminal dipeptide sequence [J]. J Biol Chem. 1980;255:401–7.

Coakley WT, James CJ. A simple linear transform for the Folin-Lowry protein calibration curve to 1.0 mg/ml. Anal Biochem. 1978;85(1):90–7.

Cushman DW, Cheung HS, Sabo EF, Ondetti MA. Design of potent competitive inhibitors of angiotensin-converting enzyme. Carboxyalkanoyl and mercaptoalkanoyl amino acids. Biochemistry. 2002;16(25):5484–91.

Deng JC. Effect of temperatures on fish alkaline protease, protein Interaction and texture quality. J Food Sci. 1981;46(1):62–5.

Douglas RG, Sturrock ED. Structure-based design of domain-selective angiotensin-converting enzyme inhibitors. Drug Discovery Afr. 2012:355–77.

Dun XP, Chen ZW. Enzyme hydrolysis preparation of rice small polypeptides from rice residue [J]. Food Sci. 2004;25(6):113–6.

Enyonam Quist E, Dixon Phillips R, Saalia FK. Angiotensin converting enzyme inhibitory activity of proteolytic digests of peanut (Arachis hypogaea L.) flour. LWT Food Sci Technol. 2009;42 (3):694–9.

Gill I, López-Fandiño R, Jorba X, et al. Biologically active peptides and enzymatic approaches to their production[J]. Enzym Microb Technol. 1996;18(3):163.

Grimble GK, Silk DBA. Intravenous protein hydrolysates-Time to turn the clock back[J]. Clin Nutr. 1990;9(1):39–45.

Guang C, Phillips RD. Purification, activity and sequence of angiotensin I converting enzyme inhibitory peptide from Alcalase hydrolysate of peanut flour[J]. J Agric Food Chem. 2009;57 (21):10102–6.

Guanhong L. Studies on angiotensin-converting enzyme inhibitory peptides derived from food proteins [D]. Wuxi: Jiangnan University; 2005.

Guangchang P et al. Theoretical basis and prospect of research progress of bioactive peptides[J]. Food Sci. 2001;22(2):80–4.

Guo H. Preparation of wheat germ protein and its antioxidant function[D]. Wuxi: Jiangnan University; 2009.

Guo HQ, Mao H, Zhao B, et al. Molecular mechanism study of targeting of two angiotensin-converting enzyme inhibitory peptides[J]. Food Sci. 2010;31(23):1–5.

Jamdar SN, Rajalakshmi V, Pednekar MD, Juan F, Yardi V, Sharma A. Influence of degree of hydrolysis on functional properties, antioxidant activity and ACE inhibitory activity of peanut protein hydrolysate. Food Chem. 2010;121(1):178–84.

Jianping Wu, Xiaolin Ding. Characterization of inhibition and stability of soy-protein-derived angiotensin I-converting enzyme inhibitory peptides. Food Res Int. 2002;35(4):367–75.

Jung WK, Mendis E, Je JY, et al. Angiotensin I-converting enzyme inhibitory peptide from yellowfin sole (Limanda aspera) frame protein and its antihypertensive effect in spontaneously hypertensive rats [J]. Food Chem. 2006;94(1):26–32.

Kannel WB. Blood pressure as a cardiovascular risk factor: prevention and treatment. JAMA: J Am Med Assoc. 1996;275(20):1571–6.

Liao Dan-kui, Sun Xiu-hua, Wan Shun-gang, Yang Ke-di, Tong Zhang-fa, Liu Xiong-min. Enzymatic Hydrolysis and Screening of Prodrugtype Angiotensin I-converting Enzyme Inhibitor from Chicken Egg Yolk [J]. Fine Chem. 2006;8:757–9.

Liu Lisheng. Hypertension[M]. Beijing: People's Health Publishing House; 2001. p. 30–4.

Liu YD, Cao YP. Prospect review on polypeptide [J]. Food Sci. 2002;23(8):319–20.

Marquez MC, Fernandez V. Enzymic hydrolysis of vegetable proteins: mechanism and kinetics [J]. Process Biochem. 1993;28:481–90.

Mo Qi. A Study on the Papain Hydrolysis of Peanut Meal [J]. Cereal Feed Ind. 1996;12:37–8.

Matsui T, Li C, Osajima Y. Preparation and characterization of novel bioactive peptides responsible for angiotensin I – converting enzyme inhibition from wheat germ [J]. J Peptide Sci. 1999;5(7):289–97.

Munilla-moran R, Saborido-rey F. Digestive enzymes in marine species. 1: Protease activities ingut from redfish (Sebastes mentella), seabream (Sparusaarata) and turbot (Scophthalmus inaximus) [J]. Comp Biochem Physiol B: Biochem Mol Biol. 1996;113(2):395–402.

Nakagomi K, Fujimura A, Ebisu H, et al. Acein-1, a novel angiotensin-I-converting enzyme inhibitory peptide isolated from tryptic hydrolysate of human plasma[J]. FEBS Lett. 1998;438(3):255–7.

Ondetti MA, Cushman DW. Design of specific inhibitors of angiotensin-converting enzyme: new class of orally active antihypertensive agents [J]. Science. 1977;196(4288):441–4.

Patchett AA, Harris E, Tristram EW, et al. A new class of angiotensin-converting enzyme inhibitors [J]. Nature. 1980;298:280–3.

Shen Beiying. Study on Antioxidant Peptides of Soybean Protein [J]. Chinese fats oils. 1996; 21(6):21–4.

Stevens RL, Micalizzi ER, Fessler DC, Pals DT. Angiotensin I converting enzyme of calf lung. Method of assay and partial purification. Biochemistry. 2002;11(16):2999–3007.

Rohrbach MS, Williams JEB, Rolstad RA, et al. Purification and substrate specificity of bovine angiotensin-converting enzyme [J]. J Biol Chem. 1981;256:225–30.

Ruiz JÁG, Ramos M, Recio I. Angiotensin converting enzyme-inhibitory activity of peptides isolated from Manchego cheese. Stability under simulated gastrointestinal digestion[J]. Int Dairy J. 2004;14(12):1075–80.

Saito Y, Wanezaki K, Kawato A, et al. Structure and activity of angiotensin I converting enzyme inhibitory peptides from sake and sake lees [J]. Biosci Biotechnol Biochem. 1994;58(10):1767–71.

Sheng GH. Function and application of soybean peptide [J]. Food Sci Technol. 1993;6:21–6.

Tello PG. Enzymatic hydrolysis of whey protein: I kinetic models [J]. Biotechnol Bioeng. 1994;44:523–8.

Vermeirssen V, Deplancke B, Tappenden KA, Van Camp J, Gaskins HR, Verstraete W. Intestinal transport of the lactokinin Ala-Leu-Pro-Met-His-Ile-Arg through a Caco-2 Bbe monolayer. J Pept Sci. 2002;8(3):95–100.

Wang Jiazheng. Protein technical handbook[M]. Beijing: Science Press; 2002.

Wang M, Gu WY. On preparation of enzyme-modified peptides from maize gluten meal [J]. Sci Technol Cereal Oils Foods. 1999;1–3.

Wang H, Ng TB. Pleureryn, a novel protease from fresh fruiting bodies of the edible mushroom Pleurotus eryngii. Biochem Biophys Res Commun. 2001;289(3):750–5.

Wang Q, Wang CY, Hu H, et al. Structural characterization and structure-activity relationship of ACE inhibitory peptides from peanut [J]. Food Sci. 2013.

Wu JP, Ding XL. Characterization of inhibition and stability of soy-protein-derived angiotensin I-converting enzyme inhibitory peptides[J]. Food Res Int. 2002;35:367–75.

Yamamoto D, Takai S, Miyazaki M. Prediction of interaction mode between a typical ACE inhibitor and MMP-9 active site. Biochem Biophys Res Commun. 2007;354(4):981–4.

Yude L, Yanping C. Development of biological protein peptide[J]. Food Sci. 2002;23(8):319–20.

Zhang XH. Preparation of soy peptide and study on the separation of purified grade [D]. Guangzhou: South China University of Technology; 2002.

Zhang YH. Study on preparation of peanut oligopeptides and its functional activities [D]. Beijing: Chinese Academy of Agricultural Sciences; 2007.

Zhang XS, Feng JC. Study on protease production of functional protein short peptides [J]. Anim Sci Vet Med. 2004;21(3):48–51.

Zhang YH, Wang Q. Peanut protein hydrolyzing by Alcalase to prepare peanut oligopeptides [J]. Trans CSAE. 2007;23(4):258–63.

Chapter 10
Oxidation Stability Improvement of Peanut Oil

The content of unsaturated fatty acids in peanut oil was high (>70%), while the contents of endogenous V_E, phytosterols, squalene, and other antioxidant substances in peanut were low and not enough to prevent the oxidation of peanut oil, so the peanut oil was easily oxidized in the storage and edible process, and the generated peroxide was quickly decomposed into aldehydes and ketones, which caused the rancidity of peanut oil (Hao Xiaoli 2003) and reduced the quality of peanut oil. Therefore, it is required to add exogenous antioxidants to improve the oxidation stability of peanut oil to ensure its storage quality and nutritional quality. Zhang Jianshu and Wang Qiang (2012) selected six antioxidants (tea polyphenol (TP), phytosterol, TBHQ, ascorbyl palmitate (AP), BHT, and V_E) and three synergists (phytic acid, citric acid, and V_C) to improve the oxidation stability of peanut oil and researched to determine the natural and synthetic antioxidants with the best antioxidant effect and the optimal ratio between the antioxidants and synergists.

1 Relationship Between Antioxidant and Stability of Peanut Oil

About 0.02% TP, V_E, phytosterol, TBHQ, AP, and BHT were added to the peanut oil to determine and compare their accelerated oxidation induction time. The results are shown in Fig. 10.1. The synthetic antioxidant with the best antioxidant effect among these six antioxidants on peanut oil was TBHQ with the induction time of 17.90 h $\pm$ 0.71 h, and the best natural antioxidant was TP with the induction time of 11.43 h $\pm$ 0.27 h; their induction time was significantly higher than that of the blank control group ($P < 0.01$), and they were 3.67 times and 2.34 times that of the blank control group, respectively; the antioxidant effect of TBHQ was significantly higher than that of tea polyphenol ($P < 0.01$), and the induction time of the former was 1.57 times that of the latter. Tea polyphenol and TBHQ belonged to the phenolic

Q. Wang, *Peanut Processing Characteristics and Quality Evaluation*,
DOI 10.1007/978-981-10-6175-2_10

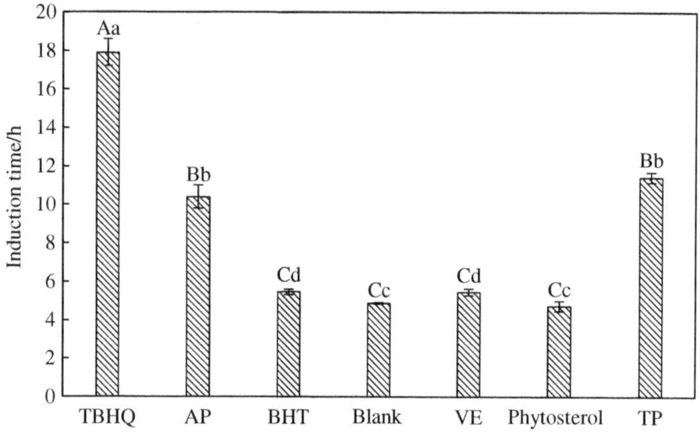

Fig. 10.1 Comparison of antioxidant effects of different antioxidants on peanut oil. Note: *A*, *B*, *C* – different letters mean highly significant difference ($P < 0.01$); *a*, *b*, *c*, *d* – different letters mean significant difference ($P < 0.05$)

Fig. 10.2 Structure drawings of TBHQ and tea polyphenol

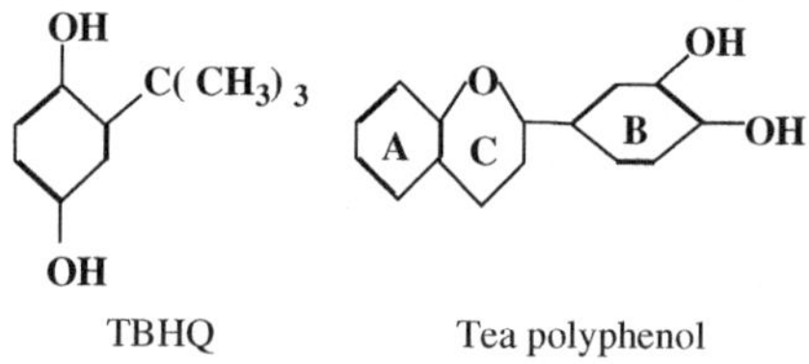

antioxidants, and they could provide hydrogen atoms to combine with oil peroxide radicals to block the chain reaction so as to terminate the oil oxidation, namely, AH + ROO·→ ROOH + A; the stabler the generated A, the better the antioxidant effect; TBHQ and tea polyphenol (the structures were shown in Fig. 10.2) reacted with the oil peroxide free radicals to generate the semiquinone free radicals, the stable structures were presented through the resonance inside the molecules, and thus the oil oxidation chain reaction was prevented. Tea polyphenol was a natural antioxidant with natural and nontoxic advantages (Liu Jian et al. 2007), and it had anticancer, antiaging, anti-radiation, blood glucose and blood lipid reduction, and other pharmacological effects (Chen Yuxiang et al. 2001); therefore, it was a recommended antioxidant with better antioxidant effects.

The induction time of peanut oil with V_E and BHT was significantly higher than that of blank control group ($P < 0.05$), but there was no significant difference in the induction time between V_E and BHT ($P > 0.05$). Therefore, V_E and BHT could significantly improve the oxidation stability of peanut oil, but there was no significant difference in the antioxidant effect of V_E and BHT on peanut oil. There was no significant difference in the induction time of peanut oil with phytosterol and that of blank control peanut oil. The antioxidant effect of AP on peanut oil was second only to that of TBHQ, and there was no significant difference between the antioxidant

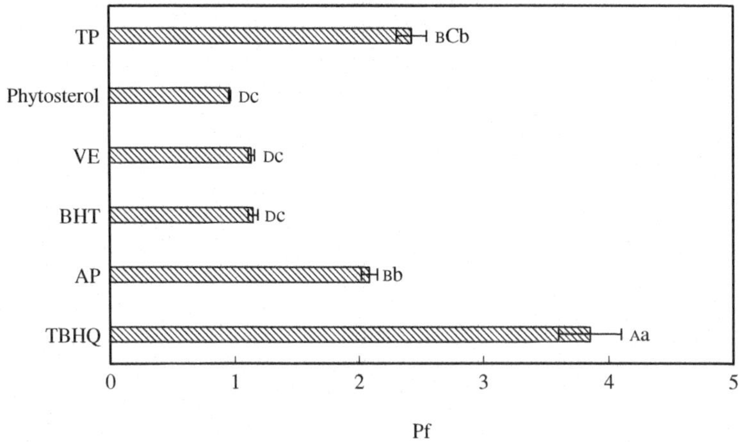

Fig. 10.3 Comparison of Pf values of different antioxidants. Note: *A*, *B*, *C*, *D* – different letters mean highly significant difference ($P < 0.01$); *a*, *b*, *c* – different letters mean significant difference ($P < 0.05$)

effect of AP and that of tea polyphenol. The antioxidant effects of six antioxidants were ranked as TBHQ>TP>AP>BHT>V_E>blank>phytosterol. The results were consistent with the research results of He Biyan et al. (1999) by using the AOM method and Yao Yunyou et al. (2004) by using the Schaal heat resistance test method.

The Pf values of the six antioxidants selected were calculated and compared by plotting, as shown in Fig. 10.3. Pf value was the ratio of induction time of sample to that of blank control group. It was generally believed that when the Pf value was between 2 and 4, the antioxidant had antioxidant activity; when Pf > 4, the antioxidant had strong antioxidant activity (Wu Hou et al. 2001; Xu Fuping et al. 2008). Among the six antioxidants tested, the Pf values of three antioxidants were between 2 and 4, and they were TP, AP, and TBHQ, which indicated that they had antioxidant activity on peanut oil (2<Pf<4), but did not have strong antioxidant activity (Pf>4), and the antioxidant activity of TBHQ was significantly higher than that of TP and AP ($P < 0.05$).

2 Impact of Usage and Usage Amount of Antioxidant on Stability of Peanut Oil

According to GB2760-2007, the maximum addition amount of antioxidants (TBHQ, BHT, and AP) allowed in China was 0.02%, and that of tea polyphenols was 0.04%. The addition amount of antioxidant was changed within the allowable range, and the accelerated oxidation induction time was determined and compared. The results are shown in Fig. 10.4.

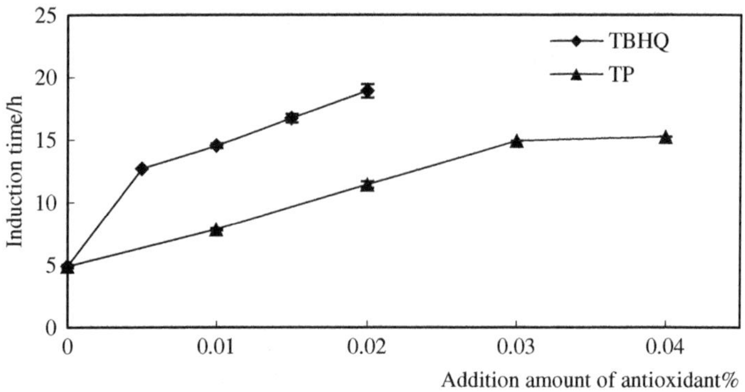

Fig. 10.4 Comparison of antioxidant effects of different TBHQ and TP addition amounts on peanut oil

The induction time of peanut oil increased with the increase of TBHQ, and when the maximum addition amount was 0.02% specified in GB2760-2007, the oxidation induction time was the maximum (18.91 h ± 0.54 h) which was significantly higher than that under other addition amounts ($P < 0.01$), which indicated that TBHQ had the best antioxidant effect on peanut oil when the addition amount was 0.02% in the limited range of national standard. With the increase of TP addition amount, the induction time of peanut oil also increased; when the maximum addition amount was 0.04% specified in GB2760-2007, the induction time was the longest (15.24 h ± 0.09 h) which did not have a significant difference from that (14.93 h ± 0.09 h) when the addition amount was 0.03% ($P > 0.05$), and the induction time when the addition amounts were 0.04% and 0.03% was significantly higher than that of TP under other addition amounts. Therefore, by considering the economic benefits and other factors comprehensively, 0.03% was selected as the optimal addition amount of TP antioxidant for peanut oil.

3 Impact of Synergistic Effect of Antioxidant and Synergist on Stability of Peanut Oil

Through the comparison of induction time of peanut oil after the antioxidants and synergists with different ratios were added (Figs. 10.5 and 10.6), it could be seen that among the three synergists (phytic acid, citric acid, and V_C), the peanut oil induction time of 0.02%TBHQ + 0.02%V_C was the longest (21.05 h ± 0.26 h) which was significantly higher than that of other groups ($P < 0.01$), so V_C was a good antioxidant synergist for TBHQ, followed by the 0.02% TBHQ + 0.02% citric acid with the induction time of 20.07 h ± 0.20 h which was significantly higher than that of 0.02%TBHQ (18.91 h ± 0.28 h) ($P < 0.05$); there was no significant difference between the induction time of 0.02%TBHQ + 0.02% phytic acid

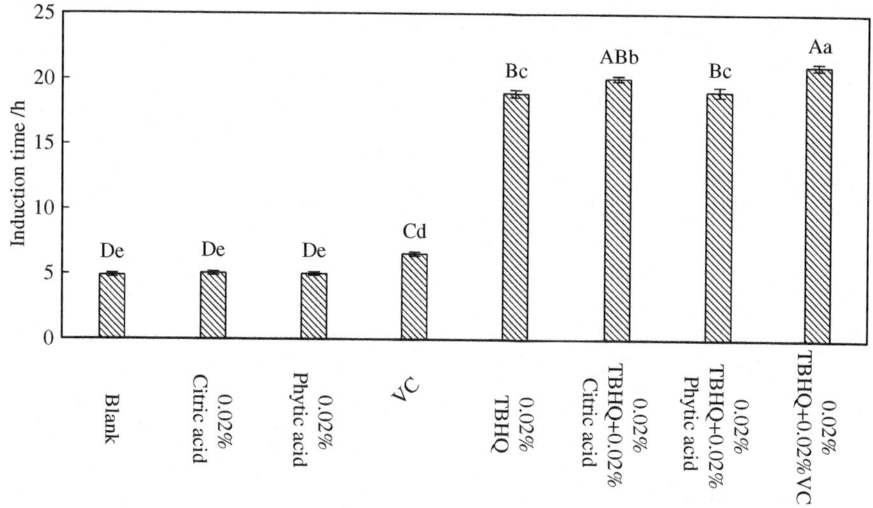

Fig. 10.5 Impact of synergistic effect of TBHQ and synergist on oxidation stability of peanut oil. Note: *A*, *B*, *C*, *D* – different letters mean highly significant difference ($P < 0.01$); *a*, *b*, *c*, *d*, *e* – different letters mean significant difference ($P < 0.05$)

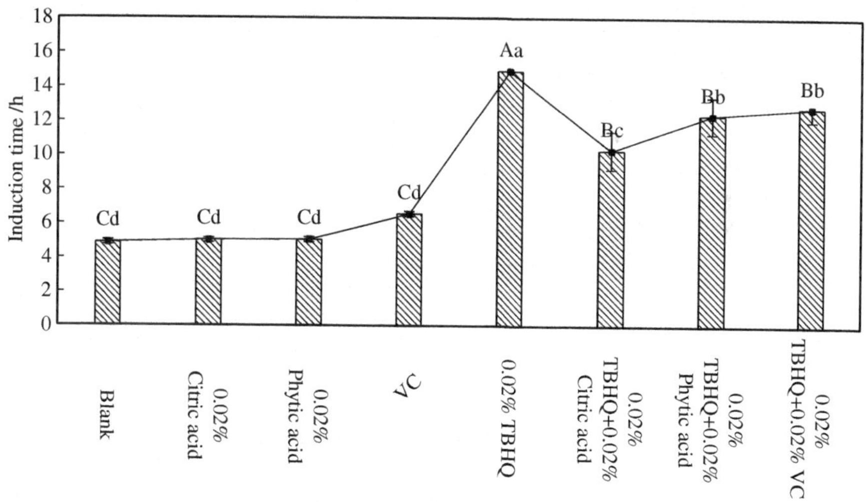

Fig. 10.6 Impact of synergistic effect of TP and synergist on oxidation stability of peanut oil. Note: *A*, *B*, *C* – different letters mean highly significant difference ($P < 0.01$); *a*, *b*, *c*, *d* – different letters mean significant difference ($P < 0.05$)

(19.07 ± 0.39 h) and that of 0.02% TBHQ group. Therefore, the synergistic effect of phytic acid on TBHQ was not significant. In addition, it could be seen from the figure that the induction time when these three synergists were added to the peanut oil individually was not significantly different from that of the blank group, indicating that the synergistic effect could be showed through the combination of synergist and antioxidant.

As an antioxidant synergist, V_C could capture the peroxide free radicals to block the oil oxidation chain reaction; with strong reducibility, V_C could react with the oxygen in the oil to reduce the concentration of oxygen in the oil (Wang Shaomei et al. 2000), and V_C also had an effect on chelating of metal ions (Kang Jiangquan 2002). Therefore, V_C could play the synergistic effect in the above three aspects as an antioxidant synergist; due to the polyhydroxyl structure, the citric acid had the antioxidant synergism effect; on the one hand, the citric acid could release protons to reduce TBHQ to restore its antioxidant capacity; on the other hand, it could chelate metal ions to form a complexing structure, so that metal ions could not play the prooxidant effect (You Jianming et al. 2004; Zhou Caiqiong et al. 2001); the pH value of system is an important factor affecting the free radical scavenging ability of V_C and phytic acid; the free radical scavenging ability of V_C increased with the decrease of pH when the pH was <5.66, and the free radical scavenging ability of phytic acid increased with the decrease of pH value when pH was within 2.0–8.0 and reached the maximum when pH was 2.0 (Jia Yan et al. 2011).

After 0.03% TP + 0.02% citric acid, 0.03% TP + 0.02% phytic acid, and 0.03% TP + 0.02% V_C were added to the peanut oil, the induction time was significantly lower than that of oil when only 0.03% TP was added (11.43 h $\pm$ 0.27 H) ($P < 0.01$), indicating that these three synergists had no significant synergistic effect on TP. Because the phytic acid, citric acid, and V_C were water-soluble substances and their solubility in oil was low, Wang Yuefei et al. (2010) pointed out that the synergistic antioxidant effect of antioxidant was related to its antioxidant capacity and effective concentration and the synergistic effect was low when the effective concentration was low; the solubility in oil and effective concentration of phytic acid, citric acid, and V_C were low, and TP was a water-soluble antioxidant with poor solubility in oil, so there was no significantly synergistic effect between TP and phytic acid, citric acid, and V_C in peanut oil.

References

Chen Yuxiang, Liu Yang, Zhou Daowei. Effect of tea polyphenols on soybean oil and lard oil [J]. Food Sci. 2001;22(11):27–9.

Hao Xiaoli, Xu Shenhong, Hang Hu. Synergistic effect of TBHQ and VE on antioxidant [J]. J Qingdao Univ. 2003;18(3):53–5.

He Biyan, Ou Guangnan. Comparative study on antioxidant activity of tea polyphenols, BHT and TBHQ [J]. J Jimei Univ. 1999;4(3):40–3.

Jia Yan, Tu Shuxin, Tang Shirong. A comparative study of roles of phytic acid and several antioxidants in eradication of free radicals. J Huazhong Agric Univ. 2011;30(5):618–23.

Kang Jiangquan. Food chemistry. Beijing: China Agric UnivPress. 2002:112–23.

Liu Jian, Meng Chunli,Yang Ping. Study on anti-oxidation ability of edible oil with tea polyphenols: J Henan Univ Technol; 2007.

You Jian Ming, Liu Da Yu. Study on antioxidant activity of tea polyphenols to palm oil [J]. Sichuan Food Ferment. 2004;40(3):43–5.

Wang Shaomei, Huang Meixia, Qing Xiaohong, et al. Study on synergistic antioxidation of Tea polyphenols and VC to lard emulsification system [J]. Food Sci. 2000;21(10):22–5.
Wang Yuefei, Xu Ping, Li Lei et al. Study on synergetic effect of tea polyphenols and several kinds of natural antioxidants [J]. Teach Sci, 2010,30 (2):109–114.
Wu Hou, Weng Xinchu. Study on antioxidant activity of natural tocopherol [J]. J Shanghai Univ. 2001;7(2):142–6.
Xu Fuping, Li Junyan, Sun Zhenhai. Study on antioxidant activity of natural tocopherols extracted from bromide slabs [J]. Acad Forum. 2008;11:55–7.
Yao Yunyou, Yu Hui. Comparative study on antioxidant effects of tea polyphenols and TBHQ on peanut oil and peanut soybean blends [J]. Food and Food Ind. 2004;4:19–21.
Zhang Jianshu, Wang Qiang. Effects of fatty acid composition and endogenous antioxidants on the stability of peanut oil in different varieties [D]. Beijing: Chinese Academy of Agricultural Sciences; 2012.
Zhou CQ, Duan ZP, Li CB. Influence of some organic acids on anti- oxidation of tea-poplyphenols in raoe salad oil. J Southwest Agric Univ. 2001;23(4):362–4.